Laser Spectroscopy 1

Wolfgang Demtröder

Laser Spectroscopy 1

Basic Principles

5th Edition

 Springer

Prof. Dr. Wolfgang Demtröder
University of Kaiserslautern
Kaiserslautern, Germany

ISBN 978-3-662-49542-1   ISBN 978-3-642-53859-9 (eBook)
DOI 10.1007/978-3-642-53859-9

Springer
© Springer-Verlag Berlin Heidelberg 2014
Softcover reprint of the hardcover 5th edition 2014

Springer is part of Springer Science+Business Media
www.springer.com

Preface to the Fifth Edition

Since the 4th edition 2008, many new achievements in Laser Spectroscopy have been successfully realized. as for example the expansion of optical comb spectroscopy into the extreme UV and its applications to metrology and astronomy, the progress in high harmonics generation and in attosecond spectroscopy, the experimental realization of entangled atoms in traps, the improvement of optical parametric oscillators with extended tuning ranges and of fibre lasers in a wide spectral range. In particular the manifold of applications of laser spectroscopy in chemistry, biology medicine and for solving technical problems has considerably increased.

The relevance of these developments has been underlined by awarding 3 times in the last 12 years Nobel prizes to researchers in the field of laser spectroscopy, namely 2001 to E. Cornell, W. Ketterle and C.E. Wieman for realizing Bose–Einstein Condensation, 2005 to J.L. Hall, Th.W. Hänsch for the development of the optical frequency comb and to R. Glauber for his outstanding contributions to the quantum theory of coherence and 2012 to S. Haroche and D. Wineland for their groundbreaking experimental methods applied to single trapped atoms and ions which they investigated with ultrahigh precision laser spectroscopy.

Some of these new developments are included in this new edition. The short representation of lasers and the discussion of the experimental tools indispensible for laser spectroscopy have been improved and extended and the references to the relevant literature have been supplemented by recent publications. However, also references to older papers or books are still kept, because they are often authentic for the development of new techniques and instrumentation and should not be forgotten.

As in the 4th edition the textbook is divided into two volumes. In the first volume the basic foundations of laser spectroscopy and the necessary experimental equipment is covered, while in Vol. 2 the different techniques of laser spectroscopy and various applications are outlined.

The author thanks many colleagues who gave permission to reproduce figures from their work or who supplied the author with papers on new developments. Many thanks go to Dr. Schneider and Ute Heuser at Springer Heidelberg, who supported this new edition and last but not least to my wife Harriet for her continuous encouragement and support.

v

The author hopes that this new edition will find a similar friendly acceptance as the former editions. He will appreciate any comments regarding mistakes, possible improvements of the representation or the inclusion of subjects not covered in this book. Any mail will be answered as soon as possible.

Wolfgang Demtröder   Kaiserslautern, December 2013

Preface to the Fourth Edition

About 50 years after the realization of the first laser in 1960, laser spectroscopy is still a very intense field of research which has expanded with remarkable progress into many areas of science, medicine and technology, and has provided an ever-increasing number of applications. The importance of laser spectroscopy and its appreciation by many people is, for instance, proved by the fact that over the last ten years three Nobel Prizes have been awarded to nine scientists in the field of laser spectroscopy and quantum optics.

This positive development is partly based on new experimental techniques, such as improvements of existing lasers and the invention of new laser types, the realization of optical parametric oscillators and amplifiers in the femtosecond range, the generation of attosecond pulses, the revolution in the measurements of absolute optical frequencies and phases of optical waves using the optical frequency comb, or the different methods developed for the generation of Bose–Einstein condensates of atoms and molecules and the demonstration of atom lasers as a particle equivalent to photon lasers.

These technical developments have stimulated numerous applications in chemistry, biology, medicine, atmospheric research, materials science, metrology, optical communication networks, and many other industrial areas.

In order to cover at least some of these new developments, a single volume would need too many pages. Therefore the author has decided to split the book into two parts. The first part contains the foundations of laser spectroscopy, i.e., the basic physics of spectroscopy, optical instruments and techniques. It furthermore provides a short introduction to the physics of lasers, and discusses the role of optical resonators and techniques for realizing tunable narrowband lasers, the working horses of laser spectroscopy. It gives a survey on the different types of tunable lasers and represents essentially the updated and enlarged edition of the first six chapters of the third edition. In order to improve its value as a textbook for students, the number of problems has been increased and their solutions are given at the end of Vol. 1. The second volume discusses the different techniques of laser spectroscopy. Compared to the third edition, it adds many new developments and tries to bring the reader up to speed on the present state of laser spectroscopy.

The author wishes to thank all of the people who have contributed to this new edition. There is Dr. Th. Schneider at Springer-Verlag, who has always supported the author and has shown patience when deadlines were not kept. Claudia Rau from LE-TeX has taken care of the layout, and many colleagues have given their permission to use figures from their research. Several readers have sent me their comments on errors or possible improvements. I thank them very much.

The author hopes that this new edition will find a similar friendly approval to the former editions and that it will enhance interest in the fascinating field of laser spectroscopy. He would appreciate any suggestions for improvement or hints about possible errors, and he will try to answer every question as soon as possible.

Kaiserslautern, February 2008   Wolfgang Demtröder

Preface to the Third Edition

Laser Spectroscopy continues to develop and expand rapidly. Many new ideas and recent realizations of new techniques based on old ideas have contributed to the progress in this field since the last edition of this textbook appeared. In order to keep up with these developments it was therefore necessary to include at least some of these new techniques in the third edition.

There are, firstly, the improvement of frequency-doubling techniques in external cavities, the realization of more reliable cw-parametric oscillators with large output power, and the development of tunable narrow-band UV sources, which have expanded the possible applications of coherent light sources in molecular spectroscopy. Furthermore, new sensitive detection techniques for the analysis of small molecular concentrations or for the measurement of weak transitions, such as overtone transitions in molecules, could be realized. Examples are Cavity Ringdown Spectroscopy, which allows the measurement of absolute absorption coefficients with great sensitivity or specific modulation techniques that push the minimum detectable absorption coefficient down to $10^{-14}\,\mathrm{cm}^{-1}$!

The most impressive progress has been achieved in the development of tunable femtosecond and subfemtosecond lasers, which can be amplified to achieve sufficiently high output powers for the generation of high harmonics with wavelengths down into the X-ray region and with pulsewidths in the attosecond range. Controlled pulse shaping by liquid crystal arrays allows coherent control of atomic and molecular excitations and in some favorable cases chemical reactions can already be influenced and controlled using these shaped pulses.

In the field of metrology a big step forward was the use of frequency combs from cw mode-locked femtosecond lasers. It is now possible to directly compare the microwave frequency of the cesium clock with optical frequencies, and it turns out that the stability and the absolute accuracy of frequency measurements in the optical range using frequency-stabilized lasers greatly surpasses that of the cesium clock. Such frequency combs also allow the synchronization of two independent femtosecond lasers.

The increasing research on laser cooling of atoms and molecules and many experiments with Bose–Einstein condensates have brought about some remarkable results and have considerably increased our knowledge about the interaction of light with matter on a microscopic scale and the interatomic interactions at very low temperatures. Also the realization of coherent matter waves (atom lasers) and investigations of interference effects between matter waves have proved fundamental aspects of quantum mechanics.

The largest expansion of laser spectroscopy can be seen in its possible and already realized applications to chemical and biological problems and its use in medicine as a diagnostic tool and for therapy. Also, for the solution of technical problems, such as surface inspections, purity checks of samples or the analysis of the chemical composition of samples, laser spectroscopy has offered new techniques.

In spite of these many new developments the representation of established fundamental aspects of laser spectroscopy and the explanation of the basic techniques are not changed in this new edition. The new developments mentioned above and also new references have been added. This, unfortunately, increases the number of pages. Since this textbook addresses beginners in this field as well as researchers who are familiar with special aspects of laser spectroscopy but want to have an overview on the whole field, the author did not want to change the concept of the textbook.

Many readers have contributed to the elimination of errors in the former edition or have made suggestions for improvements. I want to thank all of them. The author would be grateful if he receives such suggestions also for this new edition.

Many thanks go to all colleagues who gave their permission to use figures and results from their research. I thank Dr. H. Becker and T. Wilbourn for critical reading of the manuscript, Dr. H.J. Koelsch and C.-D. Bachem of Springer-Verlag for their valuable assistance during the editing process, and LE-TeX Jelonek, Schmidt and Vöckler for the setting and layout. I appreciate, that Dr. H. Lotsch, who has taken care for the foregoing editions, has supplied his computer files for this new edition. Last, but not least, I would like to thank my wife Harriet who made many efforts in order to give me the necessary time for writing this new edition.

Kaiserslautern, April 2002   Wolfgang Demtröder

Preface to the Second Edition

During the past 14 years since the first edition of this book was published, the field of laser spectroscopy has shown a remarkable expansion. Many new spectroscopic techniques have been developed. The time resolution has reached the femtosecond scale and the frequency stability of lasers is now in the millihertz range.

In particular, the various applications of laser spectroscopy in physics, chemistry, biology, and medicine, and its contributions to the solutions of technical and environmental problems are remarkable. Therefore, a new edition of the book seemed necessary to account for at least part of these novel developments. Although it adheres to the concept of the first edition, several new spectroscopic techniques such as optothermal spectroscopy or velocity-modulation spectroscopy are added.

A whole chapter is devoted to time-resolved spectroscopy including the generation and detection of ultrashort light pulses. The principles of coherent spectroscopy, which have found widespread applications, are covered in a separate chapter. The combination of laser spectroscopy and collision physics, which has given new impetus to the study and control of chemical reactions, has deserved an extra chapter. In addition, more space has been given to optical cooling and trapping of atoms and ions.

I hope that the new edition will find a similar friendly acceptance as the first one. Of course, a textbook never is perfect but can always be improved. I, therefore, appreciate any hint to possible errors or comments concerning corrections and improvements. I will be happy if this book helps to support teaching courses on laser spectroscopy and to transfer some of the delight I have experienced during my research in this fascinating field over the last 30 years.

Many people have helped to complete this new edition. I am grateful to colleagues and friends, who have supplied figures and reprints of their work. I thank the graduate students in my group, who provided many of the examples used to illustrate the different techniques. Mrs. Wollscheid who has drawn many figures, and Mrs. Heider who typed part of the corrections. Particular thanks go to Helmut Lotsch of Springer-Verlag, who worked very hard for this book and who showed much patience with me when I often did not keep the deadlines.

Last but not least, I thank my wife Harriet who had much understanding for the many weekends lost for the family and who helped me to have sufficient time to write this extensive book.

Kaiserslautern, June 1995   Wolfgang Demtröder

Preface to the First Edition

The impact of lasers on spectroscopy can hardly be overestimated. Lasers represent intense light sources with spectral energy densities which may exceed those of incoherent sources by several orders of magnitude. Furthermore, because of their extremely small bandwidth, single-mode lasers allow a spectral resolution which far exceeds that of conventional spectrometers. Many experiments which could not be done before the application of lasers, because of lack of intensity or insufficient resolution, are readily performed with lasers.

Now several thousands of laser lines are known which span the whole spectral range from the vacuum-ultraviolet to the far-infrared region. Of particular interst are the continuously tunable lasers which may in many cases replace wavelength-selecting elements, such as spectrometers or interferometers. In combination with optical frequency-mixing techniques such continuously tunable monochromatic coherent light sources are available at nearly any desired wavelength above 100 nm.

The high intensity and spectral monochromasy of lasers have opened a new class of spectroscopic techniques which allow investigation of the structure of atoms and molecules in much more detail. Stimulated by the variety of new experimental possibilities that lasers give to spectroscopists, very lively research activities have developed in this field, as manifested by an avalanche of publications. A good survey about recent progress in laser spectroscopy is given by the proceedings of various conferences on laser spectroscopy (see "Springer Series in Optical Sciences"), on picosecond phenomena (see "Springer Series in Chemical Physics"), and by several quasi-mongraphs on laser spectroscopy published in "Topics in Applied Physics".

For nonspecialists, however, or for people who are just starting in this field, it is often difficult to find from the many articles scattered over many journals a coherent representation of the basic principles of laser spectroscopy. This textbook intends to close this gap between the advanced research papers and the representation of fundamental principles and experimental techniques. It is addressed to physicists and chemists who want to study laser spectroscopy in more detail. Students who have some knowledge of atomic and molecular physics, electrodynamics, and optics should be able to follow the presentation.

The fundamental principles of lasers are covered only very briefly because many excellent textbooks on lasers already exist.

On the other hand, those characteristics of the laser that are important for its applications in spectroscopy are treated in more detail. Examples are the frequency spectrum of different types of lasers, their linewidths, amplitude and frequency stability, tunability, and tuning ranges. The optical components such as mirrors, prisms, and gratings, and the experimental equipment of spectroscopy, for example, monochromators, interferometers, photon detectors, etc., are discussed extensively because detailed knowledge of modern spectroscopic equipment may be crucial for the successful performance of an experiment.

Each chapter gives several examples to illustrate the subject discussed. Problems at the end of each chapter may serve as a test of the reader's understanding. The literature cited for each chapter is, of course, not complete but should inspire further studies. Many subjects that could be covered only briefly in this book can be found in the references in a more detailed and often more advanced treatment. The literature selection does not represent any priority list but has didactical purposes and is intended to illustrate the subject of each chapter more thoroughly.

The spectroscopic applications of lasers covered in this book are restricted to the spectroscopy of free atoms, molecules, or ions. There exists, of course, a wide range of applications in plasma physics, solid-state physics, or fluid dynamics which are not discussed because they are beyond the scope of this book. It is hoped that this book may be of help to students and researchers. Although it is meant as an introduction to laser spectroscopy, it may also facilitate the understanding of advanced papers on special subjects in laser spectroscopy. Since laser spectroscopy is a very fascinating field of research, I would be happy if this book can transfer to the reader some of my excitement and pleasure experienced in the laboratory while looking for new lines or unexpected results.

I want to thank many people who have helped to complete this book. In particular the students in my research group who by their experimental work have contributed to many of the examples given for illustration and who have spent their time reading the galley proofs. I am grateful to colleages from many laboratories who have supplied me with figures from their publications. Special thanks go to Mrs. Keck and Mrs. Ofiiara who typed the manuscript and to Mrs. Wollscheid and Mrs. Ullmer who made the drawings. Last but not least, I would like to thank Dr. U. Hebgen, Dr. H. Lotsch, Mr. K.-H. Winter, and other coworkers of Springer-Verlag who showed much patience with a dilatory author and who tried hard to complete the book in a short time.

Kaiserslautern, March 1981   Wolfgang Demtröder

Contents

Chapter 1
Introduction

Most of our knowledge about the structure of atoms and molecules is based on spectroscopic investigations. Thus spectroscopy has made an outstanding contribution to the present state of atomic and molecular physics, to chemistry, and to molecular biology. Information on molecular structure and on the interaction of molecules with their surroundings may be derived in various ways from the absorption or emission spectra generated when electromagnetic radiation interacts with matter.

Wavelength measurements of spectral lines allow the determination of energy levels of the atomic or molecular system. The *line intensity* is proportional to the transition probability, which measures how strongly the two levels of a molecular transition are coupled. Since the transition probability depends on the wave functions of both levels, intensity measurements are useful to verify the spatial charge distribution of excited electrons, which can only be roughly calculated from approximate solutions of the Schrödinger equation. The *natural linewidth* of a spectral line may be resolved by special techniques, allowing mean lifetimes of excited molecular states to be determined. Measurements of the *Doppler width* yield the velocity distribution of the emitting or absorbing molecules and with it the temperature of the sample. From *pressure broadening* and *pressure shifts* of spectral lines, information about collision processes and interatomic potentials can be extracted. *Zeemann* and *Stark splittings* by external magnetic or electric fields are important means of measuring magnetic or electric moments and elucidating the coupling of the different angular momenta in atoms or molecules, even with complex electron configurations. The *hyperfine structure* of spectral lines yields information about the interaction between the nuclei and the electron cloud and allows nuclear magnetic dipole moments, electric quadrupole moments or even higher moments, such as octupole moments to be determined. Time-resolved measurements allow the spectroscopist to follow up dynamical processes in ground-state and excited-state molecules, to investigate details of collision processes and various energy transfer mechanisms. The combination of optical excitation with femto-to attosecond laser pulses and X-ray diffraction on the same time scale allows time-resolved snapshots of molecular structure in electronically excited states not washed out by vibrations of the excited molecule, because the X-ray diffraction pattern is measured within a

W. Demtröder, *Laser Spectroscopy 1*, DOI 10.1007/978-3-642-53859-9_1,
© Springer-Verlag Berlin Heidelberg 2014

time interval, which is short compared to the vibrational period. Laser spectroscopic studies of the interaction of single atoms with a radiation field provide stringent tests of quantum electrodynamics and the realization of high-precision frequency standards allows one to check whether fundamental physical constants show small changes with time.

These examples represent only a small selection of the many possible ways by which spectroscopy provides tools to explore the microworld of atoms and molecules. However, the amount of information that can be extracted from a spectrum depends essentially on the attainable spectral or time resolution and on the detection sensitivity that can be achieved.

The application of new technologies to optical instrumentation (for instance, the production of larger and better ruled gratings in spectrographs, the use of highly reflecting dielectric coatings in interferometers, and the development of optical multichannel analyzers, CCD cameras, and image intensifiers) has certainly significantly extended the sensitivity limits. Considerable progress was furthermore achieved through the introduction of new spectroscopic techniques, such as Fourier spectroscopy, optical pumping, level-crossing techniques, and various kinds of double-resonance methods and molecular beam spectroscopy.

Although these new techniques have proved to be very fruitful, the really stimulating impetus to the whole field of spectroscopy was given by the introduction of lasers. In many cases these new spectroscopic light sources may increase spectral resolution and sensitivity by several orders of magnitude. Combined with new spectroscopic techniques, lasers are able to surpass basic limitations of classical spectroscopy. Many experiments that could not be performed with incoherent light sources are now feasible or have already been successfully completed recently. This book deals with such new techniques of laser spectroscopy and explains the necessary instrumentation. It is divided into two volumes.

The first volume contains the basic physical foundations of laser spectroscopy and the most important experimental equipment in a spectroscopic laboratory. It begins with a discussion of the fundamental definitions and concepts of classical spectroscopy, such as thermal radiation, induced and spontaneous emission, radiation power, intensity and polarization, transition probabilities, and the interaction of weak and strong electromagnetic (EM) fields with atoms. Since the coherence properties of lasers are important for several spectroscopic techniques, the basic definitions of coherent radiation fields are outlined and the description of coherently excited atomic levels is briefly discussed.

In order to understand the theoretical limitations of spectral resolution in classical spectroscopy, Chap. 3 treats the different causes of the broadening of spectral lines and the information drawn from measurements of line profiles. Numerical examples at the end of each section illustrate the order of magnitude of the different effects.

The contents of Chap. 4, which covers spectroscopic instrumentation and its application to wavelength and intensity measurements, are essential for the experimental realization of laser spectroscopy. Although spectrographs and monochromators, which played a major rule in classical spectroscopy, may be abandoned for many experiments in laser spectroscopy, there are still numerous applications where these instruments are indispensible. Of major importance for laser spectroscopists are the different kinds of interferometers. They are used not only in laser resonators ιιι ιεαllἐ οιιιѕlε moɗe operation but also for line-profile measurements of spectral lines and for very precise wavelength measurement means, Since the determination of wavelength is a central problem in spectroscopy, a whole section discusses some modern techniques for precise wavelength measurements and their accuracy.

Lack of intensity is one of the major limitations in many spectroscopic investigations. It is therefore often vital for the experimentalist to choose the proper light detector. Section 4.5 surveys several light detectors and sensitive techniques such as photon counting, which is becoming more commonly used. While Chaps. 2–4 cover subjects that are not restricted to laser spectroscopy (they are general spectroscopy concepts), Chap. 5 deals with the "working horse" of laser spectroscopy: the different kinds of lasers and their design. It treats the basic properties of lasers as spectroscopic radiation sources and starts with a short recapitulation of the fundamentals of lasers, such as threshold conditions, optical resonators, and laser modes. Only those laser characteristics that are important in laser spectroscopy are discussed here. For a more detailed treatment the reader is referred to the extensive laser literature cited in Chap. 5. Those properties and experimental techniqes that make the laser such an attractive spectroscopic light source are discussed more thoroughly. For instance, the important questions of wavelength stabilization and continuous wavelength tuning are treated, and experimental realizations of single-mode tunable lasers and limitations of laser linewidths are presented. The last part of this chapter gives a survey of the various types of tunable lasers that have been developed for different spectral ranges. Advantages and limitations of these lasers are discussed. The available spectral range could be greatly extended by optical frequency doubling and mixing processes. This interesting field of nonlinear optics is presented in Chap. 5 as far as it is relevant to spectroscopy.

The second volume presents various applications of lasers in spectroscopy and discusses the different methods that have been developed recently. The presentation relies on the general principles and the instrumentation of spectroscopy outlined in the first volume. It starts with different techniques of laser spectroscopy and also covers recent developments and the various applications of laser spectroscopy in science, technology, medicine and environmental studies.

The exciting development of ultra-short pulses in the femtosecond and attosecond range emitted by tunable lasers has opened many interesting new fields of molecular dynamics, where the forming and dissociation of molecules can be followed up in real time. This enables a basic understanding of many fast processes in Chemistry and Biology.

The vastly increasing number of applications of laser spectroscopy in basic sciences such as metrology, Chemistry, Biology, Medicine, and for the solution of technical problems will be illustrated by several examples. Even in astronomy lasers have brought about new techniques such as adaptive optics which increase the angular resolution of optical telescopes by several orders of magnitude or the exact determination of the Doppler-shift of astronomical objects, using the optical comb spectroscopy.

This book is intended as an *introduction* to the basic methods and instrumentation of spectroscopy, with special emphasis placed on laser spectroscopy. The examples in each chapter illustrate the text and may suggest other possible applications. They are mainly concerned with the spectroscopy of free atoms and molecules and are, of course, not complete, but have been selected from the literature or from our own laboratory work for didactic purposes and may not represent the priorities of publication dates. For a far more extensive survey of the latest publications in the broad field of laser spectroscopy, the reader is referred to the proceedings of various conferences on laser spectroscopy [1–11] and to textbooks or collections of articles on modern aspects of laser spectroscopy [12–32]. Since scientific achievements in laser physics have been pushed forward by a few pioneers, it is interesting to look back to the historical development and to the people who influenced it. Such a personal view can be found in [33, 34]. The reference list at the end of the book might be helpful in finding more details of a special experiment or to dig deeper into theoretical and experimental aspects of each chapter. A useful "Encyclopedia of spectroscopy" [35, 36] gives a good survey on different aspects of laser spectroscopy.

Chapter 2
Absorption and Emission of Light

This chapter deals with basic considerations about absorption and emission of electromagnetic waves interacting with matter. Especially emphasized are those aspects that are important for the spectroscopy of gaseous media. The discussion starts with thermal radiation fields and the concept of cavity modes in order to elucidate differences and connections between spontaneous and induced emission and absorption. This leads to the definition of the Einstein coefficients and their mutual relations. The next section explains some definitions used in photometry such as radiation power, intensity, spectral power density and polarization of electromagnetic waves.

It is possible to understand many phenomena in optics and spectroscopy in terms of classical models based on concepts of classical electrodynamics. For example, the absorption and emission of electromagnetic waves in matter can be described using the model of damped oscillators for the atomic electrons. In most cases, it is not too difficult to give a quantum-mechanical formulation of the classical results. The semiclassical approach will be outlined briefly in Sect. 2.8.

Many experiments in laser spectroscopy depend on the coherence properties of the radiation and on the coherent excitation of atomic or molecular levels. Some basic ideas about temporal and spatial coherence of optical fields and the density-matrix formalism for the description of coherence in atoms are therefore discussed at the end of this chapter.

Throughout this text the term "light" is frequently used for electromagnetic radiation in all spectral regions. Likewise, the term "molecule" in general statements includes atoms as well. We shall, however, restrict the discussion and most of the examples to *gaseous* media, which means essentially free atoms or molecules.

For more detailed or more advanced presentations of the subjects summarized in this chapter, the reader is referred to the extensive literature on spectroscopy [37–47]. Those interested in light scattering from solids are directed to the sequence of Topics volumes edited by Cardona and coworkers [48].

W. Demtröder, *Laser Spectroscopy 1*, DOI 10.1007/978-3-642-53859-9_2,
© Springer-Verlag Berlin Heidelberg 2014

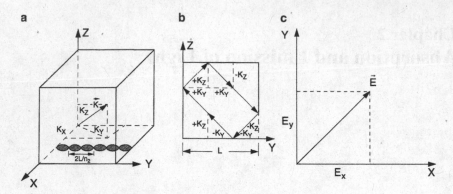

Figure 2.1 Modes of a stationary EM field in a cavity: **a** Standing waves in a cubic cavity; **b** superposition of possible **k** vectors to form standing waves, illustrated in a two-dimensional coordinate system; **c** illustration of the polarization of cavity modes

2.1 Cavity Modes

Consider a cubic cavity with the sides L at the temperature T. The walls of the cavity absorb and emit electromagnetic radiation. At thermal equilibrium the absorbed power $P_a(\omega)$ has to be equal to the emitted power $P_e(\omega)$ for all frequencies ω. Inside the cavity there is a *stationary radiation field* E, which can be described at the point r by a superposition of plane waves with the amplitudes A_p, the wave vectors k_p, and the angular frequencies ω_p as

$$E = \sum_p A_p \exp[i(\omega_p t - k_p \cdot r)] + \text{compl. conj.} \tag{2.1}$$

The waves are reflected at the walls of the cavity. For each wave vector $k = (k_x, k_y, k_z)$, this leads to eight possible combinations $k_i = (\pm k_x, \pm k_y, \pm k_z)$ that interfere with each other. A stationary-field configuration only occurs if these superpositions result in *standing waves* (Fig. 2.1a,b). This imposes boundary conditions for the wave vector, namely

$$k_x = \frac{\pi}{L} n_1 ; \quad k_y = \frac{\pi}{L} n_2 ; \quad k_z = \frac{\pi}{L} n_3 , \quad n_i = \text{integer} , \tag{2.2a}$$

which means that, for all three components, the side-length L of the cavity must be an integer multiple of $1/2$ of the wavelength $\lambda = 2\pi/k$.

The wave vector of the electromagnetic wave is then:

$$k = \frac{\pi}{L}(n_1, n_2, n_3) , \tag{2.2b}$$

where n_1, n_2, n_3 are positive integers.

Figure 2.2 Illustration of the maximum number of possible **k** vectors with $|\mathbf{k}| \leq k_{max}$ in momentum space (n_x, n_y, n_z)

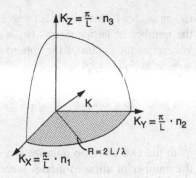

The magnitudes of the wave vectors allowed by the boundary conditions are

$$|\mathbf{k}| = \frac{\pi}{L}\sqrt{n_1^2 + n_2^2 + n_3^2}, \qquad (2.3)$$

which can be written in terms of the wavelength $\lambda = 2\pi/|\mathbf{k}|$ or the frequency $\omega = c|\mathbf{k}|$.

$$\lambda = 2L/\sqrt{n_1^2 + n_2^2 + n_3^2} \quad \text{or} \quad \omega = \frac{\pi c}{L}\sqrt{n_1^2 + n_2^2 + n_3^2}. \qquad (2.4)$$

These standing waves are called *cavity modes* (Fig. 2.1b).

Since the amplitude vector $\mathbf{A}$ of a transverse wave $\mathbf{E}$ is always perpendicular to the wave vector $\mathbf{k}$, it can be composed of two components a_1 and a_2 with the unit vectors $\hat{\mathbf{e}}_1$ and $\hat{\mathbf{e}}_2$

$$\mathbf{A} = a_1\hat{\mathbf{e}}_1 + a_2\hat{\mathbf{e}}_2 \qquad (\hat{\mathbf{e}}_1 \cdot \hat{\mathbf{e}}_2 = \delta_{12}; \quad \hat{\mathbf{e}}_1, \hat{\mathbf{e}}_2 \perp \mathbf{k}). \qquad (2.5)$$

The complex numbers a_1 and a_2 define the polarization of the standing wave. Equation (2.5) states that any arbitrary polarization can always be expressed by a linear combination of two mutually orthogonal linear polarizations. To each cavity mode defined by the wave vector $\mathbf{k}_p$ therefore belong two possible polarization states. This means that *each triple of integers* (n_1, n_2, n_3) *represents two cavity modes. Any arbitrary stationary field configuration can be expressed as a linear combination of cavity modes.*

We shall now investigate how many modes with frequencies $\omega \leq \omega_m$ are possible. Because of the boundary condition (2.4), this number is equal to the number of all integer triples (n_1, n_2, n_3) that fulfil the condition

$$c^2 k^2 = \omega^2 \leq \omega_m^2. $$

In a system with the coordinates $(\pi/L)(n_1, n_2, n_3)$ (see Fig. 2.2), each triple (n_1, n_2, n_3) represents a point in a three-dimensional lattice with the lattice constant π/L. In this system, (2.4) describes all possible frequencies within a sphere of radius ω_m/c. If this radius is large compared to π/L, which means that $2L \gg \lambda_m$, the number of lattice points (n_1, n_2, n_3) with $\omega^2 \leq \omega_m^2$ is roughly given by the volume of the octant of the sphere shown in Fig. 2.2 with $k = |\mathbf{k}| = \omega/c$. This volume is

$$V_k = \frac{1}{8}\frac{4\pi}{3}\left(\frac{k_{\max}}{\pi/L}\right)^3 . \tag{2.6a}$$

With the two possible polarization states of each mode, one therefore obtains for the number of allowed modes with frequencies between $\omega = 0$ and $\omega = \omega_m$ in a cubic cavity of volume L^3 with $L \gg \lambda$

$$N(\omega_m) = 2\frac{1}{8}\frac{4\pi}{3}\left(\frac{L\omega_m}{\pi c}\right)^3 = \frac{1}{3}\frac{L^3\omega_m^3}{\pi^2 c^3} . \tag{2.6b}$$

The spatial mode density (the number of modes per unit volume) is then

$$n(\omega_m) = N(\omega_m)/L^3 = \frac{1}{3}\frac{\omega_m^3}{\pi^2 c^3} . \tag{2.6c}$$

It is often interesting to know the number $n(\omega)\mathrm{d}\omega$ of modes per unit volume within a certain frequency interval $\mathrm{d}\omega$, for instance, within the width of a spectral line. The *spectral mode density* $n(\omega)$ can be obtained directly from (2.6a–2.6c) by differentiating $N(\omega)/L^3$ with respect to ω. $N(\omega)$ is assumed to be a continuous function of ω, which is, strictly speaking, only the case for $L \to \infty$. We get

$$n(\omega)\mathrm{d}\omega = \frac{\omega^2}{\pi^2 c^3}\mathrm{d}\omega . \tag{2.7a}$$

In spectroscopy the frequency $\nu = \omega/2\pi$ is often used instead of the angular frequency ω. With $\mathrm{d}\omega = 2\pi\mathrm{d}\nu$, the number of modes per unit volume within the frequency interval $\mathrm{d}\nu$ is then

$$n(\nu)\mathrm{d}\nu = \frac{8\pi\nu^2}{c^3}\mathrm{d}\nu . \tag{2.7b}$$

In Fig. 2.3 the spectral mode density $n(\nu)$ is plotted against the frequency ν on a double-logarithmic scale. This illustrates, that the spectral mode density (number of modes within the frequency interval $\mathrm{d}\nu = 1\,\mathrm{s}^{-1}$) in the visible range is about $n(\nu) = 10^5\,\mathrm{m}^{-3}$, which gives inside the Doppler widths of a spectral line $\mathrm{d}\nu = 10^9\,\mathrm{s}^{-1}$ the large number of 10^{14} modes/m^3.

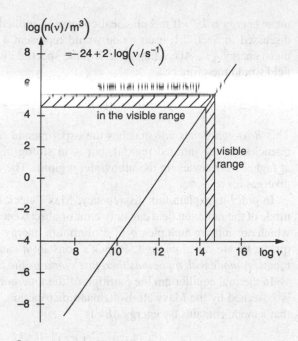

Figure 2.3 Spectral mode density $n(\nu) = N(\nu)/L^3$ as a function of the frequency ν

Example 2.1

1. In the visible part of the spectrum ($\lambda = 500\,\text{nm}$; $\nu = 6 \times 10^{14}\,\text{Hz}$), (2.7b) yields for the number of modes per m³ within the Doppler width of a spectral line ($d\nu = 10^9\,\text{Hz}$)

$$n(\nu)d\nu = 3 \times 10^{14}\,\text{m}^{-3} \;.$$

2. In the microwave region ($\lambda = 1\,\text{cm}$; $\nu = 3 \times 10^{10}\,\text{Hz}$), the number of modes per m³ within the typical Doppler width $d\nu = 10^5\,\text{Hz}$ is only $n(\nu)d\nu = 10^2\,\text{m}^{-3}$.
3. In the X-ray region ($\lambda = 1\,\text{nm}$; $\nu = 3 \times 10^{17}\,\text{Hz}$), one finds $n(\nu)d\nu = 8.4 \times 10^{21}\,\text{m}^{-3}$ within the typical natural linewidth $d\nu = 10^{11}\,\text{Hz}$ of an X-ray transition.

2.2 Thermal Radiation and Planck's Law

In classical thermodynamics each degree of freedom of a system in thermal equilibrium at a temperature T has the mean energy $kT/2$, where k is the Boltzmann constant. Since classical oscillators have kinetic as well as potential energies, their mean energy is kT. If this classical concept is applied to the electromagnetic field discussed in Sect. 2.1, each mode would represent a classical oscillator with the mean energy kT. According to (2.7b), the spectral energy density of the radiation field would therefore be

$$\rho(\nu)d\nu = n(\nu)kTd\nu = \frac{8\pi\nu^2 k}{c^3}Td\nu \ . \tag{2.8}$$

This *Rayleigh–Jeans law* matches the experimental results fairly well at low frequencies (in the infrared region), but is in strong disagreement with experiment at higher frequencies (in the ultraviolet region). The energy density $\rho(\nu)$ actually diverges for $\nu \rightarrow \infty$.

In order to explain this discrepancy, Max Planck suggested in 1900 that each mode of the radiation field can only emit or absorb energy in discrete amounts $qh\nu$, which are integer multiples q of a minimum energy quantum $h\nu$. These energy quanta $h\nu$ are called *photons*. Planck's constant h can be determined from experiments. *A mode with q photons therefore contains the energy $qh\nu$.*

In thermal equilibrium the partition of the total energy into the different modes is governed by the Maxwell–Boltzmann distribution, so that the probability $p(q)$ that a mode contains the energy $qh\nu$ is

$$p(q) = (1/Z)e^{-qh\nu/kT} \ , \tag{2.9}$$

where k is the Boltzmann constant and

$$Z = \sum_q e^{-qh\nu/kT} \tag{2.10}$$

is the partition function summed over all modes containing q photons $h \cdot \nu$. Z acts as a normalization factor which makes $\sum_q p(q) = 1$, as can be seen immediately by inserting (2.10) into (2.9). This means that a mode has to contain with certainty ($p = 1$) some number ($q = 0, 1, 2, \ldots$) of photons.

The mean energy per mode is therefore

$$\overline{W} = \sum_{q=0}^{\infty} p(q)qh\nu = \frac{1}{Z}\sum_{q=0}^{\infty} qh\nu e^{-qh\nu/kT} = \frac{\sum qh\nu e^{-qh\nu/kT}}{\sum e^{-qh\nu/kT}} \ . \tag{2.11}$$

The evaluation of the sum yields [42] (see Problem 2.1)

$$\overline{W} = \frac{h\nu}{e^{h\nu/kT} - 1} \ . \tag{2.12}$$

The thermal radiation field has the energy density $\rho(\nu)d\nu$ within the frequency interval ν to $\nu + d\nu$, which is equal to the number $n(\nu)d\nu$ of modes in the interval $d\nu$ times the mean energy $\overline{W}$ per mode. Using (2.7b, 2.12) one obtains

$$\rho(\nu)d\nu = \frac{8\pi\nu^2}{c^3}\frac{h\nu}{e^{h\nu/kT} - 1}d\nu \ . \tag{2.13}$$

Figure 2.4 Spectral distribution of the energy density $\rho_\nu(\nu)$ for different temperatures

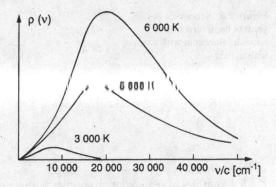

This is *Planck's* famous *radiation law* (Fig. 2.4), which predicts a spectral energy density of the thermal radiation that is fully consistent with experiments. The expression "thermal radiation" comes from the fact that the spectral energy distribution (2.13) is characteristic of a radiation field that is in thermal equilibrium with its surroundings (in Sect. 2.1 the surroundings are determined by the cavity walls).

The thermal radiation field described by its energy density $\rho(\nu)$ is *isotropic*. This means that through any transparent surface element dA of a sphere containing a thermal radiation field, the same power flux dP is emitted into the solid angle $d\Omega$ at an angle θ to the surface normal $\hat{n}$ (Fig. 2.5)

$$dP = \frac{c}{4\pi}\rho(\nu)dA d\Omega d\nu \cos\theta . \tag{2.14}$$

It is therefore possible to determine $\rho(\nu)$ experimentally by measuring the spectral distribution of the radiation penetrating through a small hole in the walls of the cavity. If the hole is sufficiently small, the energy loss through this hole is negligibly small and does not disturb the thermal equilibrium inside the cavity.

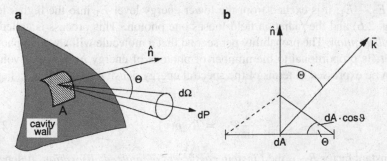

Figure 2.5 Illustration of (2.14)

Figure 2.6 Schematic diagram of the interaction of a two-level system with a radiation field

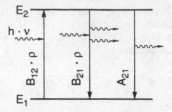

Example 2.2

1. Examples of real radiation sources with spectral energy distributions close to the Planck distribution (2.13) are the sun, the bright tungsten wire of a light bulb, flash lamps, and high-pressure discharge lamps.

2. Spectral lamps that emit discrete spectra are examples of *nonthermal* radiation sources. In these gas-discharge lamps, the light-emitting atoms or molecules may be in thermal equilibrium with respect to their translational energy, which means that their velocity distribution is Maxwellian. However, the population of the different excited atomic levels may not necessarily follow a Boltzmann distribution. There is generally no thermal equilibrium between the atoms and the radiation field. The radiation may nevertheless be isotropic.

3. Lasers are examples of nonthermal and anisotropic radiation sources (Chap. 5). The radiation field is concentrated in a few modes, and most of the radiation energy is emitted into a small solid angle. This means that the laser represents an extreme anisotropic nonthermal radiation source.

2.3   Absorption, Induced, and Spontaneous Emission

Assume that molecules with the energy levels E_1 and E_2 have been brought into the thermal radiation field of Sect. 2.2. If a molecule absorbs a photon of energy $h\nu = E_2 - E_1$, it is excited from the lower energy level E_1 into the higher level E_2 (Fig. 2.6) and the radiation field looses one photon. This process is called *induced absorption*. The probability per second that a molecule will absorb a photon, dP_{12}/dt, is proportional to the number of photons of energy $h\nu$ per unit volume and can be expressed in terms of the spectral energy density $\rho_\nu(\nu)$ of the radiation field as

$$\frac{d}{dt}P_{12} = B_{12}\rho(\nu) . \tag{2.15}$$

The constant factor B_{12} is the *Einstein coefficient of induced absorption*. It depends on the electronic structure of the atom, i.e. on its electronic wave functions in the two levels $|1\rangle$ and $|2\rangle$. Each absorbed photon of energy $h\nu$ decreases the number of photons in one mode of the radiation field by one.

The radiation field can also induce molecules in the excited state E_2 to make a transition to the lower state E_1 with simultaneous emission of a photon of energy $h\nu$. This process is called induced (or stimulated) emission. The induced photon of energy $h\nu$ is emitted into the same mode that caused the emission. This means that the number of photons in this mode is increased by one. The probability $\mathrm{d}P_{21}/\mathrm{d}t$ that one molecule emits one induced photon per second is in analogy to (2.15)

$$\frac{\mathrm{d}}{\mathrm{d}t}P_{21} = B_{21}\rho(\nu) . \qquad (2.16)$$

The constant factor B_{21} is the *Einstein coefficient of induced emission*.

An excited molecule in the state E_2 may also *spontaneously* convert its excitation energy into an emitted photon $h\nu$. This spontaneous radiation can be emitted in the arbitrary direction k and increases the number of photons in the mode with frequency ν and wave vector k by one. In the case of isotropic emission, the probability of gaining a spontaneous photon is equal for all modes with the same frequency ν but different directions k.

The probability per second $\mathrm{d}P_{21}^{\text{spont}}/\mathrm{d}t$ that a photon $h\nu = E_2 - E_1$ is spontaneously emitted by a molecule, depends on the structure of the molecule and the selected transition $|2\rangle \rightarrow |1\rangle$, *but it is independent of the external radiation field*,

$$\frac{\mathrm{d}}{\mathrm{d}t}P_{21}^{\text{spont}} = A_{21} . \qquad (2.17)$$

A_{21} is the *Einstein coefficient of spontaneous emission* and is often called the *spontaneous transition probability*.

Let us now look for relations between the three Einstein coefficients B_{12}, B_{21}, and A_{21}. The total number N of all molecules per unit volume is distributed among the various energy levels E_i of population density N_i such that $\sum_i N_i = N$. At thermal equilibrium the population distribution $N_i(E_i)$ is given by the Boltzmann distribution

$$N_i(E_i) = N\frac{g_i}{Z}\mathrm{e}^{-E_i/kT} . \qquad (2.18)$$

The statistical weight $g_i = 2J_i + 1$ gives the number of degenerate sublevels of the level $|i\rangle$ with total angular momentum J_i and the partition function

$$Z = \sum_i g_i \mathrm{e}^{-E_i/kT} ,$$

acts again as a normalization factor which ensures that $\sum_i N_i = N$.

In a stationary field the total absorption rate $N_1 B_{12}\rho(\nu)$, which gives the number of photons absorbed per unit volume per second, has to equal the total emission rate

Figure 2.7 Average number of photons per mode in a thermal radiation field as a function of temperature T and frequency ν

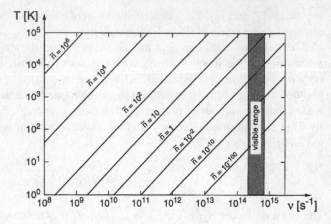

$N_2 B_{21}\rho(\nu) + N_2 A_{21}$ (otherwise the spectral energy density $\rho(\nu)$ of the radiation field would change). This gives

$$[B_{21}\rho(\nu) + A_{21}]N_2 = B_{12}N_1\rho(\nu) . \tag{2.19}$$

Using the relation

$$N_2/N_1 = (g_2/g_1)e^{-(E_2-E_1)/kT} = (g_2/g_1)e^{-h\nu/kT} ,$$

deduced from (2.18), and solving (2.19) for $\rho(\nu)$ yields

$$\rho(\nu) = \frac{A_{21}/B_{21}}{\frac{g_1}{g_2}\frac{B_{12}}{B_{21}}e^{h\nu/kT} - 1} . \tag{2.20}$$

In Sect. 2.2 we derived Planck's law (2.13) for the spectral energy density $\rho(\nu)$ of the thermal radiation field. Since both (2.13) and (2.20) must be valid for an arbitrary temperature T and all frequencies ν, comparison of the constant coefficients yields the relations

$$\boxed{B_{12} = \frac{g_2}{g_1}B_{21}} , \tag{2.21}$$

$$\boxed{A_{21} = \frac{8\pi h\nu^3}{c^3}B_{21}} . \tag{2.22}$$

Equation (2.21) states that for levels $|1\rangle$, $|2\rangle$ with equal statistical weights $g_2 = g_1$, *the probability of induced emission is equal to that of induced absorption.*

From (2.22) the following illustrative result can be extracted: since $n(\nu) = 8\pi\nu^2/c^3$ gives the number of modes per unit volume and frequency interval $d\nu = 1$ Hz, (see (2.7b)), (2.22) can be written as

$$\frac{A_{21}}{n(\nu)} = B_{21}h\nu , \tag{2.23a}$$

which means that the spontaneous emission per mode $A_{21}^* = A_{21}/n(\nu)$ equals the induced emission that is triggered by one photon. This can be generalized to: *the ratio of the induced- to the spontaneous-emission rate in an arbitrary mode is equal to the number q of photons in this mode.*

$$\frac{B_{21}\rho(\nu)}{A_{21}^*} = q \quad \text{with } \rho(\nu) = q\,h\,\nu \text{ in 1 mode.} \tag{2.23b}$$

In Fig. 2.7 the mean number of photons per mode in a thermal radiation field at different absolute temperatures is plotted as a function of frequency ν. The graphs illustrate that in the visible spectrum this number is small compared to unity at temperatures that can be realized in a laboratory. *This implies that in thermal radiation fields, the spontaneous emission per mode exceeds by far the induced emission.* If it is possible, however, to concentrate most of the radiation energy into a few modes, the number of photons in these modes may become exceedingly large and the induced emission in these modes dominates, although the total spontaneous emission into *all* modes may still be larger than the induced rate. Such a selection of a few modes is realized in a laser (Chap. 5).

Comment Note that the relations (2.21, 2.22) are valid for all kinds of radiation fields. Although they have been derived for stationary fields at thermal equilibrium, the Einstein coefficients are constants that depend only on the molecular properties and not on external fields as far as these fields do not alter the molecular properties. These equations therefore hold for arbitrary $\rho_\nu(\nu)$.

Using the angular frequency $\omega = 2\pi\nu$ instead of ν, the unit frequency interval $d\omega = 1\,\text{s}^{-1}$ corresponds to $d\nu = 1/2\pi\,\text{s}^{-1}$. The spectral energy density $\rho_\omega(\omega) = n(\omega)\hbar\omega$ is then, according to (2.7a),

$$\rho_\omega(\omega) = \frac{\omega^2}{\pi^2 c^3} \frac{\hbar\omega}{e^{\hbar\omega/kT} - 1}, \tag{2.24}$$

where $\hbar$ is Planck's constant h divided by 2π. The ratio of the Einstein coefficients

$$A_{21}/B_{21} = \frac{\hbar\omega^3}{\pi^2 c^3}, \tag{2.25a}$$

now contains $\hbar$ instead of h, and is smaller by a factor of 2π. However, the ratio $A_{21}/[B_{21}\rho_\omega(\omega)]$, which gives the ratio of the spontaneous to the induced transition probabilities, remains the same:

$$A_{21}/\left[B_{21}^{(\nu)}\rho_\nu(\nu)\right] = A_{21}/\left[B_{21}^{(\omega)}\rho_\omega(\omega)\right]. \tag{2.25b}$$

Example 2.3

1. In the thermal radiation field of a 100 W light bulb, 10 cm away from the tungsten wire, the number of photons per mode at $\lambda = 500$ nm is about 10^{-8}. If a molecular probe is placed in this field, the induced emission is therefore completely negligible.

2. In the center spot of a high-current mercury discharge lamp with very high pressure, the number of photons per mode is about 10^{-2} at the center frequency of the strongest emission line at $\lambda = 253.6$ mm. This shows that, even in this very bright light source, the induced emission only plays a minor role.

3. Inside the cavity of a HeNe laser (output power 1 mW with mirror transmittance $T = 1\%$) that oscillates in a single mode, the number of photons in this mode is about 10^7. In this example the *spontaneous* emission into this mode is completely negligible. Note, however, that the total spontaneous emission power at $\lambda = 632.2$ nm, which is emitted into all directions, is much larger than the induced emission. This spontaneous emission is more or less uniformly distributed among all modes. Assuming a volume of 1 cm^3 for the gas discharge, the number of modes within the Doppler width of the neon transition is about 10^8, which means that the total spontaneous rate is about 10 times the induced rate.

2.4   Basic Photometric Quantities

In spectroscopic applications of light sources, it is very useful to define some characteristic quantities of the emitted and absorbed radiation. This allows a proper comparison of different light sources and detectors and enables one to make an appropriate choice of apparatus for a particular experiment.

2.4.1   Definitions

The *radiant energy* W (measured in joules) refers to the total amount of energy emitted by a light source, transferred through a surface, or collected by a detector. The *radiant power* $P = dW/dt$ (often called *radiant flux* Φ [W]) is the radiant energy per second. The *radiant energy density* ρ [J/m^3] is the radiant energy per unit volume of space.

Consider a surface element dA of a light source (Fig. 2.8a). The radiant power emitted from dA into the solid angle $d\Omega$, around the angle θ against the surface

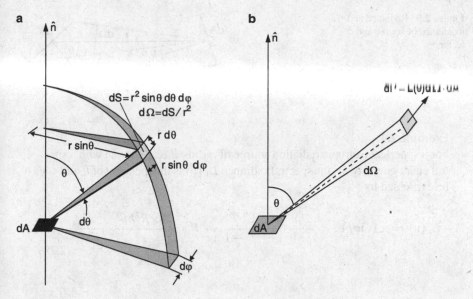

Figure 2.8   **a** Definition of solid angle $d\Omega$; **b** definition of radiance $L(\theta)$

normal $\hat{n}$ is

$$dP = L(\theta)dAd\Omega \, , \qquad (2.26a)$$

where the *radiance* L [W/m^2 sr^{-1}] is the power emitted per unit surface element $dA = 1\,m^2$ into the unit solid angle $d\Omega = 1$ sr which depends on the angle θ of the radiation direction against the surface normal.

If we assume $L(\theta) = L(0)\cos\theta$ (*Lambert's law*), where $L(0) = L(\theta = 0)$ we obtain with $d\Omega = \sin\theta \cdot d\theta \cdot d\varphi$ the total power emitted into the half-sphere above the surface element dA

$$P = L(0)\,dA \int\limits_{\theta=0}^{\pi/2} \int\limits_{\varphi=0}^{2\pi} \cos\theta \sin\theta \, d\theta \, d\varphi = \pi L(0)\,dA \qquad (2.26b)$$

The above three quantities refer to the total radiation integrated over the entire spectrum. Their spectral versions $W_\nu(\nu)$, $P_\nu(\nu)$, $\rho_\nu(\nu)$, and $L_\nu(\nu)$ are called the *spectral densities*, and are defined as the amounts of W, P, ρ, and L within the unit frequency interval $d\nu = 1\,s^{-1}$ around the frequency ν:

$$W = \int\limits_0^\infty W_\nu(\nu)d\nu \; ; \; P = \int\limits_0^\infty P_\nu(\nu)d\nu \; ; \; \rho = \int\limits_0^\infty \rho_\nu(\nu)d\nu \; ; \; L = \int\limits_0^\infty L_\nu(\nu)d\nu \, .$$

$$(2.27)$$

Figure 2.9 Radiance and
irradiance of source and detector

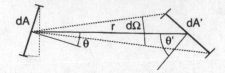

Example 2.4

For a spherical isotropic radiation source of radius R (e.g., a star) with a spectral energy density ρ_v, the spectral radiance $L_v(v)$ is independent of θ and can be expressed by

$$L_v(v) = \rho_v(v)c/4\pi = \frac{2hv^3}{c^2}\frac{1}{e^{hv/kT}-1} \rightarrow P_v = \frac{8\pi R^2 hv^3}{c^2}\frac{1}{e^{hv/kT}-1}.$$

$$(2.28)$$

A surface element dA' of a detector at distance r from the source element dA covers a solid angle $d\Omega = dA'\cos\theta'/r^2$ as seen from the source (Fig. 2.9). With $r^2 \gg dA$ and dA', the radiant flux Φ received by dA' is

$$d\Phi = L(\theta)dA\cos\theta d\Omega = L(\theta)\cos\theta dA\cos\theta' dA'/r^2, \qquad (2.29a)$$

The total flux Φ received by the surface A' and emitted by A is then

$$\Phi = \int_A\int_{A'} \frac{1}{r^2}L(\theta)\cos\theta\,\cos\theta'\,dA\,dA'. \qquad (2.29b)$$

The same flux Φ is received by A if A' is the emitter. For isotropic sources (2.29a, 2.29b) is symmetric with regard to θ and θ' or dA and dA'. The positions of detector and source may be interchanged without altering (2.29a, 2.29b). Because of this reciprocity, L may be interpreted either as the *radiance of the source* at the angle θ to the surface normal or, equally well, as the *radiance incident onto the detector at the angle θ'*.

For isotropic sources, where L is independent of θ, (2.29a, 2.29b) demonstrates that the radiant flux emitted into the unit solid angle is proportional to $\cos\theta$ (*Lambert's law*). An example for such a source is a hole with the area dA in a blackbody radiation cavity (Fig. 2.5).

The radiant flux incident on the unit detector area is called *irradiance I*, while in the spectroscopic literature it is often termed *intensity*. The flux density or intensity I [W/m²] of a plane wave $E = E_0\cos(\omega t - kz)$ traveling in vacuum in the z-

Figure 2.10 Flux densities of detectors with extended area

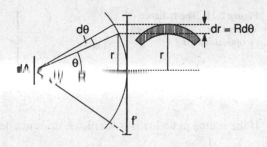

direction is given by

$$I = c \int \rho(\omega) d\omega = c\epsilon_0 E^2 = c\epsilon_0 E_0^2 \cos^2(\omega t - kz) \, . \qquad (2.30a)$$

With the complex notation

$$E = A_0 e^{i(\omega t - kz)} + A_0^* e^{-i(\omega t - kz)} \quad (|A_0| = \tfrac{1}{2} E_0) \, , \qquad (2.30b)$$

the intensity becomes

$$I = c\epsilon_0 E^2 = 4c\epsilon_0 A_0^2 \cos^2(\omega t - kz) \, . \qquad (2.30c)$$

Most detectors cannot follow the rapid oscillations of light waves with the angular frequencies $\omega \sim 10^{13}$–10^{15} Hz in the visible and near-infrared region. With a time constant $T \gg 1/\omega$ they measure, at a fixed position z, the time-averaged intensity

$$\langle I \rangle = \frac{c\epsilon_0 E_0^2}{T} \int\limits_0^T \cos^2(\omega t - kz) dt = \tfrac{1}{2} c\epsilon_0 E_0^2 = 2c\epsilon_0 A_0^2 \, . \qquad (2.31)$$

2.4.2 *Illumination of Extended Areas*

In the case of extended detector areas, the total power received by the detector is obtained by integration over all detector elements dA' (Fig. 2.10). The detector receives all the radiation that is emitted from the source element dA within the angles $-u \leq \theta \leq +u$. The same radiation passes an imaginary spherical surface in front of the detector. We choose as elements of this spherical surface circular rings with $dA' = 2\pi r dr = 2\pi R^2 \sin \theta \cos \theta d\theta$ because $r = R \sin \theta \Rightarrow dr = R \cos \theta d\theta$. From (2.29a, 2.29b) one obtains for the total flux Φ impinging onto the detector with $\cos \theta' = 1$

$$\Phi(u) = 2\pi \int\limits_0^u L dA \cos \theta \, \sin \theta d\theta \, . \qquad (2.32)$$

Figure 2.11 The radiance of
a source cannot be increased
by optical imaging

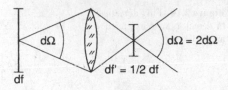

If the source radiation is isotropic, L does not depend on θ and (2.32) yields

$$\Phi = \pi L \sin^2 u \, dA \; . \tag{2.33}$$

Comment Note that it is impossible to increase the radiance of a source by any so-phisticated imaging optics. This means that the image dA^* of a radiation source dA never has a larger radiance than the source itself. It is true that the flux density can be increased by focussing the radiation. The solid angle, however, into which radia-tion from the image dA^* is emitted is also increased by the same factor. Therefore, the radiance *does not* increase. In fact, because of inevitable reflection, scattering, and absorption losses of the imaging optics, the radiance of the image dA^* is, in practice, always less than that of the source (Fig. 2.11).

A strictly parallel light beam would be emitted into the solid angle $d\Omega = 0$. With a finite radiant power this would imply an infinite radiance L, which is im-possible. This illustrates that such a light beam cannot be realized. The radiation source for a strictly parallel beam anyway has to be a point source in the focal plane of a lens. Such a point source with zero surface cannot emit any power.

For more extensive treatments of photometry see [49, 50].

Example 2.5
1. *Radiance of the sun.* An area equal to $1 \, m^2$ of the earth's surface receives at normal incidence without reflection or absorption through the atmo-sphere an incident radiant flux I_e of about $1.35 \, kW/m^2$ (solar constant). Because of the symmetry of (2.32) we may regard dA' as emitter and dA as receiver. The sun is seen from the earth under an angle of $2u = 32$ min-utes of arc. This yields $\sin u = 4.7 \times 10^{-3}$. Inserting this number into (2.33), one obtains $L_s = 2 \times 10^7 \, W/(m^2 \, sr)$ for the radiance of the sun's surface. The total radiant power Φ of the sun can be obtained from (2.32) or from the relation $\Phi = 4\pi R^2 I_e$, where $R = 1.5 \times 10^{11} \, m$ is the distance from the earth to the sun. These numbers give $\Phi = 4 \times 10^{26} \, W$.
2. *Radiance of a HeNe laser.* We assume that the output power of $1 \, mW$ is emitted from $1 \, mm^2$ of the mirror surface into an angle of 4 minutes of arc, which is equivalent to a solid angle of $1 \times 10^{-6} \, sr$. The maximum radiance in the direction of the laser beam is then $L = 10^{-3}/(10^{-6} \cdot 10^{-6}) = 10^9 \, W/(m^2 \, sr)$. This is about 50 times larger than the radiance

of the sun. For the spectral density of the radiance the comparison is even more dramatic. Since the emission of a stabilized single-mode laser is restricted to a spectral range of about 1 MHz, the laser has a spectral radiance density $L_\nu = 1 \times 10^4$ W · s/(m² sr), whereas the sun, which emits within a mean spectral range of $\approx 10^{15}$ Hz, only reaches $L_\nu = 2 \times 10^{-8}$ W · s/(m² sr).

3. Looking directly into the sun, the retina receives a radiant flux of 1 mW if the diameter of the iris is 1 mm. This is just the same flux the retina receives staring into the laser beam of Example 2.5b. There is, however, a big difference regarding the irradiance of the retina. The image of the sun on the retina is about 100 times as large as the focal area of the laser beam. This means that the power density incident on single retina cells is about 100 times larger in the case of the laser radiation.

2.5 Polarization of Light

The complex amplitude vector A_0 of the plane wave

$$E = A_0 \cdot e^{i(\omega t - kz)} \tag{2.34}$$

can be written in its component representation

$$A_0 = \begin{Bmatrix} A_{0x} e^{i\phi_x} \\ A_{0y} e^{i\phi_y} \end{Bmatrix} . \tag{2.35}$$

For unpolarized light the phases ϕ_x and ϕ_y are uncorrelated and their difference fluctuates statistically. For linearly polarized light with its electric vector in x-direction $A_{0y} = 0$. When E points into a direction α against the x axis, $\phi_x = \phi_y$ and $\tan\alpha = A_{0y}/A_{0x}$. For circular polarization $A_{0x} = A_{0y}$ and $\phi_x = \phi_y \pm \pi/2$.

The different states of polarization can be characterized by their Jones vectors, which are defined as follows:

$$E = \begin{Bmatrix} E_x \\ E_y \end{Bmatrix} = |E| \cdot \begin{Bmatrix} a \\ b \end{Bmatrix} e^{i(\omega t - kz)} \tag{2.36}$$

where the normalized vector $\{a, b\}$ is the *Jones vector*. In Table 2.1 the Jones vectors are listed for the different polarization states. For linearly polarized light with $\alpha = 45°$, for example, the amplitude A_0 can be written as

$$A_0 = \sqrt{A_{0x}^2 + A_{0y}^2}\, \frac{1}{\sqrt{2}} \begin{Bmatrix} 1 \\ 1 \end{Bmatrix} = |A_0| \frac{1}{\sqrt{2}} \begin{Bmatrix} 1 \\ 1 \end{Bmatrix} , \tag{2.37}$$

Table 2.1 Jones vectors for light traveling in the z-direction and Jones matrices for polarizers

Jones vectors	Jones matrices

Linear polarization | **Linear polarizers**

$\leftrightarrow$ $\quad$ $\updownarrow$ $\quad$ $\nearrow$ 45° $\quad$ $\nwarrow$ −45°

$$\begin{pmatrix} 1 & 0 \\ 0 & 0 \end{pmatrix} \quad \begin{pmatrix} 0 & 0 \\ 0 & 1 \end{pmatrix} \quad \frac{1}{2}\begin{pmatrix} 1 & 1 \\ 1 & 1 \end{pmatrix} \quad \frac{1}{2}\begin{pmatrix} 1 & -1 \\ -1 & 1 \end{pmatrix}$$

x-direction $\begin{pmatrix} 1 \\ 0 \end{pmatrix}$
$\leftrightarrow$

y-direction $\begin{pmatrix} 0 \\ 1 \end{pmatrix}$
$\updownarrow$

angle θ against x-axis $\begin{pmatrix} \cos^2\theta & \sin\theta\cos\theta \\ \sin\theta\cos\theta & \sin^2\theta \end{pmatrix}$

$\begin{pmatrix} \cos\alpha \\ \sin\alpha \end{pmatrix}$

$\lambda/4$-plates with fast axis in the direction of

x $\qquad\qquad\qquad\qquad\qquad$ y

$\leftrightarrow + \updownarrow$ $\begin{pmatrix} 1 \\ 1 \end{pmatrix}$

$$M_{H\lambda/4} = e^{i\pi/4}\begin{pmatrix} 1 & 0 \\ 0 & i \end{pmatrix} \qquad M_{V\lambda/4} = e^{i\pi/4}\begin{pmatrix} 1 & 0 \\ 0 & -i \end{pmatrix}$$

$\alpha = 45°$: $\frac{1}{\sqrt{2}}\begin{pmatrix} 1 \\ 1 \end{pmatrix}$

$$= \frac{1}{\sqrt{2}}\begin{pmatrix} 1+i & 0 \\ 0 & i-1 \end{pmatrix} \qquad = \frac{1}{\sqrt{2}}\begin{pmatrix} 1+i & 0 \\ 0 & 1-i \end{pmatrix}$$

$\alpha = -45°$: $\frac{1}{\sqrt{2}}\begin{pmatrix} 1 \\ -1 \end{pmatrix}$

$\lambda/2$-plates with fast axis in the direction of

x $\qquad\qquad\qquad\qquad\qquad$ y

$$M_{H\lambda/2} = e^{-i\pi/2}\begin{pmatrix} 1 & 0 \\ 0 & -1 \end{pmatrix} \qquad M_{V\lambda/2} = e^{i\pi/2}\begin{pmatrix} 1 & 0 \\ 0 & -1 \end{pmatrix}$$

$$= \begin{pmatrix} -i & 0 \\ 0 & i \end{pmatrix} \qquad\qquad = \begin{pmatrix} i & 0 \\ 0 & -i \end{pmatrix}$$

Wave plate with phaseshift $\varphi/2$ in x-direction and $-\varphi/2$ in y-direction

$$M_{\Delta\varphi} = \begin{pmatrix} e^{i\varphi/2} & 0 \\ 0 & e^{-i\varphi/2} \end{pmatrix} = e^{i\varphi/2}\begin{pmatrix} 1 & 0 \\ 0 & e^{-i\varphi} \end{pmatrix}$$

Circular polarization | **Rotator (device which turns the polarization vector by an angle θ)**

σ^+: $\frac{1}{\sqrt{2}}\begin{pmatrix} 1 \\ i \end{pmatrix}$

$$M_{rot}(\theta) = \begin{pmatrix} \cos\theta & \sin\theta \\ -\sin\theta & \cos\theta \end{pmatrix}$$

A polarizing element rotated by an angle θ

σ^-: $\frac{1}{\sqrt{2}}\begin{pmatrix} 1 \\ -i \end{pmatrix}$

$$M(\theta) = M_{rot}(-\theta)M(0)M_{rot}(\theta)$$

$\lambda/2$-plate rotated by an angle θ

$\sigma^+ + \sigma^-$: $\sqrt{2}\begin{pmatrix} 1 \\ 0 \end{pmatrix}$

$$M_{\lambda/2}(\theta) = \begin{pmatrix} \cos 2\theta & \sin 2\theta \\ \sin 2\theta & -\cos 2\theta \end{pmatrix}$$

Polarizing wave plate rotated by an angle θ

$$M(\theta) = \begin{pmatrix} \cos\varphi/2 + i\sin\varphi/2\cos 2\theta & i\sin\varphi/2\sin 2\theta \\ i\sin\varphi/2\sin 2\theta & \cos\varphi/2 - i\sin\varphi/2\cos 2\theta \end{pmatrix}$$

Figure 2.12 A polarizing element changes the input electric vector $E_0 = \{E_{x0}, E_{y0}\}$ of an electromagnetic wave travelling in the z-direction into the transmitted electric vector $E_t = \{E_{xt}, E_{yt}\}$

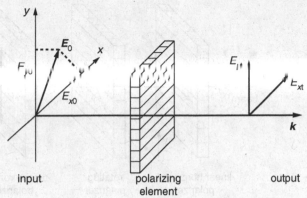

input   polarizing   output
                                   element

while for circular polarization (σ^+ or σ^- light), we obtain

$$A_0^{(\sigma^+)} = \frac{1}{\sqrt{2}} |A_0| \begin{Bmatrix} 1 \\ i \end{Bmatrix} \quad ; \quad A_0^{(\sigma^-)} = \frac{1}{\sqrt{2}} |A_0| \begin{Bmatrix} 1 \\ -i \end{Bmatrix} \tag{2.38}$$

because $\exp(-i\pi/2) = -i$.

The Jones representation shows its advantages when we consider the transmission of light through optical elements such as polarizers, $\lambda/4$ plates, or beamsplitters (Fig. 2.12). These elements can be described by 2×2 matrices (*Jones matrices*), which are compiled for some elements in Table 2.1. The polarization state of the transmitted light is then obtained by multiplication of the Jones vector of the incident wave by the Jones matrix of the optical element.

$$E_t = \begin{Bmatrix} E_{xt} \\ E_{yt} \end{Bmatrix} = \begin{pmatrix} a & b \\ c & d \end{pmatrix} \cdot \begin{Bmatrix} E_{x0} \\ E_{y0} \end{Bmatrix} , \tag{2.39}$$

If the light wave passes through several polarizing elements their Jones matrices are multiplied, where the first matrix in the product represents the last element. We will illustrate this by some examples:

1. If a polarizer which rotates the polarisation plane of the incident wave is placed between two crossed linear polarizers (Fig. 2.13) the electric vector of the input beam will be turned and the crossed polarizer transmits only a fraction of the input intensity which depends on the turning angle θ of the rotating polarizer. The Jones formalism yields the output electric vector as

$$\begin{pmatrix} E_{xt} \\ E_{yt} \end{pmatrix} = \begin{pmatrix} 0 & 0 \\ 0 & 1 \end{pmatrix} \begin{pmatrix} \cos\theta & \sin\theta \\ -\sin\theta & \cos\theta \end{pmatrix} \begin{pmatrix} 1 & 0 \\ 0 & 0 \end{pmatrix} \begin{pmatrix} E_{x0} \\ E_{y0} \end{pmatrix} = \begin{pmatrix} 0 \\ -\sin\theta E_{x0} \end{pmatrix} .$$

In Fig. 2.14 the transmitted intensity is plotted as a function of the angle θ. For $\theta = n\pi$ ($n = 0, 1, 2, \ldots$) the transmitted intensity is zero because for these angles the E-vector points into the x-direction and the wave is blocked by the polarizer in y-direction.

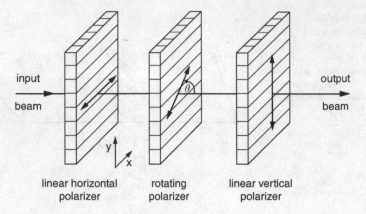

Figure 2.13 Rotating polarizer between two crossed linear polarizers

Figure 2.14 Transmitted
intensity as a function of the
tilting angle θ

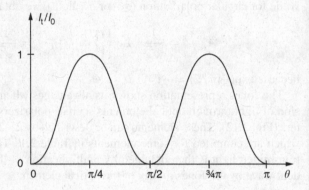

2. Incident light linearly polarized in the x-direction ($\alpha = 0°$) becomes, after transmission through a $\lambda/4$ polarizer with its slow axis in the x-direction

$$E_t = e^{i\pi/4} \begin{pmatrix} 1 & 0 \\ 0 & -i \end{pmatrix} \cdot \begin{pmatrix} 1 \\ 0 \end{pmatrix} |E_0| = e^{i\pi/4} \begin{pmatrix} 1 \\ -i \end{pmatrix} |E_0|$$

$$= e^{-i\pi/4} \begin{pmatrix} E_{x0} \\ -iE_{y0} \end{pmatrix}, \qquad (2.40)$$

a right circular polarized σ^--wave.

3. A σ^+-wave passes through a $\lambda/2$-wave plate with its fast axis in y-direction. The transmitted light is then

$$E_t = \frac{1}{\sqrt{2}} \begin{pmatrix} i & 0 \\ 0 & -i \end{pmatrix} \begin{pmatrix} 1 \\ -i \end{pmatrix} |E_0| = \frac{1}{\sqrt{2}} \begin{pmatrix} i \\ 1 \end{pmatrix} |E_0| = \frac{1}{\sqrt{2}} e^{i\pi/2} \begin{pmatrix} 1 \\ -i \end{pmatrix} |E_0| \, .$$

The transmitted light is σ^--light, where the phase factor of $\pi/2$ does not affect the state of polarization.

4. The Jones Matrix $M(\theta)$ for any polarizing element, turned by an angle θ against its original position $M(0)$ is obtained by the product

$$M(\theta) = M_{\text{rot}}(-\theta)M(0)M_{\text{rot}}(\theta) .$$

For example a linear horizontal polarizer turned by an angle θ is described by the Jones matrix

$$M^{\text{lin}}(\theta) = \begin{pmatrix} \cos^2 \theta & \sin \theta \cos \theta \\ \sin \theta \cos \theta & \sin^2 \theta \end{pmatrix}$$

which converts for $\theta = 0$ to

$$M^{\text{lin}}(0) = \begin{pmatrix} 1 & 0 \\ 0 & 0 \end{pmatrix}$$

and for $\theta = 45°$ to

$$M^{\text{lin}}(45°) = \begin{pmatrix} 1/2 & 1/2 \\ 1/2 & 1/2 \end{pmatrix} = \frac{1}{2}\begin{pmatrix} 1 & 1 \\ 1 & 1 \end{pmatrix} .$$

A $\lambda/2$-plate with its fast axis in the x-direction is rotated by an angle θ around the z-axis (direction of light propagation). If linear horizontal polarized light passes through the device the output light is

$$\begin{pmatrix} E_{xt} \\ E_{yt} \end{pmatrix} = \begin{pmatrix} \cos 2\theta & \sin 2\theta \\ \sin 2\theta & -\cos 2\theta \end{pmatrix} \begin{pmatrix} 0 \\ 1 \end{pmatrix} = \begin{pmatrix} \cos 2\theta \\ \sin 2\theta \end{pmatrix} .$$

For $\theta = 45°$ this becomes

$$\begin{pmatrix} E_{xt} \\ E_{yt} \end{pmatrix} = \begin{pmatrix} 0 \\ 1 \end{pmatrix}$$

which is the Jones vector for linear vertical polarized light.
For $\theta = 180°$ we obtain linear horizontal polarized light

$$\begin{pmatrix} E_{xt} \\ E_{yt} \end{pmatrix} = \begin{pmatrix} -1 \\ 0 \end{pmatrix} ,$$

where the E-vector has been turned by 180°.

More examples can be found in [51–53].

2.6   Absorption and Dispersion

When an electromagnetic wave $E = E_0 \cdot e^{i(\omega t - Kz)}$ with wavelength $\lambda = 2\pi/K$ passes through an absorbing medium with refractive index $n(\lambda)$, the phase velocity

v_{Ph} changes from the vacuum value c to $v_{Ph} = c/n$ (**dispersion**) and the amplitude decreases due to **absorption**. On the macroscopic scale these phenomena can be well described by a classical model which treats the atomic electrons as classical harmonic oscillators, which are forced by the electromagnetic field to damped harmonic oscillations. This model gives an illustrative picture of the relation between absorption and dispersion, leading to the *dispersion relations*. It relates the macroscopic quantities „*refractive index n*" and „*absorption coefficient α*" to the microscopic properties of the atomic electron shell. The classical results can be readily transferred to the real conditions described by quantum mechanical models.

2.6.1   Classical Model

The forced oscillation $x(t)$ of a damped harmonic oscillator with mass m is described by the equation

$$m\ddot{x} + b\dot{x} + Dx = q \cdot E_0 \cdot e^{i\omega t} . \tag{2.41}$$

Inserting the ansatz $x = x_0 \exp(i\omega t)$ into (2.41) gives with the abbreviations $\gamma = b/m$ and $\omega_0^2 = D/m$ for the amplitude x_0 the solution

$$x_0 = \frac{q \cdot E_0}{m(\omega_0^2 - \omega^2 + i\gamma\omega)} . \tag{2.42}$$

These forced oscillations of a charge q produces an induced dipole moment

$$p_{el} = q \cdot x = \frac{q^2 \cdot E_0 \cdot e^{i\omega t}}{m(\omega_0^2 - \omega^2 + i\gamma\omega)} . \tag{2.43}$$

For N oscillators per unit volume the macroscopic polarization (sum of the induced dipole moments per unit volume induced by the light wave) becomes

$$P = N \cdot q \cdot x . \tag{2.44}$$

Classical electrodynamics proves that this polarization is related to the electric field E as

$$P = \varepsilon_0(\varepsilon - 1)E . \tag{2.45}$$

The relative dielectric constant is related to the refractive index n by

$$n = \sqrt{\varepsilon} . \tag{2.46}$$

From (2.43) to (2.46) we obtain for the refractive index n the relation

$$n^2 = 1 + \frac{Nq^2}{\varepsilon_0 m(\omega_0^2 - \omega^2 + i\gamma\omega)} \tag{2.47}$$

which shows that n is a complex quantity which can be written as

$$n = n' - i\kappa \quad (n' \text{ and } \kappa \text{ are real quantities}) . \tag{2.48}$$

An electromagnetic wave with frequency $\omega = 2\pi\nu$ travelling in the z direction through a medium with refractive index n can be written as

$$E = E_0 e^{i(\omega t - Kz)} . \tag{2.49a}$$

The wavenumber $K = 2\pi/\lambda$ is $K = K_0$ in vacuum and $K_n = nK_0$ in the medium which implies $\lambda_n = \lambda_0/n$. Inserting this into (2.49a) gives with (2.48)

$$E = E_0 e^{-K_0\kappa z} \cdot e^{i(\omega t - n'K_0 z)} = E_0 e^{-2\pi\kappa z/\lambda} \cdot e^{iK_0 n' z} \cdot e^{i\omega t} . \tag{2.49b}$$

This illustrates that the imaginary part κ of the komplex refractive index describes the absorption of the electromagnetic wave. After a penetration depth $\Delta z = \lambda/(2\pi\kappa)$ the amplitude E_0 has decreased to E_0/e. The real part $n'(\omega)$ of the refractive index represents the dispersion, i.e. the dependence of the phase velocity $v_{Ph}(\omega)$ on the frequency ω of the electromagnetic wave.

For gaseous media at not too high pressures the refractive index n is nearly 1. For example for air at 1 atmosphere $n = 1.0003$. In this case we can use the approximation

$$n^2 - 1 = (n+1)(n-1) \approx 2(n-1)$$

and obtain instead of (2.47) the approximate expression

$$n = 1 + \frac{1}{2} \frac{Nq^2}{\varepsilon_0 m(\omega_0^2 - \omega^2 + i\gamma\omega)} . \tag{2.50}$$

Using (2.48) we can separate real and imaginary part of n and obtain:

$$\kappa = \frac{Nq^2}{2\varepsilon_0 m} \frac{\gamma\omega}{(\omega_0^2 - \omega^2)^2 + \gamma^2\omega^2} , \tag{2.51a}$$

$$n' = 1 + \frac{Nq^2}{2\varepsilon_0 m} \frac{\omega_0^2 - \omega^2}{(\omega_0^2 - \omega^2)^2 + \gamma^2\omega^2} . \tag{2.51b}$$

These equations are called *dispersion relations*. They link absorption and dispersion through the complex refractive index. They are valid for classical oscillators at rest and at sufficiently small densities to guarantee that $(n-1) \ll 1$.

Note, that the thermal motion of atoms in a gas results in an additional Doppler broadening which will be treated in Chap. 3.

Figure 2.15 Absorption and
Dispersion profiles around an
atomic absorption transition

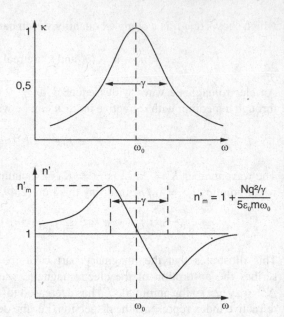

In the vicinity of an eigenfrequency ω_0, where $|\omega - \omega_0| \ll \omega_0 \Rightarrow (\omega_0^2 - \omega^2) \approx 2\omega_0(\omega_0 - \omega)$ (2.51a, 2.51b) simplifies to

$$\kappa = \frac{Nq^2}{8\varepsilon_0 m\omega_0} \frac{\gamma}{(\omega_0 - \omega)^2 + (\gamma/2)^2} , \tag{2.52a}$$

$$n' = 1 + \frac{Nq^2}{4\varepsilon_0 m\omega_0} \frac{\omega_0 - \omega}{(\omega_0 - \omega)^2 + (\gamma/2)^2} . \tag{2.52b}$$

In Fig. 2.15 the absorption profile $\kappa(\omega)$ and the dispersion profile $n'(\omega)$ are plotted in the vicinity of an eigenfrequency ω_0. The extrema of n' appear at $\omega_m = \omega_0 \pm \gamma$.

Generally the absorption of light is described by its intensity attenuation rather than its amplitude decrease since it is the intensity not the amplitude which is measured in most experiments. A plane wave with the intensity $I(\omega)$ travelling into the z-direction through an absorbing medium suffers after a pathlength dz the intensity attenuation

$$dI = -\alpha I \, dz . \tag{2.53}$$

The absorption coefficient $\alpha(\omega)$ [cm^{-1}] gives the relative intensity decrease dI/I along the absorption pathlength $dz = 1$ cm. If α is independent of the incident intensity I (linear absorption) the integration of (2.53) yields *Beer's absorption law*

$$I(\omega) = I_0 e^{-\alpha(\omega)z} . \tag{2.54}$$

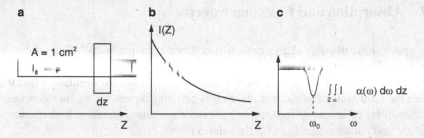

Figure 2.16 **a** Absorption of a parallel light wave with cross section A travelling into the z-direction. **b** Exponential decrease of the transmitted intensity $I(z)$. **c** Absorption profile of in incident spectral continuum with an intensity hole around the centrefrequency ω_0

Since the intensity $I = \varepsilon_0 c E^2$ is proportional to the square of the amplitude E the comparison of (2.54) with (2.49b) gives the relation

$$\boxed{\alpha = 4\pi\kappa/\lambda = 2K\kappa}\quad \text{with } K = 2\pi/\lambda . \tag{2.55}$$

In summary:

The absorption coefficient $\alpha(\omega)$ is proportional to the imaginary part κ of the refractive index $n = n' - i\kappa$. In the vicinity of an oscillator eigenfrequency ω_0 the absorption coefficient $\alpha(\omega) \sim \kappa(\omega)$ has a Lorentzian profile (2.52a). The real part $n'(\omega)$ of $n(\omega)$ describes the dispersion. For $|\omega - \omega_0| \ll \omega_0$ its profile is proportional to the derivative $d\alpha/d\omega$.

The power absorbed in the volume element $\Delta V = A \cdot \Delta z$ on the path length Δz from the incident monochromatic wave with cross section F and intensity I is

$$\Delta P(\omega) = \alpha(\omega)I(\omega)\Delta V . \tag{2.56a}$$

If the incident radiation has a spectral continuum with bandwidth $\Delta\omega$ and the absorption coefficient in the vicinity of an absorption line has the bandwidth $\delta\omega \ll \Delta\omega$ the absorbed power is

$$\Delta P = \int \alpha(\omega)I(\omega)\,d\omega\,\Delta V . \tag{2.56b}$$

The transmitted intensity $I_t(\omega)$ shows a hole with width $\delta\omega$ around the centre frequency ω_0 of the absorption profile (Fig. 2.16). For $\delta\omega > \Delta\omega$ the integral extends from $\omega_0 - \Delta\omega/2$ to $\omega_0 + \Delta\omega/2$, where ω_0 is the centre frequency of the absorbing atomic transition and $\alpha(\omega)$ does not change much within this interval. It can be therefore extracted from the integral. The total transmitted intensity decreases without a hole in the spectral distribution.

In the next section we will check how the classical model can be modified for its application to real atoms or molecules.

2.7   Absorption and Emission Spectra

The spectral distribution of the radiant flux from a source is called its emission *spectrum*. The thermal radiation discussed in Sect. 2.2 has a *continuous* spectral distribution described by its spectral energy density (2.13). *Discrete* emission spectra, where the radiant flux has distinct maxima at certain frequencies ν_{ik}, are generated by transitions of atoms or molecules between two bound states, a higher energy state E_k and a lower state E_i, with the relation

$$h\nu_{ik} = E_k - E_i \; . \tag{2.57}$$

In a spectrograph (see Sect. 4.1 for a detailed description) the entrance slit S is imaged into the focal plane B of the camera lens. Because of dispersive elements in the spectrograph, the position of this image depends on the wavelength of the incident radiation. In a discrete spectrum each wavelength λ_{ik} produces a separate line in the imaging plane, provided the spectrograph has a sufficiently high resolving power (Fig. 2.17). Discrete spectra are therefore also called *line spectra*, as opposed to *continuous spectra* where the slit images form a continuous band in the focal plane, even for spectrographs with infinite resolving power.

If radiation with a continuous spectrum passes through a gaseous molecular sample, molecules in the lower state E_i may absorb radiant power at the eigenfrequencies $\nu_{ik} = (E_k - E_i)/h$, which is thus missing in the transmitted power. The difference in the spectral distributions of incident minus transmitted power is the *absorption spectrum* of the sample. The absorbed energy $h\nu_{ik}$ brings a molecule into the higher energy level E_k. If these levels are bound levels, the resulting spectrum is a discrete absorption spectrum. If E_k is above the dissociation limit or above the ionization energy, the absorption spectrum becomes continuous. In Fig. 2.18 both cases are schematically illustrated for atoms (a) and molecules (b).

Examples of discrete absorption lines are the Fraunhofer lines in the spectrum of the sun, which appear as dark lines in the bright continuous spectrum (Fig. 2.19).

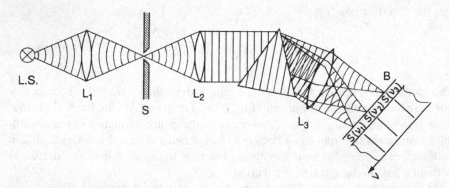

Figure 2.17 Spectral lines in a discrete spectrum as images of the entrance slit of a spectrograph

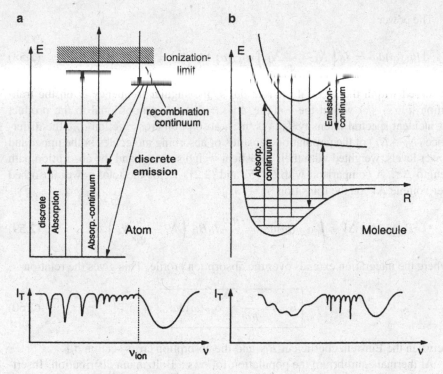

Figure 2.18 Schematic diagram to illustrate the origin of discrete and continuous absorption and emission spectra for atoms (**a**) and molecules (**b**)

Figure 2.19 Prominent Fraunhofer absorption lines within the visible and near-UV spectral range

They are produced by atoms in the sun's atmosphere that absorb at their specific eigenfrequencies the continuous blackbody radiation from the sun's photosphere.

2.7.1   Absorption Cross Section and Einstein Coefficients

A measure of the absorption strength is the absorption cross section σ_{ik}. Each photon passing through the circular area $\sigma_{ik} = \pi r_{ik}^2$ around the atom is absorbed on the transition $|i\rangle \rightarrow |k\rangle$.

The power

$$\mathrm{d}P_{ik}(\omega)\mathrm{d}\omega = I_0 \left(N_i - \frac{g_i}{g_k} N_k \right) \sigma_{ik}(\omega) A \Delta z \mathrm{d}\omega = I_0 \alpha_{ik}(\omega) \Delta V \mathrm{d}\omega , \qquad (2.58)$$

absorbed within the spectral interval $\mathrm{d}\omega$ at the angular frequency ω on the transition $|i\rangle \rightarrow |k\rangle$ within the volume $\Delta V = A \Delta z$ is proportional to the product of incident spectral intensity I_0 [Ws/m^2], absorption cross section σ_{ik}, the difference $(N_i - N_k)$ of the population densities of absorbing molecules in the upper and lower levels, weighted with their statistical weights g_i, g_k, and the absorption path length Δz. A comparison with (2.15) and (2.21) yields the total power absorbed per volume ΔV on the transition $|i\rangle \rightarrow |k\rangle$:

$$P_{ik} = I_0 \cdot \Delta V \cdot \int \alpha_{ik}(\omega)\mathrm{d}\omega = \frac{\hbar\omega}{c} I_0 B_{ik} \left(N_i - \frac{g_i}{g_k} N_k \right) \Delta V , \qquad (2.59)$$

where the integration extends over the absorption profile. This gives the relation

$$\boxed{ B_{ik} = \frac{c}{\hbar\omega} \int \sigma_{ik}(\omega)\mathrm{d}\omega , } \qquad (2.60)$$

between the Einstein coefficient B_{ik} and the absorption cross section σ_{ik}.

At thermal equilibrium the population follows a Boltzmann distribution. Inserting (2.18) yields the power absorbed within the volume $\Delta V = A \Delta z$ by a sample with molecular density N and temperature T out of an incident beam with the cross section A and intensity I

$$\begin{aligned}
P_{ik} &= (N/Z)g_i(\mathrm{e}^{-E_i/kT} - \mathrm{e}^{-E_k/kT}) A \Delta z \int I_0 \sigma_{ik}\mathrm{d}\omega , \\
&= I_0 \sigma_{ik}(\omega_0)(N/Z)g_i(\mathrm{e}^{-E_i/kT} - \mathrm{e}^{-E_k/kT})\Delta V ,
\end{aligned} \qquad (2.60a)$$

for a monochromatic laser with $I_0(\omega) = I_0 \delta(\omega - \omega_0)$.

In the far-infrared region is $\Delta E = (E_k - E_i) \ll kT$. In this case the exponential function can be approximated by $\exp(-x) \approx 1 - x$. This converts (2.60a) into

$$P_{ik} \approx I_0 \sigma_{ik}(\omega_0)g_i(N/Z) \cdot (\Delta E/kT)\Delta V . \qquad (2.60b)$$

In case of a collimated laser beam passing through the absorbing sample the volume element $\Delta V = A \cdot \Delta z$ is the product of the laser beam cross section A and the absorption path length Δz.

The absorption lines are only measurable if the absorbed power is sufficiently high, which means that the density N or the absorption path length Δz must be large enough. Furthermore, the difference in the two Boltzmann factors in (2.60a) should be sufficiently large, which means E_i should be not much larger than kT, but $E_k \gg kT$. Absorption lines in gases at thermal equilibrium are therefore only intense for transitions from low-lying levels E_i that are thermally populated.

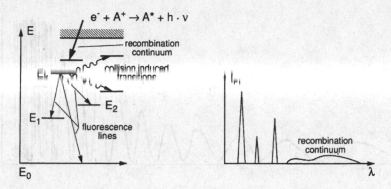

Figure 2.20 Discrete and continuous emission spectrum and the corresponding level diagram, which also shows radiationless transitions induced by inelastic collisions (*wavy lines*)

It is, however, possible to pump molecules into higher energy states by various excitation mechanisms such as optical pumping or electron excitation. This allows the measurement of absorption spectra for transition from these states to even higher molecular levels (Vol. 2, Sect. 5.3).

2.7.2 Fluorescence Spectra

The excited molecules release their energy either by spontaneous or induced emission or by collisional deactivation (Fig. 2.20). The spatial distribution of spontaneous emission depends on the orientation of the excited molecules and on the symmetry properties of the excited state E_k. If the molecules are randomly oriented, the spontaneous emission (often called *fluorescence*) is isotropic.

The fluorescence spectrum (emission spectrum) emitted from a discrete upper level E_k consists of discrete lines if the terminating lower levels E_i are bound states. A continuum is emitted if E_i belongs to a repulsive state of a molecule that dissociates. As an example, the fluorescence spectrum of the $^3\Pi \rightarrow {}^3\Sigma$ transition of the NaK molecule is shown in Fig. 2.21. It is emitted from a selectively excited bound vibrational level of the $^3\Pi$ state that has been populated by optical pumping with an argon laser. The fluorescence terminates into a repulsive $^3\Sigma$ state, which has a shallow van der Waals minimum. Transitions terminating to energies E_k above the dissociation energy form the continuous part of the spectrum, whereas transitions to lower bound levels in the van der Waals potential well produce discrete lines. The modulation of the continuum reflects the modulation of the transmission probability due to the maxima and nodes of the vibrational wave function $\psi_{\text{vib}}(R)$ in the upper bound level [54].

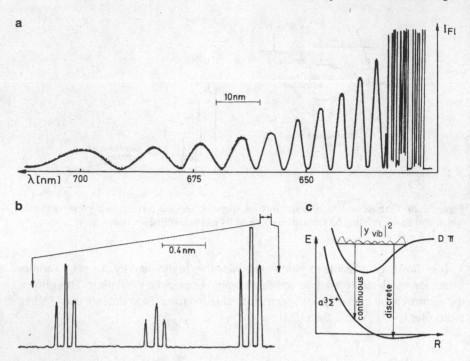

Figure 2.21 Continuous "bound–free" and discrete "bound–bound" fluorescence transitions of the NaK molecule observed upon laser excitation at $\lambda = 488$ nm: **a** part of the spectrum; **b** enlargement of three discrete vibrational bands; **c** level scheme [54]

2.7.3 Oscillator Strength

In contrast to the harmonic oscillator of our classical model for absorption, real atoms or molecules have many energy levels and therefore also many possible transitions from lower thermally populated levels. This means that there are many absorption lines at different frequencies ω_i. The magnitude of the absorption coefficient α depends on the population density of the absorbing level and on the transition probabilities for the different transitions. These transition probabilities can be only calculated by using quantum mechanical methods, but there are several experimental techniques for their determination (see next section).

The transition probability can be vividly described by the concept of **oscillator strength** f which relates the classical model with a more realistic approach. It can be explained as follows:

An atom with an electron in an outer shell (,,*Leucht-Electron*") which can be excited by absorption of a photon from a lower state E_i into higher electronic states E_k can be treated as classical oscillator with the oscillator strength $f = 1$. However, the total absorption of the atom in level E_i is due to the sum of all transitions into the levels E_k. Each of these transitions contribute only a fraction $f_{ik} < 1$ to the total absorption, which is called the oscillator strength of this transition. With other words: N **atoms absorb on a transition** $|i\rangle \rightarrow |k\rangle$ **as much as** $f_{ik} \cdot N$ **classical oscillators.**

Since the sum of all transitions from level E_i should have the absorption strength of a classical oscillator we obtain the relation

$$\sum_k f_{ik} = 1 . \tag{2.61}$$

The summation extends to all upper levels E_k which can be reached by allowed transitions from level E_i including the ionization or dissociation continuum.

When an excited level E_k is populated, also induced emission can take place which reduces the total absorption. The corresponding oscillator strengths f_{ki} for induced emission are therefore negative.

Example 2.6

1. For the transitions of the resonance lines in the sodium atom is $f(3\,S_{1/2} \rightarrow P_{3/2}) = 0.33$ and $f(3\,S_{1/2} \rightarrow 3\,P_{3/2}) = 0.66$. This illustrates that the two fine-structure components of the yellow sodium line contribute already nearly 99 % of the total absorption from the ground state $3\,S_{1/2}$. All other possible transitions carry only 1 % of the total absorption.
2. For the hydrogen Atom H is $f(1\,S_{1/2} \rightarrow 2\,p) = 0.4162$. The transition probability is $A_{ik} = 4.7 \times 10^8 \text{ s}^{-1}$.

Taking into account the oscillator strengths the equations (2.51a, 2.51b) are modified to

$$\kappa_i = \frac{N_i e^2}{2\varepsilon_0 m} \sum_k \frac{\omega f_{ik} \gamma_{ik}}{(\omega_{ik}^2 - \omega^2)^2 + \gamma_{ik}^2 \omega^2} , \tag{2.62a}$$

$$n_i' = 1 + \frac{N_i e^2}{2\varepsilon_0 m} \frac{(\omega_{ik}^2 - \omega^2) f_{ik}}{(\omega_{ik}^2 - \omega^2)^2 + \gamma_{ik}^2 \omega^2} , \tag{2.62b}$$

where N_i is the number density of atoms per volume in level $|i\rangle$ and γ_i is the full spectral width (FWHM) of the absorbing transition $|i\rangle \rightarrow |k\rangle$. In the vicinity of an absorption line ($|\omega_0 - \omega| \ll \omega_0$) the equations simplify according to (2.52a, 2.52b). The equations (2.62a, 2.62b) are valid for atoms or molecules at rest. The thermal motion of atoms broadens the spectral width γ_{ik} of the transition (Doppler broadening, see Sect. 3.2).

The oscillator strengths can be measured by several experimental techniques. One of them uses the measurement of absorption profiles or dispersion profiles. Lifetime measurement (see Sect. 2.10), which give direct values of transition probabilities and therefore information on the natural line width of a transition (see Sect. 3.1) are a valuable source for the determination of oscillator strengths.

2.7.4 Relation Between Absorption Cross Section and Einstein Coefficients

At a density N_i of molecules in level $|i\rangle$ and a negligible population N_k in the upper level of a transition the power absorbed from the incident light wave within the volume element ΔV is

$$\frac{dW_{ik}}{dt} = N_i \cdot B_{ik} \cdot \rho(\omega) \cdot \hbar\omega \cdot \Delta V . \tag{2.63}$$

The comparison with (2.56a, 2.56b) gives with $I(\omega) = c \cdot \rho(\omega)$ for a plane incident light wave

$$B_{ik} = \frac{2\pi c}{N_i \hbar \omega_{ik}} \int_{\omega_0 - \Delta\omega/2}^{\omega_0 + \Delta\omega/2} \alpha_{ik}(\omega)\, d\omega \approx \frac{2\pi c}{N_i \hbar \omega_{ik}} \int_{0}^{+\infty} \alpha_{ik}(\omega)\, d\omega . \tag{2.64}$$

The integration limits can be extended from 0 to $\pm\infty$, because the absorption coefficient outside the absorption profile is zero.

The Einstein coefficient for absorption B_{ik} is proportional to the absorption coefficient α_{ik} integrated over the absorption profile.

Inserting for $\alpha = (2\omega/c)\kappa$ the expression (2.62a) one obtains in the vicinity of the absorption centre frequency the relation between Einstein coefficient B_{ik} and oscillator strength f_{ik}

$$B_{ik}^{(\omega)} = \frac{e^2 f_{ik}}{4\hbar\omega_{ik}^2 \varepsilon_0 m} \int_{0}^{\infty} \frac{\omega \gamma_{ik}\, d\omega}{(\omega_{ik}^2 - \omega)^2 + (\gamma_{ik}/2)^2} . \tag{2.65}$$

The integral is readily solved and has the value $2\pi\omega_{ik}$. This gives finally the relation

$$\boxed{B_{ik}^{(\omega)} = \frac{\pi e^2}{2m\varepsilon_0 \hbar \omega_{ik}} f_{ik} , \quad B_{ik}^{(\nu)} = \frac{e^2}{4m\varepsilon_0 h \nu_{ik}} f_{ik} .} \tag{2.66}$$

Note, that $\rho(\nu)$ is the energy density within the spectral interval $d\nu = 1\,\mathrm{s}^{-1} \Rightarrow d\omega = 2\pi\, d\nu = 2\pi\,\mathrm{s}^{-1}$. Since the transition probability $B_{ik}^{(\nu)}\rho(\nu) = B_{ik}^{(\omega)}\rho(\omega)$ is independent of the choice of ν or ω it follows that $\rho(\nu) = 2\pi\rho(\omega)$ and therefore $B_{ik}^{(\nu)} = (1/2\pi)B_{ik}^{(\omega)}$. One has to pay attention to this difference when comparing formulas expressed in ν or in ω.

2.7.5 Integrated Absorption and Line Strength

If the Einstein coefficients are known, the oscillator strength can be derived and the absorption coefficient and the dispersion are obtained, or, vice versa, from measured oscillator strengths the Einstein coefficients can be determined.

The optical absorption coefficient σ_{ik} for a transition $|i\rangle \to |k\rangle$ in a gas with N_i absorbing atoms per cm^3 is related to the absorption coefficient α_{ik} by

$$\alpha_{ik} = N_i \cdot \sigma_{ik} .$$ (2.67)

Inserting this into (2.64) yields

$$B_{ik}^{(\omega)} = \frac{2\pi c}{\hbar\omega} \int \sigma_{ik}(\omega)\, d\omega , \quad B_{ik}^{(v)} = \frac{c}{hv} \int \sigma_{ik}(v)\, dv .$$ (2.68)

The integral

$$S_{ik} = \int \sigma_{ik}(\omega)\, d\omega = \int \sigma_{ik}(v)\, dv$$ (2.69)

is called the **line strength** of the transition $|i\rangle \to |k\rangle$.

Defining a mean absorption cross section

$$\bar{\sigma}_{ik} = \frac{1}{\Delta v} \int \sigma_{ik}(v)\, dv ,$$ (2.70)

where Δv is the full width (HWHM) of the absorption line (see Sect. 3.1), we can write the line strength as the product

$$S_{ik} = \Delta v \cdot \bar{\sigma}_{ik}$$ (2.71)

of linewidth Δv and mean absorption cross section $\bar{\sigma}_{ik}$ (averaged over the linewidth). From (2.22) and (2.68) we obtain the relation

$$\bar{\sigma}_{ik} = \frac{\lambda^2 A_{ik}}{8\pi \Delta v}$$ (2.72)

between mean absorption cross section $\bar{\sigma}_{ik}$ and Einstein Coefficient A_{ik} for spontaneous emission. If Δv is the natural linewidth Δv_n it follows (see Sect. 3.1)

$$\Delta v_n = A_{ik}/2\pi$$

and we get the interesting result for the mean absorption cross section

$$\bar{\sigma}_{ik} = (\lambda/2)^2 .$$ (2.73)

The mean absorption cross section for the absorption within the natural linewidth of a transition with oscillator strength f_{ik} is about $f_{ik} \cdot \lambda^2/4$ and depends therefore only on the wavelength λ and the oscillator strength.

Example 2.7
The mean absorption cross section of the Na D-line at $\lambda = 589\,\text{nm}$ is according to (2.73) about $\sigma_{ik} = 9 \times 10^{-10}\,\text{cm}^2$ (the measured value is $10 \times 10^{-10}\,\text{cm}^2$). At a vapour pressure of $10^{-6}\,\text{mb}$ the atom density is $N_i = 2.5 \times 10^{10}\,\text{cm}^{-3}$. The absorption coefficient is then $\alpha_{ik} = 2\,\text{cm}^{-1}$. This means that the intensity I_0 of a laser beam, tuned to the centre wavelength of the Na transition has decreased already to $(1/e)I_0$ after 0.5 cm absorption path length.

2.8   Transition Probabilities

The intensities of spectral lines depend not only on the population density of the molecules in the absorbing or emitting level but also on the transition probabilities of the corresponding molecular transitions. If these probabilities are known, the population density can be obtained from measurements of line intensities. This is very important, for example, in astrophysics, where spectral lines represent the main source of information from the extraterrestrial world. Intensity measurements of absorption and emission lines allow the concentration of the elements in stellar atmospheres or in interstellar space to be determined. Comparing the intensities of different lines of the same element (e.g., on the transitions $E_i \rightarrow E_k$ and $E_e \rightarrow E_k$ from different upper levels E_i, E_e to the same lower level E_k) furthermore enables us to derive the temperature of the radiation source from the relative population densities N_i, N_e in the levels E_i and E_e at thermal equilibrium according to (2.18). *All these experiments, however, demand a knowledge of the corresponding transition probabilities.*

There is another aspect that makes measurements of transition probabilities very attractive with regard to a more detailed knowledge of molecular structure. Transition probabilities derived from computed wave functions of upper and lower states are much more sensitive to approximation errors in these functions than are the energies of these states. Experimentally determined transition probabilities are therefore well suited to test the validity of calculated approximate wave functions. A comparison with computed probabilities allows theoretical models of electronic charge distributions in excited molecular states to be improved [55, 56].

2.8.1   Lifetimes, Spontaneous and Radiationless Transitions

The probability $\mathcal{P}_{ik}$ that an excited molecule in the level E_i makes a transition to a lower level E_k by spontaneous emission of a fluorescence quantum $h\nu_{ik} =$

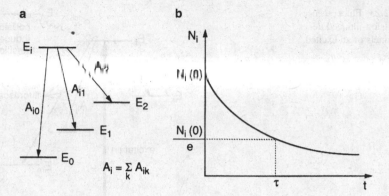

Figure 2.22   Radiative decay of the level $|i\rangle$: **a** Level scheme; **b** decay curve $N_i(t)$

$E_i - E_k$ is, according to (2.17), related to the Einstein coefficient A_{ik} by

$$d\mathcal{P}_{ik}/dt = A_{ik} \; .$$

When several transition paths from E_i to different lower levels E_k are possible (Fig. 2.22), the total transition probability is given by

$$A_i = \sum_k A_{ik} \; . \tag{2.74}$$

The decrease dN_i of the population density N_i during the time interval dt due to radiative decay is then

$$dN_i = -A_i N_i dt \; . \tag{2.75}$$

Integration of (2.75) yields

$$N_i(t) = N_{i0} e^{-A_i t} \; , \tag{2.76}$$

where N_{i0} is the population density at $t = 0$.

After the time $\tau_i = 1/A_i$ the population density N_i has decreased to $1/e$ of its initial value at $t = 0$. The time τ_i represents the *mean spontaneous lifetime* of the level E_i as can be seen immediately from the definition of the mean time

$$\bar{\tau_i} = \int\limits_0^\infty t \mathcal{P}_i(t) dt = \int\limits_0^\infty t A_i e^{-A_i t} dt = \frac{1}{A_i} = \tau_i \; , \tag{2.77}$$

where $\mathcal{P}_i(t) dt$ is the probability that one atom in the level E_i makes a spontaneous transition within the time interval between t and $t + dt$.

Figure 2.23 Fluorescence- and collision-induced decay channels of an excited level $|i\rangle$

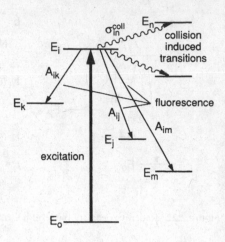

The *radiant power* emitted from N_i molecules on the transition $E_i \rightarrow E_k$ is

$$P_{ik} = N_i h \nu_{ik} A_{ik} \; . \tag{2.78}$$

If several transitions $E_i \rightarrow E_k$ from the same upper level E_i to different lower levels E_k are possible, the radiant powers of the corresponding spectral lines are proportional to the product of the Einstein coefficient A_{ik} and the photon energy $h\nu_{ik}$. The relative radiation intensities in a certain direction may also depend on the spatial distribution of the fluorescence, which can be different for the different transitions.

The level E_i of the molecule A can be depopulated not only by spontaneous emission but also by collision-induced *radiationless* transitions (Fig. 2.23). The probability $\mathrm{d}\mathcal{P}_{ik}^{\mathrm{coll}}/\mathrm{d}t$ of such a transition depends on the density N_B of the collision partner B, on the mean relative velocity $\bar{v}$ between A and B, and on the collision cross section $\sigma_{ik}^{\mathrm{coll}}$ for an inelastic collision that induces the transition $E_i \rightarrow E_k$ in the molecule A

$$\mathrm{d}\mathcal{P}_{ik}^{\mathrm{coll}}/\mathrm{d}t = \bar{v} N_B \sigma_{ik}^{\mathrm{coll}} \; . \tag{2.79}$$

When the excited molecule $A(E_i)$ is exposed to an intense radiation field, the *induced emission* may become noticeable. It contributes to the depopulation of level E_i in a transition $|i\rangle \rightarrow |k\rangle$ with the probability

$$\mathrm{d}\mathcal{P}_{ik}^{\mathrm{ind}}/\mathrm{d}t = \rho(\nu_{ik}) B_{ik} \; . \tag{2.80}$$

The total transition probability that determines the effective lifetime of a level E_i is then the sum of spontaneous, induced, and collisional contributions, and the mean lifetime τ_i^{eff} becomes

$$\frac{1}{\tau_i^{\mathrm{eff}}} = \sum_k \left[A_{ik} + \rho(\nu_{ik}) B_{ik} + N_B \sigma_{ik} \bar{v} \right] \; . \tag{2.81}$$

Figure 2.24 Two-level
system with open decay
channels into other levels
interacting with an EM field

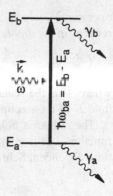

Measuring the effective lifetime τ_i^{eff} as a function of the exciting radiation intensity
and also its dependence on the density N_B of collision partners (Stern–Vollmer
plot) allows one to determine the three transition probabilities separately (Vol. 2,
Sect. 8.3).

2.8.2 Semiclassical Description: Basic Equations

In the semiclassical description, the radiation incident upon an atom is described by
a classical electromagnetic (EM) plane wave

$$E = E_0 \cos(\omega t - kz) . \tag{2.82a}$$

The atom, on the other hand, is treated quantum-mechanically. In order to simplify
the equations, we restrict ourselves to a two-level system with the eigenstates E_a
and E_b (Fig. 2.24).

Until now laser spectroscopy was performed in spectral regions where the wavelength λ was large compared to the diameter d of an atom (e.g., in the visible
spectrum λ is 500 nm, but d is only about 0.5 nm). For $\lambda \gg d$, the phase of
the EM wave does not change much within the volume of an atom because $kz =
(2\pi/\lambda)z \ll 1$ for $z \leq d$. We can therefore neglect the spatial derivatives of the
field amplitude (dipole approximation). In a coordinate system with its origin in
the center of the atom, we can assume $kz \simeq 0$ within the atomic volume, and write
(2.82a) in the form

$$E = E_0 \cos \omega t = A_0(\mathrm{e}^{\mathrm{i}\omega t} + \mathrm{e}^{-\mathrm{i}\omega t}) \quad \text{with } |A_0| = \tfrac{1}{2}E_0 . \tag{2.82b}$$

The Hamiltonian operator

$$\mathcal{H} = \mathcal{H}_0 + \mathcal{V} , \tag{2.83}$$

of the atom interacting with the light field can be written as a sum of the unperturbed Hamiltonian $\mathcal{H}_0$ of the free atom without the light field plus the perturbation

operator $\mathcal{V}$, which describes the interaction of the atom with the field and which reduces in the *dipole approximation* to

$$\mathcal{V} = \boldsymbol{p} \cdot \boldsymbol{E} = \boldsymbol{p} \cdot \boldsymbol{E}_0 \cos \omega t , \qquad (2.84)$$

where $\mathcal{V}$ is the scalar product of the dipole operator $\boldsymbol{p} = -e \cdot \boldsymbol{r}$ and the electric field $\boldsymbol{E}$ of the electromagnetic wave.

The radiation field causes transitions in the atom. This means that the eigenfunctions of the atom become time-dependent. The general solution $\psi(\boldsymbol{r}, t)$ of the time-dependent Schrödinger equation

$$\mathcal{H}\psi = i\hbar \frac{\partial \psi}{\partial t} \qquad (2.85)$$

can be expressed as a linear superposition

$$\psi(\boldsymbol{r}, t) = \sum_{n=1}^{\infty} c_n(t) u_n(\boldsymbol{r}) e^{-iE_n t/\hbar} , \qquad (2.86)$$

of the eigenfunctions of the unperturbed atom

$$\phi_n(\boldsymbol{r}, t) = u_n(\boldsymbol{r}) e^{-iE_n t/\hbar} . \qquad (2.87)$$

The spatial parts $u_n(\boldsymbol{r})$ of these eigenfunctions are solutions of the time-independent Schrödinger equation

$$\mathcal{H}_0 u_n(\boldsymbol{r}) = E_n u_n(\boldsymbol{r}) , \qquad (2.88)$$

and satisfy the orthogonality relations[1]

$$\int u_i^* u_k \mathrm{d}\tau = \delta_{ik} . \qquad (2.89)$$

For our two-level system with the eigenstates $|a\rangle$ and $|b\rangle$ and the energies E_a and E_b, (2.86) reduces to a sum of two terms

$$\psi(\boldsymbol{r}, t) = a(t) u_a e^{-iE_a t/\hbar} + b(t) u_b e^{-iE_b t/\hbar} . \qquad (2.90)$$

The coefficients $a(t)$ and $b(t)$ are the time-dependent *probability amplitudes* of the atomic states $|a\rangle$ and $|b\rangle$. This means that the value $|a(t)|^2$ gives the probability of finding the system in level $|a\rangle$ at time t. Obviously, the relation $|a(t)|^2 + |b(t)|^2 = 1$ must hold at all times t, if decay into other levels is neglected.

Substituting (2.90) and (2.83) into (2.85) gives

$$i\hbar \dot{a}(t) u_a e^{-iE_a t/\hbar} + i\hbar \dot{b}(t) u_b e^{-iE_b t/\hbar} = a \mathcal{V} u_a e^{-iE_a t/\hbar} + b \mathcal{V} u_b e^{-iE_b t/\hbar} , \quad (2.91)$$

---

[1] Note that in (2.86–2.88) a nondegenerate system has been assumed.

where the relation $\mathcal{H}_0 u_n = E_n u_n$ has been used to cancel equal terms on both sides. Multiplication with $u_n^*(n = a, b)$ and spatial integration results in the following two equations

$$\dot{a}(t) = -(i/\hbar)[a(t)V_{aa} + b(t)V_{ab}e^{-i\omega_{ba}t}] , \qquad (2.92a)$$

$$\dot{b}(t) = -(i/\hbar)[b(t)V_{bb} + a(t)V_{ba}e^{-i\omega_{ab}t}] , \qquad (2.92b)$$

with $\omega_{ab} = (E_a - E_b)/\hbar = -\omega_{ba}$ and with the spatial integral

$$V_{ab} = \int u_a^* V u_b d\tau = -eE \int u_a^* r u_b d\tau . \qquad (2.93a)$$

Since r has odd parity, the integrals V_{aa} and V_{bb} vanish when integrating over all coordinates from $-\infty$ to $+\infty$. The quantity

$$\boldsymbol{D}_{ab} = \boldsymbol{D}_{ba} = -e \int u_a^* r u_b d\tau , \qquad (2.93b)$$

is called the atomic *dipole matrix element*. It depends on the stationary wave functions u_a and u_b of the two states $|a\rangle$ and $|b\rangle$ and is determined by the charge distribution in these states.

The expectation value $\boldsymbol{D}_{ab}$ of the dipole matrix element for our two-level system should be distinguished from the expectation value of the dipole moment in a specific state $|\psi\rangle$

$$\boldsymbol{D}_n = -e \int u_n^* r u_n \, d\tau = 0 , \quad n = a, b , \qquad (2.94a)$$

which is zero because the integrand is an odd function of the coordinates. Using (2.90) and the abbreviation $\omega_{ba} = (E_b - E_a)/\hbar = -\omega_{ab}$, the general dipole matrix element $\boldsymbol{D} = -e \int \psi^* r \psi \, d\tau$ can be expressed by the coefficients $a(t)$ and $b(t)$, and by the matrix element $\boldsymbol{D}_{ab}$ as

$$\boldsymbol{D} = +\boldsymbol{D}_{ab}(a^* b e^{-i\omega_{ba}t} + a b^* e^{+i\omega_{ba}t}) = D_0 \cos(\omega_{ba}t + \varphi) , \qquad (2.94b)$$

with

$$D_0 = D_{ab}|a^* b| \quad \text{and} \quad \tan\varphi = -\frac{\Im\{a^* b\}}{\Re\{a^* b\}} .$$

Even without the external field, the expectation value of the atomic dipole moment oscillates with the eigenfrequency ω_{ba} and the amplitude $|a^* \cdot b|$ if the wavefunction of the atomic system can be represented by the superposition (2.93a, 2.93b). The time average of this oscillation's dipole moment is zero!

Using (2.82b) for the EM field and the abbreviation

$$\Omega_{ab} = D_{ab} E_0/\hbar = 2 D_{ab} A_0/\hbar = \Omega_{ba} \qquad (2.95)$$

which depends on the field amplitude E_0 and (2.92a, 2.92b) reduces to

$$\dot{a}(t) = -(i/2)\Omega_{ab}\left(e^{i(\omega-\omega_{ba})t} + e^{-i(\omega+\omega_{ba})t}\right)b(t) \,, \qquad (2.96a)$$

$$\dot{b}(t) = -(i/2)\Omega_{ab}\left(e^{-i(\omega-\omega_{ba})t} + e^{i(\omega+\omega_{ba})t}\right)a(t) \,. \qquad (2.96b)$$

where $\omega_{ba} = -\omega_{ba} > 0$.

These are the basic equations that must be solved to obtain the probability amplitudes $a(t)$ and $b(t)$. The frequency Ω_{ab} is called the **Rabi frequency**. Its physical interpretation will be discussed in Sect. 2.8.6.

2.8.3 Weak-Field Approximation

Suppose that at time $t = 0$, the atoms are in the lower state E_a, which implies that $a(0) = 1$ and $b(0) = 0$. We assume the field amplitude A_0 to be sufficiently small so that for times $t < T$ the population of E_b remains small compared with that of E_a, i.e., $|b(t < T)|^2 \ll 1$. Under this *weak-field condition* we can solve (2.96a, 2.96b) with an iterative procedure starting with $a = 1$ and $b = 0$. Using thermal radiation sources, the field amplitude A_0 is generally small enough to make the first iteration step already sufficiently accurate.

With these assumptions the first approximation of (2.96a, 2.96b) gives

$$\dot{a}(t) = 0 \,, \qquad (2.97a)$$
$$\dot{b}(t) = -(i/2)\Omega_{ba}\left(e^{i(\omega_{ba}-\omega)t} + e^{i(\omega_{ba}+\omega)t}\right) \,. \qquad (2.97b)$$

With the initial conditions $a(0) = 1$ and $b(0) = 0$, integration of (2.97a, 2.97b) from 0 to t yields

$$a(t) = a(0) = 1 \,, \qquad (2.98a)$$
$$b(t) = \left(\frac{\Omega_{ab}}{2}\right)\left(\frac{e^{i(\omega-\omega_{ba})t} - 1}{\omega - \omega_{ba}} - \frac{e^{i(\omega+\omega_{ba})t} - 1}{\omega + \omega_{ba}}\right) \,. \qquad (2.98b)$$

For $E_b > E_a$ the term $\omega_{ba} = (E_b - E_a)/\hbar$ is positive. In the transition $E_a \rightarrow E_b$, the atomic system absorbs energy from the radiation field. Noticeable absorption occurs, however, only if the field frequency ω is close to the eigenfrequency ω_{ba}. In the optical frequency range this implies that $|\omega_{ba} - \omega| \ll \omega_{ba}$. The second term in (2.98b) is then small compared to the first one and may be neglected. This is called the *rotating-wave approximation* for only that term is kept in which the atomic wave functions and the field waves with the phasors $\exp(-i\omega_{ab}t)$ and $\exp(-i\omega t)$ rotate together.

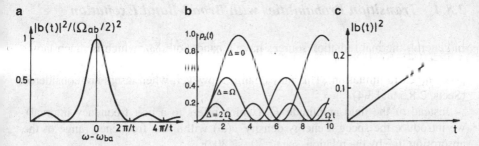

Figure 2.25 **a** Normalized transition probability for monochromatic excitation as a function of the detuning $(\omega - \omega_{ba})$ in the rotating-wave approximation; **b** probability of a transition to the upper level as a function of time for different detuning; **c** $|b(t)|^2$ under broadband excitation and weak fields

In the rotating-wave approximation we obtain from (2.98b) for the probability $|b(t)|^2$ that the system is at time t in the upper level E_b

$$|b(t)|^2 = \left(\frac{\Omega_{ab}}{2}\right)^2 \left(\frac{\sin(\omega - \omega_{ba})t/2}{(\omega - \omega_{ba})/2}\right)^2 . \tag{2.99}$$

Since we had assumed that the atom was at $t = 0$ in the lower level E_a, (2.99) *gives the transition probability for the atom to go from E_a to E_b during the time t.*

Figure 2.25a illustrates this transition probability as a function of the detuning $\Delta\omega = \omega - \omega_{ba}$. Equation (2.99) shows that $|b(t)|^2$ depends on the absolute value of the detuning $\Delta\omega = |\omega - \omega_{ba}|$ of the field frequency ω from the eigenfrequency ω_{ba}. When tuning the frequency ω into resonance with the atomic system $(\omega \to \omega_{ba})$, the second factor in (2.99) approaches the value t^2 because $\lim_{x\to 0}[(\sin^2 xt)/x^2] = t^2$. The transition probability at resonance,

$$|b(t)|^2_{\omega=\omega_{ba}} = \left(\frac{\Omega_{ab}}{2}\right)^2 t^2 , \tag{2.100}$$

increases proportionally to t^2. The approximation used in deriving (2.99) has, however, anticipated that $|b(t)|^2 \ll 1$. According to (2.100) and (2.95), this assumption for the resonance case is equivalent to

$$\left(\frac{\Omega_{ab}}{2}\right)^2 t^2 \ll 1 \quad \text{or} \quad t \ll T = \frac{2}{\Omega_{ab}} = \frac{\hbar}{D_{ab}E_0} . \tag{2.101}$$

Our small-signal approximation only holds if the interaction time t of the field (amplitude E_0) with the atom (matrix element D_{ab}) is restricted to $t \ll T = \hbar/(D_{ab}E_0)$. Because the spectral analysis of a wave with the finite detection time T gives the spectral width $\Delta\omega \simeq 1/T$ (see also Sect. 3.2), we cannot assume monochromaticity, but have to take into account the frequency distribution of the interaction term.

2.8.4   *Transition Probabilities with Broad-Band Excitation*

In general, thermal radiation sources have a bandwidth $\delta\omega$, which is much larger than the Fourier limit $\Delta\omega = 1/T$. Therefore, the finite interaction time imposes no extra limitation. This may change, however, when lasers are considered (Sects. 2.8.5 and 3.4).

Instead of the field amplitude E_0 (which refers to a unit frequency interval), we introduce the spectral energy density $\rho(\omega)$ within the frequency range of the absorption line by the relation, see (2.30a–2.30c),

$$\int \rho(\omega)d\omega = \epsilon_0 E_0^2/2 = 2\epsilon_0 A_0^2 .$$

We can now generalize (2.99) to include the interaction of broadband radiation with our two-level system by integrating (2.99) over all frequencies ω of the radiation field. This yields the total transition probability $\mathcal{P}_{ab}(t)$ within the time T. If $\boldsymbol{D}_{ab} \parallel \boldsymbol{E}_0$, we obtain with $\Omega_{ab} = D_{ab}E_0/\hbar$

$$\mathcal{P}_{ab}(t) = \int |b(t)|^2 d\omega = \frac{(\boldsymbol{D}_{ab})^2}{2\epsilon_0\hbar^2} \int \rho(\omega) \left(\frac{\sin(\omega_{ba} - \omega)t/2}{(\omega_{ba} - \omega)/2}\right)^2 d\omega . \quad (2.102)$$

For thermal light sources or broadband lasers, $\rho(\omega)$ is slowly varying over the absorption line profile. It is essentially constant over the frequency range where the factor $[\sin^2(\omega_{ba} - \omega)t/2]/[(\omega_{ba} - \omega)/2]^2$ is large (Fig. 2.25a). We can therefore replace $\rho(\omega)$ by its resonance value $\rho(\omega_{ba})$. The integration can then be performed, which gives the value $\rho(\omega_{ba})2\pi t$ for the integral because

$$\int\limits_{-\infty}^{\infty} \frac{\sin^2(xt)}{x^2} \, dx = 2\pi t .$$

For broadband excitation, the transition probability for the time interval between 0 and t

$$\mathcal{P}_{ab}(t) = \frac{\pi}{\epsilon_0\hbar^2} D_{ab}^2 \rho(\omega_{ba})t , \quad (2.103)$$

is linearly dependent on t (Fig. 2.25c).

For broadband excitation the *transition probability per second*

$$\frac{d}{dt}\mathcal{P}_{ab} = \frac{\pi}{\epsilon_0\hbar^2} D_{ab}^2 \rho(\omega_{ba}) , \quad (2.104)$$

becomes independent of time!

To compare this result with the Einstein coefficient B_{ab} derived in Sect. 2.3, we must take into account that the blackbody radiation was isotropic, whereas the EM wave (2.82a, 2.82b) used in the derivation of (2.104) propagates into one direction. For randomly oriented atoms with the dipole moment p, the averaged component of p^2 in the z-direction is $\langle p_z^2 \rangle = p^2 \langle \cos^2 \theta \rangle = p^2/3$.

In the case of isotropic radiation, the interaction term $D_{ab}^2 \rho(\omega_{ba})$ therefore has to be divided by a factor of 3. A comparison of (2.16) with the modified equation (2.104) yields

$$\frac{\mathrm{d}}{\mathrm{d}t} \mathcal{P}_{ab} = \frac{\pi}{3\epsilon_0 \hbar^2} \rho(\omega_{ba}) D_{ab}^2 = \rho(\omega_{ba}) B_{ab} \ . \tag{2.105}$$

With the definition (2.93a, 2.93b) for the dipole matrix element D_{ik}, the Einstein coefficient B_{ik} of induced absorption $E_i \rightarrow E_k$ finally becomes

$$B_{ik}^{\omega} = \frac{\pi e^2}{3\epsilon_0 \hbar^2} \left| \int u_i^* \boldsymbol{r} u_k \mathrm{d}\tau \right|^2 \quad \text{and} \quad B_{ik}^{\nu} = B_{ik}^{\omega}/2\pi \ . \tag{2.106}$$

Equation (2.106) gives the Einstein coefficient for a one-electron system where $r = (x, y, z)$ is the vector from the nucleus to the electron, and $u_n(x, y, z)$ denotes the one-electron wave functions.[2] From (2.106) we learn that the **Einstein coefficient B_{ik} is proportional to the squared transition dipole moment**.

So far we have assumed that the energy levels E_i and E_k are not degenerate, and therefore have the statistical weight factor $g = 1$. In the case of a degenerate level $|k\rangle$, the total transition probability ρB_{ik} of the transition $E_i \rightarrow E_k$ is the sum

$$\rho B_{ik} = \rho \sum_n B_{ik_n} \ ,$$

over all transitions to the sublevels $|k_n\rangle$ of $|k\rangle$. If level $|i\rangle$ is also degenerate, an additional summation over all sublevels $|i_m\rangle$ is necessary, taking into account that the population of each sublevel $|i_m\rangle$ is only the fraction N_i/g_i.

The Einstein coefficient B_{ik} for the transition $E_i \rightarrow E_k$ between the two degenerate levels $|i\rangle$ and $|k\rangle$ is therefore

$$B_{ik} = \frac{\pi}{3\epsilon_0 \hbar^2} \frac{1}{g_i} \sum_{m=1}^{g_i} \sum_{n=1}^{g_k} |D_{im k_n}|^2 = \frac{\pi}{3\epsilon_0 \hbar^2 g_i} S_{ik} \ . \tag{2.107}$$

The double sum is called the *line strength* S_{ik} of the atomic transition $|i\rangle \leftarrow |k\rangle$.

---

[2] Note that when using the frequency $\nu = \omega/2\pi$ instead of ω, the spectral energy density $\rho(\nu)$ per unit frequency interval is larger by a factor of 2π than $\rho(\omega)$ because a unit frequency interval $\mathrm{d}\nu = 1$ Hz corresponds to $\mathrm{d}\omega = 2\pi$ [Hz]. The right-hand side of (2.106) must then be divided by a factor of 2π, since $B_{ik}^{\nu} \rho(\nu) = B_{ik}^{\omega} \rho(\omega)$.

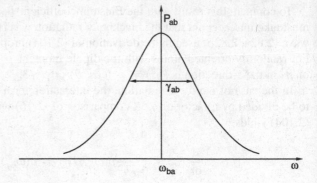

Figure 2.26 Transition probability of a damped system under weak broadband excitation

2.8.5   Phenomenological Inclusion of Decay Phenomena

So far we have neglected the fact that the levels $|a\rangle$ and $|b\rangle$ are not only coupled by transitions induced by the external field but may also decay by spontaneous emission or by other relaxation processes such as collision-induced transitions. We can include these decay phenomena in our formulas by adding phenomenological decay terms to (2.96a, 2.96b), which can be expressed by the decay constant γ_a and γ_b (Fig. 2.24). A rigorous treatment requires quantum electrodynamics [59].

In the rotating-wave approximation, for which the term with the frequency $(\omega_{ba} + \omega)$ is neglected, (2.96a, 2.96b) then becomes

$$\dot{a}(t) = -\frac{1}{2}\gamma_a a - \frac{i}{2}\Omega_{ab}e^{-i(\omega_{ba}-\omega)t}b(t) , \tag{2.108a}$$

$$\dot{b}(t) = -\frac{1}{2}\gamma_b b - \frac{i}{2}\Omega_{ab}e^{+i(\omega_{ba}-\omega)t}a(t) . \tag{2.108b}$$

When the field amplitude E_0 is sufficiently small, see (2.101), we can use the weak-signal approximation of Sect. 2.8.3. This means that $|a(t)|^2 = 1$, $|b(t)|^2 \ll 1$, and also $aa^* - bb^* \simeq 1$. With this approximation, one obtains in a similar way as in the derivation of (2.99) the transition probability

$$\mathcal{P}_{ab}(\omega) = |b(t,\omega)|^2 = \int \gamma_{ab}e^{-\gamma_{ab}t}\,|b(t)|^2\mathrm{d}t = \frac{1}{2}\frac{\Omega_{ab}^2}{(\omega_{ba}-\omega)^2 + (\frac{1}{2}\gamma_{ab})^2} . \tag{2.108c}$$

This is a Lorentzian line profile (Fig. 2.26) with a full halfwidth $\gamma_{ab} = \gamma_a + \gamma_b$.

After taking the second-time derivative of (2.94b) and using (2.108a–2.108c), the equation of motion for the dipole moment D of the atom under the influence of a radiation field, becomes

$$\ddot{D} + \gamma_{ab}\dot{D} + (\omega_{ba}^2 + \gamma_{ab}^2/4)D$$
$$= (\Omega_{ab})[(\omega_{ba}+\omega)\cos\omega t + (\gamma_{ab}/2)\sin\omega t] . \tag{2.109a}$$

The homogeneous equation

$$\ddot{D} + \gamma_{ab}\dot{D} + (\omega_{ba}^2 + \gamma_{ab}^2/4)D = 0 , \qquad (2.109b)$$

which describes the atomic dipoles without the driving field ($\Omega_{ab} = 0$), has the solution for weak damping ($\gamma_{ab} \ll \omega_{ba}$)

$$D(t) = D_0 e^{(-\gamma_{ab}/2)t} \cos \omega_{ba} t . \qquad (2.110)$$

The inhomogeneous equation (2.109a) shows that the induced dipole moment of the atom interacting with a monochromatic radiation field behaves like a driven damped harmonic oscillator with $\omega_{ba} = (E_b - E_a)/\hbar$ for the eigenfrequency and $\gamma_{ab} = (\gamma_a + \gamma_b)$ for the damping constant oscillating at the driving field frequency ω.

Using the approximation $(\omega_{ba} + \omega) \simeq 2\omega$ and $\gamma_{ab} \ll \omega_{ba}$, which means weak damping and a close-to-resonance situation, we obtain solutions of the form

$$D = D_1 \cos \omega t + D_2 \sin \omega t , \qquad (2.111)$$

where the factors D_1 and D_2 include the frequency dependence,

$$D_1 = \frac{\Omega_{ab}(\omega_{ba} - \omega)}{(\omega_{ba} - \omega)^2 + (\gamma_{ab}/2)^2} , \qquad (2.112a)$$

$$D_2 = \frac{\frac{1}{2}\Omega_{ab}\gamma_{ab}}{(\omega_{ab} - \omega)^2 + (\gamma_{ab}/2)^2} . \qquad (2.112b)$$

These two equations for D_1 and D_2 describe dispersion and absorption of the EM wave. The former is caused by the phase lag between the radiation field and the induced dipole oscillation, and the latter by the atomic transition from the lower level E_a to the upper level E_b and the resultant conversion of the field energy into the potential energy $(E_b - E_a)$.

The macroscopic polarization P of a sample with N atoms/cm^3 is related to the induced dipole moment D by $P = ND$.

2.8.6   Interaction with Strong Fields

In the previous sections we assumed weak-field conditions where the probability of finding the atom in the initial state was not essentially changed by the interaction with the field. This means that the population in the initial state remains approximately constant during the interaction time. In the case of broadband radiation, this approximation results in a *time-independent transition probability*. Also the inclusion of weak-damping terms with $\gamma_{ab} \ll \omega_{ba}$ did not affect the assumption of a constant population in the initial state.

When intense laser beams are used for the excitation of atomic transitions, the weak-field approximation is no longer valid. In this section, we therefore consider the "strong-field case." The corresponding theory, developed by Rabi, leads to a time-dependent probability of the atom being in either the upper or lower level. The representation outlined below follows that of [57].

We consider a monochromatic field of frequency ω and start from the basic equations (2.96a, 2.96b) for the probability amplitudes in the rotating wave approximation with $\omega_{ba} = -\omega_{ab}$

$$\dot{a}(t) = \frac{i}{2}\Omega_{ab}e^{-i(\omega_{ba}-\omega)t}b(t) , \tag{2.113a}$$

$$\dot{b}(t) = \frac{i}{2}\Omega_{ab}e^{+i(\omega_{ba}-\omega)t}a(t) . \tag{2.113b}$$

Inserting the trial solution

$$a(t) = e^{i\mu t} \;\Rightarrow\; \dot{a}(t) = i\mu e^{i\mu t} ,$$

into (2.113a) yields

$$b(t) = \frac{2\mu}{\Omega_{ab}}e^{i(\omega_{ba}-\omega+\mu)t} \;\Rightarrow\; \dot{b}(t) = \frac{2i\mu(\omega_{ba}-\omega+\mu)}{\Omega_{ab}}e^{i(\omega_{ba}-\omega+\mu)t} .$$

Substituting this back into (2.113b) gives the relation

$$2\mu(\omega_{ba} - \omega + \mu) = \Omega_{ab}^2/2 . \tag{2.114}$$

This is a quadratic equation for the unknown quantity μ with the two solutions

$$\mu_{1,2} = -\frac{1}{2}(\omega_{ba} - \omega) \pm \frac{1}{2}\sqrt{(\omega_{ba} - \omega)^2 + \Omega_{ab}^2} . \tag{2.115}$$

The general solutions for the amplitudes a and b are then

$$a(t) = C_1 e^{i\mu_1 t} + C_2 e^{i\mu_2 t} , \tag{2.116a}$$

$$b(t) = (2/\Omega_{ab})e^{i(\omega_{ba}-\omega)t}(C_1\mu_1 e^{i\mu_1 t} + C_2\mu_2 e^{i\mu_2 t}) . \tag{2.116b}$$

With the initial conditions $a(0) = 1$ and $b(0) = 0$, we find for the coefficients

$$C_1 + C_2 = 1 \quad \text{and} \quad C_1\mu_1 = -C_2\mu_2 ,$$

$$\Rightarrow C_1 = -\frac{\mu_2}{\mu_1 - \mu_2} \quad C_2 = +\frac{\mu_1}{\mu_1 - \mu_2} .$$

From (2.115) we obtain $\mu_1\mu_2 = -\Omega_{ab}^2/4$. With the shorthand

$$\Omega = \mu_1 - \mu_2 = \sqrt{(\omega_{ba} - \omega)^2 + \Omega_{ab}^2} , \tag{2.117}$$

we get the probability amplitude

$$b(t) = i(\Omega_{ab}/\Omega)e^{i(\omega_{ba}-\omega)t/2} \sin(\Omega t/2) . \tag{2.118}$$

The probability $|b(t)|^2 = b(t)b^*(t)$ of finding the system in level E_b is then

$$|b(t)|^2 = (\Omega_{ab}/\Omega)^2 \sin^2(\Omega t/2) , \tag{2.119}$$

where

$$\Omega = \sqrt{(\omega_{ba} - \omega)^2 + (D_{ab} \cdot E_0/\hbar)^2} \tag{2.120}$$

is called the general *"Rabi flopping frequency"* for the nonresonant case $\omega \neq \omega_{ba}$. Equation (2.119) reveals that the transition probability is a periodic function of time. Since

$$|a(t)|^2 = 1 - |b(t)|^2 = 1 - (\Omega_{ab}/\Omega)^2 \sin^2(\Omega t/2) , \tag{2.121}$$

the system oscillates with the frequency Ω between the levels E_a and E_b, where the level-flopping frequency Ω depends on the detuning $(\omega_{ba} - \omega)$, on the field amplitude E_0, and the matrix element D_{ab} (Fig. 2.25b).

> The general Rabi flopping frequency Ω gives the frequency of population oscillation in a two-level system in an electromagnetic field with amplitude E_0.

Note In the literature often the term "Rabi frequency" is restricted to the resonant case $\omega = \omega_{ba}$.

At resonance $\omega_{ba} = \omega$, and (2.119) and (2.121) reduce to

$$|a(t)|^2 = \cos^2(D_{ab} \cdot E_0 t/2\hbar) , \tag{2.122a}$$
$$|b(t)|^2 = \sin^2(D_{ab} \cdot E_0 t/2\hbar) . \tag{2.122b}$$

After a time

$$T = \pi\hbar/(D_{ab} \cdot E_0) = \pi/\Omega_{ab} , \tag{2.123}$$

the probability $|b(t)|^2$ of finding the system in level E_b becomes unity. This means that the population probability $|a(0)|^2 = 1$ and $|b(0)|^2 = 0$ of the initial system has been inverted to $|a(T)|^2 = 0$ and $|b(T)|^2 = 1$ (Fig. 2.27).

Radiation with the amplitude A_0, which resonantly interacts with the atomic system for exactly the time interval $T = \pi\hbar/(D_{ab} \cdot E_0)$, is called a *π-pulse*

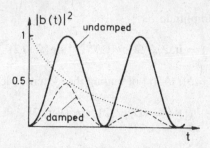

Figure 2.27 Population probability $|b(t)|^2$ of the levels E_b altering with the Rabi flopping frequency due to the interaction with a strong field. The resonant case is shown without damping and with damping due to decay channels into other levels. The decaying curve represents the factor $\exp[-(\gamma_{ab}/2)t]$

because it changes the phases of the probability amplitudes $a(t)$, $b(t)$ by π, see (2.116a, 2.116b, 2.118).

We now include the damping terms γ_a and γ_b, and again insert the trial solution

$$a(t) = e^{i\mu t} ,$$

into (2.108a, 2.108b). Similar to the procedure used for the undamped case, this gives a quadratic equation for the parameter μ with the two complex solutions

$$\mu_{1,2} = -\frac{1}{2}\left(\omega_{ba} - \omega - \frac{i}{2}\gamma_{ab}\right) \pm \frac{1}{2}\sqrt{\left(\omega_{ba} - \omega - \frac{i}{2}\gamma\right)^2 + \Omega_{ab}^2} ,$$

where

$$\gamma_{ab} = \gamma_a + \gamma_b \quad \text{and} \quad \gamma = \gamma_a - \gamma_b . \tag{2.124}$$

From the general solution

$$a(t) = C_1 e^{i\mu_1 t} + C_2 e^{i\mu_2 t} ,$$

we obtain from (2.108a) with the initial conditions $|a(0)|^2 = 1$ and $|b(0)|^2 = 0$ the transition probability

$$|b(t)|^2 = \frac{\Omega_{ab}^2 e^{(-\gamma_{ab}/2)t}[\sin(\Omega/2)t]^2}{(\omega_{ba} - \omega)^2 + (\gamma/2)^2 + \Omega_{ab}^2} . \tag{2.125}$$

This is a damped oscillation (Fig. 2.27) with the damping constant $\frac{1}{2}\gamma_{ab} = (\gamma_a + \gamma_b)/2$, the Rabi flopping frequency

$$\Omega = \mu_1 - \mu_2 = \sqrt{\left(\omega_{ba} - \omega + \frac{1}{2}\gamma\right)^2 + \Omega_{ab}^2} , \tag{2.126}$$

and the envelope $\Omega_{ab}^2 e^{-(\gamma_{ab}/2)t}/[(\omega_{ba} - \omega)^2 + (\gamma/2)^2 + \Omega_{ab}^2]$. The spectral profile of the transition probability is Lorentzian (Sect. 3.1), with a halfwidth depending

Figure 2.28 Population of level $|b\rangle$ for a closed two-level system where the relaxation channels are open only for transitions between $|a\rangle$ and $|b\rangle$

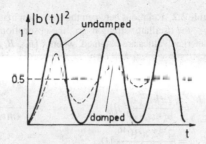

on $\gamma = \gamma_a - \gamma_b$ and on the strength of the interaction. Since $\Omega_{ab}^2 = (D_{ab} \cdot E_0/\hbar)^2$ is proportional to the intensity of the electromagnetic wave, the linewidth *increases* with increasing intensity (saturation broadening, Sect. 3.5). Note, that $|a(t)|^2 + |b(t)|^2 < 1$ for $t > 0$, because the levels a and b can decay into other levels.

In some cases the two-level system may be regarded as isolated from its environment. The relaxation processes then occur only between the levels $|a\rangle$ and $|b\rangle$, but do not connect the system with other levels. This implies $|a(t)|^2 + |b(t)|^2 = 1$. Equation (2.108a–2.108c) then must be modified as

$$\dot{a}(t) = -\frac{1}{2}\gamma_a a(t) + \frac{1}{2}\gamma_b b(t) + \frac{i}{2}\Omega_{ab}e^{-i(\omega_{ba}-\omega)t}b(t) , \qquad (2.127a)$$

$$\dot{b}(t) = -\frac{1}{2}\gamma_b b(t) + \frac{1}{2}\gamma_a a(t) + \frac{i}{2}\Omega_{ab}e^{+i(\omega_{ba}-\omega)t}a(t) . \qquad (2.127b)$$

The trial solution $a = \exp(i\mu t)$ yields, for the resonance case $\omega = \omega_{ba}$, the two solutions

$$\mu_1 = \frac{1}{2}\Omega_{ab} + \frac{i}{2}\gamma_{ab} , \quad \mu_2 = -\frac{1}{2}\Omega_{ab} ,$$

and for the transition probability $|b(t)|^2$, one obtains with $|a(0)|^2 = 1$, $|b(0)|^2 = 0$ a damped oscillation that approaches the steady-state value

$$|b(t = \infty)|^2 = \frac{1}{2}\frac{\Omega_{ab}^2 + \gamma_a\gamma_b}{\Omega_{ab}^2 + (\frac{1}{2}\gamma_{ab})^2} . \qquad (2.128)$$

This is illustrated in Fig. 2.28 for the special case $\gamma_a = \gamma_b$ where $|b(\infty)|^2 = 1/2$, which means that the two levels become equally populated.

For a more detailed treatment see [57–60].

2.8.7 Relations Between Transition Probabilities, Absorption Coefficient, and Line Strength

In this section we will summarize important relations between the different quantities discussed so far.

Table 2.2 Relations between the transition matrix element D_{ik} and the Einstein coefficients A_{ik}, B_{ik}, the oscillator strength f_{ik}, the absorption cross section σ_{ik}, and the line strength S_{ik}. The numerical values are obtained, when λ [m], B_{ik} [m^{3}s^{-2}J^{-1}], D_{ik} [As m], m_c [kg]

$$A_{ki} = \frac{1}{g_k} \frac{16\pi^3 \nu^3}{3\varepsilon_0 hc^3} |D_{ik}|^2 \qquad B_{ik}^{(\nu)} = \frac{1}{g_i} \frac{2\pi^2}{3\varepsilon_0 h^2} |D_{ik}|^2 \qquad B_{ik}^{(\omega)} = \frac{1}{g_i} \frac{\pi}{3\varepsilon_0 \hbar^2} |D_{ik}|^2$$

$$= \frac{2\pi h \nu_{ik}^2 e^2}{m \cdot \varepsilon_0 \cdot c^3} f_{ik} \qquad\qquad = \frac{e^2 f_{ik}}{4m\varepsilon_0 h\nu_{ik}} \qquad\qquad = \frac{\pi e^2 f_{ik}}{2m\varepsilon_0 \hbar \omega_{ik}}$$

$$= \frac{2.82 \times 10^{46}}{g_k \cdot \lambda^3} |D_{ik}|^2 \, \text{s}^{-1} \qquad = 6 \times 10^{31} \lambda^3 \frac{g_i}{g_k} A_{ki} \qquad = \frac{g_k}{g_i} B_{ki}$$

$$f_{ik} = \frac{1}{g_i} \frac{8\pi^2 m_e \nu}{e^2 h} |D_{ik}|^2 \qquad S_{ik} = |D_{ik}|^2 \qquad\qquad \sigma_{ik} = \frac{1}{\Delta\nu} \frac{2\pi^2 \nu}{3\varepsilon_0 chg_i} \cdot S_{ik}$$

$$= \frac{g_k}{g_i} \cdot 4.5 \times 10^4 \lambda^2 A_{ki} \qquad\qquad = (7.8 \times 10^{-21} g_i \lambda) f_{ik}$$

$$\alpha_{ik} = \sigma_{ik} \cdot N_i = \frac{2\omega}{c}\kappa \qquad A_{ik} = \frac{8\pi h\nu^3}{c^3} B_{ik}^{(\nu)} \qquad B_{ik}^{(\nu)} = \frac{c}{h\nu} \int\limits_0^\infty \sigma_{ik}(\nu)\,\mathrm{d}\nu$$

The absorption coefficient $\alpha(\omega)$ for a transition between levels $|i\rangle$ and $|k\rangle$ with population densities N_i and N_k and statistical weights g_i, g_k is related to the absorption cross section $\sigma_{ik}(\omega)$ by

$$\alpha(\omega) = [N_i - (g_i/g_k)N_k]\sigma_{ik}(\omega) . \tag{2.129}$$

The Einstein coefficient for absorption B_{ik} is given by

$$B_{ik} = \frac{c}{\hbar\omega} \int\limits_0^\infty \sigma_{ik}(\omega)\,\mathrm{d}\omega = \frac{c\,\overline{\sigma}_{ik}}{\hbar\omega} \int\limits_0^\infty g(\omega - \omega_0)\,\mathrm{d}\omega \tag{2.130}$$

where $g(\omega - \omega_0)$ is the line profile of the absorbing transition at center frequency ω_0. The transition probability per second according to (2.15) is then

$$P_{ik} = B_{ik} \cdot \varrho = \frac{c}{\hbar\omega \cdot \Delta\omega} \int \varrho(\omega) \cdot \sigma_{ik}(\omega)\,\mathrm{d}\omega , \tag{2.131}$$

where $\Delta\omega$ is the spectral linewidth of the transition.

The line strength S_{ik} of a transition is defined as the sum

$$S_{ik} = \sum_{m_i, m_k} |D_{m_i, m_k}|^2 = |D_{ik}|^2 , \tag{2.132}$$

over all dipole-allowed transitions between all subcomponents m_i, m_k of levels $|i\rangle$, $|k\rangle$. The oscillator strength f_{ik} gives the ratio of the power absorbed by a molecule on the transition $|i\rangle \rightarrow |k\rangle$ to the power absorbed by a classical oscillator on its eigenfrequency $\omega_{ik} = (E_k - E_i)/h$.

Some of these relations are compiled in Table 2.2.

Figure 2.29 The field amplitudes A_n at a point P in a radiation field as superposition of an infinite number of waves from different surface elements dS_i of an extended source

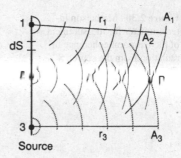

2.9 Coherence Properties of Radiation Fields

The radiation emitted by an extended source S generates a total field amplitude A at the point P that is a superposition of an infinite number of partial waves with the amplitudes A_n and the phases ϕ_n emitted from the different surface elements dS (Fig. 2.29), i.e.,

$$A(P) = \sum_n A_n(P) e^{i\phi n(P)} = \sum_n \left[A_n(0)/r_n^2 \right] e^{i(\phi n_0 + 2\pi r_n/\lambda)}, \qquad (2.133)$$

where $\phi_{n0}(t) = \omega t + \phi_n(0)$ is the phase of the nth partial wave at the surface element dS of the source. The phases $\phi_n(r_n, t) = \phi_{n,0}(t) + 2\pi r_n/\lambda$ depend on the distances r_n from the source and on the angular frequency ω.

If the phase differences $\Delta\phi_n = \phi_n(P, t_1) - \phi_n(P, t_2)$ at a *given point* P between two different times t_1, t_2 are nearly the same for all partial waves, the radiation field at P is *temporally coherent*. The maximum time interval $\Delta t = t_2 - t_1$ for which $\Delta\phi_n$ for all partial waves differ by less than π is termed the *coherence time* of the radiation source. The path length $\Delta s_c = c\Delta t$ traveled by the wave during the coherence time Δt is the *coherence length*.

If a constant time-independent phase difference $\Delta\phi = \phi(P_1) - \phi(P_2)$ exists for the total amplitudes $A = A_0 e^{i\phi}$ at two different points P_1, P_2, the radiation field is *spatially coherent*. All points P_m, P_n that fulfill the condition that for all times t, $|\phi(P_m, t) - \phi(P_n, t)| < \pi$ have nearly the same optical path difference from the source. They form the *coherence volume*.

The superposition of coherent waves results in interference phenomena that, however, can be observed directly only within the coherence volume. The dimensions of this coherence volume depend on the size of the radiation source, on the spectral width of the radiation, and on the distance between the source and observation point P.

The following examples illustrate these different expressions for the coherence properties of radiation fields.

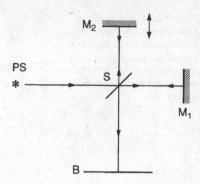

2.9.1 Temporal Coherence

Consider a point source PS in the focal plane of a lens forming a parallel light beam
that is divided by a beam splitter S into two partial beams (Fig. 2.30).

They are superimposed in the plane of observation B after reflection from the
mirrors M_1, M_2. This arrangement is called a *Michelson interferometer* (Sect. 4.2).
The two beams with wavelength λ travel different optical path lengths SM_1SB and
SM_2SB, and their path difference in the plane B is

$$\Delta s = 2(SM_1 - SM_2) .$$

The mirror M_2 is mounted on a carriage and can be moved, resulting in a continuous
change of Δs. In the plane B, one obtains maximum intensity when both amplitudes
have the same phase, which means $\Delta s = m\lambda$, and minimum intensity if $\Delta s =
(2m + 1)\lambda/2$. With increasing Δs, the contrast $V = (I_{max} - I_{min})/(I_{max} + I_{min})$
decreases (Fig. 2.31) and vanishes if Δs becomes larger than the coherence length
Δs_c (Sect. 2.9.4). Experiments show that Δs_c is related to the spectral width $\Delta\omega$ of
the incident wave by

$$\Delta s_c \simeq c/\Delta\omega = c/(2\pi\Delta\nu) . \tag{2.134}$$

This observation may be explained as follows. A wave emitted from a point
source with the spectral width $\Delta\omega$ can be regarded as a superposition of many
quasi-monochromatic components with frequencies ω_n within the interval $\Delta\omega$. The
superposition results in wave trains of finite length $\Delta s_c = c\Delta t = c/\Delta\omega$ because
the different components with slightly different frequencies ω_n come out of phase
during the time interval Δt and interfere destructively, causing the total amplitude
to decrease (Sect. 3.1). If the path difference Δs in the Michelson interferometer
becomes larger than Δs_c, the split wave trains no longer overlap in the plane B. The
coherence length Δs_c of a light source therefore becomes larger with decreasing
spectral width $\Delta\omega$.

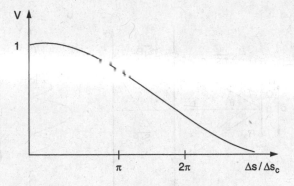

Figure 2.31 Visibility V as a function of path difference Δs for a Michelson interferometer with a light source with spectral band width $\Delta\omega$

Example 2.8

1. A low-pressure mercury spectral lamp with a spectral filter that only transmits the green line $\lambda = 546\,\text{nm}$ has, because of the Doppler width $\Delta\omega_D = 4 \times 10^9\,\text{Hz}$, a coherence length of $\Delta s_c \simeq 8\,\text{cm}$.
2. A single-mode HeNe laser with a bandwidth of $\Delta\omega = 2\pi \cdot 1\,\text{MHz}$ has a coherence length of about 50 m.

2.9.2   Spatial Coherence

The radiation from an *extended source* LS of size b illuminates two slits S_1 and S_2 in the plane A at a distance d apart (Young's double-slit interference experiment, Fig. 2.32a). The total amplitude and phase at each of the two slits are obtained by superposition of all partial waves emitted from the different surface elements df of the source, taking into account the different paths df–S_1 and df–S_2.

The intensity at the point of observation P in the plane B depends on the path difference S_1P–S_2P and on the phase difference $\Delta\phi = \phi(S_1) - \phi(S_2)$ of the total field amplitudes in S_1 and S_2. If the different surface elements df of the source emit independently with random phases (thermal radiation source), the phases of the total amplitudes in S_1 and S_2 will also fluctuate randomly. However, this would not influence the intensity in P as long as these fluctuations occur in S_1 and S_2 synchronously, because then the phase difference $\Delta\phi$ would remain constant. In this case, the two slits form two coherent sources that generate an interference pattern in the plane B.

For radiation emitted from the central part 0 of the light source, this proves to be true since the paths $0S_1$ and $0S_2$ are equal and all phase fluctuations in 0 arrive simultaneously in S_1 and S_2. For all other points Q of the source, however, path differences $\Delta s_Q = QS_1 - QS_2$ exist, which are largest for the edges R_1, R_2 of the source. From Fig. 2.32 one can infer for $b \ll r \Rightarrow \sin\theta \approx \theta$ the relation

$$\Delta s_R = \Delta s_{\max} = R_2S_1 - R_1S_1 \simeq b\sin\theta = R_1S_2 - R_1S_1 \,.$$

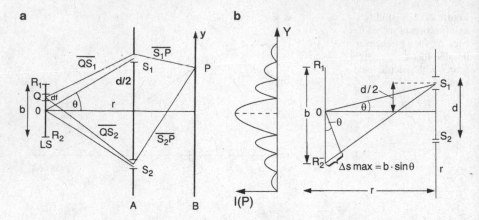

Figure 2.32 a Young's double-slit arrangement for measurements of spatial coherence; **b** path difference between a slit S_1 and different points of an extended source

For $\Delta s_{max} > \lambda/2$, the phase difference $\Delta\phi$ of the partial amplitudes in S_1 and S_2 exceeds π. With random emission from the different surface elements df of the source, the time-averaged interference pattern in the plane B will be washed out. The condition for coherent illumination of S_1 and S_2 from a light source with the dimension b is therefore

$$\Delta s = b\sin(\theta/2) < \lambda/2 .$$

because $\overline{R_2S_1} = \overline{R_1S_2}$.

With $2\sin\theta = d/r$, this condition can be written as

$$bd/r < \lambda . \tag{2.135a}$$

Extension of this coherence condition to two dimensions yields, for a source area $A_s = b^2$, the following condition for the maximum surface $A_c = d^2$ that can be illuminated coherently:

$$b^2d^2/r^2 \leq \lambda^2 . \tag{2.135b}$$

Since $d\Omega = d^2/r^2$ is the solid angle accepted by the illuminated surface $A_c = d^2$, this can be formulated as

$$\boxed{A_s d\Omega \leq \lambda^2} . \tag{2.135c}$$

The source surface $A_s = b^2$ determines the maximum solid angle $d\Omega \leq \lambda^2/A_s$ inside which the radiation field shows spatial coherence. Equation (2.135c) reveals that radiation from a point source (spherical waves) is spatially coherent within the whole solid angle $d\Omega = 4\pi$. The coherence surfaces are spheres with the source

at the center. Likewise, a plane wave produced by a point source in the focus of
a lens shows spatial coherence over the whole aperture confining the light beam.
For given source dimensions, the coherence surface $A_c = d^2$ increases with the
square of the distance from the source. Because of the vast distances to stars, the
starlight received by telescopes is spatially coherent across the telescope aperture,
in spite of the large diameter of the radiation source.

The arguments above may be summarized as follows: the coherence surface S_c
(i.e., that maximum area A_c that can be coherently illuminated at a distance r from
an extended quasi-monochromatic light source with area A_s emitting at a wavelength λ) is determined by

$$\boxed{S_c = \lambda^2 r^2 / A_s}\,. \tag{2.136}$$

2.9.3   Coherence Volume

With the coherence length $\Delta s_c = c/\Delta\omega$ in the propagation direction of the radiation with the spectral width $\Delta\omega$ and the coherence surface $S_c = \lambda^2 r^2 / A_s$, the
coherence volume $V_c = S_c \Delta s_c$ becomes

$$V_c = \frac{\lambda^2 r^2 c}{\Delta\omega A_s}\,. \tag{2.137}$$

A unit surface element of a source with the spectral radiance L_ω [W/(m^{2}sr)] emits
$L_\omega/\hbar\omega$ photons per second within the frequency interval $d\omega = 1$ Hz into the unit
solid angle 1 sr.

The mean number $\bar{n}$ of photons in the spectral range $\Delta\omega$ within the coherence
volume defined by the solid angle $\Delta\Omega = \lambda^2/A_s$ and the coherence length $\Delta s_c = c\Delta t_c$ generated by a source with area A_s is therefore

$$\bar{n} = (L_\omega/\hbar\omega) A_s \Delta\Omega \Delta\omega \Delta t_c\,.$$

With $\Delta\Omega = \lambda^2/A_s$ and $\Delta t_c \simeq 1/\Delta\omega$, this gives

$$\bar{n} = (L_\omega/\hbar\omega)\lambda^2\,. \tag{2.138}$$

Example 2.9
For a thermal radiation source, the spectral radiance for linearly polarized
light (given by (2.28) divided by a factor 2) is for $\cos\phi = 1$ and $L_\nu d\nu = L_\omega d\omega$

$$L_\nu = \frac{h\nu^3/c^2}{e^{h\nu/kT} - 1}\,.$$

The mean number of photons within the coherence volume is then with $\lambda = c/v$

$$\bar{n} = \frac{1}{e^{hv/kT} - 1} \, .$$

This is identical to the mean number of photons per mode of the thermal radiation field, as derived in Sect. 2.2.

Figure 2.7 and Example 2.3 give values of $\bar{n}$ for different conditions.

The mean number $\bar{n}$ of photons per mode is often called the *degeneracy parameter* of the radiation field. This example shows that the coherence volume is related to the modes of the radiation field. This relation can be also illustrated in the following way:

If we allow the radiation from all modes with the same direction of k to escape through a hole in the cavity wall with the area $A_s = b^2$, the wave emitted from A_s will not be strictly parallel, but will have a diffraction-limited divergence angle $\theta \simeq \lambda/b$ around the direction of k. This means that the radiation is emitted into a solid angle $d\Omega = \lambda^2/b^2$. This is the same solid angle (2.135c) that limits the spatial coherence.

The radiation with the same direction of k (which we assume to be the z direction) may still differ in $|k|$, i.e., it may have different frequencies ω. The coherence length is determined by the spectral width $\Delta\omega$ of the radiation emitted from A_s. Since $|k| = \omega/c$, the spectral width $\Delta\omega$ corresponds to an interval $\Delta k = \Delta\omega/c$ of k values. This radiation illuminates a minimum "diffraction surface"

$$A_D = r^2 d\Omega = r^2\lambda^2/A_s \, .$$

Multiplication with the coherence length $\Delta s_c = c/\Delta\omega$ yields again the coherence volume $V_c = A_D c/\Delta\omega = r^2\lambda^2 c/(\Delta\omega A_s)$ of (2.137). We shall now demonstrate that the coherence volume is identical with the spatial part of the elementary cell in the general phase space.

As is well known from atomic physics, the diffraction of light can be explained by Heisenberg's uncertainty relation. Photons passing through a slit of width Δx have the uncertainty Δp_x of the x-component p_x of their momentum p, given by $\Delta p_x \Delta x \geq \hbar$ (Fig. 2.33).

Generalized to three dimensions, the uncertainty principle postulates that the simultaneous measurements of momentum and location of a photon have the minimum uncertainty

$$\Delta p_x \Delta p_y \Delta p_z \Delta x \Delta y \Delta z \geq \hbar^3 = V_{\text{ph}} \, , \tag{2.139}$$

Figure 2.33 The uncertainty principle applied to the diffraction of light by a slit

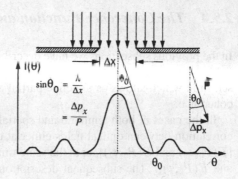

where $V_{\mathrm{ph}} = \hbar^3$ is the volume of the elementary cell in phase space. *Photons within the same cell of the phase space are indistinguishable and can be therefore regarded as identical.*

Photons that are emitted from the hole $A_{\mathrm{s}} = b^2$ within the diffraction angle $\theta = \lambda/b$ against the surface normal (Fig. 2.34), which may point into the z-direction, have the minimum uncertainty

$$\Delta p_x = \Delta p_y = |\boldsymbol{p}|\,\lambda/(2\pi b) = (\hbar\omega/c)\lambda/(2\pi b) = (\hbar\omega/c)d/(2\pi r)\,,\quad (2.140)$$

of the momentum components p_x and p_y, where the last equality follows from (2.135b).

The uncertainty Δp_z is mainly caused by the spectral width $\Delta\omega$. Since $p = \hbar\omega/c$, we find

$$\Delta p_z = (\hbar/c)\Delta\omega\,.\tag{2.141}$$

Substituting (2.140, 2.141) into (2.139), we obtain for the spatial part of the elementary phase space cell

$$\Delta x\Delta y\Delta z = \frac{\lambda^2 r^2 c}{\Delta\omega A_{\mathrm{s}}} = V_{\mathrm{c}}\,,\tag{2.142}$$

which turns out to be identical with the coherence volume defined by (2.137).

Figure 2.34 Coherence volume and phase space cell

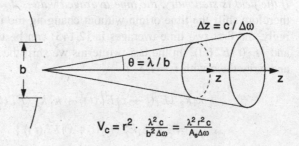

2.9.4 The Coherence Function and the Degree of Coherence

In the previous subsections we have described the coherence properties of radiation fields in a more illustrative way. We now briefly discuss a more quantitative description which allows us to define partial coherence and to measure the degree of coherence.

In the cases of both temporal and spatial coherence, we are concerned with the correlation between optical fields either at the same point P_0 but at different times $[E(P_0, t_1)$ and $E(P_0, t_2)]$, or at the same time t but at two different points $[E(P_1, t)$ and $E(P_2, t)]$. The subsequent description follows the representation in [39, 61, 62].

Suppose we have an extended source that generates a radiation field with a narrow spectral bandwidth $\Delta\omega$, which we shall represent by the complex notation of a plane wave, i.e.,

$$E(r, t) = A_0 e^{i(\omega t - k \cdot r)} + \text{c.c.}$$

The field at two points in space S_1 and S_2 (e.g., the two apertures in Young's experiment) is then $E(S_1, t)$ and $E(S_2, t)$. The two apertures serve as secondary sources (Fig. 2.32), and the resultant field at the point of observation P at time t is

$$E(P, t) = k_1 E_1(S_1, t - r_1/c) + k_2 E_2(S_2, t - r_2/c), \qquad (2.143)$$

where the imaginary numbers k_1 and k_2 depend on the size of the apertures and on the distances $r_1 = S_1 P$ and $r_2 = S_2 P$.

The resulting time averaged irradiance at P measured over a time interval which is long compared to the coherence time is

$$I_p = \epsilon_0 c \langle E(P, t) E^*(P, t) \rangle, \qquad (2.144)$$

where the brackets $\langle \ldots \rangle$ indicate the time average. Using (2.143), this becomes

$$I_p = c\epsilon_0 \big[k_1 k_1^* \langle E_1(t - t_1) E_1^*(t - t_1) \rangle + k_2 k_2^* \langle E_2(t - t_2) E_2^*(t - t_2) \rangle$$

$$+ k_1 k_2^* \langle E_1(t - t_1) E_2^*(t - t_2) \rangle + k_1^* k_2 \langle E_1^*(t - t_1) E_2(t - t_2) \rangle \big]. \qquad (2.145)$$

If the field is stationary, the time-averaged values do not depend on time. We can therefore shift the time origin without changing the irradiances (2.144). Accordingly, the first two time averages in (2.145) can be transformed to $\langle E_1(t) E_1^*(t) \rangle$ and $\langle E_2(t) E_2^*(t) \rangle$. In the last two terms we shift the time origin by an amount t_2 and write them with $\tau = t_2 - t_1$

$$k_1 k_2^* \langle E_1(t + \tau) E_2^*(t) \rangle + k_1^* k_2 \langle E_1^*(t + \tau) E_2(t) \rangle$$

$$= 2\Re\{ k_1 k_2^* \langle E_1(t + \tau) E_2^*(t) \rangle \}. \qquad (2.146)$$

The term

$$\Gamma_{12}(\tau) = \langle E_1(t + \tau)E_2^*(t)\rangle \;, \tag{2.147}$$

is called the *mutual coherence function* and describes the cross correlation of the field amplitudes at S_1 and S_2. When the amplitudes and phases of E_1 and E_2 fluctuate within a time interval $\Delta t < \tau$, the time average $\Gamma_{12}(\tau)$ will be zero if these fluctuations of the two fields at two different points and at two different times are completely uncorrelated. If the field at S_1 at time $t + \tau$ were perfectly correlated with the field at S_2 at time t, the relative phase would be unaltered despite individual fluctuations, and Γ_{12} would become independent of τ.

Inserting (2.147) into (2.145) gives for the irradiance at P (note that k_1 and k_2 are pure imaginary numbers for which $2\Re\{k_1 \cdot k_2\} = 2|k_1| \cdot |k_2|$)

$$I_p = \epsilon_0 c \left[|k_1|^2 I_{S1} + |k_2|^2 I_{S2} + 2|k_1||k_2| \Re\{\Gamma_{12}(\tau)\}\right] \;. \tag{2.148}$$

The first term $I_1 = \epsilon_0 c |k_1|^2 I_{S1}$ yields the irradiance at P when only the aperture S_1 is open ($k_2 = 0$); the second term $I_2 = \epsilon_0 c |k_2|^2 I_{S2}$ is that for $k_1 = 0$.

Let us introduce the first-order correlation functions

$$\Gamma_{11}(\tau) = \langle E_1(t + \tau)E_1^*(t)\rangle \;,$$
$$\Gamma_{22}(\tau) = \langle E_2(t + \tau)E_2^*(t)\rangle \;, \tag{2.149}$$

which correlate the field amplitude at the same point but at different times. For $\tau = 0$ the *self-coherence functions*

$$\Gamma_{11}(0) = \langle E_1(t)E_1^*(t)\rangle = I_1/(\epsilon_0 c) \;,$$
$$\Gamma_{22}(0) = I_2/(\epsilon_0 c) \;, \tag{2.150}$$

are proportional to the irradiance I at S_1 and S_2, respectively.

With the definition of the normalized form of the mutual coherence function,

$$\gamma_{12}(\tau) = \frac{\Gamma_{12}(\tau)}{\sqrt{\Gamma_{11}(0)\Gamma_{22}(0)}} = \frac{\langle E_1(t + \tau)E_2^*(t)\rangle}{\sqrt{\langle |E_1(t)|^2 |E_2(t)|^2\rangle}} \;, \tag{2.151}$$

(2.148) can be written as

$$\boxed{I_p = I_1 + I_2 + 2\sqrt{I_1 I_2}\,\Re\{\gamma_{12}(\tau)\}} \;. \tag{2.152}$$

This is the general interference law for partially coherent light; $\gamma_{12}(\tau)$ is called the *complex degree of coherence*. Its meaning will be illustrated by the following: we express the complex quantity $\gamma_{12}(\tau)$ as

$$\gamma_{12}(\tau) = |\gamma_{12}(\tau)|\,e^{i\phi_{12}(\tau)} \;, \tag{2.153}$$

where the phase angle $\phi_{12}(\tau) = \phi_1(\tau) - \phi_2(\tau)$ is related to the phases of the fields E_1 and E_2 in (2.147).

For $|\gamma_{12}(\tau)| = 1$, (2.152) describes the interference of two completely coherent waves out of phase at S_1 or S_2 by the amount $\phi_{12}(\tau)$. For $|\gamma_{12}(\tau)| = 0$, the interference term vanishes. The two waves are said to be completely incoherent. For $0 < |\gamma_{12}(\tau)| < 1$ we have *partial coherence*. $\gamma_{12}(\tau)$ is therefore a measure of the degree of coherence. We illustrate the mutual coherence function $\gamma_{12}(\tau)$ by applying it to the situations outlined in Sects. 2.9.1 and 2.9.2.

Example 2.10

In the Michelson interferometer, the incoming nearly parallel light beam is split by S (Fig. 2.30) and recombined in the plane B. If both partial beams have the same amplitude $E = E_0 e^{i\phi(t)}$, the degree of coherence becomes

$$\gamma_{11}(\tau) = \frac{\langle E(t + \tau)E^*(t)\rangle}{|E(t)|^2} = \langle e^{i\phi(t+\tau)}e^{-i\phi(t)}\rangle \ .$$

For long averaging times T we obtain with $\Delta\phi = \phi(t + \tau) - \phi(t)$,

$$\gamma_{11}(\tau) = \lim_{T\to\infty} \frac{1}{T} \int\limits_0^T (\cos\Delta\phi + i\sin\Delta\phi)dt \ . \tag{2.154}$$

For a strictly monochromatic wave with infinite coherence length Δs_c, the phase function is $\phi(t) = \omega t - \boldsymbol{k}\cdot\boldsymbol{r}$ and $\Delta\phi = +\omega\tau$ with $\tau = \Delta s/c$. This yields

$$\gamma_{11}(\tau) = \cos\omega\tau + i\sin\omega\tau = e^{i\omega\tau} \ , \quad |\gamma_{11}(\tau)| = 1 \ .$$

For a wave with spectral width $\Delta\omega$ so large that $\tau > \Delta s_c/c = 1/\Delta\omega$, the phase differences $\Delta\phi$ vary randomly between 0 and 2π and the integral averages to zero, giving $\gamma_{11}(\tau) = 0$. In Fig. 2.35 the interference pattern $I(\Delta\phi) \propto |E_1(t)\cdot E_2(t + \tau)|^2$ in the observation plane behind a Michelson interferometer is illustrated as a function of the phase difference $\Delta\phi = (2\pi/\lambda)\Delta s$ for equal intensities $I_1 = I_2$ but different values of $|\gamma_{12}(\tau)|$. For completely coherent light ($|\gamma_{12}(\tau)| = 1$) the intensity $I(\tau)$ changes between $4I_1$ and zero, whereas for $|\gamma(t)| = 0$ the interference term vanishes and the total intensity $I = 2I_1$ does not depend on τ.

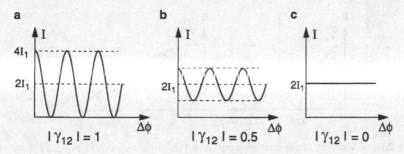

Figure 2.35 Interference pattern $I(\Delta\phi)$ of two-beam interference for different degrees of coherence

Example 2.11

For the special case of a quasi-monochromatic plane wave $E = E_0 \times \exp(i\omega t - i\mathbf{k} \cdot \mathbf{r})$, an optical path difference $(\mathbf{r}_2 - \mathbf{r}_1)$ causes a corresponding phase difference

$$\phi_{12}(\tau) = \mathbf{k} \cdot (\mathbf{r}_2 - \mathbf{r}_1) ,$$

and (2.152) can be expressed with $\Re\{\gamma_{12}(\tau)\} = |\gamma_{12}(\tau)| \cos \phi_{12}$ by

$$I_p = I_1 + I_2 + 2\sqrt{I_1 I_2} \, |\gamma_{12}(\tau)| \cos \phi_{12}(\tau) . \qquad (2.155)$$

For $|\gamma_{12}(\tau)| = 1$, the interference term causes a full modulation of the irradiance $I_p(\tau)$. For $\gamma_{12}(\tau) = 0$, the interference vanishes and the total intensity becomes independent of the time delay τ between the two beams.

Example 2.12

Referring to Young's experiment (Fig. 2.32) with a narrow bandwidth but extended source, *spatial* coherence effects will predominate. The fringe pattern in the plane B will depend on $\Gamma(S_1, S_2, \tau) = \Gamma_{12}(\tau)$. In the region around the central fringe $(r_2 - r_1) = 0$, $\tau = 0$, the values of $\Gamma_{12}(0)$ and $\gamma_{12}(0)$ can be determined from the visibility of the interference pattern.

To find the value $\gamma_{12}(\tau)$ for any point P on the screen B in Fig. 2.32, the time-averaged intensity $I(P)$ is measured when both slits are open, and also $I_1(P)$ and $I_2(P)$ when one of the slits is blocked. In terms of these observed

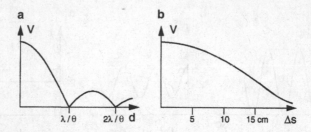

Figure 2.36 a Visibility of the interference pattern behind the two slits of Fig. 2.32 if they are illuminated by a monochromatic extended source. The abscissa gives the slit separation d in units of λ/θ. **b** Visibility of a Doppler-broadened line behind a Michelson interferometer as a function of the path difference Δs

quantities, the degree of coherence can be determined from (2.152) to be

$$\Re\{\gamma_{12}(P)\} = \frac{I(P) - I_1(P) - I_2(P)}{2\sqrt{I_1(P)I_2(P)}}.$$

This yields the desired information about the spatial coherence of the source, which depends on the size of the source and its distance from the pinholes.

The visibility of the fringes at P is defined as

$$V(P) = \frac{I_{\max} - I_{\min}}{I_{\max} + I_{\min}} = \frac{2\sqrt{I_1(P)}\sqrt{I_2(P)}}{I_1(P) + I_2(P)}\,|\gamma_{12}(\tau)|\,, \qquad (2.156)$$

where the last equality follows from (2.155). If $I_1 = I_2$ (equal size pinholes), we see that

$$V(P) = |\gamma_{12}(\tau)|\,.$$

The visibility is then equal to the degree of coherence. Figure 2.36a depicts the visibility V of the fringe pattern in P as a function of the slit separation d, indicated in Fig. 2.32, when these slits are illuminated by monochromatic light from an extended uniform source with quadratic size $b \times b$ that appears from S_1 under the angle θ. Figure 2.36b illustrates the visibility as a function of path difference Δs in a Michelson interferometer which is illuminated with the Doppler-broadened line $\lambda = 632.8\,\text{nm}$ from a neon discharge lamp.

For more detailed presentations of coherence see the textbooks [41, 62–64].

Figure 2.37 Coherent excitation of two atomic levels $|a\rangle$ and $|b\rangle$ from the same lower level $|g\rangle$ with a broadband laser pulse with $\hbar\Delta\omega > |E_b - E_a|$

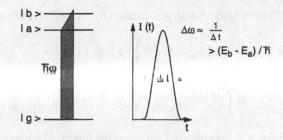

2.10 Coherence of Atomic Systems

Two levels of an atom are said to be coherently excited if their corresponding wave functions are in phase at the excitation time. With a short laser pulse of duration Δt, which has a Fourier-limited spectral bandwidth $\Delta\omega \simeq 1/\Delta t$, two atomic levels a and b can be excited simultaneously if their energy separation ΔE is smaller than $\hbar\Delta\omega$ (Fig. 2.37). The wave function of the excited atom is then a linear combination of the wave functions ψ_a and ψ_b, and the atom is said to be in a coherent superposition of the two states $|a\rangle$ and $|b\rangle$.

An ensemble of atoms is *coherently* excited if the wave functions of the excited atoms, at a certain time t, have the same phase for all atoms. This phase relation may change with time due to differing frequencies ω in the time-dependent part $\exp(i\omega t)$ of the excited-state wave functions or because of relaxation processes, which may differ for the different atoms. This will result in a "phase diffusion" and a time-dependent decrease of the degree of coherence.

The realization of such coherent systems requires special experimental preparations that, however, can be achieved with several techniques of coherent laser spectroscopy (Vol. 2, Chap. 7). An elegant theoretical way of describing observable quantities of a coherently or incoherently excited system of atoms and molecules is based on the density-matrix formalism.

2.10.1 Density Matrix

Let us assume, for simplicity, that each atom of the ensemble can be represented by a two-level system (Sect. 2.8), described by the wave function

$$\psi(r,t) = \psi_a + \psi_b = a(t)u_a e^{-iE_a t/\hbar} + b(t)u_b e^{-i[(E_b/\hbar)t - \phi]} , \qquad (2.157a)$$

where the phase ϕ is introduced because it might be different for each of the atoms. We can write ψ as the state vector

$$\begin{pmatrix} \psi_a \\ \psi_b \end{pmatrix} \quad \text{or} \quad (\psi_a, \psi_b) \qquad (2.157b)$$

The density matrix $\tilde{\rho}$ is defined by the product of the two state vectors

$$\tilde{\rho} = |\psi\rangle\langle\psi| = \begin{pmatrix} \psi_a \\ \psi_b \end{pmatrix} (\psi_a, \psi_b)$$

$$= \begin{pmatrix} |a(t)|^2 & abe^{-i[(E_a-E_b)t/\hbar+\phi]} \\ abe^{+i[(E_a-E_b)t/\hbar+\phi]} & |b(t)|^2 \end{pmatrix} = \begin{pmatrix} \rho_{aa} & \rho_{ab} \\ \rho_{ba} & \rho_{bb} \end{pmatrix}, \quad (2.158)$$

since the normalized atomic wave functions in vector notation are

$$u_a = \begin{pmatrix} 1 \\ 0 \end{pmatrix} \quad \text{and} \quad u_b = \begin{pmatrix} 0 \\ 1 \end{pmatrix}.$$

The diagonal elements ρ_{aa} and ρ_{bb} represent the probabilities of finding the atoms of the ensemble in the level $|a\rangle$ and $|b\rangle$, respectively.

If the phases ϕ of the atomic wave function (2.157a) are randomly distributed for the different atoms of the ensemble, the nondiagonal elements of the density matrix (2.158) average to zero and the incoherently excited system is therefore described by the diagonal matrix

$$\tilde{\rho}_{\text{incoh}} = \begin{pmatrix} [a(t)]^2 & 0 \\ 0 & [b(t)]^2 \end{pmatrix}. \quad (2.159)$$

If definite phase relations exist between the wave functions of the atoms, the system is in a *coherent state*. The nondiagonal elements of (2.158) describe the degree of coherence of the system and are therefore often called "coherences."

Such a coherent state can, for example, be generated by the interaction of the atomic ensemble with a sufficiently strong EM field that induces atomic dipole moments, which add up to a macroscopic oscillating dipole moment if all atomic dipoles oscillate in phase. The expectation value D of such an atomic dipole moment is

$$D = -e \int \psi^* r \psi \, d\tau . \quad (2.160)$$

With (2.94b) this becomes

$$D = -D_{ab}(a^*be^{-i\omega_{ba}t} + ab^*e^{i\omega_{ba}t}) = D_{ab}(\rho_{ab} + \rho_{ba}). \quad (2.161)$$

The nondiagonal elements of the density matrix are therefore proportional to the expectation value of the dipole moment.

2.10.2 Coherent Excitation

We saw in Sect. 2.10.1 that in a coherently excited system of atoms, well-defined phase relations exist between the time-dependent wavefunctions of the atomic

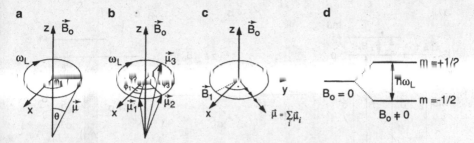

Figure 2.38 Precession of a magnetic dipole in a homogeneous magnetic field B_0 (a); Incoherent precession of the different dipoles (b); Synchronization of dipoles by a radio frequency (RF) field (c); Coherent superposition of two Zeeman sublevels (d) as the quantum-mechanical equivalent to the classical picture (c)

levels. In this section we will illustrate such coherent excitations by several examples.

- If identical paramagnetic atoms with magnetic moments μ and total angular momentum J are brought into a homogeneous magnetic field $B_0 = \{0, 0, B_z\}$, the angular momentum vectors J_i of the atoms will precess with the Lamor frequency $\omega_L = \gamma B_0$ around the z-direction, where $\gamma = \mu/|J|$ is the gyromagnetic ratio (Fig. 2.38a). The phases φ_i of this precession will be different for the different atoms and, in general, are randomly distributed. The precession occurs incoherently (Fig. 2.38b). The dipole moments μ of the N atoms add up to a macroscopic "*longitudinal magnetization*"

$$M_z = \sum_{i=1}^{N} \mu \cos \theta_i = N\mu \cos \theta , \qquad (2.162)$$

but the average "*transversal magnetization*" is zero.

When an additional radio frequency field $B_1 = B_{10} \cos \omega t$ is added with $B_1 \perp B_0$, the dipoles are forced to precess synchroneously with the RF field B_1 in the x–y-plane if $\omega = \omega_L$. This results in a macroscopic magnetic moment $M = N\mu$, which rotates with ω_L in the x–y-plane and has a phase angle $\pi/2$ against B_1 (Fig. 2.38c). The precession of the atoms becomes coherent through their coupling to the RF field. In the quantum-mechanical description, the RF field induces transitions between the Zeeman sublevels (Fig. 2.38d). If the RF field B_1 is sufficiently intense, the atoms are in a coherent superposition of the wave functions of both Zeeman levels.

- Excitation by visible or UV light may also create a coherent superposition of Zeeman sublevels. As an example, we consider the transition $6\,^1S_0 \rightarrow 6\,^3P_1$ of the Hg atom at $\lambda = 253.7\,\text{nm}$ (Fig. 2.39). In a magnetic field $B = \{0, 0, B_z\}$, the upper level $6\,^3P_1$ splits into three Zeeman sublevels with magnetic quantum numbers $m_z = 0, \pm 1$. Excitation with linear polarized light ($E \parallel B$) only

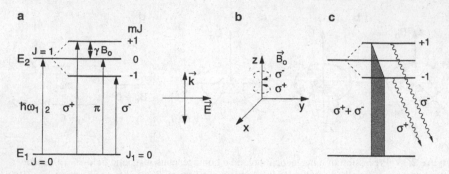

Figure 2.39   Coherent excitation of Zeeman sublevels with $m = \pm 1$ (**a**) by linear polarized light with $\boldsymbol{E} \perp \boldsymbol{B}$ (**b**). The fluorescence is a superposition of σ^+ and σ^- light (**c**)

populates the level $m_J = 0$. The fluorescence emitted from this Zeeman level is also linearly polarized.

However, if the exciting light is polarized perpendicularly to the magnetic field ($\boldsymbol{E} \perp \boldsymbol{B}$), it may be regarded as superposition of σ^+ and σ^- light traveling into the z-direction, which is chosen as the quantization axis.

In this case, the levels with $m = \pm 1$ are populated. As long as the Zeeman splitting is smaller than the homogeneous width of the Zeeman levels (e.g., the natural linewidth $\Delta\omega = 1/\tau$), both components are excited coherently (even with monochromatic light!). The wave function of the excited state is represented by a linear combination $\psi = a\psi_a + b\psi_b$ of the two wavefunctions of the Zeeman sublevels $m = \pm 1$. The fluorescence is nonisotropic, but shows an angular distribution that depends on the coefficients a, b (Vol. 2, Sect. 7.1).

• A molecule with two closely lying levels $|a\rangle$ and $|b\rangle$ that can both be reached by optical transitions from a common groundstate $|g\rangle$ can be coherently excited by a light pulse with duration ΔT, if $\Delta T < \hbar/(E_a - E_b)$, even if the levels $|a\rangle$ and $|b\rangle$ are different vibrational levels of different electronic states and their separation is larger than their homogeneous width.

The time-dependent fluorescence from these coherently excited states shows, besides the exponential decay $\exp(-t/\tau)$, a beat period $\tau_{QB} = \hbar/(E_a - E_b)$ due to the different frequencies ω_a and ω_b of the two fluorescence components (quantum beats, Vol. 2, Sect. 7.2).

2.10.3   Relaxation of Coherently Excited Systems

The time-dependent Schrödinger equation (2.85) is written in the density–matrix formalism as

$$i\hbar\dot{\tilde{\rho}} = [\mathcal{H}, \tilde{\rho}] . \tag{2.163}$$

In order to separate the different contributions of induced absorption or emission and of relaxation processes, we write the Hamiltonian $\mathcal{H}$ as the sum

$$\mathcal{H} = \mathcal{H}_0 + \mathcal{H}_1(t) + \mathcal{H}_R , \tag{2.164}$$

of the "internal" Hamiltonian of the isolated two-level system

$$\mathcal{H}_0 = \begin{pmatrix} E_a & 0 \\ 0 & E_b \end{pmatrix} ,$$

the interaction Hamiltonian of the system with an EM field $E = E_0 \cdot \cos \omega t$

$$\mathcal{H}_1(t) = -\mu E(t) = \begin{pmatrix} 0 & -D_{ab} E_0(t) \\ -D_{ba} E_0(t) & 0 \end{pmatrix} \cos \omega t , \tag{2.165}$$

and a relaxation part

$$\mathcal{H}_R = \hbar \begin{pmatrix} \gamma_a & \gamma_\varphi^a \\ \gamma_\varphi^b & \gamma_b \end{pmatrix} , \tag{2.166}$$

which describes all relaxation processes, such as spontaneous emission or collision-induced transitions. The population relaxation of level $|b\rangle$ with a decay constant γ_b causing an effective lifetime $T_b = 1/\gamma_b$ is, for example, described by

$$i\hbar \rho_{bb} \gamma_b = [\mathcal{H}_R, \tilde{\rho}]_{bb} \quad \Rightarrow \quad T_b = \frac{1}{\gamma_b} = \frac{i\hbar \rho_{bb}}{[\mathcal{H}_R, \tilde{\rho}]_{bb}} . \tag{2.167}$$

The decay of the off-diagonal elements ρ_{ab}, ρ_{ba} describes the decay of the coherence, i.e., of the phase relations between the atomic dipoles.

The dephasing rate is represented by the phase-relaxation constants γ_φ^a, γ_φ^b and the decay of the nondiagonal elements is governed by

$$\frac{i\hbar \rho_{ab}}{T_2} = -[\mathcal{H}_R, \rho]_{ab} , \tag{2.168}$$

where the "transverse" relaxation time T_2 (dephasing time) is defined by

$$\frac{1}{T_2} = \frac{1}{2}\left(\frac{1}{T_a} + \frac{1}{T_b}\right) + \gamma_\phi . \tag{2.169}$$

In general, the phase relaxation is faster than the population relaxation defined by the relaxation time T_1, which means that the nondiagonal elements decay faster than the diagonal elements (Vol. 2, Chap. 7).

For more information on coherent excitation of atomic and molecular systems see [65–67] and Vol. 2, Chap. 14.

2.11   Problems

2.1 Verify (2.12).

2.2 The angular divergence of the output from a 1-W argon laser is assumed to be 4×10^{-3} rad. Calculate the radiance L and the radiant intensity I^* of the laser beam and the irradiance I (intensity) at a surface 1 m away from the output mirror, when the laser beam diameter at the mirror is 2 mm. What is the spectral power density $\rho(\nu)$ if the laser bandwidth is 1 MHz?

2.3 Unpolarized light of intensity I_0 is transmitted through a dichroic polarizer with thickness 1 mm. Calculate the transmitted intensity when the absorption coefficients for the two polarizations are $\alpha_\| = 100\,\mathrm{cm}^{-1}$ and $\alpha_\perp = 5\,\mathrm{cm}^{-1}$.

2.4 Assume the isotropic emission of a pulsed flashlamp with spectral bandwidth $\Delta\lambda = 100$ nm around $\lambda = 400$ nm amounts to 100-W peak power out of a volume of 1 cm³. Calculate the spectral power density $\rho(\nu)$ and the spectral intensity $I(\nu)$ through a spherical surface 2 cm away from the center of the emitting sphere. How many photons per mode are contained in the radiation field?

2.5 The beam of a monochromatic laser passes through an absorbing atomic vapor with path length $L = 5$ cm. If the laser frequency is tuned to the center of an absorbing transition $|i\rangle \to |k\rangle$ with absorption cross section $\sigma_0 = 10^{-14}\,\mathrm{cm}^2$, the attenuation of the transmitted intensity is 10 %. Calculate the atomic density N_i in the absorbing level $|i\rangle$.

2.6 An excited molecular level $|E_i\rangle$ is connected with three lower levels $|n\rangle$ and the groundstate $|0\rangle$ by radiative transitions with spontaneous probabilities $A_{i0} = 4 \times 10^7\,\mathrm{s}^{-1}$, $A_{i1} = 3 \times 10^7\,\mathrm{s}^{-1}$, $A_{i2} = 1 \times 10^7\,\mathrm{s}^{-1}$, $A_{i3} = 5 \times 10^7\,\mathrm{s}^{-1}$.

1. Calculate the spontaneous lifetime τ_i and the relative population densities N_n/N_i under cw excitation of $|i\rangle$ when $\tau_1 = 500$ ns, $\tau_2 = 6$ ns, and $\tau_3 = 10$ ns.
2. Determine the Einstein coefficient B_{0i} for the excitation of $|i\rangle$ from the ground-state with $\tau_0 = \infty$ and with the statistical weights $g_0 = 1$ and $g_1 = 3$. At which spectral energy density ρ_ν is the induced absorption rate equal to the sponta-neous decay rate of level $|i\rangle$? What is the intensity of a laser with a bandwidth of 10 MHz at this radiation density?
3. How large is the absorption cross-section σ_{0i} if the absorption linewidth is solely determined by the lifetime of the upper level?

Figure 2.40 Beam-expanding telescope with an aperture in the focal plane

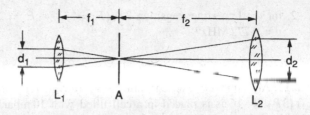

Figure 2.41 Schematic diagram of Michelson's star interferometer

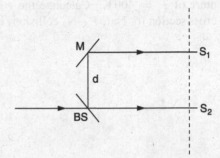

2.7 Under the conditions of Problem 2.6 there is an inversion between levels $|i\rangle$ and $|2\rangle$ which allows laser action on this transition. What is the minimum field amplitude E_0 and energy density ρ, of this transition that cause a Rabi oscillation between levels $|2\rangle$ and $|i\rangle$ with a period $T = 1/\Omega$ which is shorter than the lifetime of $|2\rangle$?

2.8 Expansion of a laser beam is accomplished by two lenses with different focal lengths (Fig. 2.40). Why does an aperture in the focal plane improve the quality of the wave fronts in the expanded beam by eliminating perturbations due to diffraction effects by dust and other imperfections on the lens surfaces?

2.9 Calculate the maximum slit separation in Young's interference experiments that still gives distinct interference fringes, if the two slits are illuminated

1. by incoherent light of $\lambda = 500$ nm from a hole with 1-mm diameter, 1 m away from the slits;
2. by a star with 10^6-km diameter, at a distance of 4 light-years;
3. by two partial beams of a He-Ne laser with a spectral width of 1 MHz (Fig. 2.41).

2.10 A sodium atom is placed in a cavity $V = 1$ cm^3 with walls at the temperature T, producing a thermal radiation field with spectral energy density $\rho(\nu)$. At what temperature T are the spontaneous and induced transition probabilities equal

1. for the transition $3P \to 3S$ ($\lambda = 589$ nm) with $\tau(3P) = 16$ ns;

2. for the hyperfine transition $3S$ $(F = 3 \rightarrow F = 2)$ with $\tau(3F) \simeq 1\,\mathrm{s}$ and $\nu = 1772\,\mathrm{MHz}$?

2.11 An optically excited sodium atom $\mathrm{Na}(3P)$ with a spontaneous lifetime $\tau(3P) = 16\,\mathrm{ns}$ is placed in a cell filled with $10\,\mathrm{mbar}$ nitrogen gas at a temperature of $T = 400\,\mathrm{K}$. Calculate the effective lifetime $\tau_{\mathrm{eff}}(3P)$ if the quenching cross section for $\mathrm{Na}(3P)$–N_2 collisions is $\sigma_q = 4 \times 10^{-15}\,\mathrm{cm}^2$.

Chapter 3
Widths and Profiles of Spectral Lines

Spectral lines in discrete absorption or emission spectra are never strictly monochromatic. Even with the very high resolution of interferometers, one observes a spectral distribution $I(\nu)$ of the absorbed or emitted intensity around the central frequency $\nu_0 = (E_i - E_k)/h$ corresponding to a molecular transition with the energy difference $\Delta E = E_i - E_k$ between upper and lower levels. The function $I(\nu)$ in the vicinity of ν_0 is called the *line profile* (Fig. 3.1). The frequency interval $\delta\nu = |\nu_2 - \nu_1|$ between the two frequencies ν_1 and ν_2 for which $I(\nu_1) = I(\nu_2) = I(\nu_0)/2$ is the *full-width at half-maximum* of the line (FWHM), often shortened to the *linewidth* or *halfwidth* of the spectral line.

The halfwidth is sometimes written in terms of the angular frequency $\omega = 2\pi\nu$ with $\delta\omega = 2\pi\delta\nu$, or in terms of the wavelength λ (in units of nm or Å) with $\delta\lambda = |\lambda_1 - \lambda_2|$. From $\lambda = c/\nu$, it follows that

$$\delta\lambda = -(c/\nu^2)\delta\nu \ . \tag{3.1}$$

The *relative* halfwidths, however, are the same in all three schemes:

$$\left|\frac{\delta\nu}{\nu}\right| = \left|\frac{\delta\omega}{\omega}\right| = \left|\frac{\delta\lambda}{\lambda}\right| \ . \tag{3.2}$$

The spectral region within the halfwidth is called the *kernel of the line*, the regions outside ($\nu < \nu_1$ and $\nu > \nu_2$) are the *line wings*.

In the following sections we discuss various origins of the finite linewidth. Several examples illustrate the order of magnitude of different line-broadening effects in different spectral regions and their importance for high-resolution spectroscopy [68–71]. Following the usual convention we shall often use the angular frequency $\omega = 2\pi\nu$ to avoid factors of 2π in the equations.

W. Demtröder, *Laser Spectroscopy 1*, DOI 10.1007/978-3-642-53859-9_3,
© Springer-Verlag Berlin Heidelberg 2014

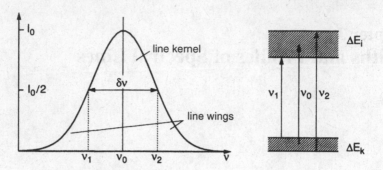

Figure 3.1 Line profile, halfwidth, kernel, and wings of a spectral line

3.1 Natural Linewidth

An excited atom can emit its excitation energy as spontaneous radiation (Sect. 2.8). In order to investigate the spectral distribution of this spontaneous emission on a transition $E_i \rightarrow E_k$, we shall describe the excited *atomic electron* by the classical model of a damped harmonic oscillator with frequency ω, mass m, and restoring force constant k. The radiative energy loss results in a damping of the oscillation described by the damping constant γ. We shall see, however, that for real atoms the damping is extremely small, which means that $\gamma \ll \omega$.

The amplitude $x(t)$ of the oscillation can be obtained by solving the differential equation of motion

$$\ddot{x} + \gamma\dot{x} + \omega_0^2 x = 0 , \qquad (3.3)$$

where $\omega_0^2 = k/m$.

The real solution of (3.3) with the initial values $x(0) = x_0$ and $\dot{x}(0) = 0$ is

$$x(t) = x_0 e^{-(\gamma/2)t}[\cos\omega t + (\gamma/2\omega)\sin\omega t] . \qquad (3.4)$$

The frequency $\omega = (\omega_0^2 - \gamma^2/4)^{1/2}$ of the damped oscillation is slightly lower than the frequency ω_0 of the undamped case. However, for small damping ($\gamma \ll \omega_0$) we can set $\omega \simeq \omega_0$ and also may neglect the second term in (3.4). With this approximation, which is still very accurate for real atoms, we obtain the solution of (3.3) as

$$x(t) = x_0 e^{-(\gamma/2)t} \cos\omega_0 t . \qquad (3.5)$$

The frequency $\omega_0 = 2\pi\nu_0$ of the oscillator corresponds to the central frequency $\omega_{ik} = (E_i - E_k)/\hbar$ of an atomic transition $E_i \rightarrow E_k$.

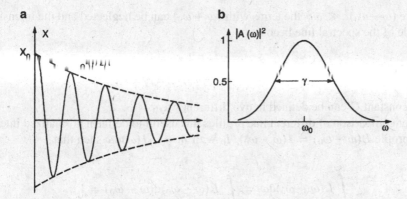

Figure 3.2 a Damped oscillation: **b** the frequency distribution $A(\omega)$ of the amplitudes obtained by the Fourier transform of $x(t)$ yields the intensity profile $I(\omega - \omega_0) \propto |A(\omega)|^2$

3.1.1 Lorentzian Line Profile of the Emitted Radiation

Because the amplitude $x(t)$ of the oscillation decreases gradually, the frequency of the emitted radiation is no longer monochromatic as it would be for an oscillation with constant amplitude. Instead, it shows a frequency distribution related to the function $x(t)$ in (3.5) by a Fourier transformation (Fig. 3.2).

The damped oscillation $x(t)$ can be described as a superposition of monochromatic oscillations $\exp(i\omega t)$ with slightly different frequencies ω and amplitudes $A(\omega)$

$$x(t) = \frac{1}{2\sqrt{2\pi}} \int_0^\infty A(\omega) e^{i\omega t} d\omega . \tag{3.6}$$

The amplitudes $A(\omega)$ are calculated from (3.5) and (3.6) as the Fourier transform

$$A(\omega) = \frac{1}{\sqrt{2\pi}} \int_{-\infty}^{+\infty} x(t) e^{-i\omega t} dt = \frac{1}{\sqrt{2\pi}} \int_0^\infty x_0 e^{-(\gamma/2)t} \cos(\omega_0 t) e^{-i\omega t} dt . \tag{3.7}$$

The lower integration limit is taken to be zero because $x(t) = 0$ for $t < 0$. Equation (3.7) can readily be integrated to give the complex amplitudes

$$A(\omega) = \frac{x_0}{\sqrt{8\pi}} \left(\frac{1}{i(\omega - \omega_0) + \gamma/2} + \frac{1}{i(\omega + \omega_0) + \gamma/2} \right) . \tag{3.8}$$

The real intensity $I(\omega) \propto A(\omega) A^*(\omega)$ contains terms with $(\omega - \omega_0)$ and $(\omega + \omega_0)$ in the denominator. In the vicinity of the central frequency ω_0 of an atomic transition where $(\omega - \omega_0)^2 \ll \omega_0^2$, the terms with $(\omega + \omega_0)$ can be neglected and the intensity profile of the spectral line becomes

$$I(\omega - \omega_0) = \frac{C}{(\omega - \omega_0)^2 + (\gamma/2)^2} \; . \tag{3.9}$$

The constant C can be defined in two different ways:

For comparison of different line profiles it is useful to define a normalized intensity profile $L(\omega - \omega_0) = I(\omega - \omega_0)/I_0$ with $I_0 = \int I(\omega)d\omega$ such that

$$\int\limits_0^\infty L(\omega - \omega_0)d\omega = \int\limits_{-\infty}^{+\infty} L(\omega - \omega_0)d(\omega - \omega_0) = 1 \; .$$

With this normalization, the integration of (3.9) yields $C = I_0\gamma/2\pi$.

$$L(\omega - \omega_0) = \frac{\gamma/2\pi}{(\omega - \omega_0)^2 + (\gamma/2)^2} \; , \tag{3.10}$$

is called the *normalized Lorentzian profile*. Its full halfwidth at half-maximum (FWHM) is

$$\delta\omega_n = \gamma \quad \text{or} \quad \delta\nu_n = \gamma/2\pi \; . \tag{3.11}$$

Any intensity distribution with a Lorentzian profile is then

$$I(\omega - \omega_0) = I_0\frac{\gamma/2\pi}{(\omega - \omega_0)^2 + (\gamma/2)^2} = I_0 L(\omega - \omega_0) \; , \tag{3.10a}$$

with a peak intensity $I(\omega_0) = 2I_0/(\pi\gamma)$.

Note Often in the literature the normalization of (3.9) is chosen in such a way that $I(\omega_0) = I_0$; furthermore, the full halfwidth is denoted by 2Γ. In this notation the line profile of a transition $|k\rangle \leftarrow |i\rangle$ is

$$I(\omega) = I_0 g(\omega - \omega_{ik}) \quad \text{with} \quad I_0 = I(\omega_0) \; ,$$

and

$$g(\omega - \omega_{ik}) = \frac{\Gamma^2}{(\omega_{ik} - \omega)^2 + \Gamma^2} \quad \text{with} \quad \Gamma = \gamma/2 \; . \tag{3.10b}$$

With $x = (\omega_{ik} - \omega)/\Gamma$ this can be abbreviated as

$$g(\omega - \omega_{ik}) = \frac{1}{1 + x^2} \quad \text{with} \quad g(\omega_{ik}) = 1 \; . \tag{3.10c}$$

In this notation the area under the line profile becomes

$$\int_0^\infty I(\omega)\,\mathrm{d}\omega = \Gamma \int_{-\infty}^{+\infty} I(\omega)\,\mathrm{d}\omega = \pi I_0 \Gamma \qquad (3.10\mathrm{d})$$

3.1.2 Relation Between Linewidth and Lifetime

The radiant power of the damped oscillator can be obtained from (3.3) if both sides of the equation are multiplied by $m\dot{x}$, which yields after rearranging

$$m\ddot{x}\dot{x} + m\omega_0^2 x\dot{x} = -\gamma m\dot{x}^2 . \qquad (3.12)$$

The left-hand side of (3.12) is the time derivative of the total energy W (sum of kinetic energy $\frac{1}{2}m\dot{x}^2$ and potential energy $kx^2/2 = m\omega_0^2 x^2/2$), and can therefore be written as

$$\frac{\mathrm{d}}{\mathrm{d}t}\left(\frac{m}{2}\dot{x}^2 + \frac{m}{2}\omega_0^2 x^2\right) = \frac{\mathrm{d}W}{\mathrm{d}t} = -\gamma m\dot{x}^2 . \qquad (3.13)$$

Inserting $x(t)$ from (3.5) and neglecting terms with γ^2 yields

$$\frac{\mathrm{d}W}{\mathrm{d}t} = -\gamma m x_0^2 \omega_0^2 \mathrm{e}^{-\gamma t} \sin^2 \omega_0 t . \qquad (3.14)$$

Because the time average $\overline{\sin^2 \omega t} = 1/2$, the time-averaged radiant power $\overline{P} = \overline{\mathrm{d}W/\mathrm{d}t}$ is

$$\frac{\overline{\mathrm{d}W}}{\mathrm{d}t} = -\frac{\gamma}{2}m x_0^2 \omega_0^2 \mathrm{e}^{-\gamma t} . \qquad (3.15)$$

Equation (3.15) shows that $\overline{P}$ and with it the intensity $I(t)$ of the spectral line decreases to $1/\mathrm{e}$ of its initial value $I(t = 0)$ after the decay time $\tau = 1/\gamma$.

In Sect. 2.9 we saw that the mean lifetime τ_i of a molecular level E_i, which decays exponentially by spontaneous emission, is related to the Einstein coefficient A_i by $\tau_i = 1/A_i$. Replacing the classical damping constant γ by the spontaneous transition probability A_i, we can use the classical formulas (3.9–3.11) as a correct description of the frequency distribution of spontaneous emission and its linewidth. The natural halfwidth of a spectral line spontaneously emitted from the level E_i is, according to (3.11),

$$\delta\nu_n = A_i/2\pi = (2\pi\tau_i)^{-1} \quad \text{or} \quad \delta\omega_n = A_i = 1/\tau_i . \qquad (3.16)$$

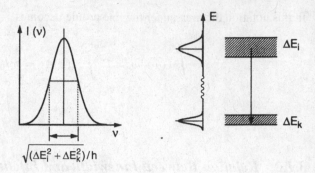

Figure 3.3 Illustration of the uncertainty principle, which relates the natural linewidth to the energy uncertainties of the upper and lower levels

The radiant power emitted from N_i excited atoms on a transition $E_i \rightarrow E_k$ is given by

$$dW_{ik}/dt = N_i A_{ik} \hbar \omega_{ik} . \tag{3.17}$$

If the emission of a source with volume ΔV is isotropic, the radiation power received by a detector of area A at a distance r through the solid angle $d\Omega = A/r^2$ is

$$P_{ik} = \left(\frac{dW_{ik}}{dt}\right) \Delta V \frac{d\Omega}{4\pi} = N_i A_{ik} \hbar \omega_{ik} \Delta V \frac{A}{4\pi r^2} . \tag{3.18}$$

This means that the density N_i of emitters can be inferred from the measured power, if A_{ik} is known (Vol. 2, Sect. 6.3).

Note Equation (3.16) can also be derived from the uncertainty principle (Fig. 3.3). With the mean lifetime τ_i of the excited level E_i, its energy E_i can be determined only with an uncertainty $\Delta E_i \simeq \hbar/\tau_i$ [72]. The frequency $\omega_{ik} = (E_i - E_k)/\hbar$ of a transition terminating in the stable ground state E_k has therefore the uncertainty

$$\delta\omega = \Delta E_i / \hbar = 1/\tau_i . \tag{3.19}$$

If the lower level E_k is not the ground state but also an excited state with the lifetime τ_k, the uncertainties ΔE_i and ΔE_k of the two levels both contribute to the linewidth. This yields for the total uncertainty

$$\Delta E = \sqrt{\Delta E_i^2 + \Delta E_k^2} \rightarrow \delta\omega_n = \sqrt{(1/\tau_i^2 + 1/\tau_k^2)} . \tag{3.20}$$

3.1.3 Natural Linewidth of Absorbing Transitions

In a similar way, the spectral profile of an *absorption line* can be derived for atoms *at rest*: the intensity I of a plane wave passing in the z direction through an absorbing

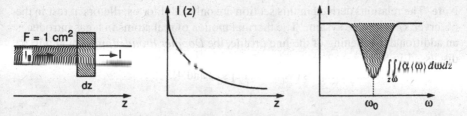

Figure 3.4 Absorption of a parallel light beam passing through an optically thin absorbing layer

sample decreases along the distance dz by

$$dI = -\alpha I \, dz \ . \tag{3.21}$$

The absorption coefficient α_{ik} [cm^{-1}] for a transition $|i\rangle \to |k\rangle$ depends on the population densities N_i, N_k of the lower and upper levels, and on the optical absorption cross section σ_{ik} [cm^2] of each absorbing atom, see (2.58):

$$\alpha_{ik}(\omega) = \sigma_{ik}(\omega)[N_i - (g_i/g_k)N_k] \ , \tag{3.22}$$

which reduces to $\alpha_{ik} = \sigma N_i$ for $N_k \ll N_i$ (Fig. 3.4). For sufficiently small intensities I, the induced absorption rate is small compared to the refilling rate of level $|i\rangle$ and the population density N_i does not depend on the intensity I (linear absorption). Integration of (3.21) then yields Beer's law

$$I = I_0 e^{-\alpha(\omega)z} = I_0 e^{-\sigma_{ik} N_i z} \ . \tag{3.23}$$

The absorption profile $\alpha(\omega)$ can be obtained from our classical model of a damped oscillator with charge q under the influence of a driving force qE caused by the incident wave with amplitude $E = E_0 e^{i\omega t}$ (see Sect. 2.6).

In Sect. 2.6 it was shown, that in the neighborhood of a molecular transition frequency ω_0 where $|\omega_0 - \omega| \ll \omega_0$, the dispersion relations (2.51a, 2.51b) reduce with $q = e$ and $\omega_0^2 - \omega^2 = (\omega_0 + \omega)(\omega_0 - \omega) \approx 2\omega_0(\omega_0 - \omega)$ to

$$\alpha(\omega) = \frac{Ne^2}{4\epsilon_0 mc} \frac{\gamma}{(\omega_0 - \omega)^2 + (\gamma/2)^2} \ , \tag{3.24a}$$

$$n' = 1 + \frac{Ne^2}{4\epsilon_0 m\omega_0} \frac{\omega_0 - \omega}{(\omega_0 - \omega)^2 + (\gamma/2)^2} \ . \tag{3.24b}$$

The absorption profile $\alpha(\omega)$ is Lorentzian with a FWHM of $\Delta\omega_n = \gamma$, which equals for free atoms at rest the natural linewidth. The difference $n' - n_0 = n' - 1$ between the refractive indices in a gas and in vacuum yields a dispersion profile.

Figure 2.15 shows the frequency dependence of $\alpha(\omega)$ and $n'(\omega)$ in the vicinity of the eigenfrequency ω_0 of an atomic transition.

Note The relations derived in this section are only valid for oscillators at rest in the observer's coordinate system. The thermal motion of real atoms in a gas introduces an additional broadening of the line profile, the *Doppler broadening*, which will be discussed in Sect. 3.2. The profiles (3.24a, 3.24b) can therefore be observed only with Doppler-free techniques (Vol. 2, Chaps. 2 and 4).

Example 3.1

1. The natural linewidth of the sodium D_1 line at $\lambda = 589.1$ nm, which corresponds to a transition between the $3\,P_{3/2}$ level ($\tau = 16$ ns) and the $3\,S_{1/2}$ ground state, is

$$\delta \nu_n = \frac{10^9}{16 \times 2\pi} = 10^7 \, \text{s}^{-1} = 10 \, \text{MHz} \,.$$

Note that with a central frequency $\nu_0 = 5 \times 10^{14}$ Hz and a lifetime of 16 ns, the damping of the corresponding classical oscillator is extremely small. Only after 8×10^6 periods of oscillation has the amplitude decreased to $1/e$ of its initial value.

2. The natural linewidth of a molecular transition between two vibrational levels of the electronic ground state with a wavelength in the infrared region is very small because of the long spontaneous lifetimes of vibrational levels. For a typical lifetime of $\tau = 10^{-3}$ s, the natural linewidth becomes $\delta \nu_n = 160$ Hz.

3. Even in the visible or ultraviolet range, atomic or molecular electronic transitions with very small transition probabilities exist. In a dipole approximation these are "forbidden" transitions. One example is the 2s ↔ 1s transition for the hydrogen atom. The upper level 2s cannot decay by electric dipole transition, but a two-photon transition to the 1s ground state is possible. The natural lifetime is $\tau = 0.12$ s and the natural linewidth of such a two-photon line is therefore $\delta \nu_n = 1.3$ Hz.

3.2 Doppler Width

Generally, the Lorentzian line profile with the natural linewidth $\delta \nu_n$, as discussed in Sect. 3.1, cannot be observed without special techniques, because it is completely concealed by other broadening effects. One of the major contributions to the spectral linewidth in gases at low pressures is the Doppler width, which is due to the thermal motion of the absorbing or emitting molecules.

Consider an excited molecule with a velocity $v = \{v_x, v_y, v_z\}$ relative to the rest frame of the observer. The central frequency of a molecular emission line that is ω_0

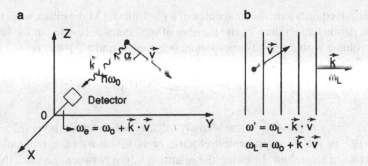

Figure 3.5 **a** Doppler shift of a monochromatic emission line and **b** absorption line

in the coordinate system of the molecule is Doppler shifted to

$$\omega_e = \omega_0 + \boldsymbol{k} \cdot \boldsymbol{v} , \qquad (3.25)$$

for an observer looking toward the emitting molecule (that is, against the direction of the wave vector $\boldsymbol{k}$ of the emitted radiation; Fig. 3.5a). For the observer, the apparent emission frequency ω_e is increased if the molecule moves toward the observer ($\boldsymbol{k} \cdot \boldsymbol{v} > 0$), and decreased if the molecule moves away ($\boldsymbol{k} \cdot \boldsymbol{v} < 0$).

Similarly, one can see that the absorption frequency ω_0 of a molecule moving with the velocity $\boldsymbol{v}$ across a plane EM wave $\boldsymbol{E} = \boldsymbol{E}_0 \exp(i\omega t - \boldsymbol{k} \cdot \boldsymbol{r})$ is shifted. The wave frequency ω in the rest frame appears in the frame of the moving molecule as

$$\omega' = \omega - \boldsymbol{k} \cdot \boldsymbol{v} .$$

The molecule can only absorb if ω' coincides with its eigenfrequency ω_0. The absorption frequency $\omega = \omega_a$ is then

$$\omega_a = \omega_0 + \boldsymbol{k} \cdot \boldsymbol{v} . \qquad (3.26a)$$

As in the emission case, the absorption frequency ω_a is increased for $\boldsymbol{k} \cdot \boldsymbol{v} > 0$ (Fig. 3.5b). This happens, for example, if the molecule moves parallel to the wave propagation. It is decreased if $\boldsymbol{k} \cdot \boldsymbol{v} < 0$, e.g., when the molecule moves against the light propagation. If we choose the $+z$-direction to coincide with the light propagation, with $\boldsymbol{k} = \{0, 0, k_z\}$ and $|k| = 2\pi/\lambda$, (3.26a) becomes

$$\omega_a = \omega_0(1 + v_z/c) . \qquad (3.26b)$$

Note Equations (3.25) and (3.26a, 3.26b) describe the *linear* Doppler shift. For higher accuracies, the quadratic Doppler effect must also be considered (Vol. 2, Sect. 9.1).

At thermal equilibrium, the molecules of a gas follow a Maxwellian velocity distribution. At the temperature T, the number of molecules $n_i(v_z)dv_z$ in the level E_i per unit volume with a velocity component between v_z and $v_z + dv_z$ is

$$n_i(v_z)dv_z = \frac{N_i}{v_p\sqrt{\pi}} e^{-(v_z/v_p)^2} dv_z \; , \tag{3.27}$$

where $N_i = \int n_i(v_z)dv_z$ is the density of all molecules in level E_i, $v_p = (2kT/m)^{1/2}$ is the most probable velocity, m is the mass of a molecule, and k is Boltzmann's constant. Inserting the relation (3.26b) between the velocity component and the frequency shift with $dv_z = (c/\omega_0)d\omega$ into (3.27) gives the number of molecules with absorption frequencies shifted from ω_0 into the interval from ω to $\omega + d\omega$

$$n_i(\omega)d\omega = N_i \frac{c}{\omega_0 v_p \sqrt{\pi}} \exp\left[-\left(\frac{c(\omega - \omega_0)}{\omega_0 v_p}\right)^2\right] d\omega \; . \tag{3.28}$$

Since the emitted or absorbed radiant power $P(\omega)d\omega$ is proportional to the density $n_i(\omega)d\omega$ of molecules emitting or absorbing in the interval $d\omega$, the intensity profile of a Doppler-broadened spectral line becomes

$$I(\omega) = I_0 \exp\left[-\left(\frac{c(\omega - \omega_0)}{\omega_0 v_p}\right)^2\right] \; . \tag{3.29}$$

This is a Gaussian profile with a full halfwidth

$$\delta\omega_D = 2\sqrt{\ln 2}\,\omega_0 v_p/c = \left(\frac{\omega_0}{c}\right)\sqrt{8kT\ln 2/m} \; , \tag{3.30a}$$

which is called the *Doppler width*. Inserting v_p from (3.30a) into (3.29) with $1/(4\ln 2) = 0.36$ yields

$$\boxed{I(\omega) = I_0 \exp\left(-\frac{(\omega - \omega_0)^2}{0.36\delta\omega_D^2}\right) \; .} \tag{3.31}$$

Note that $\delta\omega_D$ increases linearly with the frequency ω_0 and is proportional to $(T/m)^{1/2}$. The largest Doppler width is thus expected for hydrogen ($M = 1$) at high temperatures and a large frequency ω for the Lyman α line.

Equation (3.30a) can be written more conveniently in terms of the Avogadro number N_A (the number of molecules per mole), the mass of a mole, $M = N_A m$, and the gas constant $R = N_A k$. Inserting these relations into (3.30a) for the Doppler width gives

$$\delta\omega_D = (2\omega_0/c)\sqrt{2RT\ln 2/M} \; . \tag{3.30b}$$

Figure 3.6 Comparison between Lorentzian (L) and Gaussian (G) line profiles of equal halfwidths

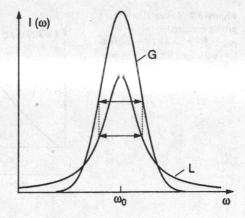

or, in frequency units, using the values for c and R,

$$\delta \nu_D = 7.16 \times 10^{-7} \nu_0 \sqrt{T/M} \quad [\text{Hz}] . \tag{3.30c}$$

Example 3.2

1. Vacuum ultraviolet: for the Lyman α line (2p $\rightarrow$ 1s transition in the H atom) in a discharge with temperature $T = 1000\,\text{K}$, $M = 1$, $\lambda = 121.6\,\text{nm}$, $\nu_0 = 2.47 \times 10^{15}\,\text{s}^{-1} \rightarrow \delta \nu_D = 5.6 \times 10^{10}\,\text{Hz}$, $\delta \lambda_D = 2.8 \times 10^{-3}\,\text{nm}$.

2. Visible spectral region: for the sodium D line (3p $\rightarrow$ 3s transition of the Na atom) in a sodium-vapor cell at $T = 500\,\text{K}$, $\lambda = 589.1\,\text{nm}$, $\nu_0 = 5.1 \times 10^{14}\,\text{s}^{-1} \rightarrow \delta \nu_D = 1.7 \times 10^9\,\text{Hz}$, $\delta \lambda_D = 1 \times 10^{-3}\,\text{nm}$.

3. Infrared region: for a vibrational transition $(J_i, v_i) \leftrightarrow (J_k, v_k)$ between two rovibronic levels with the quantum numbers J, v of the CO_2 molecule in a CO_2 cell at room temperature $(T = 300\,\text{K})$, $\lambda = 10\,\mu\text{m}$, $\nu = 3 \times 10^{13}\,\text{s}^{-1}$, $M = 44 \rightarrow \delta \nu_D = 5.6 \times 10^7\,\text{Hz}$, $\delta \lambda_D = 1.9 \times 10^{-2}\,\text{nm}$.

These examples illustrate that in the visible and UV regions, the Doppler width exceeds the natural linewidth by about two orders of magnitude. Note, however, that the intensity I approaches zero for large arguments $(\nu - \nu_0)$ much faster for a Gaussian line profile than for a Lorentzian profile (Fig. 3.6). It is therefore possible to obtain information about the Lorentzian profile from the extreme line wings, even if the Doppler width is much larger than the natural linewidth (see below).

More detailed consideration shows that a Doppler-broadened spectral line cannot be strictly represented by a pure Gaussian profile as has been assumed in the foregoing discussion, since not all molecules with a definite velocity component v_z emit or absorb radiation at the same frequency $\omega' = \omega_0(1 + v_z/c)$. Because of the finite lifetimes of the molecular energy levels, the frequency response of these

Figure 3.7 Lorentzian profile centered at $\omega' = \omega_0 + \boldsymbol{k} \cdot \boldsymbol{v} = \omega_0(1 + v_z/c)$ for molecules with a definite velocity component v_z

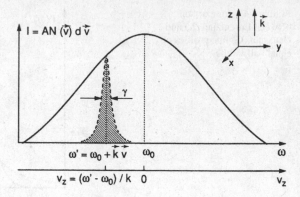

molecules is represented by a Lorentzian profile, see (3.10)

$$L(\omega - \omega') = \frac{\gamma/2\pi}{(\omega - \omega')^2 + (\gamma/2)^2} ,$$

with a central frequency ω' (Fig. 3.7). Let $n(\omega')d\omega' = n(v_z)dv_z$ be the number of molecules per unit volume with velocity components within the interval v_z to $v_z + dv_z$. The spectral intensity distribution $I(\omega)$ of the total absorption or emission of all molecules at the transition $E_i \rightarrow E_k$ is then

$$I(\omega) = I_0 \int n(\omega')L(\omega - \omega')d\omega' . \tag{3.32}$$

Inserting (3.10) for $L(\omega - \omega')d\omega'$ and (3.28) for $n(\omega')$, we obtain

$$I(\omega) = C \int\limits_0^\infty \frac{\exp\{-[(c/v_{\mathrm{p}})(\omega_0 - \omega')/\omega_0]^2\}}{(\omega - \omega')^2 + (\gamma/2)^2} d\omega' \tag{3.33}$$

with

$$C = \frac{\gamma N_i c}{2 v_{\mathrm{p}} \pi^{3/2} \omega_0} .$$

This intensity profile, which is a convolution of Lorentzian and Gaussian profiles (Fig. 3.8), is called a *Voigt profile*. Voigt profiles play an important role in the spectroscopy of stellar atmospheres, where accurate measurements of line wings allow the contributions of Doppler broadening and natural linewidth or collisional line broadening to be separated (see [73] and Sect. 3.3). From such measurements the temperature and pressure of the emitting or absorbing layers in the stellar atmospheres may be deduced [74].

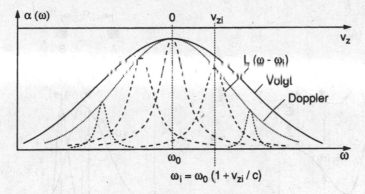

Figure 3.8 Voigt profile as a convolution of Lorentzian line shapes $L(\omega_0 - \omega_i)$ of molecules with different velocity components v_{zi} and central absorption frequencies $\omega_i = \omega_0(1 + v_{zi}/c)$

3.3 Collisional Broadening of Spectral Lines

When an atom A with the energy levels E_i and E_k approaches another atom or molecule B, the energy levels of A are shifted because of the interaction between A and B. This shift depends on the electron configurations of A and B and on the distance $R(A, B)$ between both collision partners, which we define as the distance between the centers of mass of A and B.

The energy shifts ΔE are, in general, different for the levels E_i and E_k and may be positive as well as negative. The energy shift ΔE is positive if the interaction between A and B is repulsive, and negative if it is attractive. When plotting the energy $E(R)$ for the different energy levels as a function of the interatomic distance R typical potential curves of Fig. 3.9 are obtained.

This mutual interaction of both partners at distances $R \leq R_c$ is called a *collision* and radius R_c is the *collision radius*. If no internal energy of the collision partners is transferred during the collision by nonradiative transitions, the collision is termed *elastic*. Without additional stabilizing mechanisms (recombination), the partners will separate again after the collision time $\tau_c \simeq R_c/v$, which depends on the relative velocity v.

Example 3.3
At thermal velocities of $v = 5 \times 10^2$ m/s and a typical collision radius of $R_c = 1$ nm, we obtain the collision time $\tau_c = 2 \times 10^{-12}$ s. During this time the electronic charge distribution generally follows the perturbation "adiabatically", which justifies the potential curve model of Fig. 3.9.

The *collisional shift* of a spectral line depends on the energy difference $\Delta E(R) = E_k(R) - E_i(R)$ around the collision radius R, compared to the difference at $R = \infty$.

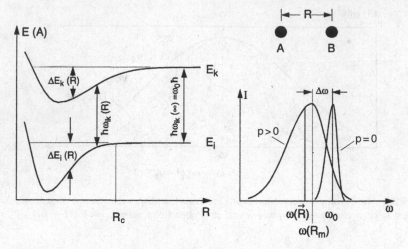

Figure 3.9 Illustration of collisional line broadening explained with the potential curves of the collision pair AB

The collisional broadening stems from two effects: The shortening of the upper state lifetime by inelastic collisions and the distribution of the energy differences $\Delta E(R)$ by elastic collisions (see below).

3.3.1 Phenomenological Description

If atom A undergoes a *radiative* transition between levels E_i and E_k during the collision time, the frequency

$$\omega_{ik} = |E_i(R) - E_k(R)| / \hbar \tag{3.34}$$

of absorbed or emitted radiation depends on the distance $R(t)$ at the time of the transition. We assume that the radiative transition takes place in a time interval that is short compared to the collision time, so that the distance R does not change during the transition. In Fig. 3.9 this assumption leads to vertical radiative transitions.

In a gas mixture of atoms A and B, the mutual distance $R(A, B)$ shows random fluctuations with a distribution around a mean value R that depends on pressure and temperature. According to (3.34), the fluorescence yields a corresponding frequency distribution around a most probable value $\omega_{ik}(R_m)$, which may be shifted against the frequency ω_0 of the unperturbed atom A. The shift $\Delta\omega = \omega_0 - \omega_{ik}$ depends on how differently the two energy levels E_i and E_k are shifted at a distance $R_m(A, B)$ where the emission probability has a maximum. The intensity profile $I(\omega)$ of the collision-broadened and shifted emission line can be obtained from

$$I(\omega) \propto \int A_{ik}(R) P_{col}(R)[E_i(R) - E_k(R)]dR , \qquad (3.35)$$

where $A_{ik}(R)$ is the spontaneous transition probability, which depends on R because the electronic wave functions of the collision pair (AB) depend on R, and $P_{col}(R)$ is the probability per unit time that the distance between A and B lies in the range from R to $R + dR$.

From (3.35) it can be seen that the intensity profile of the collision-broadened line reflects the difference of the potential curves

$$E_i(R) - E_k(R) = V[A(E_i), B] - V[A(E_k), B] . \qquad (3.36)$$

Let $V(R)$ be the interaction potential between the ground-state atom A and its collision partner B. The probability that B has a distance between R and $R + dR$ is proportional to $4\pi R^2 dR$ and (in thermal equilibrium) to the Boltzmann factor $\exp[-V(R)/kT]$. The number $N(R)$ of collision partners B with distance R from A is therefore

$$N(R)dR = N_0 4\pi R^2 e^{-V(R)/kT} dR , \qquad (3.37)$$

where N_0 is the average density of atoms B. Because the intensity of an absorption line is proportional to the density of absorbing atoms while they are forming collision pairs, the intensity profile of the absorption line can be written as

$$I(\omega)d\omega = C^* \left\{ R^2 \exp\left(-\frac{V_i(R)}{kT}\right) \frac{d}{dR}[V_i(R) - V_k(R)] \right\} dR , \qquad (3.38)$$

where $\hbar\omega(R) = [V_i(R) - V_k(R)] \rightarrow \hbar d\omega/dR = d[V_i(R) - V_k(R)]/dR$ has been used. Measuring the line profile as a function of temperature yields

$$\frac{dI(\omega, T)}{dT} = \frac{V_i(R)}{kT^2} I(\omega, T) , \qquad (3.39)$$

and therefore the ground-state potential $V_i(R)$ separately.

Frequently, different *spherical model potentials* $V(R)$ are substituted in (3.38), such as the Lennard–Jones potential

$$V(R) = a/R^{12} - b/R^6 , \qquad (3.40)$$

The coefficients a, b are adjusted for optimum agreement between theory and experiment [75–83].

The line shift caused by elastic collisions corresponds to an energy shift $\Delta E = \hbar\Delta\omega$ between the excitation energy $\hbar\omega_0$ of the free atom A* and the photon energy $\hbar\omega$. It is supplied from the kinetic energy of the collision partners. This means that in case of positive shifts ($\Delta\omega > 0$), the kinetic energy is smaller after the collision than before.

Besides elastic collisions, inelastic collisions may also occur in which the excitation energy E_i of atom A is either partly or completely transferred into internal energy of the collision partner B, or into translational energy of both partners. Such inelastic collisions are often called *quenching collisions* because they decrease the number of excited atoms in level E_i and therefore quench the fluorescence intensity. The total transition probabiltiy A_i for the depopulation of level E_i is a sum of radiative and collision-induced probabilities (Fig. 2.20)

$$A_i = A_i^{\text{rad}} + A_i^{\text{coll}} \quad \text{with} \quad A_i^{\text{coll}} = N_B\sigma_i v . \tag{3.41}$$

Inserting the relations

$$v = \sqrt{\frac{8kT}{\pi\mu}} , \quad \mu = \frac{M_A \cdot M_B}{M_A + M_B} , \quad p_B = N_B kT , \tag{3.42}$$

between the mean relative velocity v, the responsible pressure p_B, and the gas temperature T into (3.41) gives the total transition probability

$$A_i = \frac{1}{\tau_{\text{sp}}} + ap_B \quad \text{with} \quad a = 2\sigma_{ik}\sqrt{\frac{2}{\pi\mu kT}} . \tag{3.43}$$

It is evident from (3.16) that this pressure-dependent transition probability causes a corresponding pressure-dependent linewidth $\delta\omega$, which can be described by a sum of two damping terms

$$\delta\omega = \delta\omega_n + \delta\omega_{\text{col}} = \gamma_n + \gamma_{\text{col}} = \gamma_n + ap_B . \tag{3.44}$$

The collision-induced additional line broadening ap_B is therefore often called *pressure broadening*.

From the derivation in Sect. 3.1, one obtains a Lorentzian profile (3.9) with a halfwidth $\gamma = \gamma_n + \gamma_{\text{col}}$ for the line broadened by inelastic collisions:

$$I(\omega) = \frac{C}{(\omega - \omega_0)^2 + [(\gamma_n + \gamma_{\text{col}})/2]^2} . \tag{3.45}$$

The *elastic collisions* do not change the amplitude, but the *phase* of the damped oscillator is changed due to the frequency shift $\Delta\omega(R)$ during the collisions. They are often termed *phase-perturbing collisions* (Fig. 3.10).

When taking into account line shifts $\Delta\omega$ caused by elastic collisions, the line profile for cases where it still can be described by a Lorentzian becomes

$$I(\omega) = \frac{C^*}{(\omega - \omega_0 - \Delta\omega)^2 + (\gamma/2)^2} , \tag{3.46}$$

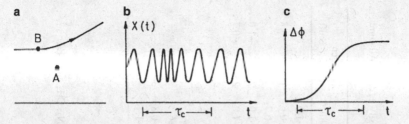

Figure 3.10 Phase perturbation of an oscillator by collisions: **a** classical path approximation of colliding particles; **b** frequency change of the oscillator $A(t)$ during the collision; **c** resulting phase shift

Figure 3.11 Shift and broadening of a Lorentzian line profile by collisions

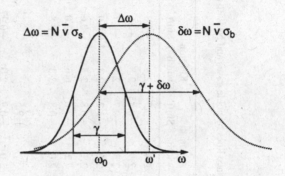

where the line shift

$$\Delta\omega = N_B \cdot \bar{v} \cdot \sigma_s \tag{3.47a}$$

and the line broadening

$$\gamma = \gamma_n + N_B \cdot \bar{v} \cdot \sigma_b \tag{3.47b}$$

are determined by the number density N_B of collision parameters B and by the collision cross sections σ_s for line shifts and σ_b for broadening (Fig. 3.11). The constant $C^* = (I_0/2\pi)(\gamma + N_B\bar{v}\sigma_b)$ becomes $I_0\gamma/2\pi$ for $N_B = 0$, when (3.46) becomes identical to (3.10).

Note The real collision-induced line profile depends on the interaction potential between A and B. In most cases it is no longer Lorentzian, but has an asymmetric profile because the transition probability depends on the internuclear distance and because the energy difference $\Delta E(R) = E_i(R) - E_k(R)$ is generally not a uniformly rising or falling function but may have extrema.

Figure 3.12 depicts as examples pressure broadening and shifts in $[\text{cm}^{-1}]$ of the lithium resonance line perturbed by different noble gas atoms. Table 3.1 compiles pressure-broadening and line shift data for different alkali resonance lines.

Table 3.1 Broadening (fullhalfwidth γ/n) and shift $\Delta\omega/n$ of atomic alkali resonance lines by noble gases and N_2 with number density n/cm^3. (All numbers are given in units of $10^{-20}\,cm^{-1}/cm^{-3} \approx 10\,MHz/torr$ at $T = 300\,K$)

Transition	λ (nm)	Selfbroadening	Helium width	Helium Shift	Neon width	Neon Shift	Argon width	Argon Shift	Krypton width	Krypton Shift	Xenon width	Xenon Shift	Nitrogen width	Nitrogen Shift
Li $2S$–$2P$	670.8	2.5×10^2	2.2	−0.08	1.5	−0.2	2.4	−0.7	2.9	−0.8	3.3	−1.0		
Na $3S_{1/2}$–$3P_{1/2}$	598.6	1.6×10^2	1.6	0.00	1.3	−0.3	2.9	−0.85	2.8	−0.6	3.0	−0.6	1.8	−0.8
–$P_{3/2}$	598.0	2.7×10^2	3.0	−0.06	1.5	−0.75	2.3	−0.7	2.5	−0.7	2.5	−0.7		
K $4S_{1/2}$–$4P_{1/2}$	769.9	3.2×10^2	1.5	+0.24	0.9	−0.22	2.6	−1.2	2.4	−0.9	2.9	−1.0	2.6	−1.0
–$4P_{3/2}$	766.5	2.2×10^2	2.1	+0.13	1.2	−0.33	2.1	−0.8	2.5	−0.6	2.9	−1.0	2.6	−0.7
–$5P_{1/2}$	404.7	0.8×10^1	3.8	+0.74	1.6	0.0	7.2	−2.0	6.6	−2.0	6.6			
Rb $5S_{1/2}$–$5P_{1/2}$	794.7	3.7×10^2	2.0	1.0	1.0	−0.04	2.0	−0.8	2.3	−0.8				
–$6P_{1/2}$	421.6	1.6×10^1												
–$10P_{1/2}$	315.5	0.4×10^1	5.0					−9.5				−6		
Cs $6S_{1/2}$–$6P_{1/2}$	894.3		2.0	+0.67	1.0	−0.29	2.0	−0.9	2.0	−0.27	2.1	−0.8	3.1	−0.7
–$7P_{1/2}$	459.0		8.8	+1.0	3.5	0.0	8.6	−1.6	2.3	−1.5	6.3	−1.7		

Note: The values differ quite substantially in the literature. Therefore some average values were used in Table 3.1.

References: N. Allard, J. Kielkopf: Rev. Mod. Phys. **54**, 1103 (1982)
M.J. O'Callaghan, A. Gallagher: Phys. Rev. **39**, 6190 (1989)
E. Schüler, W. Behmenburg: Phys. Rep. **12C**, 274 (1974)
M.D. Rotondaro, G.P. Perram: Collisional broadening and shift of the rubidium D_1 and D_2 lines by rare gases, H_2, D_2, N_2, CH_4 and CF_4.
J. Quant. Spectrosc. Rad. Transf. **57**, 497 (1997)

Figure 3.12 Pressure broadening (*left scale*) and shifts
(*right scale*) of the lithium
resonance line by different
noble gases [84]

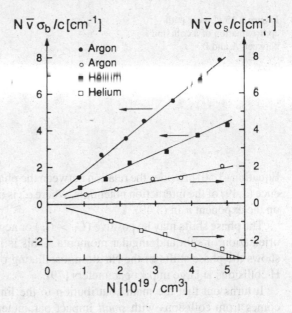

3.3.2   Relations Between Interaction Potential, Line Broadening, and Shifts

In order to gain more insight into the physical meaning of the cross sections σ_s and σ_b, we have to discover the relation between the phase shift $\eta(R)$ and the potential $V(R)$. Assume potentials of the form

$$V_i(R) = C_i/R^n , \quad V_k(R) = C_k/R^n ,\tag{3.48}$$

between the atom in level E_i or E_k and the perturbing atom B. The frequency shift $\Delta\omega$ for the transition $E_i \rightarrow E_k$ is then

$$\hbar\Delta\omega(R) = \frac{C_i - C_k}{R^n} .\tag{3.49}$$

The line broadening comes from two contributions:

1. The phase shift, due to the frequency shift of the oscillator during the collision
2. the quenching collisions which shorten the effective lifetime of the upper level of A.

The corresponding phase shift of the oscillator A due to a collision with impact parameter R_0, where we neglect the scattering of B and assume that the path of B is not deflected but follows a straight line (Fig. 3.13), is

$$\Delta\phi(R_0) = \int\limits_{-\infty}^{+\infty} \Delta\omega dt = \frac{1}{\hbar} \int\limits_{-\infty}^{+\infty} \frac{(C_i - C_k)dt}{[R_0^2 + \bar{v}^2(t - t_0)^2]^{n/2}} = \frac{\alpha_n(C_i - C_k)}{v R_0^{n-1}} .\tag{3.50}$$

Figure 3.13 Linear path
approximation of a collision
between A and B

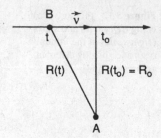

Equation (3.50) provides the relation between the phase shift $\Delta\phi(R_0)$ and the difference (3.49) of the interaction potentials, where α_n is a numerical constant depending
on the exponent n in (3.49).

The phase shifts may be positive ($C_i > C_k$) or negative depending on the relative
orientation of spin and angular momenta. This is illustrated by Fig. 3.14, which
shows the phase shifts of the Na atom, oscillating on the 3s–3p transition for Na–
H collisions at large impact parameters [79].

It turns out that the main contribution to the line broadening cross section σ_b
comes from collisions with *small* impact parameters, whereas the lineshift cross
section σ_s still has large values for *large* impact parameters. This means that elastic

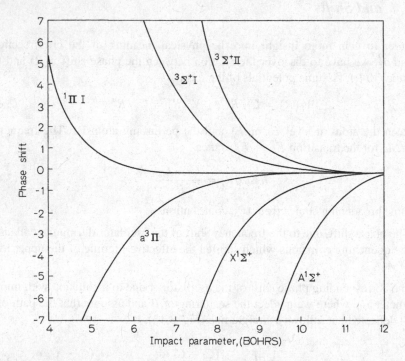

Figure 3.14 Phase shift of the Na*(3p) oscillation for Na*–H collisions versus impact parameter.
The various adiabatic molecular states for Na*H are indicated [79]

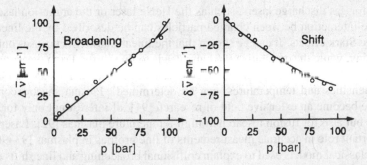

Figure 3.15 Broadening and shift of the Cs resonance line at $\lambda = 894.3$ nm by argon

Figure 3.16 Satellites in the pressure-broadened line profile of the cesium transition $6s \rightarrow 9p_{3/2}$ for Cs–Xe collisions at different xenon densities [atoms/cm^3] [81]

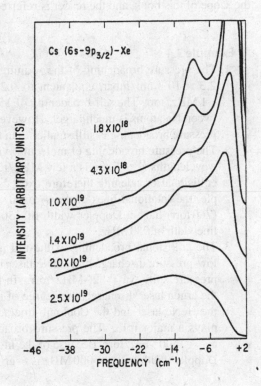

collisions at large distances do not cause noticeable broadening of the line, but can still very effectively shift the line center [85]. Figure 3.15 exhibits broadening and shift of the Cs resonance line by argon atoms.

Nonmonotonic interaction potentials $V(R)$, such as the Lennard–Jones potential (3.40), cause satellites in the wings of the broadened profiles (Fig. 3.16) From the satellite structure the interaction potential may be deduced [86].

Because of the long-range Coulomb interactions between charged particles (electrons and ions) described by the potential (3.48) with $n = 1$, pressure broadening and shift is particularly large in plasmas and gas discharges [87, 88]. This is of interest for gas discharge lasers, such as the HeNe laser or the argon-ion laser [89, 90]. The interaction between charged particles can be described by the linear and quadratic Stark effects. It can be shown that the linear Stark effect causes only line broadening, while the quadratic effect also leads to line shifts. From measurements of line profiles in plasmas, very detailed plasma characteristics, such as electron or ion densities and temperatures, can be determined. Plasma spectroscopy has therefore become an extensive field of research [91], of interest not only for astro-physics, but also for fusion research in high-temperature plasmas [92]. Lasers play an important role in accurate measurements of line profiles in plasmas [93–96].

The classical models used to explain collisional broadening and line shifts can be improved by using quantum mechanical calculations. These are, however, beyond the scope of this book, and the reader is referred to the literature [68, 81, 89–101].

Example 3.4

1. The pressure broadening of the sodium D line $\lambda = 589$ nm by argon is 2.3×10^{-5} nm/mbar, equivalent to 0.228 MHz/Pa. The shift is about -1 MHz/torr. The self-broadening of 150 MHz/torr due to collisions be-tween Na atoms is much larger. However, at pressures of several torr, the pressure broadening is still smaller than the Doppler width.

2. The pressure broadening of molecular vibration–rotation transitions with wavelengths $\lambda \simeq 5\,\mu$m is a few MHz/torr. At atmospheric pressure, the collisional broadening therefore exceeds the Doppler width. For exam-ple, the rotational lines of the ν_2 band of H_2O in air at normal pressure (760 torr) have a Doppler width of 150 MHz, but a pressure-broadened linewidth of 930 MHz.

3. The collisional broadening of the red neon line at $\lambda = 633$ nm in the low-pressure discharge of a HeNe laser is about $\delta\nu = 150$ MHz/torr; the pressure shift $\Delta\nu = 20$ MHz/torr. In high-current discharges, such as the argon laser discharge, the degree of ionization is much higher than in the HeNe laser and the Coulomb interaction between ions and electrons plays a major role. The pressure broadening is therefore much larger: $\delta\nu = 1500$ MHz/torr. Because of the high temperature in the plasma, the Doppler width $\delta\nu_D \simeq 5000$ MHz is even larger [90].

3.3.3 Collisional Narrowing of Lines

In the infrared and microwave ranges, collisions may sometimes cause a narrow-ing of the linewidth instead of a broadening (*Dicke narrowing*) [102]. This can be explained as follows: if the lifetime of the upper molecular level (e.g., an excited vibrational level in the electronic ground state) is long compared to the mean time between successive collisions, the velocity of the oscillator is often altered by elastic

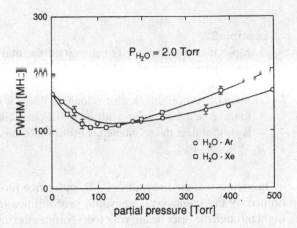

Figure 3.17 Dicke narrowing and pressure broadening of a rotational line of a vibrational transition in H_2O at $1871\ cm^{-1}\ (\lambda = 5.34\ \mu m)$ as a function of Ar and Xe pressure [103]

collisions and the mean velocity component is smaller than without these collisions, resulting in a smaller Doppler shift. When the Doppler width is larger than the pressure-broadened width, this effect causes a narrowing of the Doppler-broadened lines, if the mean-free path is smaller than the wavelength of the molecular transition [103]. Figure 3.17 illustrates this Dicke narrowing for a rotational transition of the H_2O molecule at $\lambda = 5.34\ \mu m$. The linewidth decreases with increasing pressure up to pressures of about 100–150 torr, depending on the collision partner, which determines the mean-free path Λ. For higher pressures, the pressure broadening overcompensates the Dicke narrowing, and the linewidth increases again.

There is a second effect that causes a collisional narrowing of spectral lines. In the case of very long lifetimes of levels connected by an EM transition, the linewidth is not determined by the lifetimes but by the diffusion time of the atoms out of the laser beam (Sect. 3.4). Inserting a noble gas into the sample cell decreases the diffusion rate and therefore increases the interaction time of the sample atoms with the laser field, which results in a decrease of the linewidth with pressure [104] until the pressure broadening overcompensates the narrowing effect.

3.4   Transit-Time Broadening

In many experiments in laser spectroscopy, the interaction time of molecules with the radiation field is small compared with the spontaneous lifetimes of excited levels. Particularly for transitions between rotational–vibrational levels of molecules with spontaneous lifetimes in the millisecond range, the transit time $T = d/|v|$ of molecules with a mean thermal velocity v passing through a laser beam of diameter d may be smaller than the spontaneous lifetime by several orders of magnitude.

Example 3.5
1. Molecules in a molecular beam with thermal velocities $|v| = 5 \times 10^4$ cm/s passing through a laser beam of 0.1-cm diameter have the mean transit time $T = 2 \, \mu$s.
2. For a beam of fast ions with velocities $\overline{v} = 3 \times 10^8$ cm/s, the time required to traverse a laser beam with $d = 0.1$ cm is already below 10^{-9} s, which is shorter than the spontaneous lifetimes of most atomic levels.

In such cases, the linewidth of a Doppler-free molecular transition is no longer limited by the spontaneous transition probabilities (Sect. 3.1), but by the time of flight through the laser beam, which determines the interaction time of the molecule with the radiation field. This can be seen as follows: consider an undamped oscillator $x = x_0 \cos \omega_0 t$ that oscillates with constant amplitude during the time interval T and then suddenly stops oscillating. Its frequency spectrum is obtained from the Fourier transform

$$A(\omega) = \frac{1}{\sqrt{2\pi}} \int\limits_0^T x_0 \cos(\omega_0 t) e^{-i\omega t} \, dt \; . \tag{3.51}$$

The spectral intensity profile $I(\omega) = A^* A$ is, for $(\omega - \omega_0) \ll \omega_0$,

$$I(\omega) = C \frac{\sin^2[(\omega - \omega_0) T/2]}{(\omega - \omega_0)^2} \; , \tag{3.52}$$

according to the discussion in Sect. 3.1. This is a function with a full halfwidth $\delta \omega_T = 5.6/T$ around its central maximum ω_0 (Fig. 3.18a) and a full width $\delta \omega_b = 4\pi/T \simeq 12.6/T$ between the zero points on both sides of the central maximum.

This example can be applied to an atom that traverses a laser beam with a rectangular intensity profile (Fig. 3.18a). The oscillator amplitude $x(t)$ is proportional to the field amplitude $E = E_0(r) \cos \omega t$. If the interaction time $T = d/v$ is small compared to the damping time $T = 1/\gamma$, the oscillation amplitude can be regarded as constant during the time T. The full halfwidth of the absorption line is then $\delta \omega = 5.6 v/d \rightarrow \delta v \simeq v/d$.

In reality, the field distribution across a laser beam that oscillates in the fundamental mode is given by (Sect. 5.3)

$$E = E_0 e^{-r^2/w^2} \cos \omega t \; , \tag{3.53}$$

in which $2w$ gives the diameter of the Gaussian beam profile across the points where $E = E_0/e$. Substituting the forced oscillator amplitude $x = \alpha E$ into (3.51), one obtains instead of (3.52) a Gaussian line profile (Fig. 3.18b)

$$I(\omega) = I_0 \exp\left(-(\omega - \omega_0)^2 \frac{w^2}{2v^2}\right) \; , \tag{3.54}$$

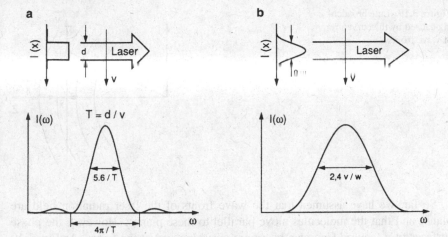

Figure 3.18 Transition probability $\mathcal{P}(\omega)$ of an atom traversing a laser beam **a** with a rectangular intensity profile $I(x)$; and **b** with a Gaussian intensity profile for the case $\gamma < 1/T = v/d$. The intensity profile $I(\omega)$ of an absorption line is proportional to $\mathcal{P}(\omega)$

with a transit-time limited halfwidth (FWHM)

$$\delta\omega_{tt} = 2(v/w)\sqrt{2\ln(2)} \simeq 2.4v/w \rightarrow \delta v \simeq 0.4v/w . \qquad (3.55)$$

The quantity $w = (\lambda R/2\pi)^{1/2}$ (see Sect. 5.2.3) is called the beam waist of the Gaussian beam profile.

There are two possible ways of reducing the transit-time broadening: one may either enlarge the laser beam diameter $2w$, or one may decrease the molecular velocity v. Both methods have been verified experimentally and will be discussed in Vol. 2, Sects. 2.3 and 9.2. The most efficient way is to directly reduce the atomic velocity by optical cooling (Vol. 2, Chap. 9).

Example 3.6

1. A beam of NO_2 molecules with $\overline{v} = 600\,\text{m/s}$ passes through a focused laser beam with $w = 0.1\,\text{mm}$. Their transit time broadening $\delta v = \delta\omega/2\pi \simeq 400\,\text{kHz}$ is large compared to their natural linewidth $\delta v_n \simeq 10\,\text{kHz}$ of NO_2 transitions in the visible region.

2. For frequency standards the rotational–vibrational transition of CH_4 at $\lambda = 3.39\,\mu\text{m}$ is used (Vol. 2, Sect. 2.3). In order to reduce the transit-time broadening for CH_4 molecules with $\overline{v} = 7 \times 10^4\,\text{cm/s}$ below their natural linewidth of $\delta v = 10\,\text{kHz}$, the laser-beam diameter must be enlarged to $2w \geq 6\,\text{cm}$.

Figure 3.19 Line broadening caused by the curvature of wave fronts

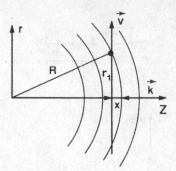

So far, we have assumed that the wave fronts of the laser radiation field are planes and that the molecules move parallel to these planes. However, the phase surfaces of a focused Gaussian beam are curved except at the focus. As Fig. 3.19 illustrates, an atom moving along the r-direction perpendicular to the laser beam z-axis experiences a maximum phase shift $\Delta\phi = x2\pi/\lambda$, between the points $r = 0$ and $r = r_1$. With $r^2 = R^2 - (R - x)^2$ we obtain the approximation $x \simeq r^2/2R$ for $x \ll R$. This gives for the phase shifts

$$\Delta\phi = kr^2/2R = \omega r^2/(2cR) , \qquad (3.56)$$

where $k = \omega/c$ is the magnitude of the wave vector, and R is the radius of curvature of the wave front. This phase shift depends on the location of an atom and is therefore different for the different atoms, and causes additional line broadening (Sect. 3.3.1). The calculation [105] yields for the transit-time broadened halfwidth, including the wave-front curvature,

$$\delta\omega = \frac{2v}{w}\sqrt{2\ln 2}\left[1 + \left(\frac{\pi w^2}{R\lambda}\right)^2\right]^{1/2}$$

$$= \delta\omega_{tt}\left[1 + \left(\frac{\pi w^2}{R\lambda}\right)^2\right]^{1/2} \approx \delta\omega_{tt}(1 + \Delta\phi^2)^{1/2} . \qquad (3.57)$$

In order to minimize this additional broadening, the radius of curvature has to be made as large as possible. If $\Delta\phi \ll \pi$ for a distance $r = w$, the broadening by the wave-front curvature is small compared to the transit-time broadening. This imposes the condition $R \gg w^2/\lambda$ on the radius of curvature.

Example 3.7
For a wave with $\lambda = 1\,\mu m \rightarrow \omega = 2 \times 10^{15}\,Hz$. With $w = 1\,cm$, this gives, according to (3.56), a maximum phase shift $\Delta\phi = 2\times10^{15} / (6\times10^{10}R\,[cm])$. In order to keep $\Delta\phi \ll 2\pi$, the radius of curvature should be $R \gg 5\times10^3\,cm$.

For $R = 5 \times 10^3$ cm $\rightarrow \Delta\phi = 2\pi$ and the phase-front curvature causes an additional broadening by a factor of about 6.5.

3.5   Homogeneous and Inhomogeneous Line Broadening

If the probability $\mathcal{P}_{ik}(\omega)$ of absorption or emission of radiation with frequency ω causing a transition $E_i \rightarrow E_k$ is equal for all the molecules of a sample that are in the same level E_i, we call the spectral line profile of this transition *homogeneously broadened*. Natural line broadening is an example that yields a homogeneous line profile. In this case, the probability for emission of light with frequency ω on a transition $E_i \rightarrow E_k$ with the normalized Lorentzian profile $L(\omega - \omega_0)$ and central frequency ω_0 is given by

$$\mathcal{P}_{ik}(\omega) = A_{ik} L(\omega - \omega_0) .$$

It is equal for all atoms in level E_i.

The standard example of inhomogeneous line broadening is Doppler broadening. In this case, the probability of absorption or emission of monochromatic radiation $E(\omega)$ is not equal for all molecules, but depends on their velocity $\overline{v}$ (Sect. 3.2). We divide the molecules in level E_i into subgroups such that all molecules with a velocity component within the interval v_z to $v_z + \Delta v_z$ belong to one subgroup. If we choose Δv_z to be $\delta\omega_n/k$ where $\delta\omega_n$ is the natural linewidth, we may consider the frequency interval $\delta\omega_n$ to be homogeneously broadened inside the much larger inhomogeneous Doppler width. That is to say, all molecules in the subgroup can absorb or emit radiation with wave vector k and frequency $\omega = \omega_0 + v_z|k|$ (Fig. 3.7), because in the coordinate system of the moving molecules, this frequency is within the natural width $\delta\omega_n$ around ω_0 (Sect. 3.2).

Collisions alter the line profile in a more complex way. In Sect. 3.3 we saw that the spectral line profile is altered by two kinds of collisions: Inelastic and elastic collisions. *Inelastic* collisions cause additional damping, resulting in a shortening of the excited state lifetime and a pure broadening of the Lorentzian line profile. This broadening by inelastic collisions brings about a homogeneous Lorentzian line profile. The *elastic* collisions could be described as phase-perturbing collisions. The Fourier transform of the oscillation trains with random phase jumps again yields a Lorentzian line profile, as derived in Sect. 3.3. Summarizing, we can state that elastic and inelastic collisions that only perturb the phase or amplitude of an oscillating atom without changing its velocity cause homogeneous line broadening.

So far, we have neglected the fact that collisions also change the velocity of both collision partners. If the velocity component v_z of a molecule is altered by an amount u_z during the collision, the molecule is transferred from one subgroup $(v_z \pm \Delta v_z)$ within the Doppler profile to another subgroup $(v_z + u_z \pm \Delta v_z)$.

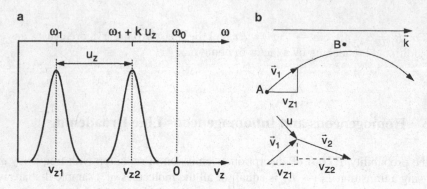

Figure 3.20 a Effect of velocity-changing collisions on the frequency shift of homogeneous sub-groups within a Doppler-broadened line profile, **b** velocity change of atom A during a collision with atom B

This causes a shift of its absorption or emission frequency from ω to $\omega + ku_z$ (Fig. 3.20). This shift should not be confused with the line shift caused by phase-perturbing elastic collisions that also occurs when the velocity of the oscillator does not noticeably change.

At thermal equilibrium, the changes u_z of v_z by velocity-changing collisions are randomly distributed. Therefore, the whole Doppler profile will, in general, not be affected and the effect of these collisions is canceled out in Doppler-limited spectroscopy. In Doppler-free laser spectroscopy, however, the velocity-changing collisions may play a non-negligible role. They cause effects that depend on the ratio of the mean time $T = \Lambda/\overline{v}$ between collisions (λ = mean free pathlength between successive collisions) to the interaction time τ_c with the radiation field. For $T > \tau_c$, the redistribution of molecules by velocity-changing collisions causes only a small change of the population densities $n_i(v_z)dv_z$ within the different subgroups, without noticeably changing the homogeneous width of this subgroup. If $T \ll \tau_c$, the different subgroups are uniformly mixed. This results in a broadening of the homogeneous linewidth associated with each subgroup. The effective interaction time of the molecules with a monochromatic laser field is shortened because the velocity-changing collisions move a molecule out of resonance with the field. The resultant change of the line shape can be monitored using saturation spectroscopy (Vol. 2, Sect. 2.3).

Under certain conditions, if the mean free path Λ of the molecules is smaller than the wavelength of the radiation field, velocitychanging collisions may also result in a narrowing of a Doppler-broadening line profile (Dicke narrowing, Sect. 3.3.3).

3.6 Saturation and Power Broadening

At sufficiently large laser intensities, the optical pumping rate on an absorbing transition becomes larger than the relaxation rates. This results in a noticeable decrease of the population in the absorbing levels. This saturation of the population den-

Figure 3.21 Two-level system with no relaxation into
other levels

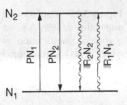

sities also causes additional line broadening. The spectral line profiles of such
partially saturated transitions are different for homogeneously and for inhomogeneously broadened lines [106]. Here we treat the homogeneous case, while the
saturation of inhomogeneous line profiles is discussed in Vol. 2, Chap. 2.

3.6.1   Saturation of Level Population by Optical Pumping

The effect of optical pumping on the saturation of population densities is illustrated
by a two-level system with population densities N_1 and N_2. The two levels are
coupled to each other by absorption or emission and by relaxation processes, but
have no transitions to other levels (Fig. 3.21). Such a "true" two-level system is
realized by many atomic resonance transitions without hyperfine structure.

With the probability $\mathfrak{P}_{12} = B_{12}\rho(\omega)$ for a transition $|1\rangle \rightarrow |2\rangle$ by absorption of
photons $\hbar\omega$ and the relaxation probability R_i for level $|i\rangle$, the rate equation for the
level population is

$$\frac{dN_1}{dt} = -\frac{dN_2}{dt} = -\mathfrak{P}_{12}N_1 - R_1 N_1 + \mathfrak{P}_{12}N_2 + R_2 N_2 , \qquad (3.58)$$

where we have assumed nondegenerate levels with statistical weight factors $g_1 =
g_2 = 1$. Under stationary conditions ($dN_i/dt = 0$) we obtain with $N_1 + N_2 = N$
from (3.58) with the abbreviation $\mathfrak{P}_{12} = P$

$$(P + R_1)N_1 = (P + R_2)(N - N_1) \Rightarrow N_1 = N \frac{P + R_2}{2P + R_1 + R_2} \qquad (3.59a)$$

$$(P + R_2)N_2 = (P + R_1)(N - N_2) \Rightarrow N_2 = N \frac{P + R_1}{2P + R_1 + R_2} . \qquad (3.59b)$$

When the pump rate P becomes much larger than the relaxation rates R_i, the population N_1 approaches $N/2$, i.e., $N_1 = N_2$. This means that the absorption coefficient
$\alpha = \sigma(N_1 - N_2)$ approaches zero for $P \rightarrow \infty$ (Fig. 3.22). The medium becomes
completely transparent.

Without a radiation field ($P = 0$), the population densities at thermal equilibrium according to (3.59a, 3.59b) are

$$N_{10} = \frac{R_2}{R_1 + R_2}N ; \quad N_{20} = \frac{R_1}{R_1 + R_2}N . \qquad (3.59c)$$

Figure 3.22 Saturation of population density N_1 and absorption coefficient $\alpha = \sigma(N_1 - N_2)$ as functions of the saturation parameter S (see text)

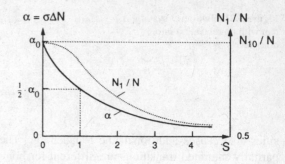

With the abbreviations

$$\Delta N = N_1 - N_2 \quad \text{and} \quad \Delta N_0 = N_{10} - N_{20}$$

we obtain from (3.59a–3.59c)

$$\Delta N = N\frac{R_2 - R_1}{2P + R_1 + R_2}$$

$$\Delta N_0 = N\frac{R_2 - R_1}{R_2 + R_1}$$

which gives:

$$\Delta N = \frac{\Delta N_0}{1 + 2P/(R_1 + R_2)} = \frac{\Delta N_0}{1 + S} . \tag{3.59d}$$

The *saturation parameter*

$$S = 2P/(R_1 + R_2) = P/\overline{R} = B_{12}\rho(\omega)/\overline{R} \tag{3.60}$$

represents the ratio of pumping rate P to the average relaxation rate $\overline{R} = (R_1 + R_2)/2$. If the spontaneous emission of the upper level $|2\rangle$ is the only relaxation mechanism, we have $R_1 = 0$ and $R_2 = A_{21}$. Since the pump rate due to a monochromatic wave with intensity $I(\omega)$ is $P = \sigma_{12}(\omega)I(\omega)/\hbar\omega$, we obtain for the saturation parameter

$$S = \frac{2\sigma_{12}I(\omega)}{\hbar\omega A_{12}} . \tag{3.61}$$

The saturated absorption coefficient $\alpha(\omega) = \sigma_{12}\Delta N$ is, according to (3.59d),

$$\boxed{\alpha = \frac{\alpha_0}{1 + S}} , \tag{3.62}$$

where α_0 is the unsaturated absorption coefficient without pumping.

3.6.2 Saturation Broadening of Homogeneous Line Profiles

According to (2.15) and (3.59d), the power absorbed per unit volume on the transition $|1\rangle \rightarrow |2\rangle$ by atoms with the population densities N_1, N_2 in a radiation field with a broad spectral profile and spectral energy density ρ is

$$\frac{dW_{12}}{dt} = \hbar\omega B_{12}\rho(\omega)\Delta N = \hbar\omega B_{12}\rho(\omega)\frac{\Delta N_0}{1+S} \, . \tag{3.63}$$

With $S = B_{12}\rho(\omega)/\overline{R}$, see (3.60), this can be written as

$$\frac{dW_{12}}{dt} = \hbar\omega\overline{R}\frac{\Delta N_0}{1+S^{-1}} \, . \tag{3.64}$$

Since the absorption profile $\alpha(\omega)$ of a homogeneously broadened line is Lorentzian, see (3.24a), the induced absorption probability of a monochromatic wave with frequency ω follows a Lorentzian line profile $B_{12}\rho(\omega) \cdot L(\omega - \omega_0)$ (see 3.10). We can therefore introduce a frequency-dependent spectral saturation parameter S_ω for the transition $E_1 \rightarrow E_2$,

$$S_\omega = \frac{B_{12}\rho(\omega)}{\overline{R}}L(\omega - \omega_0) \, . \tag{3.65}$$

We can assume that the mean relaxation rate $\overline{R}$ is independent of ω within the frequency range of the line profile. With the definition (3.24a) of the Lorentzian profile $L(\omega - \omega_0)$, we obtain for the spectral saturation parameter S_ω

$$S_\omega = S_0\frac{(\gamma/2)^2}{(\omega - \omega_0)^2 + (\gamma/2)^2} \, , \quad \text{with} \quad S_0 = S_\omega(\omega_0) \, . \tag{3.66}$$

Substituting (3.66) into (3.64) yields the frequency dependence of the absorbed radiation power per unit frequency interval $d\omega = 1\,\text{s}^{-1}$

$$\frac{d}{dt}W_{12}(\omega) = \frac{\hbar\omega\overline{R}\Delta N_0 S_0(\gamma/2)^2}{(\omega - \omega_0)^2 + (\gamma/2)^2(1+S_0)} = \frac{C}{(\omega - \omega_0)^2 + (\gamma_s/2)^2} \, . \tag{3.67}$$

This a Lorentzian profile with the increased halfwidth

$$\boxed{\gamma_s = \gamma\sqrt{1 + S_0} \, .} \tag{3.68}$$

The halfwidth $\gamma_s = \delta\omega_s$ of the saturation-broadened line increases with the saturation parameter S_0 at the line center ω_0. If the induced transition rate at ω_0 equals the total relaxation rate $\overline{R}$, the saturation parameter $S_0 = [B_{12}\rho(\omega_0)]/\overline{R}$ becomes $S_0 = 1$, which increases the linewidth by a factor $\sqrt{2}$, compared to the unsaturated linewidth $\delta\omega_0$ for weak radiation fields ($\rho \rightarrow 0$).

Figure 3.23 Saturation broadening of a homogeneous line profile

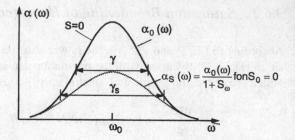

Since the power dW_{12}/dt absorbed per unit volume equals the intensity decrease per centimeter, $dI = -\alpha_s I$, of an incident wave with intensity I, we can derive the absorption coefficient α from (3.67). With $I = c\rho$ and S_ω from (3.66) we obtain

$$\alpha_s(\omega) = \alpha_0(\omega_0) \frac{(\gamma/2)^2}{(\omega - \omega_0)^2 + (\gamma_s/2)^2} = \frac{\alpha_0(\omega)}{1 + S_\omega}, \tag{3.69}$$

where the unsaturated absorption profile is

$$\alpha_0(\omega) = \frac{\alpha_0(\omega_0)(\gamma/2)^2}{(\omega - \omega_0)^2 + (\gamma/2)^2} \tag{3.70}$$

with $\alpha_0(\omega_0) = 2\hbar\omega B_{12}\Delta N_0/\pi c\gamma$.

This shows that the saturation decreases the absorption coefficient $\alpha(\omega)$ by the factor $(1 + S_\omega)$. At the line center, this factor has its maximum value $(1 + S_0)$, while it decreases for increasing $(\omega - \omega_0)$ to 1, see (3.66). The saturation is therefore strongest at the line center, and approaches zero for $(\omega - \omega_0) \to \infty$ (Fig. 3.23). This is the reason why the line broadens. For a more detailed discussion of saturation broadening, see Vol. 2, Chap. 9 and [105–107].

3.6.3   Power Broadening

The broadening of homogeneous line profiles by intense laser fields can also be regarded from another viewpoint compared to Sect. 3.6.2. When a two-level system is exposed to a radiation field $E = E_0 \cos \omega t$, the population probability of the upper level $|b\rangle$ is, according to (2.95) and (2.119),

$$|b(\omega, t)|^2 = \frac{D_{ab}^2 E_0^2}{\hbar^2(\omega_{ab} - \omega)^2 + D_{ab}^2 E_0^2}$$
$$\times \sin^2 \left[\frac{1}{2} \sqrt{(\omega_{ab} - \omega)^2 + (D_{ab}E_0/\hbar)^2} \cdot t \right], \tag{3.71}$$

an oscillating function of time, which oscillates at exact resonance $\omega = \omega_{ab}$ with the Rabi flopping frequency $\Omega_R = \Omega_{ab} = D_{ab}E_0/\hbar$.

If the upper level $|b\rangle$ can decay by spontaneous processes with a relaxation constant γ, its mean population probability is

$$\mathcal{P}_b(\omega) = \overline{|b(\omega,t)|}' = \int_0^\infty \gamma u \,\Gamma'' \,|b(\omega,t)|^2 \,dt \quad (3.72)$$

Inserting (3.71) and integrating yields

$$\mathcal{P}_b(\omega) = \frac{1}{2} \frac{D_{ab}^2 E_0^2/\hbar^2}{(\omega_{ab} - \omega)^2 + \gamma^2(1 + S)}, \quad (3.73)$$

with $S = D_{ab}^2 E_0^2/(\hbar^2\gamma^2)$. Since $\mathcal{P}_b(\omega)$ is proportional to the absorption line profile, we obtain as in (3.67) a power-broadened Lorentzian line profile with the linewidth

$$\gamma_S = \gamma\sqrt{1 + S}. \quad (3.74)$$

Since the induced absorption rate within the spectral interval γ is, according to (2.57) and (2.105)

$$B_{12}\rho\gamma = B_{12}I\gamma/c \simeq D_{12}^2 E_0^2/\hbar^2, \quad (3.75)$$

the quantity S in (3.73) turns out to be identical with the saturation parameter S in (3.60).

If both levels $|a\rangle$ and $|b\rangle$ decay with the relaxation constants γ_a and γ_b, respectively, the line profile of the homogeneously broadened transition $|a\rangle \to |b\rangle$ is again described by (3.73), where now (Vol. 2, Sect. 2.1 and [107])

$$\gamma = \tfrac{1}{2}(\gamma_a + \gamma_b) \quad \text{and} \quad S = D_{ab}^2 E_0^2/(\hbar^2\gamma_a\gamma_b). \quad (3.76)$$

If a strong pump wave is tuned to the center $\omega_0 = \omega_{ab}$ of the transition and the absorption profile is probed by a tunable weak probe wave, the absorption profile looks different: due to the population modulation with the Rabi flopping frequency Ω, sidebands are generated at $\omega_0 \pm \Omega$ that have the homogeneous linewidth γ_S. The superposition of these sidebands (Fig. 3.24) gives a line profile that depends on the ratio Ω/γ_S of the Rabi flopping frequency Ω and the saturated linewidth γ_S. For a sufficiently strong pump wave ($\Omega > \gamma_S$), the separation of the sidebands becomes larger than their width and a dip appears at the center ω_0.

3.7 Spectral Line Profiles in Liquids and Solids

Many different types of lasers use liquids or solids as amplifying media. Since the spectral characteristics of such lasers play a significant role in applications of laser spectroscopy, we briefly outline the spectral linewidths of optical transitions

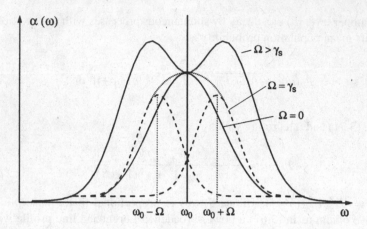

Figure 3.24 Absorption profile of a homogeneous transition pumped by a strong pump wave kept at ω_0 and probed by a weak tunable probe wave for different values of the ratio Ω/γ_s of the Rabi frequency Ω to the linewidth γ_s

in liquids and solids. Because of the large densities compared with the gaseous state, the mean relative distances $R(A, B_j)$ between an atom or molecule A and its surrounding partners B_j are very small (typically a few tenths of a nanometer), and the interaction between A and the adjacent partners B_j is accordingly large.

In general, the atoms or molecules used for laser action are diluted to small concentrations in liquids or solids. Examples are the dye laser, where dye molecules are dissolved in organic solutions at concentrations of 10^{-4} to 10^{-3} moles/liter, or the ruby laser, where the relative concentration of the active Cr^{3+} ions in Al_3O_3 is on the order of 10^{-3}. The optically pumped laser molecules A^* interact with their surrounding host molecules B. The resulting broadening of the excited levels of A^* depends on the total electric field produced at the location of A by all adjacent molecules B_j, and on the dipole moment or the polarizability of A^*. The linewidth $\Delta\omega_{ik}$ of a transition $A^*(E_i) \rightarrow A^*(E_k)$ is determined by the difference in the level shifts ($\Delta E_i - \Delta E_k$).

In liquids, the distances $R_j(A^*, B_j)$ show random fluctuations analogous to the situation in a high-pressure gas. The linewidth $\Delta\omega_{ik}$ is therefore determined by the probability distribution $P(R_j)$ of the mutal distances $R_j(A^*, B_j)$ and the correlation between the phase perturbations at A^* caused by elastic collisions during the lifetime of the levels E_i, E_k (see the analogous discussion in Sect. 3.3).

Inelastic collisions of A^* with molecules B of the liquid host may cause radiationless transitions from the level E_i populated by optical pumping to lower levels E_n. These radiationless transitions shorten the lifetime of E_i and cause collisional line broadening. In liquids the mean time between successive inelastic collisions is of the order of 10^{-11} to 10^{-13} s. Therefore the spectral line $E_i \rightarrow E_k$ is greatly broadened with a homogeneously broadened profile. When the line broadening becomes larger than the separation of the different spectral lines, a broad

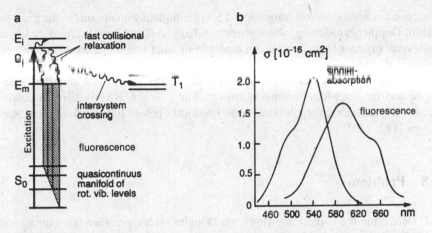

Figure 3.25 a Schematic level diagram illustrating radiative and radiationless transitions. **b** Absorption and emission cross section of rhodamine 6G dissolved in ethanol

continuum arises. In the case of molecular spectra with their many closely spaced rotational–vibrational lines within an electronic transition, such a continuum inevitably appears since the broadening at liquid densities is always much larger than the line separation.

Examples of such continuous absorption and emission line profiles are the optical dye spectra in organic solvents, such as the spectrum of Rhodamine 6G shown in Fig. 3.25b, together with a schematic level diagram [108]. The optically pumped level E_i is collisionally deactivated by radiationless transitions to the lowest vibrational level E_m of the excited electronic state. The fluorescence starts therefore from E_m instead of E_i and ends on various vibrational levels of the electronic ground state (Fig. 3.25a). The emission spectrum is therefore shifted to larger wavelengths compared with the absorption spectrum (Fig. 3.25b).

In crystalline solids the electric field $E(R)$ at the location R of the excited molecule A* has a symmetry depending on that of the host lattice. Because the lattice atoms perform vibrations with amplitudes depending on the temperatur T, the electric field will vary in time and the time average $\langle E(T, t, R) \rangle$ will depend on temperature and crystal structure [109–111]. Since the oscillation period is short compared with the mean lifetime of A*(E_i), these vibrations cause homogeneous line broadening for the emission or absorption of the atom A. If all atoms are placed at completely equivalent lattice points of an ideal lattice, the total emission or absorption of all atoms on a transition $E_i \rightarrow E_k$ would be homogeneously broadened.

However, in reality it often happens that the different atoms A are placed at nonequivalent lattice points with nonequal electric fields. This is particularly true in amorphous solids or in supercooled liquids such as glass, which have no regular lattice structure. For such cases, the line centers ω_{0j} of the homogeneously broadened lines for the different atoms A_j are placed at different frequencies. *The total emission or absorption forms an inhomogeneously broadened line profile*, which is composed of homogeneous subgroups. This is completely analogous to the gaseous case of Doppler broadening, although the resultant linewidth in solids may be larger by several orders of magnitude. An example of such inhomogeneous line broadening is the emission of excited neodymium ions in glass, which is used in the Nd-glass laser. At sufficiently low temperatures, the vibrational amplitudes decrease and the linewidth becomes narrower. For $T < 4\,\mathrm{K}$ it is possible to obtain, even in solids under favorable conditions, linewidths below $10\,\mathrm{MHz}$ for optical transitions [112, 113].

3.8   Problems

3.1 Determine the natural linewidth, the Doppler width, pressure broadening and shifts for the neon transition $3s_2 \rightarrow 2p_4$ at $\lambda = 632.8\,\mathrm{nm}$ in a HeNe discharge at $p_{He} = 2\,\mathrm{mbar}$, $p_{Ne} = 0.2\,\mathrm{mbar}$ at a gas temperature of $400\,\mathrm{K}$. The relevant data are: $\tau(3s_2) = 58\,\mathrm{ns}$, $\tau(2p_4) = 18\,\mathrm{ns}$, $\sigma_B(\mathrm{Ne-He}) \stackrel{\wedge}{=} 6 \times 10^{-14}\,\mathrm{cm}^2$, $\sigma_S(\mathrm{Ne-He}) \simeq 1 \times 10^{-14}\,\mathrm{cm}^2$, $\sigma_B(\mathrm{Ne-Ne}) = 1 \times 10^{-13}\,\mathrm{cm}^2$, $\sigma_S(\mathrm{Ne-Ne}) = 1 \times 10^{-14}\,\mathrm{cm}^2$.

3.2 What is the dominant broadening mechanism for absorption lines in the following examples:

1. The output from a CO_2 laser with $50\,\mathrm{W}$ at $\lambda = 10\,\mu\mathrm{m}$ is focussed into a sample of SF_6 molecules at the pressure p. The laser beam waist w in the focal plane is $0.25\,\mathrm{mm}$. Use the numerical parameters $T = 300\,\mathrm{K}$, $p = 1\,\mathrm{mbar}$, the broadening cross section $\sigma_b = 5 \times 10^{-14}\,\mathrm{cm}^2$ and the absorption cross section $\sigma_a = 10^{-14}\,\mathrm{cm}^2$.
2. Radiation from a star passes through an absorbing interstellar cloud of H-atoms, which absorb on the hfs-transition at $\lambda = 21\,\mathrm{cm}$ and on the Lyman-α transition $1S \rightarrow 2P$ at $\lambda = 121.6\,\mathrm{nm}$. The Einstein coefficient for the $\lambda = 21\,\mathrm{cm}$ line is $A_{ik} = 4 \times 10^{-15}\,\mathrm{s}^{-1}$, that for the Lyman-$\alpha$ transition is $A_{ik} = 1 \times 10^{9}\,\mathrm{s}^{-1}$. The atomic density of H atoms is $n = 10\,\mathrm{cm}^{-3}$ and the temperature $T = 10\,\mathrm{K}$. At which path lengths has the radiation decreased to $10\,\%$ of I_0 for the two transitions?
3. The expanded beam from a HeNe laser at $\lambda = 3.39\,\mu\mathrm{m}$ with $10\,\mathrm{mW}$ power is sent through a methane cell ($T = 300\,\mathrm{K}$, $p = 0.1\,\mathrm{mbar}$, beam diameter: $1\,\mathrm{cm}$). The absorbing CH_4 transition is from the vibrational ground state ($\tau \simeq \infty$) to an excited vibrational level with $\tau \simeq 20\,\mu\mathrm{s}$. Give the ratios of Doppler width to transit-time width to natural width to pressure-broadened linewidth for a collision cross section $\sigma_b = 10^{-16}\,\mathrm{cm}^2$.
4. Calculate the minimum beam diameter that is necessary to bring about the transit-time broadening in Example 3.2c below the natural linewidth. Is saturation broadening important, if the absorption cross section is $\sigma = 10^{-10}\,\mathrm{cm}^2$?

3.3 The sodium D-line at $\lambda = 589$ nm has a natural linewidth of 10 MHz.

1. How far away from the line center do the wings of the Lorentzian line profile exceed the Doppler profile at $T = 500$ K if both profiles are normalized to $I(\omega_0) - I_0$?
2. Calculate the intensity $I(\omega - \omega_0)$ of the Lorentzian which equals that of the Gaussian profile at this frequency ω_c relative to the line center ω_0.
3. Compare the intensities of both profiles normalized to 1 at $\omega = \omega_0$ at a distance $0.1(\omega_0 - \omega_c)$ from the line center.
4. At what laser intensity is the power broadening equal to half of the Doppler width at $T = 500$ K, when the laser frequency is tuned to the line center ω_0 and pressure broadening can be neglected?

3.4 Estimate the collision broadened width of the Li D line at $\lambda = 670.8$ nm due to

1. Li–Ar collisions at $p(\text{Ar}) = 1$ bar (Fig. 3.12);
2. Li–Li collisions at $p(\text{Li}) = 1$ mbar. This resonance broadening is due to the interaction potential $V(r) \sim 1/r^3$ and can be calculated as $\gamma_{\text{res}} = Ne^2 f_{ik}/(4\pi\epsilon_0 m\omega_{ik})$, where the oscillator strength f_{ik} is 0.65. Compare with numbers in Table 3.1.

3.5 An excited atom with spontaneous lifetime τ suffers quenching collisions. Show that the line profile stays Lorentzian and doubles its linewidth if the mean time between two collisions is $\bar{T}_c = \tau$. Calculate the pressure of N_2 molecules at $T = 400$ K for which $\bar{T}_c = \tau$ for collisions Na* + N_2 with the quenching cross section $\sigma_a = 4 \times 10^{-15}$ cm^2.

3.6 A cw laser with 100 MHz output power excites K atoms at low potassium pressures in a cell with 10 mbar neon as a buffer gas at a temperature $T = 350$ K. Estimate the different contributions to the total linewidth. At which laser intensities does the power broadening at low pressures exceeds the pressure broadening at 10 mbar (the lifetime of the upper level is $\tau_{\text{sp}} = 25$ ns) and how strong has the laser beam to be focused that power broadening at 10 mbar exceeds the Doppler width?

Chapter 4
Spectroscopic Instrumentation

This chapter is devoted to a discussion of instruments and techniques that are of fundamental importance for the measurements of wavelengths and line profiles, or for the sensitive detection of radiation. The optimum selection of proper equipment or the application of a new technique is often decisive for the success of an experimental investigation. Since the development of spectroscopic instrumentation has shown great progress in recent years, it is most important for any spectroscopist to be informed about the state-of-the-art regarding sensitivity, spectral resolving power, and signal-to-noise ratios attainable with modern equipment.

At first we discuss the basic properties of *spectrographs* and *monochromators*. Although for many experiments in laser spectroscopy these instruments can be replaced by monochromatic tunable lasers (Chap. 5 and Vol. 2, Chap. 1), they are still indispensible for the solution of quite a number of problems in spectroscopy.

Probably the most important instruments in laser spectroscopy are *interferometers*, which are applicable in various modifications to numerous problems. We therefore treat these devices in somewhat more detail. Recently, new techniques of measuring laser wavelengths with high accuracy have been developed; they are mainly based on interferometric devices. Because of their relevance in laser spectroscopy they will be discussed in a separate section.

Great progress has also been achieved in the field of low-level signal detection. Apart from new photomultipliers with an extended spectral sensitivity range and large quantum efficiencies, new detection instruments have been developed such as image intensifiers, infrared detectors, charge-coupled devices (CCDs) or optical multichannel analyzers, which could move from classified military research into the open market. For many spectroscopic applications they prove to be extremely useful.

W. Demtröder, *Laser Spectroscopy 1*, DOI 10.1007/978-3-642-53859-9_4,
© Springer-Verlag Berlin Heidelberg 2014

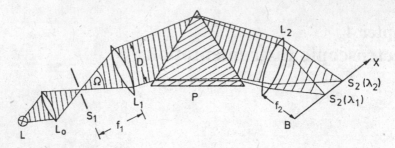

Figure 4.1 Prism spectrograph

4.1   Spectrographs and Monochromators

Spectrographs, the first instruments for measuring wavelengths, still hold their position in spectroscopic laboratories, particularly when equipped with modern accessories such as computerized microdensitometers or optical multichannel analyzers. Spectrographs are optical instruments that form images $S_2(\lambda)$ of the entrance slit S_1; the images are laterally separated for different wavelengths λ of the incident radiation (Fig. 2.17). This lateral dispersion is due to either spectral dispersion in prisms or diffraction on plane or concave reflection gratings.

Figure 4.1 depicts the schematic arrangement of optical components in a *prism spectrograph*. The light source L illuminates the entrance slit S_1, which is placed in the focal plane of the collimator lens L_1. Behind L_1 the parallel light beam passes through the prism P, where it is diffracted by an angle $\theta(\lambda)$ depending on the wavelength λ. The camera lens L_2 forms an image $S_2(\lambda)$ of the entrance slit S_1. The position $x(\lambda)$ of this image in the focal plane of L_2 is a function of the wavelength λ. The *linear dispersion* $dx/d\lambda$ of the spectrograph depends on the spectral dispersion $dn/d\lambda$ of the prism material and on the focal length of L_2.

When a reflecting diffraction grating is used to separate the spectral lines $S_2(\lambda)$, the two lenses L_1 and L_2 are commonly replaced by two spherical mirrors M_1 and M_2, which image the entrance slit either onto the exit slit S_2, or via the mirror M onto a CCD array in the plane of observation (Fig. 4.2). Both systems can use either photographic or photoelectric recording. According to the kind of detection, we distinguish between *spectrographs* and *monochromators*.

In spectrographs a charge-coupled device (CCD) diode array is placed in the focal plane of L_2 or M_2. The whole spectral range $\Delta\lambda = \lambda_1(x_1) - \lambda_2(x_2)$ covered by the lateral extension $\Delta x = x_1 - x_2$ of the diode array can be simultaneously recorded. The cooled CCD array can accumulate the incident radiant power over long periods (up to 20 h). CCD detection can be employed for both pulsed and cw light sources. The spectral range is limited by the spectral sensitivity of available CCD materials and covers the region between about 200–1000 nm.

Monochromators, on the other hand, use photoelectric recording of a selected small spectral interval. An exit slit S_2, selecting an interval Δx_2 in the focal plane B,

Figure 4.2 Grating monochromator

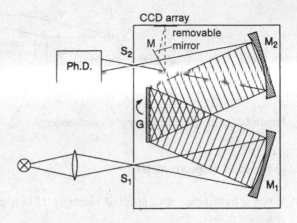

lets only the limited range $\Delta\lambda$ through to the photoelectric detector. Different spectral ranges can be detected by shifting S_2 in the x-direction. A more convenient solution (which is also easier to construct) turns the prism or grating by a gear-box drive, which allows the different spectral regions to be tuned across the fixed exit slit S_2. Modern devices uses a direct drive of the grating axis by step motors and measure the turning angle by electronic angle decoders. This avoids backlash of the driving gear. Unlike the spectrograph, different spectral regions are not detected simultaneously but successively. The signal received by the detector is proportional to the product of the area $h\Delta x_2$ of the exit slit with height h with the spectral intensity $\int I(\lambda)\mathrm{d}\lambda$, where the integration extends over the spectral range dispersed within the width Δx_2 of S_2.

Whereas the spectrograph allows the simultaneous measurement of a large region with moderate time resolution, photoelectric detection allows high time resolution but permits, for a given spectral resolution, only a small wavelength interval $\Delta\lambda$ to be measured at a time. With integration times below some minutes, photoelectric recording shows a higher sensitivity, while for very long detection times of several hours, photoplates may still be more convenient, although cooled CCD arrays currently allow integration times up to several hours.

In spectroscopic literature the name *spectrometer* is often used for both types of instruments. We now discuss the basic properties of spectrometers, relevant for laser spectroscopy. For a more detailed treatment see for instance [114–123].

4.1.1 Basic Properties

The selection of the optimum type of spectrometer for a particular experiment is guided by some basic characteristics of spectrometers and their relevance to the particular application. The basic properties that are important for all dispersive optical instruments may be listed as follows:

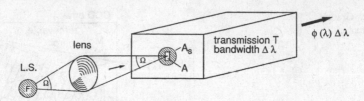

Figure 4.3 Light-gathering power of a spectrometer

a) Speed of a Spectrometer

When a light beam with spectral intensity $I^*(\lambda)$, cross section A_s and spectral radiation power

$$P_0(\lambda)\,d\lambda = I_\lambda^* \cdot A_s \cdot d\lambda \qquad (4.1a)$$

within the solid angle $d\Omega = 1$ sr falls onto the entrance slit of a spectrometer with slit area $A < A_s$ and acceptance angel Ω, the power transmitted by the spectrometer is

$$P_t(\lambda)\,d\lambda = P_0(A/A_s) \cdot T(\lambda) \cdot \Omega \cdot d\lambda = I^*(\lambda) \cdot A \cdot T(\lambda) \cdot \Omega \cdot d\lambda , \qquad (4.1b)$$

where $T(\lambda)$ is the wavelength dependent transmission of the spectrometer.

The product $U = A\Omega$ is often named *étendue*. For the prism spectrograph the maximum solid angle of acceptance, $\Omega = F/f_1^2$, is limited by the effective area $F = hD$ of the parallel light beam transmitted through the prism, which represents the limiting aperture with height h and width D for the light beam (Fig. 4.1). For the grating spectrometer the sizes of the grating and mirrors limit the acceptance solid angle Ω.

Example 4.1
For a prism with height $h = 6$ cm, $D = 6$ cm, $f_1 = 30$ cm $\rightarrow D/f = 1:5$ and $\Omega = 0.04$ sr. With an entrance slit of 5×0.1 mm^2, the étendue is $U = 5 \times 10^{-3} \times 4 \times 10^{-2} = 2 \times 10^{-4}$ cm^2 sr.

In order to utilize the optimum speed, it is advantageous to image the light source onto the entrance slit in such a way that the acceptance angle Ω is fully used (Fig. 4.4). Although more radiant power from an extended source can pass the entrance slit by using a converging lens to reduce the source image on the entrance slit, the divergence is increased. The radiation outside the acceptance angle Ω cannot be detected, but may increase the background by scattering from lens holders and spectrometer walls.

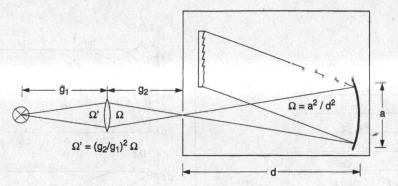

Figure 4.4 Optimized imaging of a light source onto the entrance slit of a spectrometer is achieved when the solid angle Ω' of the incoming light matches the acceptance angle $\Omega = (a/d)^2$ of the spectrometer

Figure 4.5 a Imaging of an extended light source onto the entrance slit of a spectrometer with $\Omega^* = \Omega$. **b** Correct imaging optics for laser wavelength measurements with a spectrometer. The laser light, scattered by the ground glass, forms the source that is imaged onto the entrance slit

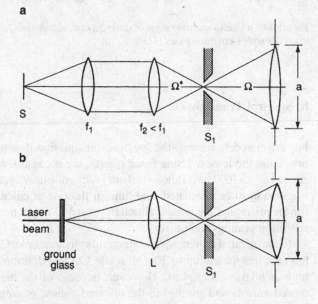

Often the wavelength of lasers is measured with a spectrometer. In this case, it is not recommended to direct the laser beam directly onto the entrance slit, because the prism or grating would be not uniformly illuminated. This decreases the spectral resolution. Furthermore, the symmetry of the optical path with respect to the spectrometer axis is not guaranteed with such an arrangement, resulting in systematic errors of wavelengths measurements if the laser beam does not exactly coincide with the spectrometer axis. It is better to illuminate a ground-glass plate with the laser and to use the incoherently scattered laser light as a secondary source, which is imaged in the usual way (Fig. 4.5).

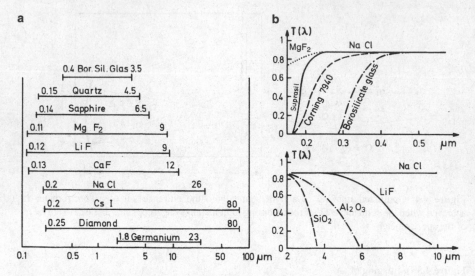

Figure 4.6 a Useful spectral ranges of different optical materials; and **b** transmittance of different materials with 1-cm thicknesses [118]b

b) Spectral Transmission

For prism spectrometers, the spectral transmission depends on the material of the prism and the lenses. Using fused quartz, the accessible spectral range spans from about 180 to 3000 nm. Below 180 nm (vacuum-ultraviolet region), the whole spectrograph must be evacuated, and lithium fluoride or calcium fluoride must be used for the prism and the lenses, although most VUV spectrometers are equipped with reflection gratings and mirrors.

In the infrared region, several materials (for example, CaF_2, NaCl, and KBr crystals) are transparent up to 30 μm, while CsI and diamond are transparent up to as high as 80 μm. (Fig. 4.6). However, because of the high reflectivity of metallic coated mirrors and gratings in the infrared region, grating spectrometers with mirrors are preferred over prism spectrographs.

Many vibrational–rotational transitions of molecules such as H_2O or CO_2 fall within the range 3–10 μm, causing selective absorption of the transmitted radiation. Infrared spectrometers therefore have to be either evacuated or filled with dry nitrogen. Because dispersion and absorption are closely related (see Sect. 2.6), prism materials with low absorption losses also show low dispersion, resulting in a limited spectral resolving power (see below).

Since the ruling or holographic production of high-quality gratings has reached a high technological standard, most spectrometers used today are equipped with diffraction gratings rather than prisms. The spectral transmission of grating spectrometers reaches from the VUV region into the far infrared. The design and the

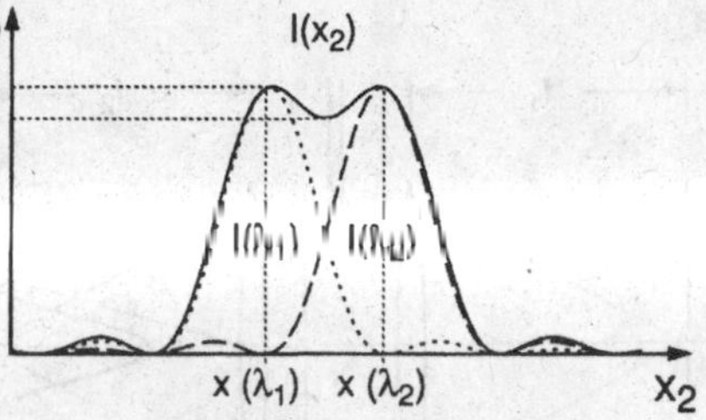

Figure 4.7 Rayleigh's criterion for the resolution of two nearly overlapping lines

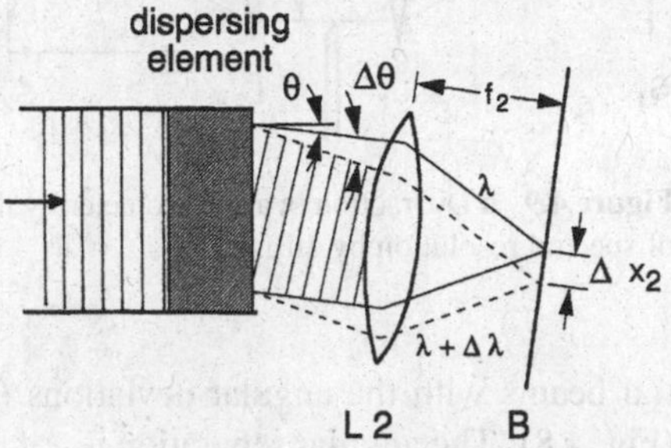

Figure 4.8 Angular dispersion of a parallel beam

coatings of the optical components as well as the geometry of the optical arrangement are optimized according to the specified wavelength region.

c) Spectral Resolving Power

The spectral resolving power of any dispersing instrument is defined by the expression

$$R = |\lambda/\Delta\lambda| = |\nu/\Delta\nu| \,, \tag{4.2}$$

where $\Delta\lambda = \lambda_1 - \lambda_2$ stands for the minimum separation of the central wavelengths λ_1 and λ_2 of two closely spaced lines that are considered to be just resolved. It is possible to recognize that an intensity distribution is composed of two lines with the intensity profiles $I_1(\lambda - \lambda_1)$ and $I_2(\lambda - \lambda_2)$ if the total intensity $I(\lambda) = I_1(\lambda - \lambda_1) + I_2(\lambda - \lambda_2)$ shows a pronounced dip between two maxima (Fig. 4.7). The intensity distribution $I(\lambda)$ depends, of course, on the ratio I_1/I_2 and on the profiles of both components. Therefore, the minimum resolvable interval $\Delta\lambda$ will differ for different profiles.

Lord Rayleigh introduced a criterion of resolution for diffraction-limited line profiles, where two lines are considered to be just resolved if the central diffraction maximum of the profile $I_1(\lambda - \lambda_1)$ coincides with the first minimum of $I_2(\lambda - \lambda_2)$ [116].

Let us consider the attainable spectral resolving power of a spectrometer. When passing the dispersing element (prism or grating), a parallel beam composed of two monochromatic waves with wavelengths λ and $\lambda + \Delta\lambda$ is split into two par-

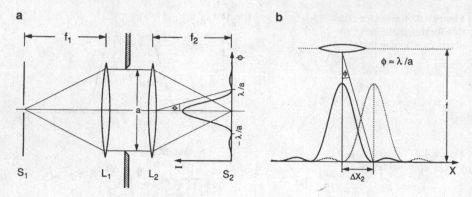

Figure 4.9 a Diffraction in a spectrometer by the limiting aperture with diameter a. **b** Limitation of spectral resolution by diffraction

tial beams with the angular deviations θ and $\theta + \Delta\theta$ from their initial direction (Fig. 4.8). The angular separation is

$$\Delta\theta = (d\theta/d\lambda)\Delta\lambda \,, \tag{4.3}$$

where $d\theta/d\lambda$ is called the *angular dispersion* [rad/nm]. Since the camera lens with focal length f_2 images the entrance slit S_1 into the plane B (Fig. 4.1), the distance Δx_2 between the two images $S_2(\lambda)$ and $S_2(\lambda + \Delta\lambda)$ is, according to Fig. 4.8,

$$\Delta x_2 = f_2 \Delta\theta = f_2 \frac{d\theta}{d\lambda}\Delta\lambda = \frac{dx}{d\lambda}\Delta\lambda \,. \tag{4.4}$$

The factor $dx/d\lambda$ is called the *linear dispersion* of the instrument. It is generally measured in mm/nm. In order to resolve two lines at λ and $\lambda + \Delta\lambda$, the separation Δx_2 in (4.4) has to be at least the sum $\delta x_2(\lambda) + \delta x_2(\lambda + \Delta\lambda)$ of the widths of the two slit images. Since the width δx_2 is related to the width δx_1 of the entrance slit according to geometrical optics by

$$\delta x_2 = (f_2/f_1)\delta x_1 \,, \tag{4.5}$$

the resolving power $\lambda/\Delta\lambda$ can be increased by decreasing δx_1. Unfortunately, there is a theoretical limitation set by diffraction. Because of the fundamental importance of this resolution limit, we discuss this point in more detail.

When a parallel light beam passes a limiting aperture with diameter a, a Fraunhofer diffraction pattern is produced in the plane of the focusing lens L_2 (Fig. 4.9). The intensity distribution $I(\phi)$ as a function of the angle ϕ with the optical axis of the system is given for $\phi \ll \pi/2 \Rightarrow \sin\phi \approx \phi$ by the well-known formula [116]

$$I(\phi) = I_0 \left(\frac{\sin(\pi \cdot (a/\lambda)\sin\phi)}{(a\pi\sin\phi)/\lambda} \right)^2 \simeq I_0 \left(\frac{\sin(a\pi\phi/\lambda)}{a\pi\phi/\lambda} \right)^2 \,. \tag{4.6}$$

The first two diffraction minima at $\phi = \pm\lambda/a \ll \pi$ are symmetrical to the central maximum (zeroth diffraction order) at $\phi = 0$. The intensity of the central diffraction maximum

$$I^{(0)} = \int_{-\lambda/a}^{+\lambda/a} I(\Phi)\, d\Phi$$

contains about 90 % of the total intensity.

Even an infinitesimally small entrance slit therefore produces a slit image of width

$$\delta x_s^{\text{diffr}} = f_2(\lambda/a)\,, \tag{4.7}$$

defined as the distance between the central diffraction maximum and the first minimum, which is approximately equal to the FWHM of the central maximum.

According to the Rayleigh criterion, two equally intense spectral lines with wavelengths λ and $\lambda + \Delta\lambda$ are just resolved if the central diffraction maximum of $S_2(\lambda)$ coincides with the first minimum of $S_2(\lambda + \Delta\lambda)$ (see above). This means that their maxima are just separated by $\delta x_s^{\text{diffr}} = f_2(\lambda/a)$. From (4.6) one can compute that, in this case, both lines partly overlap with a dip of $(8/\pi^2)I_{max} \approx 0.8 I_{max}$ between the two maxima. The distance between the centers of the two slit images is then obtained from (4.7) (see Fig. 4.9b) as

$$\Delta x_2 = f_2(\lambda/a)\,. \tag{4.8a}$$

The separation of the two lines by dispersion (4.4) $\Delta x_2 = f_2(d\theta/d\lambda)\Delta\lambda$ has to be larger than this limit. This gives the fundamental limit on the resolving power

$$\boxed{|\lambda/\Delta\lambda| \le a(d\theta/d\lambda)\,,} \tag{4.9}$$

which clearly depends only on the size a of the limiting aperture and on the angular dispersion of the instrument.

For a finite entrance slit with width b the separation Δx_2 between the central peaks of the two images $I(\lambda - \lambda_1)$ and $I(\lambda - \lambda_2)$ must be larger than (4.8a). We now obtain

$$\Delta x_2 \ge f_2\frac{\lambda}{a} + b\frac{f_2}{f_1}\,, \tag{4.8b}$$

in order to meet the Rayleigh criterion (Fig. 4.10). With $\Delta x_2 = f_2(d\theta/d\lambda)\Delta\lambda$, the smallest resolvable wavelength interval $\Delta\lambda$ is then

$$\Delta\lambda \ge \left(\frac{\lambda}{a} + \frac{b}{f_1}\right)\left(\frac{d\theta}{d\lambda}\right)^{-1}\,. \tag{4.10}$$

Figure 4.10 Intensity profiles of two monochromatic lines measured in the focal plane of L_2 with an entrance slit width $b \gg f_1 \cdot \lambda/a$ and a magnification factor f_2/f_1 of the spectrograph. *Solid line*: without diffraction; *dashed line*: with diffraction. The minimum resolvable distance between the line centers is $\Delta x_2 = f_2(b/f_1 + \lambda/a)$

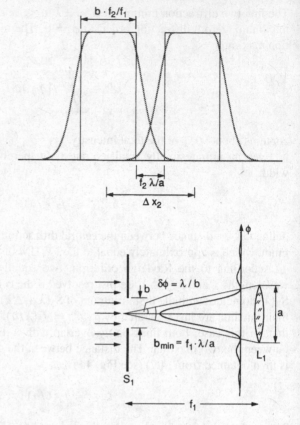

Figure 4.11 Diffraction by the entrance slit

Note **The spectral resolution is limited, *not* by the diffraction due to the entrance slit, but by the diffraction caused by the much larger aperture a, determined by the size of the prism or grating.**

Although it does not influence the spectral resolution, the much larger diffraction by the entrance slit imposes a limitation on the transmitted intensity at small slit widths. This can be seen as follows: when illuminated with parallel light, the entrance slit with width b produces a Fraunhofer diffraction pattern analogous to (4.6) with a replaced by b. The central diffraction maximum extends between the angles $\delta\phi = \pm\lambda/b$ (Fig. 4.11) and can completely pass the limiting aperture a only if $2\delta\phi$ is smaller than the acceptance angle a/f_1 of the spectrometer. This imposes a lower limit to the useful width $b_{\min}$ of the entrance slit,

$$b_{\min} \geq 2\lambda f_1/a . \tag{4.11}$$

In all practical cases, the incident light is divergent, which demands that the sum of the divergence angle and the diffraction angle has to be smaller than a/f and the minimum slit width b correspondingly larger.

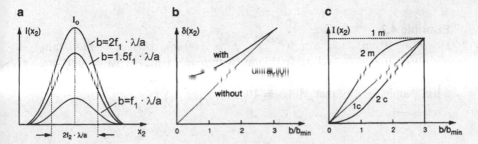

Figure 4.12   a Diffraction limited intensity distribution $I(x_2)$ in the plane B for different widths b of the entrance slit. **b** The width $\delta x_2(b)$ of the entrance slit image $S_2(x_2)$ with and without diffraction by the aperture a. **c** Intensity $I(x_2)$ in the observation plane as a function of entrance slit width b for a spectral continuum c and for a monochromatic spectral line (m) with diffraction (*solid curves* 2c and 2m) and without diffraction (*dashed curves* 1c and 1m)

Figure 4.12a illustrates the intensity distribution $I(x)$ in the plane B for different slit widths b. Figure 4.12b shows the dependence of the width $\Delta x_2(b)$ of the slit image S_2 on the entrance slit width b, taking into account the diffraction caused by the aperture a. This demonstrates that the resolution cannot be increased much by decreasing b below b_{min}. The peak intensity $I(b)$ in the plane B is plotted in Fig. 4.12c as a function of the slit width b. According to (4.1b), the transmitted radiation flux $\phi(\lambda)$ depends on the product $U = A\Omega$ of the entrance slit area A and the acceptance angle $\Omega = (a/f_1)^2$. The flux in B would therefore depend linearly on the slit width b if diffraction were not present. This means that for monochromatic radiation the peak intensity [W/m^2] in the plane B should then be constant (*curve 1m*) although the transmitted power would increase linearly with b. For a spectral continuum it should decrease linearly with decreasing slit width (*curve 1c*). Because of the diffraction by S_1, the intensity decreases with the slit width b both for monochromatic radiation $(2m)$ and for a spectral continuum $(2c)$. Note the steep decrease for $b < b_{min}$.

Substituting $b = b_{min} = 2f\lambda/a$ into (4.10) yields the practical limit for $\Delta\lambda$ imposed by diffraction by S_1 and by the limiting aperture with width a

$$\Delta\lambda = 3f(\lambda/a)d\lambda/dx .\tag{4.12}$$

Instead of the theoretical limit (4.9) given by the diffraction through the aperture a, a smaller practically attainable resolving power is obtained from (4.12), which takes into account a finite minimum entrance slit width b_{min} imposed by intensity considerations and which yields:

$$\boxed{R = \lambda/\Delta\lambda = (a/3)d\theta/d\lambda .}\tag{4.13}$$

Example 4.2

For $a = 10\,\text{cm}$, $\lambda = 5 \times 10^{-5}\,\text{cm}$, $f = 100\,\text{cm}$, $d\lambda/dx = 1\,\text{nm/mm}$, with $b = 10\,\mu\text{m}$, $\rightarrow \Delta\lambda = 0.015\,\text{nm}$; with $b = 5\,\mu\text{m}$, $\rightarrow \Delta\lambda = 0.01\,\text{nm}$. However, from Fig. 4.12 one can see that the transmitted intensity with $b = 5\,\mu\text{m}$ is only 25 % of that with $b = 10\,\mu\text{m}$.

Note For photographic detection of line spectra, it is actually better to really use the lower limit b_{min} for the width of the entrance slit, because the density of the developed photographic layer depends only on the time-integrated spectral irradiance [W/m^2] rather than on the radiation power [W]. Increasing the slit width beyond the diffraction limit b_{min}, in fact, does not significantly increase the density contrast on the plate, but does decrease the spectral resolution.

Using photoelectric recording, the detected signal depends on the radiation power $\phi_\lambda \, d\lambda$ transmitted through the spectrometer and therefore increases with increasing slit width. In the case of completely resolved line spectra, this increase is proportional to the slit width b since $\phi_\lambda \propto b$. For continuous spectra it is even proportional to b^2 because the transmitted spectral interval $d\lambda$ also increases proportional to b and therefore $\phi_\lambda \, d\lambda \propto b^2$. Using diode arrays as detectors, the image $\Delta x_2 = (f_2/f_1)b$ should have the same width as one diode in order to obtain the optimum signal at maximum resolution.

The obvious idea of increasing the product of ΩA without loss of spectral resolution by keeping the width b constant but increasing the height h of the entrance slit is of limited value because imaging defects of the spectrometer cause a curvature of the slit image, which again decreases the resolution. Rays from the rim of the entrance slit pass the prism at a small inclination to the principal axis. This causes a larger angle of incidence α_2, which exceeds that of miniumum deviation. These rays are therefore refracted by a larger angle θ, and the image of a straight slit becomes curved toward shorter wavelengths (Fig. 4.13). Since the deviation in the plane B is equal to $f_2\theta$, the radius of curvature is of the same order of magnitude as the focal length of the camera lens and increases with increasing wavelength because of the decreasing spectral dispersion. In grating spectrometers, curved images of straight slits are caused by astigmatism of the spherical mirrors. The astigmatism can be partly compensated by using curved entrance slits [122]. Another solution is based on astigmatism-corrected imaging by using an asymmetric optical setup where the first mirror M_1 in Fig. 4.2 is placed at a distance $d_1 < f_1$ from the entrance slit and the exit slit at a distance $d_2 > f_2$ from M_2. In this arrangement [124] the grating is illuminated with slightly divergent light.

When the spectrometer is used as a monochromator with an entrance slit width b_1 and an exit slit width b_2, the power $P(t)$ recorded as a function of time while the grating is uniformly turned has a trapezoidal shape for $b_1 \gg b_{\text{min}}$ (Fig. 4.14) with

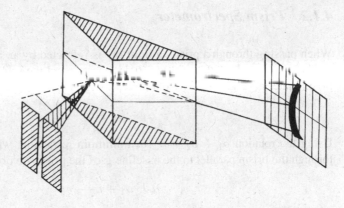

Figure 4.13 Curvature of the image of a straight entrance slit caused by astigmatic imaging errors

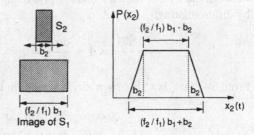

Figure 4.14 Signal profile $P(t) \propto P(x_2(t))$ at the exit slit of a monochromator with $b \gg b_{min}$ and $b_2 < (f_2/f_1)b_1$ for monochromatic incident light with uniform turning of the grating

a baseline $(f_2/f_1)b_1 + b_2$. Optimum resolution at maximum transmitted power is achieved for $b_2 = (f_2/f_1)b_1$. The line profile $P(t) = P(x_2)$ then becomes a triangle.

d) Free Spectral Range

The free spectral range of a spectrometer is the wavelength interval $\delta\lambda$ of the incident radiation for which a one-valued relation exists between λ and the position $x(\lambda)$ of the entrance slit image. Two spectral lines with wavelengths λ_1 and $\lambda_2 = \lambda_1 \pm \delta\lambda$ cannot be distinguished without further information. This means that the wavelength λ measured by the instrument must be known beforehand with an uncertainty $\Delta\lambda < \delta\lambda$. While for prism spectrometers the free spectral range covers the whole region of normal dispersion of the prism material, for grating spectrometers $\delta\lambda$ is determined by the diffraction order m and decreases with increasing m (Sect. 4.1.3).

Interferometers, which are generally used in very high orders ($m = 10^4$–10^8), have a high spectral resolution but a small free spectral range $\delta\lambda$. For unambiguous wavelength determination they need a preselector, which allows one to measure the wavelength within the free spectral range $\delta\lambda$ of the high-resolution instrument (Sect. 4.2.4).

4.1.2 Prism Spectrometer

When passing through a prism, a light ray is refracted by an angle θ that depends on the prism angle ε, the angle of incidence α_1, and the refractive index n of the prism material (Fig. 4.15). We obtain from Fig. 4.15

$$\theta = \alpha_1 - \beta_1 + \alpha_2 - \beta_2 . \tag{4.14a}$$

Using the relation $\beta_1 + \beta_2 = \varepsilon$ for minimum refraction, where the light passes through the prism parallel to the baseline g of the prism we obtain

$$\theta = \alpha_1 + \alpha_2 - \varepsilon \tag{4.14b}$$

between the total deviation θ and the prism angle ε, we find the minimum refraction by differentiating:

$$\frac{d\theta}{d\alpha_1} = 1 + \frac{d\alpha_2}{d\alpha_1} = 0 \Rightarrow d\alpha_1 = -d\alpha_2 . \tag{4.14c}$$

From Snellius' law $\sin \alpha = n \sin \beta$ we obtain the derivatives:

$$\cos \alpha_1 \, d\alpha_1 = n \, \cos \beta_1 \, d\beta_1 \tag{4.14d}$$

$$\cos \alpha_2 \, d\alpha_2 = n \, \cos \beta_2 \, d\beta_2 . \tag{4.14e}$$

Because $\beta_1 + \beta_2 = \varepsilon \Rightarrow d\beta_1 = -d\beta_2$, the division of (4.14d) by (4.14e) yields

$$\frac{\cos \alpha_1 \, d\alpha_1}{\cos \alpha_2 \, d\alpha_2} = \frac{\cos \beta_1}{\cos \beta_2} .$$

For the minimum deviation θ with $d\alpha_1 = -d\alpha_2$ we get the result:

$$\frac{\cos \alpha_1}{\cos \alpha_2} = -\frac{\cos \beta_1}{\cos \beta_2} = -\left(\frac{1 - \sin^2 \beta_1}{1 - \sin^2 \beta_2} \right)^{1/2} . \tag{4.14f}$$

Squaring the equation yields

$$\frac{1 - \sin^2 \alpha_1}{1 - \sin^2 \alpha_2} = \frac{n^2 - \sin^2 \alpha_1}{n^2 - \sin^2 \alpha_2} \tag{4.14g}$$

Figure 4.15 Refraction of light by a prism at minimum deviation where $\alpha_1 = \alpha_2 = \alpha$ and $\theta = 2\alpha - \varepsilon$

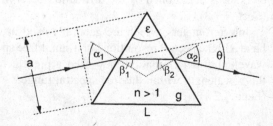

Figure 4.16 Limiting aperture in a prism spectrometer

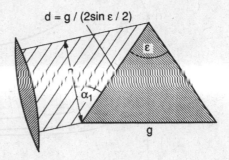

$$d = g / (2\sin \varepsilon / 2)$$

which can only be fulfilled for $n \neq 1$ if $\alpha_1 = \alpha_2$. **The minimum deviation θ is obtained for symmetrical rays with $\alpha_1 = \alpha_2 = \alpha$.** The minimum deviation

$$\theta_{\min} = 2\alpha - \varepsilon \qquad (4.14\text{h})$$

is obtained when the ray passes the prism parallel to the base g. In this case, we derive from Snellius' law:

$$\sin\left(\frac{\theta_{\min} + \varepsilon}{2}\right) = \sin \alpha = n \sin \beta = n \sin(\varepsilon/2) \qquad (4.14\text{i})$$

$$\sin\left(\frac{\theta + \varepsilon}{2}\right) = n \sin(\varepsilon/2) . \qquad (4.14\text{j})$$

From (4.14j) the derivation $d\theta/dn = (dn/d\theta)^{-1}$ is

$$\frac{d\theta}{dn} = \frac{2\sin(\varepsilon/2)}{\cos[(\theta + \varepsilon)/2]} = \frac{2\sin(\varepsilon/2)}{\sqrt{1 - n^2 \sin^2(\varepsilon/2)}} . \qquad (4.15)$$

The *angular dispersion* $d\theta/d\lambda = (d\theta/dn)(dn/d\lambda)$ is therefore

$$\boxed{\frac{d\theta}{d\lambda} = \frac{2\sin(\varepsilon/2)}{\sqrt{1 - n^2 \sin^2(\varepsilon/2)}} \frac{dn}{d\lambda}} . \qquad (4.16)$$

This shows that the angular dispersion increases with the prism angle ε, *but does not depend on the size of the prism.*

For the deviation of laser beams with small beam diameters, small prisms can therefore be used without losing angular dispersion. In a prism spectrometer, however, the size of the prism determines the limiting aperture a and therefore the diffraction; it has to be large in order to achieve a large spectral resolving power (see previous section). For a given angular dispersion, an equilateral prism with $\varepsilon = 60°$ uses the smallest quantity of possibly expensive prism material. Because $\sin 30° = 1/2$, (4.16) then reduces to

$$\frac{d\theta}{d\lambda} = \frac{dn/d\lambda}{\sqrt{1 - (n/2)^2}} . \qquad (4.17)$$

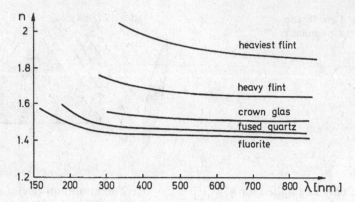

Figure 4.17 Refractive index $n(\lambda)$ for some prism materials

The diffraction limit for the resolving power $\lambda/\Delta\lambda$ according to (4.9) is

$$\lambda/\Delta\lambda \leq a(\mathrm{d}\theta/\mathrm{d}\lambda) .$$

The diameter a of the limiting aperture in a prism spectrometer is (Fig. 4.16)

$$a = \mathrm{d}\cos\alpha_1 = \frac{g\cos\alpha}{2\sin(\varepsilon/2)} . \tag{4.18}$$

Substituting $\mathrm{d}\theta/\mathrm{d}\lambda$ from (4.16) gives

$$\lambda/\Delta\lambda = \frac{g\cos\alpha_1}{\sqrt{1 - n^2\sin^2(\varepsilon/2)}}\frac{\mathrm{d}n}{\mathrm{d}\lambda} . \tag{4.19}$$

At minimum deviation, (4.14a)–(4.14j) gives $n\sin(\varepsilon/2) = \sin(\theta + \varepsilon)/2 = \sin\alpha_1$ and therefore (4.19) reduces to

$$\lambda/\Delta\lambda = g(\mathrm{d}n/\mathrm{d}\lambda) . \tag{4.20a}$$

According to (4.20a), the theoretical maximum resolving power depends solely on the base length g and on the spectral dispersion of the prism material. Because of the finite slit width $b \geq b_{min}$, the resolution reached in practice is somewhat lower. The corresponding resolving power can be derived from (4.11) to be at most

$$\boxed{R = \frac{\lambda}{\Delta\lambda} \leq \frac{1}{3}g\left(\frac{\mathrm{d}n}{\mathrm{d}\lambda}\right) .} \tag{4.20b}$$

The spectral dispersion $\mathrm{d}n/\mathrm{d}\lambda$ is a function of prism material and wavelength λ. Figure 4.17 shows dispersion curves $n(\lambda)$ for some materials commonly used for prisms. Since the refractive index increases rapidly in the vicinity of absorption

Table 4.1 Refractive index and dispersion of some materials used in prism spectrometers

Material	Useful spectral range [μm]	Refractive index n	Dispersion $-dn/d\lambda$ [nm^{-1}]		
(Glass) BK 7)	(1.35–3.5)	1.516	4.6×10^{-5}	at	589 nm
		1.53	1.1×10^{-4}	at	400 nm
Heavy flint	0.4–2	1.755	1.4×10^{-4}	at	589 nm
		1.81	4.4×10^{-4}	at	400 nm
Fused quartz	0.15–4.5	1.458	3.4×10^{-5}	at	589 nm
		1.470	1.1×10^{-4}	at	400 nm
NaCl	0.2–26	1.79	6.3×10^{-3}	at	200 nm
		1.38	1.7×10^{-5}	at	20 μm
LiF	0.12–9	1.44	6.6×10^{-4}	at	200 nm
		1.09	8.6×10^{-5}	at	10 μm

lines, glass has a larger disperison in the visible and near-ultraviolet regions than quartz, which, on the other hand, can be used advantageously in the UV down to 180 nm. In the vacuum-ultraviolet range CaF, MgF, or LiF prisms are sufficiently transparent. Table 4.1 gives a summary of the optical characteristics and useful spectral ranges of some prism materials.

If achromatic lenses (which are expensive in the infrared and ultraviolet region) are not employed, the focal length of the two lenses decreases with the wavelength. This can be partly compensated by inclining the plane B against the principal axis in order to bring it at least approximately into the focal plane of L_2 for a large wavelength range (Fig. 4.1).

In Summary: The advantage of a prism spectrometer is the unambiguous assignment of wavelengths, since the position $S_2(\lambda)$ is a monotonic function of λ. Its drawback is the moderate spectral resolution. It is mostly used for survey scans of extended spectral regions.

Example 4.3

a) Suprasil (fused quartz) has a refractive index $n = 1.47$ at $\lambda = 400$ nm and $dn/d\lambda = 1100$ cm^{-1}. This gives $d\theta/d\lambda = 1.6 \times 10^{-4}$ rad/nm. With a slitwidth $b_{min} = 2f\lambda/a$ and $g = 5$ cm we obtain from (4.20b) $\lambda/\Delta\lambda \leq 1830$. At $\lambda = 500$ nm $\Rightarrow \Delta\lambda \geq 0.27$ nm.

b) For heavy flint glass at 400 nm $n = 1.81$ and $dn/d\lambda = 4400$ cm^{-1}, giving $d\theta/d\lambda = 1.0 \times 10^{-3}$ rad/nm. This is about six times larger than that for quartz. With a focal length $f = 100$ cm for the camera lens, one achieves a linear dispersion $dx/d\lambda = 1$ mm/nm with a flint prism, but only 0.15 mm/nm with a quartz prism.

4.1.3   Grating Spectrometer

In a grating spectrometer (Fig. 4.2) the collimating lens L_1 is replaced by a spherical mirror M_1 with the entrance slit S_1 in the focal plane of M_1. The collimated parallel light is reflected by M_1 onto a reflection grating consisting of many straight grooves (about 10^5) parallel to the entrance slit. The grooves have been ruled onto an optically smooth glass substrate or have been produced by holographic techniques [125–131]. The whole grating surface is coated with a highly reflecting layer (metal or dielectric film). The light reflected from the grating is focused by the spherical mirror M_2 onto the exit slit S_2 or onto a photographic plate in the focal plane of M_2.

a) Basic Considerations

The many grooves, which are illuminated coherently, can be regarded as small radiation sources, each of them diffracting the light incident onto this small groove with a width $d \approx \lambda$ into a large range $\Delta r \approx \lambda/d$ of angles r around the direction of geometrical reflection (Fig. 4.18a). The total reflected light consists of a coherent superposition of these many partial contributions. Only in those directions where all partial waves emitted from the different grooves are in phase will constructive interference result in a large total intensity, while in all other directions the different contributions cancel by destructive interference.

Figure 4.18b depicts a parallel light beam incident onto two adjacent grooves. At an angle of incidence α to the grating normal (which is normal to the grating surface, but not necessarily to the grooves) one obtains constructive interference for those directions β of the reflected light for which the path difference $\Delta s = \Delta s_1 - \Delta s_2$ is an integer multiple m of the wavelength λ. With $\Delta s_1 = d \, \sin \alpha$ and $\Delta s_2 = d \, \sin \beta$

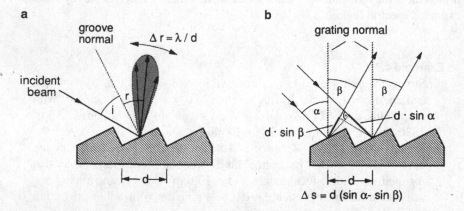

Figure 4.18 a Reflection of incident light from a single groove into the diffraction angle λ/d around the specular reflection angle $r = i$. **b** Illustration of the grating equation (4.21)

Figure 4.19 Illustration of
the blaze angle θ

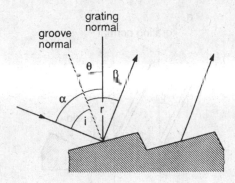

this yields the grating equation

$$d(\sin\alpha \pm \sin\beta) = m\lambda , \qquad (4.21)$$

the plus sign has to be taken if β and α are on the same side of the grating normal; otherwise the minus sign, which is the case shown in Fig. 4.18b.

The reflectivity $R(\beta,\theta)$ of a ruled grating depends on the diffraction angle β and on the blaze angle θ of the grating, which is the angle between the groove normal and the grating normal (Fig. 4.19). If the diffraction angle β coincides with the angle r of specular reflection from the groove surfaces, $R(\beta,\theta)$ reaches its optimum value R_0, which depends on the reflectivity of the groove coating. From Fig. 4.19 one infers for the case where α and β are on opposite sides of the grating normal, $i = \alpha - \theta$ and $r = \theta + \beta$, which yields, for specular reflection $i = r$, the condition for the optimum blaze angle θ

$$\theta = (\alpha - \beta)/2 . \qquad (4.22)$$

Because of the diffraction of each partial wave into a large angular range, the reflectivity $R(\beta)$ will not have a sharp maximum at $\beta = \alpha - 2\theta$, but will rather show a broad distribution around this optimum angle. The angle of incidence α is determined by the particular construction of the spectrometer, while the angle β for which constructive interference occurs depends on the wavelength λ. Therefore the blaze angle θ has to be specified for the desired spectral range and the spectrometer type.

In laser-spectroscopic applications the case $\alpha = \beta$ often occurs, which means that the light is reflected back into the direction of the incident light. For such an arrangement, called a *Littrow-grating mount* (shown in Fig. 4.20), the grating equation (4.21) for constructive interference reduces to

$$2d \sin\alpha = m\lambda . \qquad (4.21a)$$

Maximum reflectivity of the Littrow grating is achieved for $i = r = 0 \rightarrow \theta = \alpha$ (Fig. 4.20b). The Littrow grating acts as a wavelength-selective reflector because light is only reflected if the incident wavelength satisfies the condition (4.21a).

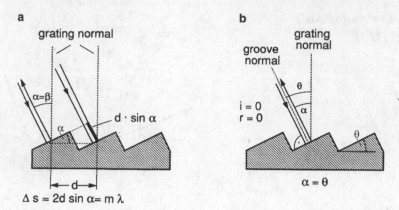

Figure 4.20 **a** Littrow mount of a grating with $\beta = \alpha$. **b** Illustration of blaze angle for a Littrow grating

For Littrow gratings used as wavelength-selective reflectors, it is desirable to have a high reflectivity in a selected order m and low reflections for all other orders. This can be achieved by selecting the width of the grooves and the blaze angle correctly. Because of diffraction by each groove with a width d, light can only reach angles β within the interval $\beta_0 \pm \lambda/d$ (Fig. 4.18a).

Example 4.4
With a blaze angle $\theta = \alpha = \beta = 30°$ and a step height $h = \lambda$, the grating can be used in second order, while the third order appears at $\beta = \beta_0 + 37°$. With $d = \lambda / \tan \theta = 2\lambda$, the central diffraction lobe extends only to $\beta_0 \pm 30°$, the intensity in the third order is very small.

b) Intensity Distribution of Reflected Light

We now examine the intensity distribution $I(\beta)$ of the reflected light when a monochromatic plane wave is incident onto an arbitrary grating.

According to (4.21) the path difference between partial waves reflected by adjacent grooves is $\Delta s = d(\sin \alpha \pm \sin \beta)$ and the corresponding phase difference is

$$\phi = \frac{2\pi}{\lambda} \Delta s = \frac{2\pi}{\lambda} d (\sin \alpha \pm \sin \beta) . \tag{4.23}$$

The superposition of the amplitudes reflected from all N grooves in the direction β gives the total reflected amplitude

$$A_R = \sqrt{R} \sum_{m=0}^{N-1} A_g e^{im\psi} = \sqrt{R} A_g \frac{1 - e^{iN\phi}}{1 - e^{i\psi}}$$ (4.24)

where $R(\beta)$ is the reflectivity of the grating, which depends on the reflection angle β, and A_g is the amplitude of the partial wave incident onto each groove. Because the intensity of the reflected wave is related to its amplitude by $I_R = \epsilon_0 c A_R A_R^*$, see (2.30c), we find, with $e^{ix} = \cos x + i \sin x$, from (4.24),

$$I_R = R I_0 \frac{\sin^2(N\phi/2)}{\sin^2(\phi/2)} \quad \text{with } I_0 = c\epsilon_0 A_g A_g^* .$$ (4.25)

This intensity distribution is plotted in Fig. 4.21 for two different values of the total groove number N. Note that for real optical gratings $N \approx 10^5$! The principal maxima occur for $\phi = 2m\pi$, which is, according to (4.23), equivalent to the grating equation (4.21) and means that at a fixed angle α the path difference between partial beams from adjacent grooves is for certain angles β_m an integer multiple of the wavelength, where the integer m is called the *order of the interference*. The function (4.25) has $(N - 1)$ minima with $I_R = 0$ between two successive principal maxima. These minima occur at values of ϕ for which $N\phi/2 = \ell\pi$, $\ell = 1, 2, \ldots, N - 1$, and mean that for each groove of the grating, another one can be found that emits light into the direction β with a phase shift π, such that all pairs of partial waves just cancel.

The line profile $I(\beta)$ of the principal maximum of order m around the diffraction angle β_m can be derived from (4.25) by substituting $\beta = \beta_m + \epsilon$. Because for large N, $I(\beta)$ is very sharply centered around β_m, we can assume $\epsilon \ll \beta_m$. With the relation

$$\sin(\beta_m + \epsilon) = \sin \beta_m \cos \epsilon + \cos \beta_m \sin \epsilon \sim \sin \beta_m + \epsilon \cos \beta_m ,$$

and because $(2\pi d/\lambda)(\sin\alpha + \sin\beta_m) = 2m\pi$, we obtain from (4.23)

$$\phi(\beta) = 2m\pi + 2\pi(d/\lambda)\epsilon \cos \beta_m = 2m\pi + \delta_1$$ (4.26)

with

$$\delta_1 = 2\pi(d/\lambda)\epsilon \cos \beta_m \ll 1 .$$

Furthermore, (4.25) can be written as

$$I_R = R I_0 \frac{[\sin(Nm\pi + N\delta_1/2)]^2}{[\sin(m\pi + \delta_1/2)]^2} = R I_0 \frac{\sin^2(N\delta_1/2)}{\sin^2(\delta_1/2)} \simeq R I_0 N^2 \frac{\sin^2(N\delta_1/2)}{(N\delta_1/2)^2} .$$ (4.27)

The first two minima with $I_R = 0$ on both sides of the central maximum at β_m are at

$$N\delta_1 = \pm 2\pi \Rightarrow \delta_1 = \pm 2\pi/N \ . \tag{4.28a}$$

From (4.26) we can now calculate the angular width $\Delta\beta$ of the central maximum around β_m: The first two minima on both sides of the intensity maximum appear for

$$\frac{2\pi d}{\lambda}\epsilon \cos\beta_m = \delta_1 = \pm\frac{2\pi}{N} \Rightarrow \tag{4.28b}$$

$$\epsilon_{1,2} = \frac{\pm\lambda}{Nd\cos\beta_m} \Rightarrow \Delta\beta = \frac{2\lambda}{Nd\cos\beta_m} \ . \tag{4.28c}$$

The intensity maximum of mth order therefore has a line profile (4.27) with a base full width $\Delta\beta = 2\lambda/(Nd\cos\beta_m)$. This corresponds to a diffraction pattern produced by an aperture with width $b = Nd\cos\beta_m$, which is just the size of the whole grating projected onto a plane, normal to the direction of β_m (Fig. 4.18).

Example 4.5
For $N \cdot d = 10\,\text{cm}$, $\lambda = 5 \times 10^{-5}\,\text{cm}$, $\beta_m = 45° \Rightarrow \cos\beta_m = \frac{1}{2}\sqrt{2} \Rightarrow$
$\varepsilon_{1/2} = 7 \times 10^{-6}\,\text{rad} = 4 \times 10^{-4}\,°$.

Note According to (4.28a)–(4.28c) the full angular halfwidth $\Delta\beta = 2\epsilon$ of the interference maxima decreases as $1/N$, while according to (4.27) the peak intensity increases with the number of illuminated grooves proportional to $N^2 I_0$, where I_0 is the power incident onto a single groove. The area under the main maxima is therefore proportional to NI_0, which is due to the increasing concentration of light into the directions β_m. Of course, the incident power per groove decreases as $1/N$. The total reflected power is therefore independent of N.

The intensity of the $N - 2$ small side maxima, which are caused by incomplete destructive interference, decreases proportional to $1/N$ with increasing groove number N. Figure 4.21 illustrates this point for $N = 5$ and $N = 11$. For gratings used in practical spectroscopy with groove numbers of about 10^5, the reflected intensity $I_R(\lambda)$ at a given wavelength λ has very sharply defined maxima only in those directions β_m, as defined by (4.21). The small side maxima are completely negligible at such large values of N, provided the distance d between the grooves is exactly constant over the whole grating area.

Figure 4.21 Intensity distribution $I(\beta)$ for two different
numbers N of illuminated
grooves. Note the different
scales of the ordinate!

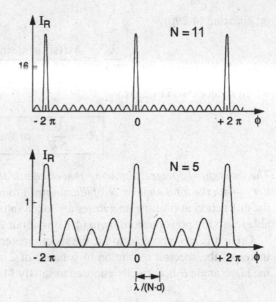

c) Spectral Resolving Power

Differentiating the grating equation (4.21) with respect to λ, we obtain at a given
angle α the angular dispersion

$$\frac{d\beta}{d\lambda} = \frac{m}{d\cos\beta} . \tag{4.29a}$$

Substituting $m/d = (\sin\alpha \pm \sin\beta)/\lambda$ from (4.21), we find

$$\frac{d\beta}{d\lambda} = \frac{\sin\alpha \pm \sin\beta}{\lambda\cos\beta} . \tag{4.29b}$$

This illustrates that the angular dispersion is determined solely by the angles
α and β and *not by the number of grooves!* For the Littrow mount with $\alpha = \beta$ and
the + sign in (4.29b), we obtain

$$\frac{d\beta}{d\lambda} = \frac{2\tan\alpha}{\lambda} . \tag{4.29c}$$

The spectral resolving power can be immediately derived from (4.29a) and the base
halfwidth $\Delta\beta = \epsilon = \lambda/(Nd\cos\beta)$ of the principal diffraction maximum (4.28a)–
(4.28c), if we apply the Rayleigh criterion (see above) that two lines λ and $\lambda + \Delta\lambda$
are just resolved when the maximum of $I(\lambda)$ falls into the adjacent minimum for
$I(\lambda + \Delta\lambda)$. This is equivalent to the condition

$$\frac{d\beta}{d\lambda}\Delta\lambda = \frac{\lambda}{Nd\cos\beta} ,$$

or, inserting (4.29b):

$$\frac{\lambda}{\Delta\lambda} = \frac{Nd(\sin\alpha \pm \sin\beta)}{\lambda},$$

(4.30)

which reduces with (4.21) to

$$R = \frac{\lambda}{\Delta\lambda} = mN .$$

(4.31)

The theoretical spectral resolving power equals the product of the diffraction order m with the total number N of illuminated grooves. If the finite slit width b_1 and the diffraction at limiting aperatures are taken into account, the practically achievable resolving power according to (4.13) is about 2–3 times lower.

Often it is advantageous to use the spectrometer in second order ($m = 2$), which increases the spectral resolution by a factor of 2 without losing much intensity, if the blaze angle θ is correctly choosen to satisfy (4.21) and (4.22) with $m = 2$.

Example 4.6
A grating with a ruled area of $10 \times 10\,\text{cm}^2$ and 10^3 grooves/mm allows in second order ($m = 2$) a theoretical spectral resolution of $R = 2 \times 10^5$. This means that at $\lambda = 500\,\text{nm}$ two lines that are separated by $\Delta\lambda = 2.5 \times 10^{-3}\,\text{nm}$ should be resolvable. Because of diffraction, the practical limit is $\Delta\lambda \approx 5 \times 10^{-3}\,\text{nm}$. The dispersion for $\alpha = \beta = 30°$ and a focal length $f = 1\,\text{m}$ is $dx/d\lambda = f\,d\beta/d\lambda = 2\,\text{mm/nm}$. With a slit width $b_1 = b_2 = 50\,\mu\text{m}$ a spectral resolution of $\Delta\lambda = 0.025\,\text{nm}$ can be achieved. In order to decrease the slit image width to $5 \times 10^{-3}\,\text{mm}$, the entrance slit width b has to be narrowed to $10\,\mu\text{m}$.

Lines around $\lambda = 1\,\mu\text{m}$ in the spectrum would appear in 1st order at the same angles β as lines with $\lambda = 500\,\text{mm}$ in 2nd order. They have to be suppressed by filters.

A special design is the so-called *echelle grating*, which has very widely spaced grooves forming right-angled steps (Fig. 4.22). The light is incident normal to the

Figure 4.22 Echelle grating

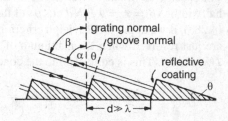

small side of the grooves. The path difference between two reflected partial beams incident on two adjacent grooves with an angle of incidence $\alpha = 90° - \theta$ is $\Delta s = 2d \cos \theta$. The grating equation (4.21) gives for the angle β of the mth diffraction order

$$d(\cos \theta + \sin \beta) \approx 2d \cos \theta = m\lambda , \qquad (4.32)$$

where β is close to $\alpha = 90° - \theta$.

With $d \gg \lambda$ the grating is used in a very high order ($m \simeq 10$–100) and the resolving power is very high according to (4.31). Because of the larger distance d between the grooves, the achievable relative ruling accuracy is higher and large gratings (up to 30 cm) can be ruled. The disadvantage of the echelle is the small free spectral range $\delta\lambda = \lambda/m$ between successive diffraction orders.

> **Example 4.7**
> $N = 3 \times 10^4$, $d = 10\,\mu\text{m}$, $\theta = 30°$, $\lambda = 500\,\text{nm}$, $m = 34$. The spectral resolving power is $R = 10^6$, but the free spectral range is only $\delta\lambda = 15\,\text{nm}$. This means that the wavelengths λ and $\lambda + \delta\lambda$ overlap in the same direction β.

d) Grating Ghosts

Minute deviations of the distance d between adjacent grooves, caused by inaccuracies during the ruling process, may result in constructive interference from parts of the grating for "wrong" wavelengths. Such unwanted maxima, which occur for a given angle of incidence α into "wrong" directions β, are called *grating ghosts*. Although the intensity of these ghosts is generally very small, intense incident radiation at a wavelength λ_i may cause ghosts with intensities comparable to those of other weak lines in the spectrum. This problem is particularly serious in laser spectroscopy when the intense light at the laser wavelength, which is scattered by cell walls or windows, reaches the entrance slit of the monochromator.

In order to illustrate the problematic nature of achieving the ruling accuracy that is required to avoid these ghosts, let us assume that the carriage of the ruling engine expands by only 1 μm during the ruling of a $10 \times 10\,\text{cm}^2$ grating, e.g., due to temperature drifts. The groove distance d in the second half of the grating differs therefore from that of the first half by $5 \times 10^{-6}d$. With $N = 10^5$ grooves, the waves from the second half are then completely out of phase with those from the first half. The condition (4.21) is then fulfilled for different wavelengths in both parts of the grating, giving rise to unwanted wavelengths at the wrong positions β. Such ghosts are particularly troublesome in laser Raman spectroscopy (Vol. 2, Chap. 3) or low-level fluorescence spectroscopy, where very weak lines have to be detected in the presence of extremely strong excitation lines. The ghosts from these excitation lines may overlap with the fluorescence or Raman lines and complicate the assignment of the spectrum.

e) Holographic Gratings

Although modern ruling techniques with interferometric length control have greatly improved the quality of ruled gratings [125–128] the most satisfactory way of producing completely ghost-free gratings is with holography. The production of holographic gratings proceeds as follows: a photosensitive layer on the grating's blank surface in the (x, y) plane is illuminated by two coherent plane waves with the wave vectors $\mathbf{k}_1$ and $\mathbf{k}_2$ ($|\mathbf{k}_1| = |\mathbf{k}_2|, \mathbf{k} = \{k_x, 0, k_z\}$), which form the angles α and β against the surface normal (Fig. 4.23). The intensity distribution of the superposition in the plane $z = 0$ of the photolayer consists of parallel dark and bright fringes imprinting an ideal grating into the layer, which becomes visible after developing the photoemulsion. The grating constant

$$d = \frac{\lambda/2}{\sin\alpha + \sin\beta}$$

depends on the wavelength $\lambda = 2\pi/|\mathbf{k}|$ and on the angles α and β. Such holographic gratings are essentially free of ghosts. Their reflectivity R, however, is lower than that of ruled gratings and is furthermore strongly dependent on the polarization of the incident wave. This is due to the fact that holographically produced grooves are no longer planar, but have a sinusoidal surface and the "blaze angle" θ varies across each groove [130].

Summary: Summarizing the considerations above, we find that the grating acts as a wavelength-selective mirror, reflecting light of a given wavelength only into definite directions β_m, called the mth diffraction orders, which are defined by (4.21). The intensity profile of a diffraction order corresponds to the diffraction profile of a slit with width $b = Nd\cos\beta_m$ representing the size of the whole grating projection as seen in the direction β_m. *The spectral resolution* $\lambda/\Delta\lambda = mN =$

Figure 4.23 **a** Photographic production of a holographic grating; **b** surface of a holographic grating

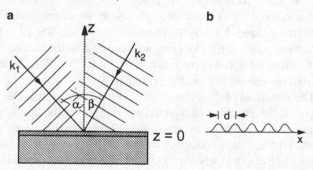

$Nd(\sin\alpha + \sin\beta)/\lambda$ *is therefore limited by the effective size of the grating measured in units of the wavelength.*

For a more detailed discussion of special designs of grating monochromators, such as the concave gratings used in VUV spectroscopy, the reader is referred to the literature on this subject [125–131]. An excellent account of the production and design of ruled gratings can be found in [125].

4.2 Interferometers

For the investigation of the various line profiles discussed in Chap. 3, interferometers are preferentially used because, with respect to the spectral resolving power, they are superior even to large spectrometers. In laser spectroscopy the different types of interferometers not only serve to measure emission – or absorption – line profiles, but they are also essential devices for narrowing the spectral width of lasers, monitoring the laser linewidth, and controlling and stabilizing the wavelength of single-mode lasers (Chap. 5).

In this section we discuss some basic properties of interferometers with the aid of some illustrating examples. The characteristics of the different types of interferometers that are essential for spectroscopic applications are discussed in more detail. Since laser technology is inconceivable without dielectric coatings for mirrors, interferometers, and filters, an extra section deals with such dielectric multilayers. The extensive literature on interferometers [133–136] informs about special designs and applications.

4.2.1 Basic Concepts

The basic principle of all interferometers may be summarized as follows (Fig. 4.24). The indicent lightwave with intensity I_0 is divided into two or more partial beams with amplitudes A_k, which pass different optical path lengths $s_k = nx_k$ (where n is the refractive index) before they are again superimposed at the exit of the interferometer. Since all partial beams come from the same source, they are co-

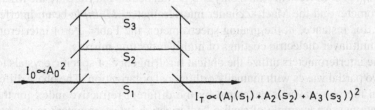

Figure 4.24 Schematic illustration of the basic principle for all interferometers

herent as long as the maximum path difference does not exceed the coherence length (Sect. 2.9). The total amplitude of the transmitted wave, which is the superposition of all partial waves, depends on the amplitudes A_k and on the phases $\phi_k = \phi_0 + 2\pi s_k/\lambda$ of the partial waves. *It is therefore sensitively dependent on the wavelength* λ.

The maximum transmitted intensity is obtained when all partial waves interfere constructively. This gives the condition for the optical path difference $\Delta s_{ik} = s_i - s_k$, namely

$$\Delta s_{ik} = m\lambda , \quad m = 1, 2, 3, \dots . \tag{4.33}$$

The condition (4.33) for maximum transmission of the interferometer applies not only to a single wavelength λ but to all λ_m for which

$$\lambda_m = \Delta s/m , \quad m = 1, 2, 3, \dots .$$

The wavelength interval

$$\delta\lambda = \lambda_m - \lambda_{m+1} = \frac{\Delta s}{m} - \frac{\Delta s}{m+1} = \frac{\Delta s}{m^2 + m} \tag{4.34a}$$

is called the *free spectral range* of the interferometer. With the mean wavelength $\bar{\lambda} = \frac{1}{2}(\lambda_m + \lambda_{m+1}) = \frac{1}{2}\Delta s(\frac{1}{m} + \frac{1}{m+1})$, we can write the free spectral range as:

$$\delta\lambda = \frac{2\bar{\lambda}}{2m+1} . \tag{4.34b}$$

It is more conveniently expressed in terms of frequency. With $\nu = c/\lambda$, (4.33) yields $\Delta s = mc/\nu_m$ and the free spectral frequency range

$$\boxed{\delta\nu = \nu_{m+1} - \nu_m = c/\Delta s ,} \tag{4.34c}$$

becomes independent of the order m.

Note It is important to realize that from one interferometric measurement alone one can only determine λ modulo $m \cdot \delta\lambda$ because all wavelengths $\lambda = \lambda_0 + m\delta\lambda$ are equivalent with respect to the transmission of the interferometer. One therefore has at first to measure λ within one free spectral range using other techniques before the absolute wavelength can be obtained with an interferometer.

Examples of devices in which only *two* partial beams interfere are the Michelson interferometer and the Mach–Zehnder interferometer. *Multiple*-beam interference is used, for instance, in the grating spectrometer, the Fabry–Perot interferometer, and in multilayer dielectric coatings of highly reflecting mirrors.

Some interferometers utilize the optical birefringence of specific crystals to produce two partial waves with mutually orthogonal polarization. The phase difference between the two waves is generated by the different refractive index for the two polarizations. An example of such a "polarization interferometer" is the *Lyot filter* [137] used in dye lasers to narrow the spectral linewidth (Sect. 4.2.11).

4.2.2 Michelson Interferometer

The basic principle of the Michelson interferometer (MI) is illustrated in Fig. 4.25. The incident plane wave

$$E = A_0 e^{i(\omega t - kx)}$$

is split by the beam splitter S (with reflectivity R and transmittance T) into two waves

$$E_1 = A_1 \exp[i(\omega t - kx + \phi_1)] \quad \text{and} \quad E_2 = A_2 \exp[i(\omega t - ky + \phi_2)] .$$

If the beam splitter has negligible absorption ($R + T = 1$), the amplitudes A_1 and A_2 are determined by $A_1 = \sqrt{T} A_0$ and $A_2 = \sqrt{R} A_0$ with $A_0^2 = A_1^2 + A_2^2$.

After being reflected at the plane mirrors M_1 and M_2, the two waves are superimposed in the plane of observation B. In order to compensate for the dispersion that beam 1 suffers by passing twice through the glass plate of beam splitter S, often an appropriate compensation plate P is placed in one side arm of the interferometer. The amplitudes of the two waves in the plane B are $\sqrt{TR} A_0$, because each wave has been transmitted and reflected once at the beam splitter surface S. The phase difference ϕ between the two waves is

$$\phi = \frac{2\pi}{\lambda} 2(\text{SM}_1 - \text{SM}_2) + \Delta\phi , \qquad (4.35)$$

where $\Delta\phi$ accounts for additional phase shifts that may be caused by reflection. The total complex field amplitude in the plane B is then

$$E = \sqrt{RT} A_0 e^{i(\omega t + \phi_0)} (1 + e^{i\phi}) . \qquad (4.36)$$

The detector in B cannot follow the rapid oscillations with frequency ω but measures the time-averaged intensity $\bar{I}$, which is, according to (2.30c),

$$\bar{I} = \tfrac{1}{2} c\epsilon_0 A_0^2 RT (1 + e^{i\phi})(1 + e^{-i\phi}) = c\epsilon_0 A_0^2 RT (1 + \cos\phi)$$

$$= \tfrac{1}{2} I_0 (1 + \cos\phi) \quad \text{for } R = T = \tfrac{1}{2} \text{ and } I_0 = \tfrac{1}{2} c\epsilon_0 A_0^2 . \qquad (4.37)$$

Figure 4.25 Two-beam interference in a Michelson interferometer

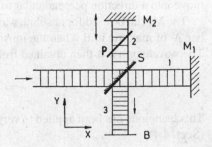

Figure 4.26 Intensity transmitted through the Michelson
interferometer as a function
of the phase difference ϕ
between the two interfering
beams for $R = T = 0.5$

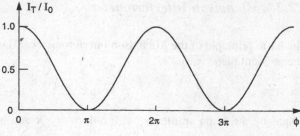

Figure 4.27 Circular fringe
pattern produced by the MI
with divergent incident light

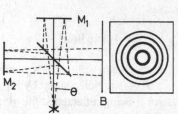

If mirror M_2 (which is mounted on a carriage) moves along a distance Δy, the
optical path difference changes by $\Delta s = 2n\Delta y$ (n is the refractive index between
S and M_2) and the phase difference ϕ changes by $2\pi\Delta s/\lambda$. Figure 4.26 shows the
intensity $I_T(\phi)$ in the plane B as a function of ϕ for a monochromatic incident plane
wave. For the maxima at $\phi = 2m\pi$ ($m = 0, 1, 2, \ldots$), the transmitted intensity I_T
becomes equal to the incident intensity I_0, which means that the transmission of
the interferometer is $T_I = 1$ for $\phi = 2m\pi$. In the minima for $\phi = (2m + 1)\pi$
the transmitted intensity I_T is zero! The incident plane wave is being reflected back
into the source.

This illustrates that the MI can be regarded either as a wavelength-dependent filter for the transmitted light, or as a wavelength-selective reflector. In the latter function it is often used for mode selection in lasers (Fox–Smith selector, Sect. 5.4.3).

For divergent incident light the path difference between the two waves depends
on the inclination angle (Fig. 4.27). In the plane B an interference pattern of circular fringes, concentric to the symmetry axis of the system, is produced. Moving
the mirror M_2 causes the ring diameters to change. The intensity behind a small
aperture in the plane B, centered around the interferometer axis still follows approximately the function $I(\phi)$ in Fig. 4.26. With parallel incident light but slightly
tilted mirrors M_1 or M_2, the interference pattern consists of parallel fringes, which
move into a direction perpendicular to the fringes when Δs is changed.

The MI can be used for absolute wavelength measurements by counting the number N of maxima in B when the mirror M_2 is moved along a known distance Δy.
The wavelength λ is then obtained from

$$\lambda = 2n\Delta y/N \ .$$

This technique has been applied to very precise determinations of laser wavelengths
(Sect. 4.4).

The MI may be described in another equivalent way, which is quite instructive. Assume that the mirror M_2 in Fig. 4.25 moves with a constant velocity $v = \Delta y / \Delta t$. A wave with frequency ω and wave vector k incident perpendicularly on the moving mirror suffers a Doppler shift

$$\Delta\omega = \omega - \omega' = 2k \cdot v = (4\pi/\lambda)v , \tag{4.38}$$

on reflection.

Inserting the path difference $\Delta s = \Delta s_0 + 2vt$ and the corresponding phase difference $\phi = (2\pi/\lambda)\Delta s$ into (4.37) gives, with (4.38) and $\Delta s_0 = 0$,

$$\bar{I} = \tfrac{1}{2}I_0(1 + \cos\Delta\omega t) \quad \text{with } \Delta\omega = 2\omega v/c . \tag{4.39}$$

We recognize (4.39) as the time-averaged beat signal, obtained from the superposition of two waves with frequencies ω and $\omega' = \omega - \Delta\omega$, giving the averaged intensity of

$$\bar{I} = I_0(1 + \cos\Delta\omega t)\overline{\cos^2[(\omega' + \omega)t/2]}x = \tfrac{1}{2}I_0(1 + \cos\Delta\omega t) .$$

Note that the frequency $\omega = (c/v)\Delta\omega/2$ of the incoming wave can be measured from the beat frequency $\Delta\omega$, provided the velocity v of the moving mirror is known. The MI with uniformly moving mirror M_2 can be therefore regarded as a device that transforms the high frequency ω (10^{14}–10^{15} s^{-1}) of an optical wave into an easily accessible rf-range $(v/c)\omega$.

Example 4.8

$v = 3\,\text{cm/s} \rightarrow (v/c) = 10^{-10}$. The frequency $\omega = 3 \times 10^{15}\,\text{Hz}$ ($\lambda = 0.6\,\mu\text{m}$) is transformed to $\Delta\omega = 6 \times 10^5\,\text{Hz} \simeq \Delta\nu \sim 100\,\text{kHz}$.

The maximum path difference Δs that still gives interference fringes in the plane B is limited by the coherence length of the incident radiation (Sect. 2.9). Using spectral lamps, the coherence length is limited by the Doppler width of the spectral lines and is typically a few centimeters. With stabilized single-mode lasers, however, coherence lengths of several kilometers can be achieved. In this case, the maximum path difference in the MI is, in general, not restricted by the source but by technical limits imposed by laboratory facilities.

The attainable path difference Δs can be considerably increased by an *optical delay line*, placed in one arm of the interferometer (Fig. 4.28). It consists of a pair of mirrors, M_3, M_4, which reflect the light back and forth many times. In order to keep diffraction losses small, spherical mirrors, which compensate by collimation the divergence of the beam caused by diffraction, are preferable. With a stable mounting

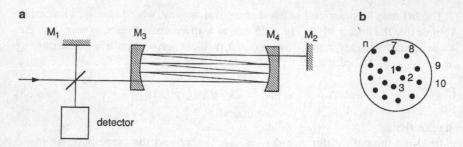

Figure 4.28 Michelson interferometer with optical delay line allowing a large path difference between the two interfering beams: **a** schematic arrangement; **b** spot positions of the reflected beams on mirror M3

of the whole interferometer, optical path differences up to 350 m could be realized [138], allowing a spectral resolution of $v/\Delta v \simeq 10^{11}$. This was demonstrated by measuring the linewidth of a HeNe laser oscillating at $v = 5\times10^{14}$ Hz as a function of discharge current. The accuracy obtained was better than 5 kHz.

For gravitational-wave detection [139], a MI with side arms of about 1-km length has been built where the optical path difference can be increased to $\Delta s > 100$ km by using highly reflective spherical mirrors and an ultrastable solid-state laser with a coherence length of $\Delta s_c \gg \Delta s$ (see Vol. 2, Sect. 9.8, Sect. 9.8.3) [140].

4.2.3   Fourier Spectroscopy

When the incoming wave consists of two components with frequencies ω_1 and ω_2, the interference pattern varies with time according to

$$\bar{I}(t) = \tfrac{1}{2}\bar{I}_{10}[1 + \cos 2\omega_1(v/c)t] + \tfrac{1}{2}I_{20}[1 + \cos 2\omega_2(v/c)t]$$
$$= \bar{I}_0\{1 + \cos[(\omega_1 - \omega_2)vt/c]\cos[(\omega_1 + \omega_2)vt/c]\} , \qquad (4.40)$$

where we have assumed $I_{10} = I_{20} = I_0$. This is a beat signal, where the amplitude of the interference signal at $(\omega_1 + \omega_2)(v/c)$ is modulated at the difference frequency $(\omega_1 - \omega_2)v/c$ (Fig. 4.29). From the sum

$$(\omega_1 + \omega_2) + (\omega_1 - \omega_2) = 2\omega_1$$

we obtain the frequency ω_1, and from the difference

$$(\omega_1 + \omega_2) - (\omega_1 - \omega_2) = 2\omega_2$$

the frequency ω_2.

The spectral resolution can roughly be estimated as follows: if Δy is the path difference traveled by the moving mirror in Fig. 4.25, the number of interference

Figure 4.29 Interference signal behind the MI with uniformly moving mirror M_2 when the incident wave consists of two components with frequencies ω_1 and ω_2 and equal amplitudes

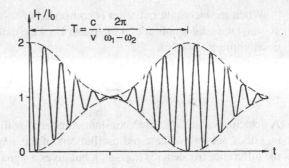

maxima that are counted by the detector is $N_1 = 2\Delta y / \lambda_1$ for an incident wave with the wavelength λ_1, and $N_2 = 2\Delta y / \lambda_2$ for $\lambda_2 < \lambda_1$. The two wavelengths can be clearly distinguished when $N_2 \geq N_1 + 1$. This yields with $\lambda_1 = \lambda_2 + \Delta\lambda$ and $\Delta\lambda \ll \lambda$ for the spectral resolving power

$$\frac{\lambda}{\Delta\lambda} = \frac{2\Delta y}{\lambda} = N = \frac{\Delta s}{\lambda} \quad \text{with } \lambda = (\lambda_1 + \lambda_2)/2 \text{ and } N = \tfrac{1}{2}(N_1 + N_2) .$$

$$(4.41a)$$

The equivalent consideration in the frequency domain follows. In order to determine the two frequencies ω_1 and ω_2, one has to measure at least over one modulation period

$$T = \frac{c}{v} \frac{2\pi}{\omega_1 - \omega_2} = \frac{c}{v} \frac{1}{\nu_1 - \nu_2} .$$

The frequency difference that can be resolved is then

$$\Delta\nu = \frac{c}{vT} = \frac{c}{\Delta s} = \frac{c}{N\lambda} \quad \Rightarrow \quad \frac{\Delta\nu}{c/\lambda} = \frac{1}{N} \text{ or } \frac{\nu}{\Delta\nu} = N = \frac{\Delta s}{\lambda} . \qquad (4.41b)$$

The spectral resolving power $\lambda/\Delta\lambda$ of the Michelson interferometer equals the maximum path difference $\Delta s/\lambda$ measured in units of the wavelength λ.

Example 4.9

a) $\Delta y = 5\,\text{cm}, \lambda = 10\,\mu\text{m} \rightarrow N = 10^4 \Rightarrow \nu/\Delta\nu = 10^4$,

b) $\Delta y = 100\,\text{cm}, \lambda = 0.5\,\mu\text{m} \rightarrow N = 4 \times 10^6 \Rightarrow \Delta\nu = 2.5 \times 10^{-7}\nu$
where the latter example can be realized only with lasers that have a sufficiently large coherence length (Sect. 4.4).

c) $\lambda_1 = 10\,\mu\text{m}, \lambda_2 = 9.8\,\mu\text{m} \rightarrow (\nu_2 - \nu_1) = 6 \times 10^{11}\,\text{Hz}$; with $v = 1\,\text{cm/s} \rightarrow T = 50\,\text{ms}$. The minimum measuring time for the resolution of the two spectral lines is 50 ms, and the minimum path difference $\Delta s = vT = 5 \times 10^{-2}\,\text{cm} = 500\,\mu\text{m}$.

When the incoming radiation is composed of several components with frequencies ω_k, the total amplitude in the plane B of the detector is the sum of all interference amplitudes (4.36),

$$E = \sum_k A_k e^{i(\omega_k t + \phi_{0k})} (1 + e^{i\phi_k}) \ . \tag{4.42a}$$

A detector with a large time constant compared with the maximum period $1/(\omega_i - \omega_k)$ does not follow the rapid oscillations of the amplitude at frequencies ω_k or at the difference frequencies $(\omega_i - \omega_k)$, but gives a signal proportional to the sum of the intensities I_k in (4.37). We therefore obtain for the time-dependent total intensity

$$\bar{I}(t) = \sum_k \tfrac{1}{2} \bar{I}_{k0} (1 + \cos \phi_k) = \sum_k \tfrac{1}{2} \bar{I}_{k0} (1 + \cos \Delta \omega_k t) \ , \tag{4.42b}$$

where the audio frequencies $\Delta \omega_k = 2 \omega_k v / c$ are determined by all differences between the frequencies ω_k of the components and by the velocity v of the moving mirror. Measurements of these frequencies $\Delta \omega_k$ allows one to reconstruct the spectral components of the incoming wave with frequencies ω_k (Fourier transform spectroscopy [141, 142]).

Since the path-difference $\Delta s(t) = v \cdot t$ is a continuous function of the time t, the sum in (4.42b) can be replaced by the integral

$$I(t) = \int I(\omega) \cos(\omega \, \Delta s / c) \, d\omega \ . \tag{4.42c}$$

The Fourier-transform of the measured intensity $I(t)$ in (4.42c) gives the wanted spectrum

$$I(\omega) = \int I(t) \cdot \cos(\omega \cdot vt / c) \, dt \ . \tag{4.42d}$$

The main advantage of Fourier spectroscopy is the fact, that all spectral intervals $d\omega$ with the intensity $I(\omega) \, d\omega$ are measured simultaneously in contrast to classical spectroscopy with a monochromator where the different spectral intervals are measured subsequently. If a spectrum consisting of N spectral intervals $\Delta \omega$ (where $\Delta \omega$ is the spectral interval which can be resolved by the wavelength-selecting instrument) is measured in a time T with a monochromator, Fourier spectroscopy can obtain this spectrum in the shorter time $T / \sqrt{N}$ with the same signal-to-noise ratio.

4.2.4   Mach–Zehnder Interferometer

Analogous to the Michelson interferometer, the Mach–Zehnder interferometer is based on the two-beam interference by amplitude splitting of the incoming wave.

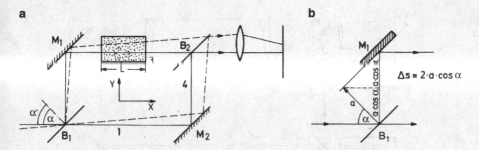

Figure 4.30 Mach–Zehnder interferometer: **a** schematic arrangement, **b** path difference between the two parallel beams

The two waves travel along different paths with a path difference $\Delta s = 2a \cos \alpha$ (Fig. 4.30b). Inserting a transparent object into one arm of the interferometer alters the optical path difference between the two beams. This results in a change of the interference pattern, which allows a very accurate determination of the refractive index of the sample and its local variation. The Mach–Zehnder interferometer may be regarded therefore as a sensitive refractometer.

If the beam splitters B_1, B_2 and the mirrors M_1, M_2 are all strictly parallel, the path difference between the two split beams does not depend on the angle of incidence α because the path difference between the beams 1 and 3 is exactly compensated by the same path length of beam 4 between M_2 and B_2 (Fig. 4.30a). This means that the interfering waves in the symmetric interferometer (without sample) experience the same path difference on the solid path as on the dashed path in Fig. 4.30a. Without the sample, the total path difference is therefore zero; it is $\Delta s = (n - 1)L$ *with* the sample having the refractive index n in one arm of the interferometer.

Expanding the beam on path 3 gives an extended interference-fringe pattern, which reflects the local variation of the refractive index. Using a laser as a light source with a large coherence length, the path lengths in the two interferometer arms can be made different without losing the contrast of the interference pattern (Fig. 4.31). With a beam expander (lenses L_1 and L_2), the laser beam can be expanded up to 10–20 cm and large objects can be tested. The interference pattern can either be photographed or may be viewed directly with the naked eye or with a television camera [143]. Such a laser interferometer has the advantage that the laser beam diameter can be kept small everywhere in the interferometer, except between the two expanding lenses. Since the illuminated part of the mirror surfaces should not deviate from an ideal plane by more than $\lambda/10$ in order to obtain good interferograms, smaller beam diameters are advantageous.

The Mach–Zehnder interferometer has found a wide range of applications. Density variations in laminar or turbulent gas flows can be seen with this technique and the optical quality of mirror substrates or interferometer plates can be tested with high sensitivity [143, 144].

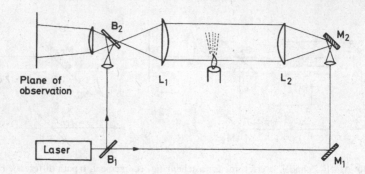

Figure 4.31 Laser interferometer for sensitive measurements of local variations of the index of refraction in extended samples, for example, in air above a candle flame

In order to get quantitative information of the local variation of the optical path through the sample, it is useful to generate a fringe pattern for calibration purposes by slightly tilting the plates B_1, M_1 and B_2, M_2 in Fig. 4.31, which makes the interferometer slightly asymmetric. Assume that B_1 and M_1 are tilted clockwise around the z-direction by a small angle β and the pair B_2, M_2 is tilted counterclockwise by the same angle β. The optical path between B_1 and M_1 is then $\Delta_1 = 2a\cos(\alpha + \beta)$, whereas $B_2 M_2 = \Delta_2 = 2a\cos(\alpha - \beta)$. After being recombined, the two beams therefore have the path difference

$$\Delta = \Delta_2 - \Delta_1 = 2a[\cos(\alpha - \beta) - \cos(\alpha + \beta)] = 4a\sin\alpha\sin\beta , \qquad (4.43)$$

which depends on the angle of incidence α. In the plane of observation, an interference pattern of parallel fringes with path differences $\Delta = m \cdot \lambda$ is observed with an angular separation $\Delta\epsilon$ between the fringes m and $m + 1$ given by $\Delta\epsilon = \alpha_m - \alpha_{m+1} = \lambda/(4a\sin\beta\cos\alpha)$.

A sample in path 3 introduces an additional path difference

$$\Delta s(\beta) = (n - 1)L/\cos\beta$$

depending on the local refractive index n and the path length L through the sample. The resulting phase difference shifts the interference pattern by an angle $\gamma = (n - 1)(L/\lambda)\Delta\varepsilon$. Using a lens with a focal length f, which images the interference pattern onto the plane O, gives the spatial distance $\Delta y = f\Delta\varepsilon$ between neighboring fringes. The additional path difference caused by the sample shifts the interference pattern by $N = (n - 1)(L/\lambda)$ fringes.

Figure 4.32 shows for illustration the interferogram of the convection zone of hot air above a candle flame, placed below one arm of the laser interferometer in Fig. 4.31. It can be seen that the optical path through this zone changes by many wavelengths.

The Mach–Zehnder interferometer has been used in spectroscopy to measure the refractive index of atomic vapors in the vicinity of spectral lines (Sect. 3.1). The

Figure 4.32 Interferogram
of the density profile in
the convection zone above
a candle flame (H. Rotten-
Kᴵᴵᴵᴵᴵᴵᴵᴵ ᴵ ᴵ ᴵ4ᴵᴵ

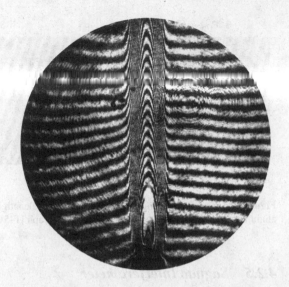

experimental arrangement (Fig. 4.33) consists of a combination of a spectrograph
and an interferometer, where the plates B_1, M_1 and B_2, M_2 are tilted in such a direction that without the sample the parallel interference fringes with the separation
$\Delta y(\lambda) = f\Delta\varepsilon$ are perpendicular to the entrance slit, which is parallel to the y-
direction. The spectrograph disperses the fringes with different wavelengths λ_i in
the z-direction. Because of the wavelength-dependent refractive index $n(\lambda)$ of the
atomic vapor (Sect. 3.1.3), the fringe shift follows a dispersion curve in the vicinity
of the spectral line (Fig. 4.34). The dispersed fringes look like hooks around an
absorption line, which gave this technique the name *hook method*. To compensate
for background shifts caused by the windows of the absorption cell, a compensating plate is inserted into the second arm. This technique was developed in 1912
by Rozhdestvenski [146] in St. Petersburg. For more details of the Hook method,
see [144–146].

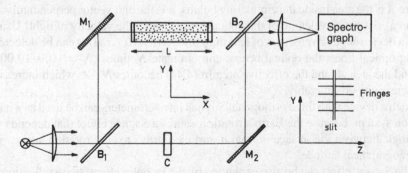

Figure 4.33 Combination of Mach–Zehnder interferometer and spectrograph used for the hook
method

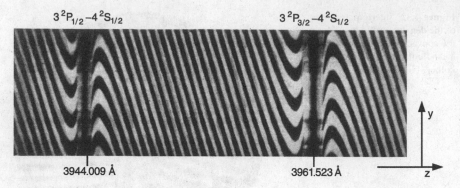

Figure 4.34 Position of fringes as a function of wavelength around the absorption line doublet of aluminium atoms, as observed behind the spectrograph [145]

4.2.5   *Sagnac Interferometer*

In the Sagnac interferometer (Fig. 4.35), the beam splitter BS splits the incoming beam into a transmitted beam and a reflected beam. The two beams circulate in opposite directions in the x, y-plane through the ring interferometer. If the whole interferometer rotates clockwise around an axis in the z-direction through the center of the $x-y$ area around which the beams circulate, the optical path for the clockwise-circulating beam becomes longer than that for the counterclockwise running beam (the Sagnac effect). This causes a phase difference between the two beams and the intensity of the interfering beams as measured in the observation plane changes with the angular speed of rotation Ω. The phase shift between the two partial waves is

$$\Delta\phi = 8\pi A \cdot \boldsymbol{n} \cdot \boldsymbol{\Omega}/(\lambda \cdot c) \tag{4.44}$$

where A is the area inside the circulating beams, $\boldsymbol{n}$ is the unit vector perpendicular to the area A, λ the wavelength of the optical waves, and c the velocity of light. Using such a device angular velocities of less than $0.1°/h$ $(5 \times 10^{-7}$ rad/s) can be detected. Using optical fibers the optical beams can circulate N times $(N = 100–10{,}000)$ around the area A, and the effective area in (4.44) becomes $N \cdot A$, which increases the sensitivity considerably.

Such a device with three orthogonal Sagnac interferometers can be used as a navigation system, because the Earth's rotation causes a Sagnac effect that depends on the angle between the surface normal $\boldsymbol{n}$ and the Earth's axis of rotation $\boldsymbol{\omega}$; i.e., on the geographical latitude.

The Sagnac effect can be also explained by the Doppler effect: upon reflection at a mirror moving at a velocity v, the frequency ν of the reflected beam is shifted by $\Delta\nu = 2v \cdot v/c$. The frequencies of the two waves circulating in opposite directions

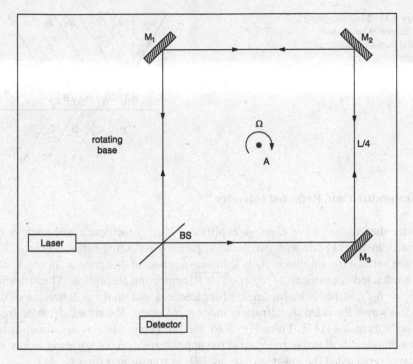

Figure 4.35 Sagnac interferometer

are therefore shifted away from each other by

$$\Delta \nu = 4A/(L \cdot \lambda)n \cdot \boldsymbol{\Omega} \qquad (4.45)$$

where L is the path length for one round trip in the ring interferometer. Since $\Delta \phi = (2\pi L/c)\Delta \nu$, both equations are equivalent, although the detection technique is different. The determination of the phase shift is based on measuring the intensity change at the detector, while the beat frequency $\Delta \nu$ can be directly counted with high precision [147].

4.2.6 Multiple-Beam Interference

In a grating spectrometer, the interfering partial waves emitted from the different grooves of the grating all have the same amplitude. In contrast, in multiple-beam interferometers these partial waves are produced by multiple reflection at plane or curved surfaces and their amplitude decreases with increasing number of reflections. The resultant total intensity therefore differs from (4.25).

Figure 4.36 Multiple-beam interference at two plane-parallel partially reflecting surfaces

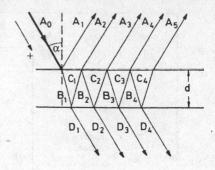

a) Transmitted and Reflected Intensity

Assume that a plane wave $E = A_0 \exp[i(\omega t - kx)]$ is incident at the angle α on a plane transparent plate with two parallel, partially reflecting surfaces (Fig. 4.36). At each surface the amplitude A_i is split into a reflected component $A_R = A_i \sqrt{R}$ and a refracted component $A_T = A_i \sqrt{1 - R}$, neglecting absorption. The reflectivity $R = I_R / I_i$ depends on the angle of incidence α and on the polarization of the incident wave. Provided the refractive index n is known, R can be calculated from Fresnel's formulas [116]. From Fig. 4.36, the following relations are obtained for the amplitudes A_i of waves reflected at the upper surface, B_i of refracted waves, C_i of waves reflected at the lower surface, and D_i of transmitted waves

$$
\begin{aligned}
&|A_1| = \sqrt{R}\,|A_0|\,, & &|B_1| = \sqrt{1 - R}\,|A_0|\,, \\
&|C_1| = \sqrt{R(1 - R)}\,|A_0|\,, & &|D_1| = (1 - R)\,|A_0|\,, \\
&|A_2| = \sqrt{1 - R}\,|C_1| = (1 - R)\sqrt{R}\,|A_0|\,, & &|B_2| = R\sqrt{1 - R}\,|A_0|\,, \\
&|C_2| = R\sqrt{R(1 - R)}\,|A_0|\,, & &|D_2| = R(1 - R)\,|A_0|\,, \\
&|A_3| = \sqrt{1 - R}\,|C_2| = R^{3/2}(1 - R)\,|A_0|\,, & & \dots\,.
\end{aligned}
\tag{4.46}
$$

This scheme can be generalized to the equations

$$
|A_{i+1}| = R|A_i|\,, \quad i \geq 2\,,
\tag{4.47a}
$$

$$
|D_{i+1}| = R|D_i|\,, \quad i \geq 1\,.
\tag{4.47b}
$$

Two successively reflected partial waves E_i and E_{i+1} have the optical path difference (Fig. 4.37)

$$
\Delta s = (2nd / \cos\beta) - 2d\,\tan\beta\,\sin\alpha\,.
$$

Because $\sin\alpha = n\sin\beta$, this can be reduced to

$$
\Delta s = 2nd\cos\beta = 2nd\sqrt{1 - \sin^2\beta} = 2d\sqrt{n^2 - \sin^2\alpha}\,,
\tag{4.48a}
$$

Figure 4.37 Optical path difference between two beams being reflected from the two surfaces of a plane-parallel plate

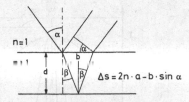

if the refractive index within the plane-parallel plate is $n > 1$ and outside the plate $n = 1$. This path difference causes a corresponding phase difference

$$\phi = 2\pi \Delta s/\lambda + \Delta\phi \, , \tag{4.48b}$$

where $\Delta\phi$ takes into account possible phase changes caused by the reflections. For instance, the incident wave with amplitude A_1 suffers the phase jump $\Delta\phi = \pi$ while being reflected at the medium with $n > 1$. Including this phase jump, we can write

$$A_1 = \sqrt{R}A_0 \exp(\mathrm{i}\pi) = -\sqrt{R}A_0 \, .$$

The total amplitude A of the reflected wave is obtained by summation over all partial amplitudes A_i, taking into account the different phase shifts,

$$A = \sum_{m=1}^{p} A_m \mathrm{e}^{\mathrm{i}(m-1)\phi} = -\sqrt{R}A_0 + \sqrt{R}A_0(1-R)\mathrm{e}^{\mathrm{i}\phi} + \sum_{m=3}^{p} A_m \mathrm{e}^{\mathrm{i}(m-1)\phi}$$

$$= -\sqrt{R}A_0 \left[1 - (1-R)\mathrm{e}^{\mathrm{i}\phi} \sum_{m=0}^{p-2} R^m \mathrm{e}^{\mathrm{i}m\phi} \right] \, . \tag{4.49}$$

For vertical incidence ($\alpha = 0$), or for an infinitely extended plate, we have an infinite number of reflections. The geometrical series in (4.49) has the limit $(1 - R\mathrm{e}^{\mathrm{i}\phi})^{-1}$ for $p \to \infty$. We obtain for the total amplitude

$$A = -\sqrt{R}A_0 \frac{1 - \mathrm{e}^{\mathrm{i}\phi}}{1 - R\mathrm{e}^{\mathrm{i}\phi}} \, . \tag{4.50}$$

The intensity $I = 2c\epsilon_0 A A^*$ of the reflected wave is then, with $I_0 = 2c\varepsilon_0 A_0 A_0^*$,

$$\boxed{I_R = I_0 R \frac{4\sin^2(\phi/2)}{(1-R)^2 + 4R\sin^2(\phi/2)} \, .} \tag{4.51a}$$

In an analogous way, we find for the total transmitted amplitude

$$D = \sum_{m=1}^{\infty} D_m \mathrm{e}^{\mathrm{i}(m-1)\phi} = (1-R)A_0 \sum_{0}^{\infty} R^m \mathrm{e}^{\mathrm{i}m\phi} \, ,$$

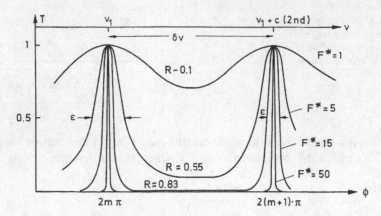

Figure 4.38 Transmittance of an absorption-free multiple-beam interferometer as a function of the phase difference ϕ for different values of the finesse F^*

which gives the total transmitted intensity

$$I_T = I_0 \frac{(1 - R)^2}{(1 - R)^2 + 4R \sin^2(\phi/2)} \cdot \tag{4.52a}$$

Equations (4.51a, 4.52a) are called the *Airy formulas*. Since we have neglected absorption, we should have $I_R + I_T = I_0$, as can easily be verified from (4.51a, 4.52a).

The abbreviation $F = 4R/(1 - R)^2$ is often used, which allows the Airy equations to be written in the form

$$I_R = I_0 \frac{F \sin^2(\phi/2)}{1 + F \sin^2(\phi/2)} \,, \tag{4.51b}$$

$$I_T = I_0 \frac{1}{1 + F \sin^2(\phi/2)} \cdot \tag{4.52b}$$

Figure 4.38 illustrates (4.52b) for different values of the reflectivity R. The maximum transmittance is $T = 1$ for $\phi = 2m\pi$. At these maxima $I_T = I_0$, therefore the reflected intensity I_R is zero. The minimum transmittance is

$$T^{\min} = \frac{1}{1 + F} = \left(\frac{1 - R}{1 + R} \right)^2 \cdot$$

Example: For $R = 0.98 \Rightarrow T^{\min} = 10^{-4}$.
$\qquad$ For $R = 0.90 \Rightarrow T^{\min} = 2.8 \times 10^{-3}$.

The ratio

$$C = I_{max}/I_{min} = 1 + \Gamma = \left(\frac{1+R}{1-\nu}\right)^2$$

is named the *contrast* of the interferometer.

b) Free Spectral Range and Finesse

The frequency range $\delta\nu$ between two maxima is the *free spectral range of the interferometer*. Interference maxima occur for $\Delta s = m \cdot \lambda$ ($m = 1, 2, 3, \ldots$) which corresponds to a phase difference $\Delta\Phi = m \cdot 2\pi$. The free spectral range $\delta\lambda$ is obtained from

$$\Phi_1 - \Phi_2 = 2\pi\Delta s/\lambda_1 - 2\pi\Delta s/\lambda_2 = 2(m+1)\pi - 2m\pi = 2\pi$$
$$\rightarrow \Delta s \cdot (\lambda_2 - \lambda_1) = \lambda_1 \cdot \lambda_2 \approx \lambda^2 \Rightarrow \delta\lambda = \lambda^2/\Delta s .$$

Wegen $\nu = c/\lambda$ folgt

$$\delta\nu = -\left(c/\lambda^2\right)\delta\lambda = c/\Delta s = c/\left(2d\sqrt{n^2 - \sin^2\alpha}\right) . \tag{4.53a}$$

For vertical incidence ($\alpha = 0$), the free spectral range becomes

$$\boxed{|\delta\nu|_{\alpha=0} = \frac{c}{2nd} .} \tag{4.53b}$$

The full halfwidth $\epsilon = |\phi_1 - \phi_2|$ with $I(\phi_1) = I(\phi_2) = I_0/2$ of the transmission maxima in Fig. 4.38 expressed in phase differences is calculated from (4.52a), (4.52b) as

$$\epsilon = 4\arcsin\left(\frac{1-R}{2\sqrt{R}}\right) , \tag{4.54a}$$

which reduces for $R \approx 1 \Rightarrow (1-R) \ll R$ to

$$\epsilon = \frac{2(1-R)}{\sqrt{R}} = \frac{4}{\sqrt{F}} . \tag{4.54b}$$

In frequency units, the free spectral range $\delta\nu$ corresponds to a phase difference $\delta\phi = 2\pi$. Therefore the halfwidth $\Delta\nu$ becomes

$$\Delta\nu = \frac{\epsilon}{2\pi}\delta\nu \simeq \frac{2\delta\nu}{\pi\sqrt{F}}$$

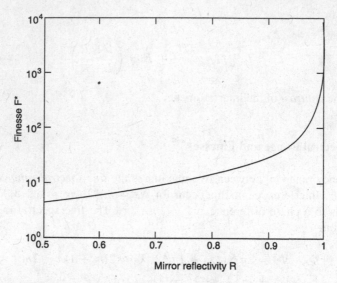

Figure 4.39 Finesse F_R^* of a Fabry–Perot interferometer as a function of the mirror reflectivity R

which yields for vertical incidence with (4.53b)

$$\Delta v = \frac{c}{2nd} \frac{1-R}{\pi\sqrt{R}} . \tag{4.54c}$$

The ratio $\delta v/\Delta v$ of free spectral range δv to the halfwidth Δv of the transmission maxima is called the *finesse* F^* of the interferometer. From (4.53b) and (4.54c) we obtain for the "reflectivity finesse" F_R^*

$$\boxed{F_R^* = \frac{\delta v}{\Delta v} = \frac{\pi\sqrt{R}}{1-R} = \frac{\pi}{2}\sqrt{F} .} \tag{4.55a}$$

The full halfwidth of the transmission peaks is then

$$\Delta v = \frac{\delta v}{F_R^*} . \tag{4.55b}$$

The finesse is a measure for the effective number of interfering partial waves in the interferometer. This means that for vertical incidence the maximum path difference between interfering waves is $\Delta s_{max} = F^* 2nd$. Figure 4.39 shows the finesse F_R^* as a function of the mirror reflectivity.

Since we have assumed an ideal plane-parallel plate with a perfect surface quality, the finesse (4.55a) is determined only by the reflectivity R of the surfaces. In practice, however, deviations of the surfaces from an ideal plane and slight inclinations of the two surfaces cause imperfect superposition of the interfering waves.

This results in a decrease and a broadening of the transmission maxima, which decreases the total finesse. If, for instance, a reflecting surface deviates by the amount
λ/q from an ideal plane the finesse cannot be larger than q. One can define the
total finesse F^* of an interferometer by

$$\frac{1}{F^{*2}} = \sum_i \frac{1}{F_i^{*2}} , \tag{4.55c}$$

where the different terms F_i^* give the contributions to the decrease of the finesse
caused by the different imperfections of the interferometer.

If, for instance, the surface of the mirror shows a parabolic deviation from a plane
surface, i.e.,

$$S(r, \varphi) = S_0 + \alpha r^2$$

the finesse becomes (with $k = 2\pi/\lambda$ [148])

$$F^* = \frac{\pi}{[(1 - R)^2/R + k^2\alpha^2]^{1/2}} \tag{4.56}$$

which yields

$$\frac{1}{F^{*2}} = \frac{(1 - R)^2}{\pi^2 R} + \frac{4\alpha^2}{\lambda^2} = \frac{1}{F_R^{*2}} + \frac{1}{F_f^{*2}} \tag{4.57}$$

where F_p is the finesse determined by the curvature of the mirror surface.

Example 4.10
A plane, nearly parallel plate has a diameter $D = 5\,\text{cm}$, a thickness $d = 1\,\text{cm}$,
and a wedge angle of $0.2''$. The two reflecting surfaces have the reflectivity
$R = 95\%$. The surfaces are flat to within $\lambda/50$, which means that no point
of the surface deviates from an ideal plane by more than $\lambda/50$. The different
contributions to the finesse are:

- Reflectivity finesse: $F_R^* = \pi\sqrt{R}/(1 - R) \simeq 60$;
- Surface finesse: $F_S \simeq 50$;
- Wedge finesse: with a wedge angle of $0.2''$ the optical path between the
  two reflecting surfaces changes by about $0.1\lambda (\lambda = 0.5\,\mu\text{m})$ across the
  diameter of the plate. For a monochromatic incident wave this causes imperfect interference and broadens the maxima corresponding to a finesse
  of about 20.

The total finesse is then $F^{*2} = 1/(1/60^2 + 1/50^2 + 1/20^2) \rightarrow F^* \simeq 17.7$.

This illustrates that high-quality optical surfaces are necessary to obtain
a high total finesse [148]. It makes no sense to increase the reflectivity without

Figure 4.40 Transmitted intensity $I_T(v)$ for two closely spaced spectral lines at the limit of spectral resolution where the linespacing equals the halfwidth of the lines

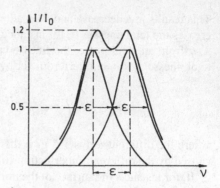

a corresponding increase of the surface finesse. In our example the imperfect parallelism was the main cause for the low finesse. Decreasing the wedge angle to $0.1''$ increases the wedge finesse to 40 and the total finesse to 27.7.

A much larger finesse can be achieved using spherical mirrors, because the demand for parallelism is dropped. With sufficiently accurate alignment and high reflectivities, values of $F^* > 50,000$ are possible (Sect. 4.2.10).

c) Spectral Resolution

The spectral resolution, $v/\Delta v$ or $\lambda/\Delta\lambda$, of an interferometer is determined by the free spectral range δv and by the finesse F^*. Two incident waves with frequencies v_1 and $v_2 = v_1 + \Delta v$ can still be resolved if their frequency separation Δv is larger than $\delta v/F^*$, which means that their peak separation should be larger than their full halfwidth.

Quantitatively this can be seen as follows: assume the incident radiation consists of two components with the intensity profiles $I_1(v - v_1)$ and $I_2(v - v_2)$ and equal peak intensities $I_1(v_1) = I_2(v_2) = I_0$. For a peak separation $v_2 - v_1 = \delta v/F^* = 2\delta v/\pi\sqrt{F}$, the total transmitted intensity $I(v) = I_1(v) + I_2(v)$ is obtained from (4.52a) as

$$I(v) = I_0 \left(\frac{1}{1 + F\sin^2(\pi v/\delta v)} + \frac{1}{1 + F\sin^2[\pi(v + \delta v/F^*)/\delta v]} \right) , \quad (4.58)$$

where the phase shift $\phi = 2\pi\Delta s/\lambda = 2\pi\Delta s(v/c) = 2\pi v/\delta v$ in (4.52b) has been expressed by the free spectral range $\delta v = c/2nd = c/\Delta s$, where Δs is the optical path difference between two successive partial waves in Fig. 4.36 for $\alpha = 0$. The function $I(v)$ is plotted in Fig. 4.40 around the frequency $v = (v_1 + v_2)/2$. For $v = v_1 = mc/2nd$, the first term in (4.58) becomes 1 and the second term can be derived with $\sin[\pi(v_1 + \delta v/F^*)/\delta v] = \sin \pi/F^* \simeq \pi/F^*$ and $F(\pi/F^*)^2 = 4$

to become 0.2. Inserting this into (4.58) yields $I(v = v_1) = 1.2I_0$, $I(v = (v_1 + v_0)/2) \simeq I_0$, and $I(v = v_2) = 1.2I_0$. This just corresponds to the Rayleigh criterion for the resolution of two spectral lines. The spectral resolving power of the interferometer is therefore

$$v/\Delta v = (v/\delta v)F^* \rightarrow \Delta v = \delta v/F^* . \tag{4.59}$$

This can be also expressed by the optical path differences Δs between two successive partial waves

$$\frac{v}{\Delta v} = \frac{\lambda}{\Delta \lambda} = F^* \frac{\Delta s}{\lambda} . \tag{4.60}$$

The resolving power of an interferometer is the product of finesse F^* and optical path difference $\Delta s/\lambda$ in units of the wavelength λ.

A comparison with the resolving power $v/\Delta v = mN = N\Delta s/\lambda$ of a grating spectrometer with N grooves shows that the finesse F^* can indeed be regarded as the effective number of interfering partial waves and $F^*\Delta s$ can be regarded as the maximum path difference between these waves.

The spectral resolution $v/\Delta v = \lambda/\Delta \lambda$ equals the maximum path-difference $\Delta s_{max}/\lambda$ in units of the wavelength λ.

Example 4.11
$d = 1\,cm$, $n = 1.5$, $R = 0.98$, $\lambda = 500\,nm$. An interferometer with negligible wedge and high-quality surfaces, where the finesse is mainly determined by the reflectivity, achieves with $F^* = \pi\sqrt{R}/(1 - R) = 155$ a resolving power of $\lambda/\Delta \lambda = 10^7$. This means that the instrument's linewidth is about $\Delta \lambda \sim 5 \times 10^{-5}$ nm or, in frequency units, $\Delta v = 60\,MHz$.

d) Influence of Absorption Losses

Taking into account the absorption $A = (1 - R - T)$ of each reflective surface, the transmitted intensity (4.52a), (4.52b) must be modified to

$$I_T = I_0 \frac{T^2}{(A + T)^2} \frac{1}{[1 + F\sin^2(\delta/2)]} , \tag{4.61a}$$

where $T^2 = T_1 T_2$ is the product of the transmittance of the two reflecting surfaces. The absorption causes three effects:

a) The maximum transmittance is decreased by the factor

$$\frac{I_T}{I_0} = \frac{T^2}{(A+T)^2} = \frac{T^2}{(1-R)^2} < 1 \ . \tag{4.61b}$$

Note that even a small absorption of each reflecting surface results in a drastic reduction of the total transmittance. For $A = 0.05$, $R = 0.9 \to T = 0.05 \Rightarrow$ $T^2/(1-R)^2 = 0.25$. This illustrates, that even for a small absorption of $A = 5\,\%$ the transmitted intensity drops to $25\,\%$ of the absorption-free transmittance.

b) For a given transmission factor T, the reflectivity $R = 1 - A - T$ decreases with increasing absorption. The quantity

$$F = \frac{4R}{(1-R)^2} = \frac{4(1-T-A)}{(T+A)^2} \tag{4.61c}$$

decreases with increasing A. For the example above we obtain $F = 360$. This makes the transmission peaks broader because of the decreasing number of interfering partial waves. The *contrast*

$$C = \frac{I_T^{max}}{I_T^{min}} = 1 + F = \left(\frac{1+R}{1-R}\right)^2 \tag{4.61d}$$

of the transmitted intensity also decreases. If absorption is taken into account, we can insert $R = 1 - A - T$ and obtain for the contrast

$$C = I_{T_{max}}/I_{T_{min}} = (2 - A - T)^2/(A + T)^2 \ .$$

For $A > 0$ but constant transmission T the reflectivity R and therefore the contrast C decreases.

Example 4.12
$R = 0.95$, $A = 0.03$, $T = 0.02 \Rightarrow C = 1521$.
Without absorption is $A = 0$, $T = 0.02$ and $R = 0.98 \Rightarrow C = 9801$.

c) The absorption causes a phase shift $\Delta\phi$ at each reflection, which depends on the wavelength λ, the polarization, and the angle of incidence α [116]. This effect causes a wavelength-dependent *shift* of the maxima.

4.2.7   Plane Fabry–Perot Interferometer

A practical realization of the multiple beam-interference discussed in this section may use either a solid plane-parallel glass or fused quartz plate with two coated reflecting surfaces (Fabry–Perot etalon, Fig. 4.41a) or two separate plates, where one

Figure 4.41 Two realizations of a Fabry–Perot Interferometer: **a** solid etalon; **b** air-spaced plane-parallel reflecting surfaces

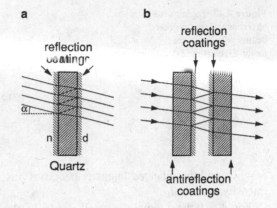

surface of each plate is coated with a reflection layer. The two reflecting surfaces are opposed and are aligned to be as parallel as achievable (Fabry–Perot interferometer (FPI), Fig. 4.41b). The outer surfaces are coated with antireflection layers in order to avoid reflections from these surfaces that might overlap the interference pattern. Furthermore, they have a slight angle against the inner surfaces (wedge).

Both devices can be used for parallel as well as for divergent incident light. We now discuss them in more detail, first considering their illumination with *parallel* light.

a) The Plane FPI as a Transmission Filter

In laser spectroscopy, etalons are mainly used as wavelength-selective transmission filters within the laser resonator to narrow the laser bandwidth (Sect. 5.4). The wavelength λ_m or frequency ν_m for the transmission maximum of mth order, where the optical path between successive beams is $\Delta s = m\lambda$, can be deduced from (4.48a) and Fig. 4.37 to be

$$\lambda_m = \frac{2d}{m}\sqrt{n^2 - \sin^2\alpha} = \frac{2nd}{m}\cos\beta \ , \tag{4.62a}$$

$$\nu_m = \frac{mc}{2nd\cos\beta} \ . \tag{4.62b}$$

For all wavelengths $\lambda = \lambda_m$ ($m = 1, 2, \ldots$) in the incident light, the phase difference between the transmitted partial waves becomes $\delta = 2m\pi$ and the transmitted intensity is, according to (4.61a)–(4.61d),

$$I_T = \frac{T^2}{(1-R)^2}I_0 = \frac{T^2}{(A+T)^2}I_0 \ , \tag{4.63}$$

where $A = 1 - T - R$ is the absorption of the etalon (substrate absorption plus absorption of one reflecting surface). The reflected waves interfere destructively for

Figure 4.42 Incomplete interference of two reflected beams with finite diameter D, causing a decrease of the maximum transmitted intensity

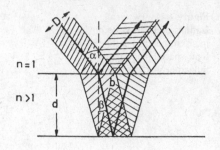

$\lambda = \lambda_m$ and the *reflected* intensity becomes zero for $A = 0$ while the *transmitted* intensity is $I_T = I_0$.

For $A > 0$ there remains a small residual reflected intensity and $I_T/I_0 < 1$.

Note, however, that this is only true for $A \ll 1$ and infinitely extended plane waves, where the different reflected partial waves completely overlap. If the incident wave is a laser beam with the finite diameter D, the different reflected partial beams do *not* completely overlap because they are laterally shifted by $\Delta = b \cos \alpha$ with $b = 2d \tan \beta$ (Fig. 4.42). For a rectangular intensity profile of the laser beam, the fraction Δ/D of the reflected partial amplitudes does not overlap and cannot interfere destructively. This means that, even for maximum transmission, the reflected intensity is not zero but a background reflection remains, which is missing in the transmitted light. For small angles α, one obtains for the intensity loss per transit due to reflection [149] for a rectangular beam profile

$$\frac{I_R}{I_0} = \frac{4R}{(1-R)^2}\left(\frac{2\alpha d}{nD}\right)^2 . \tag{4.64a}$$

For a Gaussian beam profile the calculation is more difficult, and the solution can only be obtained numerically. The result for a Gaussian beam with the radius w (Sect. 5.3) is [150]

$$\frac{I_R}{I_0} \simeq \frac{8R}{(1-R)^2}\left(\frac{2d\alpha}{nw}\right)^2 . \tag{4.64b}$$

A parallel light beam with the diameter D passing a plane-parallel plate with the angle of incidence α therefore suffers reflection losses in addition to the eventual absorption losses. The reflection losses increase with α^2 and are proportional to the ratio $(d/D)^2$ of the etalon thickness d and the beam diameter D (walk-off losses).

Example 4.13

$d = 1\,\text{cm}$, $D = 0.2\,\text{cm}$, $n = 1.5$, $R = 0.3$, $\alpha = 1° \cong 0.017\,\text{rad} \rightarrow I_R/I_0 = 0.05$, which means 5 % walk-off losses.

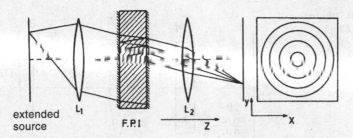

Figure 4.43 The interference ring system of the transmitted intensity may be regarded as wavelength-selective imaging of corresponding ring areas of an extended light source

The transmission peak λ_m of the etalon can be shifted by tilting the etalon. According to (4.62a), (4.62b) the wavelength λ_m *decreases* with increasing angle of incidence α. The walk-off losses, however, limit the tuning range of tilted etalons within a laser resonator. With increasing angle α, the losses may become intolerably large.

b) Illumination with Divergent Light

Illuminating the FPI with divergent monochromatic light (e.g., from an extended source or from a laser beam behind a diverging lens), a continuous range of incident angles α is offered to the FPI, which transmits, for a wavelength λ_m, those directions α_m that obey (4.62a). We then observe an interference pattern of bright rings in the transmitted light (Fig. 4.43). Since the reflected intensity $I_R = I_0 - I_T$ is complementary to the transmitted one, a corresponding system of dark rings appears in the reflected light at the same angles of incidence α_m.

When β is the angle of inclination to the interferometer axis inside the FPI, the transmitted intensity is maximum, according to (4.62a), (4.62b), for

$$m\lambda = 2nd \cos \beta \, , \tag{4.65}$$

where n is the refractive index between the reflecting planes. Let us number the rings by the integer p, beginning with $p = 0$ for the central ring. With $m = m_0 - p$, we can rewrite (4.65) for small angles β_p as

$$(m_0 - p)\lambda = 2nd \cos \beta_p \sim 2nd(1 - \beta_p^2/2) = 2nd \left[1 - \frac{1}{2} \left(\frac{n_0 \alpha_p}{n} \right)^2 \right] , \tag{4.66}$$

where n_0 is the refractive index of air, and Snell's law $\sin \alpha \simeq \alpha = (n/n_0)\beta$ has been used (Fig. 4.44).

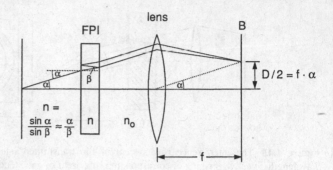

Figure 4.44 Illustration of (4.67a), (4.67b)

When the interference pattern is imaged by a lens with the focal length f into the plane of the photoplate, we obtain for the ring diameters $D_p = 2f\alpha_p$ the relations

$$(m_0 - p)\lambda = 2nd \left[1 - (n_0/n)^2 D_p^2/(8f^2)\right] , \qquad (4.67a)$$

$$(m_0 - p - 1)\lambda = 2nd \left[1 - (n_0/n)^2 D_{p+1}^2/(8f^2)\right] . \qquad (4.67b)$$

Subtracting the second equation from the first one yields

$$D_{p+1}^2 - D_p^2 = \frac{4nf^2}{n_0^2 d}\lambda . \qquad (4.68)$$

For the smallest ring with $p = 0$, (4.66) becomes

$$m_0\lambda = 2nd\left(1 - \beta_0^2/2\right) \Rightarrow m_0\lambda + nd\beta_0^2 = 2nd , \qquad (4.69)$$

which can be written as

$$(m_0 + \epsilon)\lambda = 2nd . \qquad (4.70)$$

The "excess" $\epsilon < 1$, also called *fractional interference order*, can be obtained from a comparison of (4.69) and (4.70) as

$$\epsilon = nd\beta_0^2/\lambda = (n_0/n)d\alpha_0^2/\lambda . \qquad (4.71)$$

Inserting ϵ from (4.70) into (4.67a) yields the relation

$$D_p^2 = \frac{8n^2 f^2}{n_0^2(m_0 + \epsilon)}(p + \epsilon) . \qquad (4.72)$$

A linear fit of the squares D_p^2 of the measured ring diameters versus the ring number p yields the excess ϵ and therefore from (4.70) the wavelength λ, provided the refractive index n and the value of d of the plate separation are known from a previous calibration of the interferometer. However, the wavelength is determined by

Figure 4.45 Determination
of the access ϵ from the plot
of D_p^2 versus p

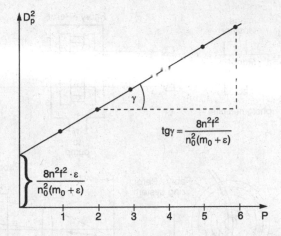

(4.70) only modulo a free spectral range $\delta\lambda = \lambda^2/(2nd)$. This means that all wavelengths λ_m differing by m free spectral ranges produce the same ring systems. For an absolute determination of λ, the integer order m_0 must be known.

When illuminated with strictly parallel light parallel to the interferometer axis the plane of observation behind the FPI is completely dark for $\lambda \neq 2nd \cos\beta/m$ and uniformly illuminated for $\lambda = 2nd \cos\beta/m$. The reflected intensity is zero for $\lambda = 2nd \cos\beta/m$, otherwise the reflection coefficient is $R = 1$ which means that the total incident intensity is reflected back into the source.

Illumination with divergent light produces a ring system of bright rings with a dark background. In the reflected light a dark ring system on a bright background appears. If the light beam is divergent in the x-direction but parallel in the y-direction the ring system changes into a system of parallel straight bright or dark lines.

The experimental scheme for the absolute determination of λ utilizes a combination of FPI and spectrograph in a so-called *crossed arrangement* (Fig. 4.46), where the ring system of the FPI is imaged onto the entrance slit of a spectrograph. The spectrograph disperses the slit images $S(\lambda)$ with a medium dispersion in the x-direction (Sect. 4.1), the FPI provides high dispersion in the y-direction. The resolution of the spectrograph must only be sufficiently high to separate the images of two wavelengths differing by one free spectral range of the FPI. Figure 4.47 shows, for illustration, a section of the Na_2 fluorescence spectrum excited by an argon laser line. The ordinate corresponds to the FPI dispersion and the abscissa to the spectrograph dispersion [151].

The angular dispersion $d\beta/d\lambda$ of the FPI can be deduced from (4.66)

$$\frac{d\beta}{d\lambda} = \left(\frac{d\lambda}{d\beta}\right)^{-1} = m/(2nd \sin\beta) = \frac{1}{\lambda_m \sin\beta} \quad \text{with } \lambda_m = 2nd/m . \quad (4.73)$$

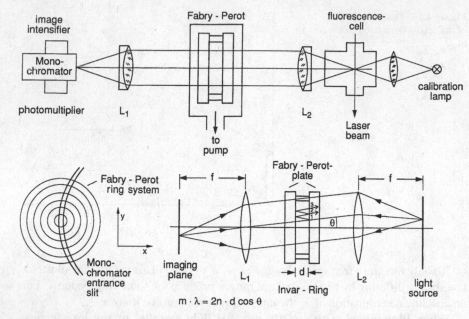

Figure 4.46 Combination of FPI and spectrograph for the unambiguous determination of the integral order m_0

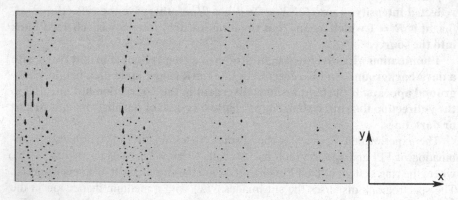

Figure 4.47 Section of the argon laser-excited fluorescence spectrum of Na_2 obtained with the arrangement of crossed FPI and spectrograph shown in Fig. 4.46 [151]

Equation (4.73) shows that the angular dispersion becomes infinite for $\beta \to 0$. The linear dispersion of the ring system on the photoplate is

$$\frac{dD}{d\lambda} = f\frac{d\beta}{d\lambda} = \frac{f}{\lambda_m \sin\beta} \, . \tag{4.74}$$

Example 4.14

$f = 30\,\text{cm}$, $\lambda = 0.5\,\mu\text{m}$. At a distance of 1 mm from the ring center is $\beta = 0.1/50$ and we obtain a linear dispersion of $d\lambda/d\lambda = 500\,\text{mm/nm}$. This is at least one order of magnitude larger than the dispersion of a large spectrograph.

c) The Air-Spaced FPI

Different from the solid etalon, which is a plane-parallel plate coated on both sides with reflecting layers, the plane FPI consists of two wedged plates, each having one high-reflection and one antireflection coating (Fig. 4.41b). The finesse of the FPI critically depends, apart from the reflectivity R and the optical surface quality, on the parallel alignment of the two reflecting surfaces. The advantage of the air-spaced FPI, that any desired free spectral range can be realized by choosing the corresponding plate separation d, must be paid for by the inconvenience of careful alignment. Instead of changing the angle of incidence α, wavelength tuning can be also achieved for $\alpha = 0$ by variation of the optical path difference $\Delta s = 2nd$, either by changing d with piezoelectric tuning of the plate separation, or by altering the refractive index by a pressure change in the container enclosing the FPI.

The tunable FPI is used for high-resolution spectroscopy of line profiles. The transmitted intensity $I_T(p)$ as a function of the optical path difference nd is given by the convolution

$$I_T(\nu) = I_0(\nu)T(nd, \lambda),$$

where the transmission of the FPI $T(nd, \lambda) = T(\phi)$ can be obtained from (4.52a), (4.52b).

With photoelectric recording (Fig. 4.48), the large dispersion at the ring center can be utilized. The light source LS is imaged onto a small pinhole P1, which serves as a point source in the focal plane of L1. The parallel light beam passes the FPI, and the transmitted intensity is imaged by L2 onto another pinhole P2 in front of the detector. All light rays within the cone $\cos \beta_0 \leq m_0\lambda/(nd)$, where β is the angle against the interferometer axis, contribute according to (4.66) to the central fringe. If the optical path length nd is tuned, the different transmission orders with $m = m_0, m_0 + 1, m_0 + 2, \ldots$ are successively transmitted for a wavelength λ according to $m\lambda = 2nd$. Light sources that come close to being a point source, can be realized when a focused laser beam crosses a sample cell and the laser-induced fluorescence emitted from a small section of the beam length is imaged through the FPI onto the entrance slit of a monochromator, which is tuned to the desired wavelength interval $\Delta\lambda$ around λ_m (Fig. 4.46). If the spectral interval $\Delta\lambda$ resolved by the monochromator is smaller then the free spectral range $\delta\lambda$ of the

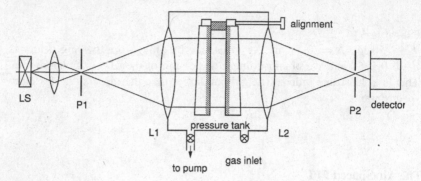

Figure 4.48 Use of a plane FPI for photoelectric recording of the spectrally resolved transmitted intensity $I_T(n \cdot d, \lambda)$ emitted from a point source

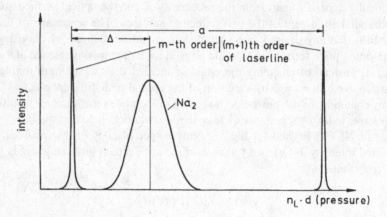

Figure 4.49 Photoelectric recording of a Doppler-broadened laser-excited fluorescence line of Na_2 molecules in a vapor cell and the Doppler-free scattered laser line. The pressure scan $\Delta p = a$ corresponds to one free spectral range of the FPI

FPI, an unambigious determination of λ is possible. For illustration, Fig. 4.49 shows a Doppler-broadened fluorescence line of Na_2 molecules excited by a single-mode argon laser at $\lambda = 488$ nm, together with the narrow line profile of the scattered laser light. The pressure change $\Delta p \cong 2d \Delta n_L = a$ corresponds to one free spectral range of the FPI, i.e., $2d \Delta n_L = \lambda$.

For Doppler-free resolution of fluorescence lines (Vol. 2, Chap. 4), the laser-induced fluorescence of molecules in a collimated molecular beam can be imaged through a FPI onto the entrance slit of the monochromator (Fig. 4.50). In this case, the crossing point of laser and molecular beam, indeed, represents nearly a point source.

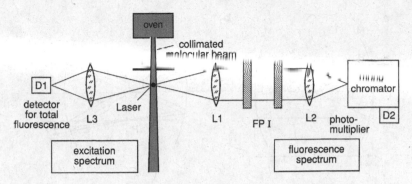

Figure 4.50 Experimental arrangement for photoelectric recording of high-resolution fluorescence lines excited by a single-mode laser in a collimated molecular beam and observed through FPI plus monochromator

4.2.8 Confocal Fabry–Perot Interferometer

A confocal interferometer, sometimes called incorrectly a *spherical FPI*, consists of two spherical mirrors M_1, M_2 with equal curvatures (radius r) that are opposed at a distance $d = r$ (Fig. 4.51a) [152, 153]. These interferometers have gained great importance in laser physics as high-resolution spectrum analyzers for detecting the mode structure and the linewidth of lasers [154–156], and, in the nearly confocal form, as laser resonators (Sect. 5.2).

Neglecting spherical aberration, all light rays entering the interferometer parallel to its axis would pass through the focal point F and would reach the entrance point P1 again after having passed the confocal FPI four times (Fig. 4.51a). Figure 4.51b illustrates the general case of a ray which enters the confocal FPI at a small inclination θ and passes the successive points P1, A, B, C, P1, shown in Fig. 4.51d in a projection. Angle θ is the skew angle of the entering ray.

Because of spherical aberration, rays with different distances ρ_1 from the axis will not all go through F but will intersect the axis at different positions F′ depending on ρ_1 and θ. Also, each ray will not exactly reach the entrance point P_1 after four passages through the confocal FPI since it is slightly shifted at successive passages. However, it can be shown [152, 155] that for sufficiently small angles θ, all rays intersect at a distance $\rho(\rho_1, \theta)$ from the axis in the vicinity of the two points P and P′ located in the central plane of the confocal FPI (Fig. 4.51b).

The optical path difference Δs between two successive rays passing through P can be calculated from geometrical optics. For $\rho_1 \ll r$ and $\theta \ll 1$, one obtains for the near confocal case $d \approx r$ [155]

$$\Delta s = 4d + \rho_1^2 \rho_2^2 \cos 2\theta / r^3 + \text{higher-order terms}. \tag{4.75}$$

An incident light beam with diameter $D = 2\rho_1$ therefore produces, in the central plane of a confocal FPI, an interference pattern of concentric rings. Analogous of

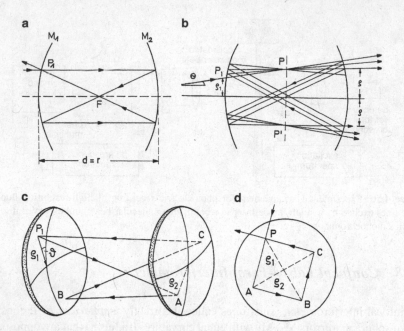

Figure 4.51 Trajectories of rays in a confocal FPI: **a** incident beam parallel to the FPI axis; **b** inclined incident beam; **c** perspective view for illustrating the skew angle; **d** projection of the skewed rays onto the mirror surfaces

the treatment in Sect. 4.2.7, the intensity $I(\rho, \lambda)$ is obtained by adding all amplitudes with their correct phases $\delta = \delta_0 + (2\pi/\lambda)\Delta s$. According to (4.52a), (4.52b) we get

$$I(\rho, \lambda) = \frac{I_0 T^2}{(1 - R)^2 + 4R \sin^2[(\pi/\lambda)\Delta s]} , \tag{4.76}$$

where $T = 1 - R - A$ is the transmission of each of the two mirrors. The intensity has maxima for $\delta = 2m\pi$, which is equivalent to

$$4d + \rho^4/r^3 = m\lambda , \tag{4.77}$$

when we neglect the higher-order terms in (4.75) and set $\theta = 0$ and $\rho^2 = \rho_1\rho_2$.

The free spectral range $\delta\nu$, i.e., the frequency separation between successive interference maxima, is for the near-confocal FPI with $\rho \ll d$

$$\delta\nu = \frac{c}{4d + \rho^4/r^3} , \tag{4.78}$$

which is *different* from the expression $\delta\nu = c/2d$ for the plane FPI.

The radius ρ_m of the mth-order interference ring is obtained from (4.77),

$$\rho_m = [(m\lambda - 4d)r^3]^{1/4} , \tag{4.79}$$

which reveals that ρ_m depends critically on the separation d of the spherical mirrors. Changing d by a small amount ϵ from $d = r$ to $d = r + \epsilon$ changes the path difference to

$$\Delta s = 4(r + \epsilon) + \rho^4/(r + \epsilon)^3 \sim 4(r + \epsilon) + \rho^4/r^3 . \qquad (4.00)$$

For a given wavelength λ, the value of ϵ can be chosen such that $4(r + \epsilon) = m_0\lambda$. In this case, the radius of the central ring becomes zero. We can number the outer rings by the integer p and obtain with $m = m_0 + p$ for the radius of the pth ring the expression

$$\rho_p = (p\lambda r^3)^{1/4} . \qquad (4.81)$$

The radial dispersion deduced from (4.79),

$$\frac{d\rho}{d\lambda} = \frac{mr^3/4}{[(m\lambda - 4d)r^3]^{3/4}} , \qquad (4.82)$$

becomes infinite for $m\lambda = 4d$, which occurs according to (4.79) at the center with $\rho = 0$.

This large dispersion can be used for high-resolution spectroscopy of narrow line profiles with a scanning confocal FPI and photoelectric recording (Fig. 4.52).

If the central plane of the near-confocal FPI is imaged by a lens onto a circular aperture with sufficiently small radius $b < (\lambda r^3)^{1/4}$ only the central interference order is transmitted to the detector while all other orders are stopped. Because of the large radial dispersion for small ρ one obtains a high spectral resolving power. With this arrangement not only spectral line profiles but also the instrumental bandwidth can be measured, when an incident monochromatic wave (from a stabilized single-mode laser) is used. The mirror separation $d = r + \epsilon$ is varied by the small amount ϵ and the power

$$P(\lambda, b, \epsilon) = 2\pi \int_{\rho=0}^{b} \rho I(\rho, \lambda, \epsilon) d\rho , \qquad (4.83)$$

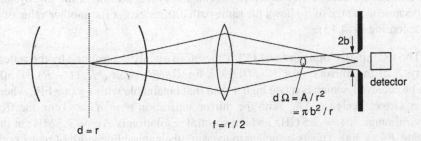

Figure 4.52 Photoelectric recording of the spectral light power transmitted of a scanning confocal FPI

Figure 4.53 Illustration of the larger sensitivity against misalignment for the plane FPI compared with the spherical FPI

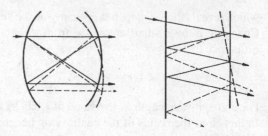

transmitted through the aperture is measured as a function of ϵ at fixed values of λ and b.

The integrand $I(\rho, \lambda, \epsilon)$ can be obtained from (4.76), where the phase difference $\delta(\epsilon) = 2\pi \Delta s/\lambda$ is deduced from (4.80).

The optimum choice for the radius b of the aperture is based on a compromise between spectral resolution and transmitted intensity. When the interferometer has the finesse F^*, the spectral halfwidth of the transmission peak is $\delta\nu/F^*$, see (4.55b), and the maximum spectral resolving power becomes $F^* \Delta s/\lambda$ (4.60). For the radius $b = (r^3\lambda/F^*)^{1/4}$ of the aperture, which is just $(F^*)^{1/4}$ times the radius ρ_1 of a fringe with $p = 1$ in (4.81), the spectral resolving power is reduced to about 70 % of its maximum value. This can be verified by inserting this value of b into (4.83) and calculating the halfwidth of the transmission peak $P(\lambda_1, F^*, \epsilon)$.

The total finesse of the confocal FPI is, in general, higher than that of a plane FPI for the following reasons:

- The alignment of spherical mirrors is far less critical than that of plane mirrors, because tilting of the spherical mirrors does not change (to a first approximation) the optical path length $4r$ through the confocal FPI, which remains approximately the same for all incident rays (Fig. 4.53). For the plane FPI, however, the path length increases for rays below the interferometer axis, but decreases for rays above the axis.
- Spherical mirrors can be polished to a higher precision than plane mirrors. This means that the deviations from an ideal sphere are less for spherical mirrors than those from an ideal plane for plane mirrors. Furthermore, such deviations do not wash out the interference structure but cause only a distortion of the ring system because a change of d allows the same path difference Δs for another value of ρ according to (4.75).

The total finesse of a confocal FPI is therefore mainly determined by the reflectivity R of the mirrors. For $R = 0.99$, a finesse $F^* = \pi\sqrt{R/(1-R)} \approx 300$ can be achieved, which is much higher than that obtainable with a plane FPI, where other factors decrease F^*. With the mirror separation $r = d = 3$ cm, the free spectral range is $\delta = 2.5$ GHz and the spectral resolution is $\Delta\nu = 7.5$ MHz at the finesse $F^* = 300$. This is sufficient to measure the natural linewidth of many optical transitions. With modern high-reflection coatings, values of $R = 0.9995$ can be obtained and confocal FPI with a finesse $F^* \geq 10^4$ have been realized [157].

From Fig. 4.52 we see that the solid angle accepted by the detector behind the aperture with radius b is $\Omega = \pi b^2/r^2$. The light power transmitted to the detector is proportional to the product of the solid angle Ω and area A in the central plane, which is imaged by the lens onto the aperture (often called the *étendue U*). With the aperture radius $b = (r^3\lambda/F^*)^{1/4}$ (see above) the étendue becomes

$$U = A\Omega = \pi^2 b^4/r^2 = \pi^2 r\lambda/F^* . \tag{4.84}$$

For a given finesse F^*, the étendue of the confocal FPI increases with the mirror separation $d = r$. The spectral resolving power

$$\frac{\nu}{\Delta\nu} = 4F^*\frac{r}{\lambda} , \tag{4.85}$$

of the confocal FPI is proportional to the product of finesse F^* and the ratio of mirror separation $r = d$ to the wavelength λ. With a given étendue $U = \pi^2 r\lambda/F^*$, we can insert $r = UF^*/(\pi^2\lambda)$ into (4.84) and obtain for the spectral resolving power

$$\frac{\nu}{\Delta\nu} = \left(\frac{2F^*}{\pi\lambda}\right)^2 U , \quad \text{(confocal FPI)} . \tag{4.86}$$

Let us compare this with the case of a plane FPI with the plate diameter D and the separation d, which is illuminated with nearly parallel light (Fig. 4.48). According to (4.66), the path difference between a ray parallel to the axis and a ray with a small inclination β is, given by $\Delta s = 2nd(1 - \cos\beta) \approx nd\beta^2$.

To achieve a finesse F^* with photoelectric recording, this variation of the path length for the different rays through the interferometer should not exceed λ/F^*, which restricts the solid angle $\Omega = \beta^2$ acceptable by the detector to $\Omega \leq \lambda/(d \cdot F^*)$. The étendue is therefore

$$U = A\Omega = \pi\frac{D^2}{4}\frac{\lambda}{d \cdot F^*} . \tag{4.87}$$

Inserting the value of d given by this equation into the spectral resolving power $\nu/\Delta\nu = 2dF^*/\lambda$, we obtain

$$\frac{\nu}{\Delta\nu} = \frac{\pi D^2}{2U} , \quad \text{(plane FPI)} . \tag{4.88}$$

While the spectral resolving power is proportional to U for the confocal FPI, it is *inversely proportional to U for the plane FPI*. This is because the étendue increases with the mirror separation d for the confocal FPI but decreases proportional to $1/d$ for the plane FPI. For a mirror radius $r > D^2/4d$, the étendue of the confocal FPI is larger than that of a plane FPI with equal spectral resolution. This means that the transmitted power is larger for the confocal FPI for $r > D^2/4d$.

Example 4.15

A confocal FPI with $r = d = 5$ cm has for $\lambda = 500$ nm the étendue $U = (2.47 \times 10^{-3}/F^*)\,\text{cm}^2/\text{sr}$. This is the same étendue as that of a plane FPI with $d = 5$ cm and $D = 10$ cm. However, the diameter of the spherical mirrors can be much smaller (less than 5 mm). With a finesse $F^* = 100$, the étendue is $U = 2.5 \times 10^{-5}$ [cm^2 sr] and the spectral resolving power is $\nu/\Delta\nu = 4 \times 10^7$. With this étendue the resolving power of the plane FPI is 6×10^6, provided the whole plane mirror surface has a surface quality to allow a surface finesse of $F^* \geq 100$. In practice, this is difficult to achieve for a flat plane with $D = 10$ cm diameter, while for the small spherical mirrors even $F^* > 10^4$ is feasible.

This example shows that for a given light-gathering power, the confocal FPI can have a much higher spectral resolving power than the plane FPI.

More detailed information on the history, theory, practice, and application of plane and spherical Fabry–Perot interferometers may be found in [158–160].

4.2.9 Multilayer Dielectric Coatings

The constructive interference found for the reflection of light from plane-parallel interfaces between two regions with different refractive indices can be utilized to produce highly reflecting, essentially absorption-free mirrors. The improved technology of such dielectric mirrors has greatly supported the development of visible and ultraviolet laser systems.

The reflectivity R of a plane interface between two regions with complex refractive indices $n_1 = n_1' - i\kappa_1$ and $n_2 = n_2' - i\kappa_2$ can be calculated from Fresnel's formulas [129]. It depends on the angle of incidence α and on the direction of polarization. For the polarization component with the electric field vector E parallel to the plane of incidence (defined by the incident and the reflected beam), the reflectivity is

$$R_p = \left(\frac{n_2 \cos\alpha - n_1 \cos\beta}{n_2 \cos\alpha + n_1 \cos\beta} \right)^2 = \left[\frac{\tan(\alpha - \beta)}{\tan(\alpha + \beta)} \right]^2 \tag{4.89a}$$

where β is the refraction angle ($\sin\beta = (n_1/n_2) \sin\alpha$). For the vertical component (E perpendicular to the plane of incidence), one obtains:

$$R_s = \left(\frac{n_1 \cos\alpha - n_2 \cos\beta}{n_1 \cos\alpha + n_2 \cos\beta} \right)^2 = \left[\frac{\sin(\alpha - \beta)}{\sin(\alpha + \beta)} \right]^2 . \tag{4.89b}$$

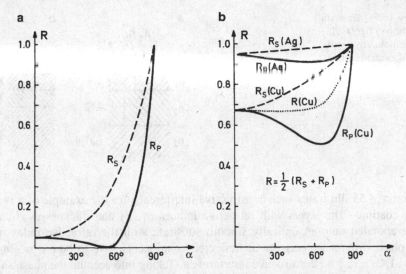

Figure 4.54 Reflectivities R_p and R_s for the two polarization components parallel and perpendicular to the plane of incidence as a function of the angle of incidence α: **a** air–glass boundary ($n_1 = 1$, $n_2 = 1, 5$); **b** air–metal boundary for Cu($n' = 0.76$, $\kappa = 3.32$) and Ag($n' = 0.055$, $\kappa = 3.32$)

The reflectivities R_p and R_s are illustrated in Fig. 4.54 for three different materials for incident light polarized parallel (R_p) and perpendicular (R_s) to the plane of incidence.

For vertical incidence ($\alpha = 0$, $\beta = 0$), one obtains from Fresnel's formulas for both polarizations

$$R|_{\alpha=0} = \left(\frac{n_1 - n_2}{n_1 + n_2}\right)^2 . \tag{4.89c}$$

Since this case represents the most common situation for laser mirrors, we shall restrict the following discussion to vertical incidence.

To achieve maximum reflectivities, the numerator $(n_1 - n_2)^2$ should be maximized and the denominator minimized. Since n_1 is always larger than one, this implies that n_2 should be as large as possible. Unfortunately, the dispersion relations (3.24a, 3.24b) imply that a large value of n also causes large absorption. For instance, highly polished metal surfaces have a maximum reflectivity of $R = 0.95$ in the visible spectral range. The residual 5 % of the incident intensity are absorbed and therefore lost.

The situation can be improved by selecting reflecting materials with low absorption (which then necessarily also have low reflectivity), but using many layers with alternating high and low refractive index n. Choosing the proper optical thickness nd of each layer allows constructive interference between the different reflected amplitudes to be achieved. Reflectivities of up to $R = 0.9999$ have been reached [161–164].

Figure 4.55 Maximum reflection of light with wavelength λ by a two-layer dielectric coating: **a** $n_1 > n_2 > n_3$; **b** $n_1 > n_2 < n_3$

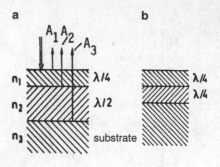

Figure 4.55 illustrates such constructive interference for the example of a two-layer coating. The layers with refractive indices n_1, n_2 and thicknesses d_1, d_2 are evaporated onto an optically smooth substrate with the refractive index n_3. The phase differences between all reflected components have to be $\delta_m = 2m\pi$ ($m = 1, 2, 3, \ldots$) for constructive interference. Taking into account the phase shift $\delta = \pi$ at reflection from an interface with a larger refractive index than that of the foregoing layer, we obtain the conditions

$$n_1 d_1 = \lambda/4 \quad \text{and} \quad n_2 d_2 = \lambda/2 \quad \text{for} \quad n_1 > n_2 > n_3 , \tag{4.90a}$$

and

$$n_1 d_1 = n_2 d_2 = \lambda/4 \quad \text{for} \quad n_1 > n_2, n_3 > n_2 . \tag{4.90b}$$

The reflected amplitudes can be calculated from Fresnel's formulas. The total reflected intensity is obtained by summation over all reflected amplitudes taking into account the correct phase. The refractive indices are now selected such that $\sum A_i$ becomes a maximum. The calculation is still feasible for our example of a two-layer coating and yields for the three reflected amplitudes (double reflections are neglected)

$$A_1 = \sqrt{R_1} A_0 ; \quad A_2 = \sqrt{R_2}(1 - \sqrt{R_1}) A_0 ,$$
$$A_3 = \sqrt{R_3}(1 - \sqrt{R_2})(1 - \sqrt{R_1}) A_0 ,$$

where the reflectivities R_i are given by (4.89a)–(4.89c).

Example 4.16

$|n_1| = 1.6$, $|n_2| = 1.2$, $|n_3| = 1.45$; $A_1 = 0.231 A_0$, $A_2 = 0.143 A_0$, $A_3 = 0.094 A_0$. $A_R = \sum A_i = 0.468 A_0 \to I_R = 0.22 I_0 \to R = 0.22$, provided the path differences have been choosen correctly.

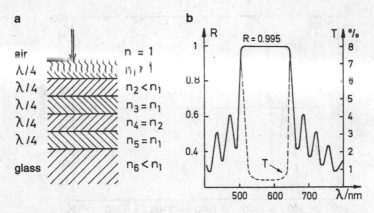

Figure 4.56 The dielectric multilayer mirror: **a** Composition of multilayers; **b** Reflectivity of a high-reflectance multilayer mirror with 17 layers as a function of the incident wavelength λ

This example illustrates that for materials with low absorption, many layers are necessary to achieve a high reflectivity. Figure 4.56a depicts schematically the composition of a dielectric multilayer mirror. The calculation and optimization of multilayer coatings with up to 20 layers becomes very tedious and time consuming, and is therefore performed using computer programs [162, 164]. Figure 4.56b illustrates the reflectivity $R(\lambda)$ of a high-reflectance mirror with 17 layers.

By proper selection of different layers with slightly different optical path lengths, one can achieve a high reflectivity over an extended spectral range. Currently, "broad-band" reflectors are available with reflectivity of $R \geq 0.99$ within the spectral range ($\lambda_0 \pm 0.2\lambda_0$), while the absorption losses are less than 0.2 % [161, 163]. At such low absorption losses, the scattering of light from imperfect mirror surfaces may become the major loss contribution. When total losses of less than 0.5 % are demanded, the mirror substrate must be of high optical quality (better than $\lambda/20$), the dielectric layers have to be evaporated very uniformly, and the mirror surface must be clean and free of dust or dirty films [164]. The best mirrors are produced by ion implantation techniques. Such dielectric mirrors with alternating $\lambda/4$-layers of materials with high and low refractive indices are often called "Bragg mirrors" because they work in a similar way to the Bragg reflection of X-rays at perfect crystal planes. With very pure materials of extremely low absorption, they reach reflectivities of $R > 0.99999$ [165]. The reflectivity $R(\lambda)$ of a Bragg mirror for vertical incidence around $\lambda = 1000$ nm is shown in Fig. 4.57.

Instead of maximizing the reflectivity of a dielectric multilayer coating through *constructive* interference, it is, of course, also possible to minimize it by destructive interference. Such *antireflection coatings* are commonly used to minimize unwanted reflections from the many surfaces of multiple-lens camera objectives, which would otherwise produce an annoying background illumination of the photomaterial. In laser spectroscopy such coatings are important for minimizing reflection losses of optical components inside the laser resonator and for avoiding

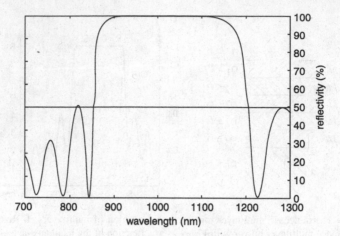

Figure 4.57 Bragg mirror with eight alternating layers of TiO_2 and SiO_2

reflections from the back surface of output mirrors, which would introduce undesirable couplings, thereby causing frequency instabilities of single-mode lasers.

Using a single layer (Fig. 4.58a), the reflectivity reaches a minimum only for a selected wavelength λ (Fig. 4.59). We obtain $I_R = 0$ for $\delta = (2m+1)\pi$, if the two amplitudes A_1 and A_2 reflected by the interfaces (n_1, n_2) and (n_2, n_3) are equal. For vertical incidence this gives the condition

$$R_1 = \left(\frac{n_1 - n_2}{n_1 + n_2}\right)^2 = R_2 = \left(\frac{n_2 - n_3}{n_2 + n_3}\right)^2 , \tag{4.91}$$

which can be reduced to

$$n_2 = \sqrt{n_1 n_3} . \tag{4.92}$$

For a single layer on a glass substrate the values are $n_1 = 1$ and $n_3 = 1.5$. According to (4.92), n_2 should be $n_2 = \sqrt{1.5} = 1.23$. Durable coatings with such low refractive indices are not available. One often uses MgF_2 with $n_2 = 1.38$, giving a reduction of reflection from 4 % to 1.2 % (Fig. 4.59).

With multilayer antireflection coatings the reflectivity can be decreased below 0.2 % for an extended spectral range [164]. For instance, with three $\lambda/4$ layers

Figure 4.58 Antireflection coating: **a** single layer; **b** multilayer coating

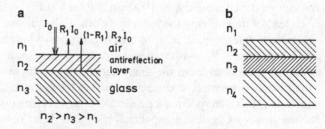

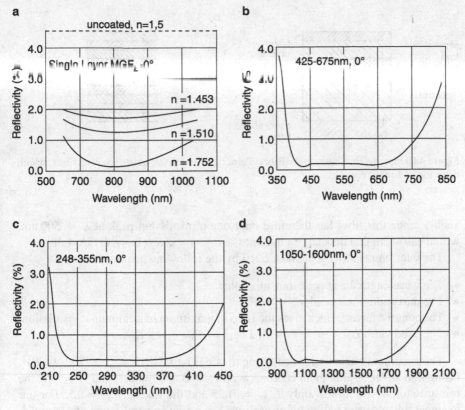

Figure 4.59 Antireflection coatings. **a** Single layer MgF$_2$ on substrates with different refractive index n; **b–d** broadband multilayer AR-coatings, optimized for different spectral ranges

(MgF$_2$, SiO, and CeF$_3$) the reflection drops to below 1 % for the whole range between 420 nm and 840 nm [161, 166, 167].

4.2.10 Interference Filters

Interference filters are used for selective transmission in a narrow spectral range. Incident radiation of wavelengths outside this transmission range is either reflected or absorbed. One distinguishes between line filters and bandpass filters.

A line filter is essentially a Fabry–Perot etalon with a very small optical path nd between the two reflecting surfaces. The technical realization uses two highly reflecting coatings (either silver coatings or dielectric multilayer coatings) that are separated by a nonabsorbing layer with a low refractive index (Fig. 4.60). For instance, for $nd = 0.5\,\mu$m the transmission maxima for vertical incidence are obtained from (4.62a) at $\lambda_1 = 1\,\mu$m, $\lambda_2 = 0.5\,\mu$m, $\lambda_3 = 0.33\,\mu$m, etc. In the

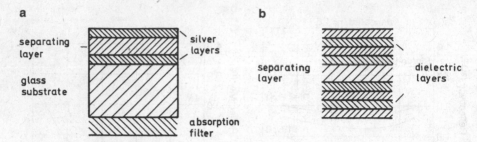

Figure 4.60 Interference filters of the Fabry–Perot type: **a** with two single layers of silver; **b** with dielectric multilayer coatings

visible range this filter has therefore only one transmission peak at $\lambda = 500\,$nm, with a halfwidth that depends on the finesse $F^* = \pi \sqrt{R}/(1 - R)$ (Fig. 4.38).

The interference filter is characterized by the following quantities:

- The wavelength λ_m at peak transmission;
- The maximum transmission;
- The contrast factor, which gives the ratio of maximum to minimum transmission;
- The bandwidth $\Delta v = v_1 - v_2$ with $T(v_1) = T(v_2) = \frac{1}{2}T_{\max}$.

The maximum transmission according to (4.61a)–(4.61d) is $T_{\max} = T^2/(1-R)^2$. Using thin silver or aluminum coatings with $R = 0.8$, $T = 0.1$, and $A = 0.1$, the transmission of the filter is only $T_{\max} = 0.25$ and the finesse $F^* = 15$. For our example this means a halfwidth of $660\,$cm^{-1} at a free spectral range of $10^4\,$cm^{-1}. At $\lambda = 500\,$nm this corresponds to a free spectral range of $250\,$nm and a halfwidth of about $16\,$nm. For many applications in laser spectroscopy, the low peak transmission of interference filters with absorbing metal coatings is not tolerable. One has to use absorption-free dielectric multilayer coatings (Fig. 4.60b) with high reflectivity, which allows a large finesse and therefore a smaller bandwidth and a larger peak transmission (Fig. 4.61).

Example 4.17

With $R = 0.95$, $A = 0.01$ and $T = 0.04$, according to (4.61a)–(4.61d) we obtain a peak transmission of $64\,\%$, which increases with $A = 0.005$, $T = 0.045$ to $81\,\%$. The contrast becomes $\gamma = I_{\mathrm{T}}^{\max}/I_{\mathrm{T}}^{\min} = (1 + F) = 1 + 4F^{*2}/\pi^2 = 1520$. With a thickness $nd = 5\,\mu$m of the separating layer, the free spectral range is $\delta v = 3 \times 10^{13}\,$Hz $\hat{=} 25\,$nm at $\lambda = 500\,$nm.

A higher finesse F^* due to larger reflectivities of the reflecting films not only decreases the bandwidth but also increases the contrast factor. With $R = 0.98 \rightarrow F = 4R/(1 - R)^2 = 9.8 \times 10^3$, which means that the intensity at the transmission minimum is only about 10^{-4} of the peak transmission.

Figure 4.61 Spectral transmission of interference filters. *Solid curve*: line filter. *Dashed curve*: bandpass filter. Note the logarithmic scale

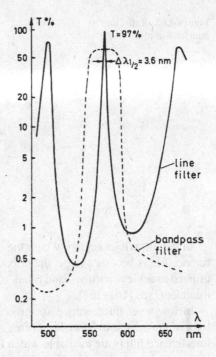

The bandwidth can be further decreased by using two interference filters in series. However, it is preferable to construct a double filter that consists of three highly-reflecting surfaces, separated by two nonabsorbing layers of the same optical thickness. If the thickness of these two layers is made slightly different, a bandpass filter results that has a flat transmission curve but steep slopes to both sides. Commercial interference filters are currently available with a peak transmission of at least 90 % and a bandwidth of less than 2 nm [162, 168]. Special narrow-band filters even reach 0.3 nm, however, with reduced peak transmission.

The wavelength λ_m of the transmission peak can be shifted to lower values by tilting the interference filter, which increases the angle of incidence α, see (4.62a). The tuning range is, however, restricted, because the reflectivity of the multilayer coatings also depends on the angle α and is, in general, optimized for $\alpha = 0$. For divergent incident light, the transmission bandwidth increases with the divergence angle. From (4.62a), we obtain for the wavelength $\lambda(\alpha)$ of a tilted filter

$$\lambda = \frac{2nd}{m} \cos \beta = \lambda_0 \cos \beta \approx \lambda_0 \left(1 - \frac{\beta^2}{2}\right) \approx \lambda_0 \left(1 - \frac{\alpha^2}{2n^2}\right). \qquad (4.93)$$

Example 4.18
$\lambda_0 = 1500\,\text{nm}, n = 1.5, \alpha = 150° \triangleq 0.25\,\text{rad} \Rightarrow \lambda(\alpha) = 1389\,\text{nm} \Rightarrow$
$\Delta\lambda = \lambda_0 - \lambda(\alpha) = 111\,\text{nm}$

Figure 4.62 Reflection interference filter

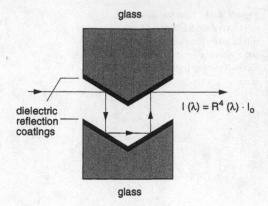

In the ultraviolet region, where the absorption of most materials used for interference filters becomes large, the selective *reflectance* of interference filters can be utilized to achieve narrow-band filters with low losses (Fig. 4.62). For more detailed treatment, see [161–168].

In low-level fluorescence spectroscopy or Raman spectroscopy, the scattered light of the intense exciting laser often overlaps the fluorescence lines. Here special interference filters are available which have a narrow minimum transmission at the laser wavelength (line-blocking filter) but a high transmission in the other spectral ranges.

Since temperature drifts cause a change of the spacing d, the wavelength λ_p at peak transmission also shifts with temperature. Typical values are $d\lambda_p/dT = 0.02 \, \text{nm/K}$. A temperature change of $10 \, \text{K}$ therefore shifts the peak transmission by $0.2 \, \text{nm}$. This is only relevant for filters with a very narrow transmission bandwidth.

4.2.11 Birefringent Interferometer

The basic principle of the birefringent interferometer or *Lyot filter* [137, 169] is founded on the interference of polarized light that has passed through a birefringent crystal. Assume that a linearly polarized plane wave

$$E = A \cdot \cos(\omega t - kx) \,,$$

with

$$A = \{0, A_y, A_z\}, \quad A_y = |A| \sin \alpha \,, \quad A_z = |A| \cos \alpha \,,$$

is incident on the birefringent crystal (Fig. 4.63). The electric vector E makes an angle α with the optical axis, which points into the z-direction. Within the crystal, the wave is split into an ordinary beam with the wave number $k_o = n_o k$

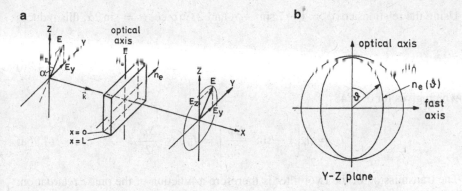

Figure 4.63 Lyot filter: **a** schematic arrangement; **b** index ellipsoid of the birefringent crystal

and the phase velocity $v_o = c/n_o$, and an extraordinary beam with $k_e = n_e k$ and $v_e = c/n_e$. The partial waves have mutually orthogonal polarization in directions parallel to the z- and y-axis, respectively. Let the crystal with length L be placed between $x = 0$ and $x = L$. Because of the different refractive indices n_0 and n_e for the ordinary and the extraordinary beams, the two partial waves at $x = L$

$$E_y(L) = A_y \cos(\omega t - k_e L) \quad \text{and} \quad E_z(L) = A_z \cos(\omega t - k_0 L) ,$$

show a phase difference of

$$\Delta\phi = k(n_0 - n_e)L = (2\pi/\lambda)\Delta n L \quad \text{with } \Delta n = n_0 - n_e . \tag{4.94}$$

The superposition of these two waves results, in general, in elliptically polarized light, where the principal axis of the ellipse is turned by an angle $\beta = \phi/2$ against the direction of A_0.

For phase differences $\Delta\phi = 2m\pi$, linearly polarized light with $E(L) \parallel E(0)$ is obtained. However, for $\Delta\phi = (2m + 1)\pi$ and $\alpha = 45°$, the transmitted wave is also linearly polarized, but now $E(L) \perp E(0)$.

The elementary Lyot filter consists of a birefringent crystal placed between two linear polarizers (Fig. 4.63a). Assume that the two polarizers are both parallel to the electric vector $E(0)$ of the incoming wave. The second polarizer parallel to $E(0)$ transmits only the projection

$$E = E_y \sin\alpha + E_z \cos\alpha$$
$$= A[\sin^2\alpha \cos(\omega t - k_e L) + \cos^2\alpha \cos(\omega t - k_0 L)] ,$$

of the amplitudes, which yields with (4.91) the transmitted time averaged intensity

$$\bar{I}_T = \tfrac{1}{2} c\epsilon_0 \overline{E}^2 = \bar{I}_0(\sin^4\alpha + \cos^4\alpha + 2\sin^2\alpha \cos^2\alpha \cos\Delta\phi) . \tag{4.95}$$

Using the relations $\cos\phi = 1 - 2\sin^2\frac{1}{2}\phi$, and $2\sin\alpha\cos\alpha = \sin 2\alpha$, this reduces to

$$\bar{I}_T = I_0[1 - \sin^2\tfrac{1}{2}\Delta\phi\sin^2(2\alpha)] ,\qquad(4.96)$$

which gives for $\alpha = 45°$

$$I_T = I_0\left[1 - \sin^2\frac{\Delta\phi}{2}\right] = I_0\cos^2\frac{\Delta\phi}{2} .\qquad(4.96a)$$

The transmission of the Lyot filter is therefore a function of the phase retardation, i.e.,

$$\boxed{T(\lambda) = \frac{I_T}{I_0} = T_0\cos^2\left(\frac{\pi\Delta n L}{\lambda}\right)}\qquad(4.97)$$

which depends on the wavelength λ.

Note According to (4.96) the maximum modulation of the transmittance with $T_{max} = T_0$ and $T_{min} = 0$ is only achieved for $\alpha = 45°$!

Taking into account absorption and reflection losses, the maximum transmission $I_T/I_0 = T_0 < 1$ becomes less than 100 %. Within a small wavelength interval, the difference $\Delta n = n_0 - n_e$ can be regarded as constant. Therefore (4.97) gives the wavelength-dependent transmission function, $\cos^2\phi$, typical of a two-beam interferometer (Fig. 4.26). For extended spectral ranges the different dispersion of $n_0(\lambda)$ and $n_e(\lambda)$ has to be considered, which causes a wavelength dependence, $\Delta n(\lambda)$.

The free spectral range $\delta\nu$ is obtained from (4.97) as

$$\frac{\Delta n\cdot L}{\lambda_1} - \frac{\Delta n\cdot L}{\lambda_2} = 1 .$$

With $\nu = c/\lambda$, this becomes

$$\delta\nu = \frac{c}{(n_0 - n_e)L} .\qquad(4.98)$$

Example 4.19
For a crystal of potassium dihydrogen phosphate (KDP), $n_e = 1.51$, $n_0 = 1.47 \rightarrow \Delta n = 0.04$ at $\lambda = 600$ nm. A crystal with $L = 2$ cm then has a free spectral range $\delta\nu = 3.75\times 10^{11}$ Hz $\hat{=} \delta\bar{\nu} = 12.5$ cm$^{-1} \rightarrow \Delta\lambda = 0.45$ nm at $\lambda = 600$ nm.

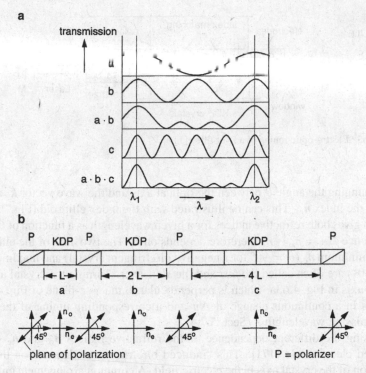

Figure 4.64 **a** Transmitted intensity $I_T(\lambda)$ of a Lyot filter composed of three birefringent crystals with lengths L, $2L$, and $4L$ between polarizers. **b** Arrangement of the crystals and the state of polarization of the transmitted wave

If N elementary Lyot filters with different lengths L_m are placed in series, the total transmission T is the product of the different transmissions T_m, i.e.,

$$T(\lambda) = \prod_{m=1}^{N} T_{0m} \cos^2\left(\frac{\pi \Delta n L_m}{\lambda}\right). \qquad (4.99)$$

Figure 4.64 illustrates a possible experimental arrangement and the corresponding transmission for a Lyot filter composed of three components with the lengths $L_1 = L$, $L_2 = 2L$, and $L_3 = 4L$. The free spectral range $\delta\nu$ of this filter equals that of the shortest component; the halfwidth $\Delta\nu$ of the transmission peaks is, however, mainly determined by the longest component. When we define, analogous to the Fabry–Perot interferometer, the finesse F^* of the Lyot filter as the ratio of the free spectral range $\delta\nu$ to the halfwidth $\Delta\nu$, we obtain, for a composite Lyot filter with N elements of lengths $L_m = 2^{m-1} L_1$, a finesse that is approximately $F^* = 2^N$.

The wavelength of the transmission peak can be tuned by changing the difference $\Delta n = n_0 - n_e$. This can be realized in two different ways:

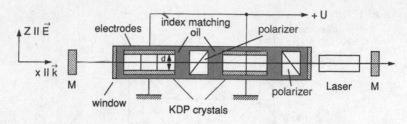

Figure 4.65 Electro-optic tuning of a Lyot filter [170]

- By changing the angle θ between the optical axis and the wave vector $\boldsymbol{k}$, which alters the index n_e. This can be illustrated with the index ellipsoid (Fig. 4.63b), which gives both refractive indices for a given wavelength as a function of θ. The difference $\Delta n = n_o - n_e$ therefore depends on θ. The two axes of the ellipsoid with minimum n_e ($\theta = 90°$ for a negative birefringent crystal) and maximum n_o ($\theta = 0°$) are often called the *fast* and the *slow* axes. Turning the crystal around the x-axis in Fig. 4.63a, which is perpendicular to the y–z-plane of Fig. 4.63b, results in a continuous change of Δn and a corresponding tuning of the peak transmission wavelength λ (Sect. 5.7.4).
- By using the different dependence of the refractive indices n_o and n_e on an applied electric field [171]. This "induced birefringence" depends on the orientation of the crystal axis in the electric field. A common arrangement employs a potassium dihydrogen phosphate (KDP) crystal with an orientation where the electric field is parallel to the optical axis (z-axis) and the wave vector $\boldsymbol{k}$ of the incident wave is perpendicular to the z-direction (transverse electro-optic effect, Fig. 4.65). Two opposite sides of the rectangular crystal with the side length d are coated with gold electrodes and the electric field $E = U/d$ is controlled by the applied voltage.

In the external electric field the uniaxial crystal becomes biaxial. In addition to the natural birefringence of the uniaxial crystal, a field-induced birefringence is generated, which is approximately proportional to the field strength E [172]. The changes of n_o or n_e by the electric field depend on the symmetry of the crystal, the direction of the applied field, and on the magnitude of the electro-optic coefficients. For the KDP crystal only one electro-optic coefficient $d_{36} = -10.7 \times 10^{-12}$ [m/V] (see Sect. 6.1) is effective if the field is applied parallel to the optical axis.

The difference $\Delta n = n_o - n_e$ then becomes

$$\Delta n(E_z) = \Delta n(E = 0) + \tfrac{1}{2} n_1^3 d_{36} E_z \ . \tag{4.100}$$

Maximum transmittance is obtained for

$$\Delta n L = m\lambda \quad (m = 0, 1, 2\ldots) \ ,$$

which gives the wavelength λ at the maximum transmittance

$$\lambda - (\Delta_\| (\Gamma - 0) + 0.5 n_1{}^3 d_{36} E_z) L/m , \qquad (4.101)$$

as a function of the applied field.

While this electro-optic tuning of the Lyot filter allows rapid switching of the peak transmission, for many applications, where a high tuning speed is not demanded, mechanical tuning is more convenient and easier to realize.

4.2.12 Tunable Interferometers

For many applications in laser spectroscopy it is advantageous to have a high-resolution interferometer that is able to scan, in a given time interval Δt, through a limited spectral range $\Delta \nu$. The scanning speed $\Delta \nu / \Delta t$ depends on the method used for tuning, while the spectral range $\Delta \nu$ is limited by the free spectral range $\delta \nu$ of the instrument. All techniques for tuning the wavelength $\lambda_m = 2nd/m$ at the transmission peak of an interferometer are based on a continuous change of the optical path difference between successive interfering beams [171, 173]. This can be achieved in different ways:

a) Change the refractive index n by altering the pressure between the reflecting plates of a FPI (pressure-scanned FPI);
b) Change the distance d between the plates with piezoelectric or magnetostrictive elements;
c) Tilt the solid etalons with a given thickness d against the direction of the incoming plane wave;
d) Change the optical path difference $\Delta s = \Delta n L$ in birefringent crystals by electro-optic tuning or by turning the optical axis of the crystal (Lyot filter).

While method (a) is often used for high-resolution fluorescence spectroscopy with slow scan rates or for tuning pulsed dye lasers, method (b) is realized in a scanning confocal FPI (used as an optical spectrum analyzer) for monitoring the mode structure of lasers.

With a commercial spectrum analyzer, the transmitted wavelength λ can be repetitively scanned over more than one free spectral range with a saw-tooth voltage (Fig. 4.66) applied to the piezoelectric distance holder [154, 173]. Scanning rates up to several kilohertz are possible. Although the finesse of such devices may exceed 10^3, the hysteresis of piezoelectric crystals limits the accuracy of absolute wavelength calibration. Here a pressure-tuned FPI may be advantageous. The pressure change has to be sufficiently slow to avoid turbulence and temperature drifts. With a digitally pressure-scanned FPI, where the pressure of the gas in the interferometer chamber is changed by small, discrete steps, repetitive scans are reproduced within about 10^{-3} of the free spectral range [174].

Figure 4.66 Scanning confocal FPI with transmission peaks of a fundamental laser mode and sawtooth voltage at the piezo on one mirror

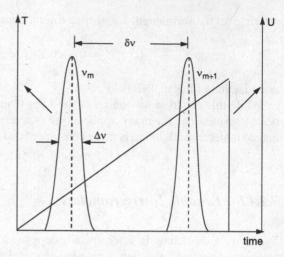

For fast wavelength tuning of dye lasers, Lyot filters with electro-optic tuning are employed within the laser resonator. A tuning range of a few nanometers can be repetitively scanned with rates up to 10^5 per second [175].

4.3   Comparison Between Spectrometers and Interferometers

When comparing the advantages and disadvantages of different dispersing devices for spectroscopic analysis, the characteristic properties of the instruments discussed in the foregoing sections, such as *spectral resolving power, étendue, spectral transmission*, and *free spectral* range, are important for the optimum choice. Of equal significance is the question of how accurately the wavelengths of spectral lines can be measured. To answer this question, further specifications are necessary, such as the backlash of monochromator drives, imaging errors in spectrographs, and hysteresis in piezo-tuned interferometers. In this section we shall treat these points in a comparison for different devices in order to give the reader an impression of the capabilities and limitations of these instruments.

4.3.1   Spectral Resolving Power

The spectral resolving power discussed for the different instruments in the previous sections can be expressed in a more general way, which applies to all devices with spectral dispersion based on interference effects. Let Δs_m be the maximum path difference between interfering waves in the instrument, e.g., between the rays from the first and the last groove of a grating (Fig. 4.67a) or between the direct beam and a beam reflected m times in a Fabry–Perot interferometer (Fig. 4.67b). Two wave-

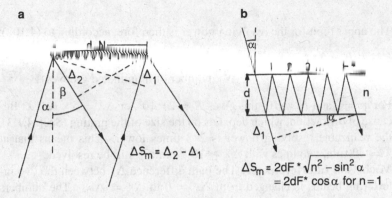

Figure 4.67 Maximum optical path difference and spectral resolving power: **a** in a grating spectrometer; **b** in a Fabry–Perot interferometer

lengths λ_1 and $\lambda_2 = \lambda_1 + \Delta\lambda$ can still be resolved if the number of wavelengths over this maximum path difference

$$\Delta s_m = 2m\lambda_2 = (2m+1)\lambda_1 , \quad m = \text{integer} ,$$

differs for the two wavelengths by at least one unit. In this case, an interference maximum for λ_1 coincides with the first minimum for λ_2. From the above equation we obtain the theoretical upper limit for the resolving power

$$\boxed{\frac{\lambda}{\Delta\lambda} = \frac{\Delta s_m}{\lambda} ,} \tag{4.102}$$

which is equal to the maximum path difference measured in units of the wavelength λ.

With the maximum time difference $\Delta T_m = \Delta s_m / c$ for traversing the two paths with the path difference Δs_m, we obtain with $\nu = c/\lambda$ from (4.102) for the minimum resolvable interval $\Delta\nu = -(c/\lambda^2)\Delta\lambda$,

$$\Delta\nu = 1/\Delta T_m \quad \Rightarrow \quad \Delta\nu \cdot \Delta T_m = 1 . \tag{4.103}$$

The product of the minimum resolvable frequency interval $\Delta\nu$ *and the maximum difference in transit times through the spectral apparatus is equal to 1.*

Example 4.20

a) **Grating Spectrometer:** The maximum path difference is, according to (4.30) and Fig. 4.67,

$$\Delta s_m = Nd(\sin\alpha - \sin\beta) = mN\lambda .$$

The upper limit for the resolving power is therefore, according to (4.102),

$$R = \lambda/\Delta\lambda = mN \quad (m : \text{diffraction order},$$
$$N : \text{number of illuminated grooves}).$$

For $m = 2$ and $N = 10^5$ this gives $R = 2 \times 10^5$, or $\Delta\lambda = 5 \times 10^{-6}\lambda$. Because of diffraction, which depends on the size of the grating (Sect. 4.1.3), the realizable resolving power is 2–3 times lower. This means that at $\lambda = 500$ nm, two lines with $\Delta\lambda \geq 10^{-2}$ nm can still be resolved.

b) **Michelson Interferometer:** The path difference Δs between the two interfering beams is changed from $\Delta s = 0$ to $\Delta s = \Delta s_m$. The numbers of interference maxima are counted for the two components λ_1 and λ_2 (Sect. 4.2.4). A distinction between λ_1 and λ_2 is possible if the number $m_1 = \Delta s/\lambda_1$ differs by at least 1 from $m_2 = \Delta s/\lambda_2$; this immediately gives (4.102). With a modern design, maximum path differences Δs up to several meters have been realized for wavelength measurements of stabilized lasers (Sect. 4.5.3). For $\lambda = 500$ nm and $\Delta s = 1$ m, we obtain $\lambda/\Delta\lambda = 2 \times 10^6$, which is one order of magnitude better than for the grating spectrometer.

c) **Fabry–Perot Interferometer:** The path difference is determined by the optical path difference $2nd$ between successive partial beams times the effective number of reflections, which can be expressed by the reflectivity finesse $F^* = \pi\sqrt{R}/(1-R)$. With ideal reflecting planes and perfect alignment, the maximum path difference would be $\Delta s_m = 2ndF^*$ and the spectral resolving power, according to (4.102), would be

$$\lambda/\Delta\lambda = F^*2nd/\lambda .$$

Because of imperfections of the alignment and deviations from ideal planes, the effective finesse is lower than the reflectivity finesse. With a value of $F^*_{\text{eff}} = 50$, which can be achieved, we obtain for $nd = 1$ cm

$$\lambda/\Delta\lambda = 2 \times 10^6 ,$$

which is comparable with the Michelson interferometer having $\Delta s_m = 100$ cm. However, with a confocal FPI, a finesse of $F^*_{\text{eff}} = 1000$ can be achieved. With $r = d = 4$ cm we then obtain

$$\lambda/\Delta\lambda = F^*4d/\lambda \approx 5 \times 10^8 ,$$

which means that for $\lambda = 500$ nm, two lines with $\Delta\lambda = 1 \times 10^{-6}$ nm ($\Delta\nu = 1$ MHz at $\nu = 5 \times 10^{14}\,\text{s}^{-1}$) are still resolvable, provided that their linewidth is sufficiently small. With high-reflection mirror coatings a finesse of $F^*_{\text{eff}} = 10^5$ has been realized. With $r = d = 1$ m this yields $\lambda/\Delta\lambda = 8 \times 10^{11}$ [160].

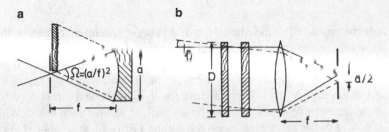

Figure 4.68 Acceptance angle of a spectrometer (**a**); and a Fabry–Perot interferometer (**b**)

4.3.2 Light-Gathering Power

The *light-gathering power*, or *étendue*, has been defined in Sect. 4.1.1 as the product $U = A\Omega$ of entrance area A and solid angle of acceptance Ω of the spectral apparatus. For most spectroscopic applications it is desirable to have an étendue U as large as possible to gain intensity. An equally important goal is to reach a maximum resolving power R. However, the two quantitites U and R are not independent of each other but are related, as can be seen from the following examples.

Example 4.21

a) **Spectrometer:** The area of the entrance slit with width b and height h is $A = b \cdot h$. The acceptance angle $\Omega = (a/f)^2$ is determined by the focal length f of the collimating lens or mirror and the diameter a of the limiting aperture in the spectrometer (Fig. 4.68a). We can write the étendue,

$$U = bha^2/f^2 ,$$

as the product of the area $A = bh$ and the solid angle $\Omega = (a/f)^2$. Using typical figures for a medium-sized spectrometer ($b = 10\,\mu$m, $h = 0.5$ cm, $a = 10$ cm, $f = 100$ cm) we obtain $\Omega = 0.01$, $A = 5 \times 10^{-4}$ cm$^2 \to U = 5 \times 10^{-6}$ cm^2 sr. With the resolving power $R = mN$, the product

$$RU = mNA\Omega \approx mN\frac{bha^2}{f^2} , \qquad (4.104a)$$

increases with the diffraction order m, the size a of the grating, the number of illuminated grooves N, and the slit area bh (as long as imaging errors can be neglected). For $m = 1$, $N = 10^5$, and the above figures for h, b, a, and f, we obtain $RU = 0.5$ cm^2 sr.

---

b) **Interferometer:** For the Michelson and Fabry–Perot interferometers, the allowable acceptance angle for photoelectric recording is limited by the aperture in front of the detector, which selects the central circular fringe. From Figs. 4.52 and 4.68b we see that the fringe images at the center and at the edge of the limiting aperture with diameter a are produced by incoming beams that are inclined by an angle ϑ against each other. With $a/2 = f\vartheta$, the solid angle accepted by the FPI is $\Omega = a^2/(4f^2)$. For a plate diameter D the étendue is then $U = \pi(D^2/4)\Omega$. According to (4.88) the spectral resolving power $R = \nu/\Delta\nu$ of a plane FPI is correlated with the étendue U by $R = \pi D^2(2U)^{-1}$. The product

$$RU = \pi D^2/2 \,, \tag{4.104b}$$

is, for a plane FPI, therefore solely determined by the plate diameter. For $D = 5\,\mathrm{cm}$, RU is about $40\,\mathrm{cm^2}$ sr, and therefore two orders of magnitude larger than for a grating spectrometer.

In Sect. 4.2.12 we saw that for a given resolving power the spherical FPI has a larger étendue for mirror separations $r > D^2/4d$. For Example 4.21 with $D = 5\,\mathrm{cm}$, $d = 1\,\mathrm{cm}$, the confocal FPI therefore gives the largest product RU of all interferometers for $r > 6\,\mathrm{cm}$. Because of the higher total finesse, however, the confocal FPI may be superior to all other instruments even for smaller mirror separations.

In summary, we can say that at comparable resolving power interferometers have a larger lightgathering power than spectrometers.

4.4 Accurate Wavelength Measurements

One of the major tasks for spectroscopists is the measurement of wavelengths of spectral lines. This allows the determination of molecular energy levels and of molecular structure. The attainable accuracy of wavelength measurements depends not only on the spectral resolution of the measuring device but also on the achievable signal-to-noise ratio and on the reproducibility of measured absolute wavelength values.

With the ultrahigh resolution, which can, in principle, be achieved with single-mode tunable lasers (Vol. 2, Chaps. 1–5), the accuracy of absolute wavelength measurements attainable with conventional techniques may not be satisfactory. New methods have been developed that are mainly based on interferometric measurements of laser wavelengths. For applications in molecular spectroscopy, the laser can be stabilized on the center of a molecular transition. Measuring the wavelength of such a stabilized laser yields simultaneously the wavelength of the molecular transition with a comparable accuracy. We shall briefly discuss some of these devices, often called *wavemeters*, that measure the unknown laser wavelength by comparison with a reference wavelength λ_R of a stabilized reference laser. Most proposals use for reference a HeNe laser, stabilized on a hyperfine component of a molecular iodine line, which has been measured by direct comparison with the primary wavelength standard to an accuracy of better than 10^{-10} [176].

Another method measures the absolute frequency ν_L of a stabilized laser and deduces the wavelength λ_L from the relation $\lambda_L = c/\nu_L$ using the *best* average of experimental values for the speed of light [177–179], which has been chosen to *define* the meter and thus the wavelength λ by the definition: 1 m *is the distance traveled by light in vacuum during the time* $\Delta t = 1/299,792,458 \, s^{-1}$. *This defines the speed of light* as

$$c = 299,792,458 \, m/s \,. \tag{4.105}$$

Such a scheme reduces the determination of lengths to the measurements of times or frequencies, which can be measured much more accurately than lengths [180]. Recently, the direct comparison of optical frequencies with the Cs standard in the microwave region has become possible with broadband frequency combs generated by visible femtosecond lasers. These frequency combs represent equidistant frequencies, separated by about 100 MHz, which span a wide frequency range, typically over 10^{14} Hz. They allow absolute frequency measurements. This method will be discussed in Vol. 2, Sect. 14.7.

4.4.1 Precision and Accuracy of Wavelength Measurements

Resolving power and light-gathering power are not the only criteria by which a wavelength-dispersing instrument should be judged. A very important question is the attainable *precision* and *accuracy* of absolute wavelength measurements.

To measure a physical quantity means to *compare* it with a reference standard. This comparison involves statistical and systematic errors. Measuring the same quantity n times will yield values X_i that scatter around the mean value

$$\overline{X} = \frac{1}{n} \sum_{i=1}^{n} X_i \,.$$

The attainable **precision** for such a set of measurements is determined by statistical errors and is mainly limited by the signal-to-noise ratio for a single measurement and by the number n of measurements (i.e., by the total measuring time). The precision can be characterized by the *standard deviation* [181, 182],

$$\sigma = \left(\sum_{i=1}^{n} \frac{(\overline{X} - X_i)^2}{n} \right)^{1/2} \,. \tag{4.106}$$

The adopted mean value $\overline{X}$, averaged over many measured values X_i, is claimed to have a certain *accuracy*, which is a measure of the reliability of this value, expressed by its probable deviation $\Delta\overline{X}$ from the unknown "true" value X. A stated accuracy of $\overline{X}/\Delta\overline{X}$ means a certain confidence that the true value X is within $\overline{X} \pm \Delta\overline{X}$. Since the accuracy is determined not only by statistical errors but, particularly, by systematic errors of the apparatus and measuring procedure, it is always lower than the precision. It is also influenced by the precision with which the reference standard can be measured and by the accuracy of its comparison with the value $\overline{X}$. Although the attainable accuracy depends on the experimental efforts and expenditures, the skill, imagination, and critical judgement of the experimentalist always have a major influence on the ultimate achieved and stated accuracy.

We shall characterize precision and accuracy by the relative uncertainties of the measured quantity X, expressed by the ratios

$$\frac{\sigma}{X} \quad \text{or} \quad \frac{\Delta\overline{X}}{X} \,,$$

respectively. A series of measurements with a standard deviation $\sigma = 10^{-8}\overline{X}$ has a relative uncertainty of 10^{-8} or a precision of 10^8. Often one says that the precision is 10^{-8}, although this statement has the disadvantage that a high precision is expressed by a small number.

Let us now briefly examine the attainable precision and accuracy of wavelength measurements with the different instruments discussed above. Although both quantities are correlated with the resolving power and the attainable signal-to-noise ratio, they are furthermore influenced by many other instrumental conditions, such as backlash of the monochromator drive, or asymmetric line profiles caused by imaging errors, or shrinking of the photographic film during the developing process. Without such additional error sources, the precision could be much higher than the resolving power, because the center of a symmetric line profile can be measured to a small fraction ϵ of the halfwidth. The value of ϵ depends on the attainable signal-to-noise ratio, which is determined, apart from other factors, by the étendue of the spectrometer. We see that for the precision of wavelength measurements, the product of resolving power R and étendue U, RU, discussed in the previous section, plays an important role.

For scanning monochromators with photoelectric recording, the main limitation for the attainable accuracy is the backlash of the grating-drive and nonuniformities of the gears, which limits the reliability of linear extrapolation between two calibration lines. Carefully designed monochromators have errors due to the drive that are less than $0.1\,\text{cm}^{-1}$, allowing a relative uncertainty of 10^{-5} or an accuracy of about 10^5 in the visible range.

In absorption spectroscopy with a tunable laser, the accuracy of line positions is also limited by the nonuniform scan speed $d\lambda/dt$ of the laser (Sect. 5.6). One has to record reference wavelength marks simultaneously with the spectrum in order to correct for the nonuniformities of $d\lambda/dt$.

A serious source of error with scanning spectrometers or scanning lasers is the distortion of the line profile and the shift of the line center caused by the time constant of the recording device. If the time constant τ is comparable with the time $\Delta t = \Delta\lambda/v_{sc}$ needed to scan through the halfwidth $\Delta\lambda$ of the line profile (which depends on the spectral resolution), the line becomes broadened, the maximum decreases, and the center wavelength is shifted. The line shift $\delta\lambda$ depends on the scanning speed v_{sc} [nm/min] and is approximately $\delta\lambda = v_{sc}\tau = (d\lambda/dt)\tau$ [122].

Example 4.22
With a scanning speed $v_{sc} = 10\,\text{nm/min}$ and a time constant of the recorder $\tau = 1\,\text{s}$ the line shift is already $\delta\lambda = 0.15\,\text{nm}$!

Because of the additional line broadening, the resolving power is reduced. If this reduction is to be less than $10\,\%$, the scanning speed must be below $v_{sc} < 0.24\Delta\lambda/\tau$. With $\Delta\lambda = 0.02\,\text{nm}$, $\tau = 1\,\text{s} \to v_{sc} < 0.3\,\text{nm/min}$.

Photographic recording avoids these problems and therefore allows a more accurate wavelength determination at the expense of an inconvenient developing process of the photoplate and the subsequent measuring procedure to determine the line positions. A typical figure for the standard deviation for a 3-m spectrograph is $0.01\,\text{cm}^{-1}$. Imaging errors causing curved lines, asymmetric line profiles due to misalignment, and backlash of the microdensitometer used for measuring the line positions on the photoplate are the main sources of errors.

Modern devices use photodiodes or CCD arrays (Sect. 4.5.2) instead of photoplates. With a diode width of $25\,\mu\text{m}$, the peak of a symmetric line profile extending over 3–5 diodes can be determined by a least-squares fit to a model profile within 1–5 μm, depending on the S/N ratio. When the array is placed behind a spectrometer with a dispersion of $1\,\text{mm/nm}$, the center of the line can be determined within $10^{-3}\,\text{nm}$. Since the signals are read electronically, there are no moving parts in the device and any mechanical error source (backlash) is eliminated.

The highest accuray (i.e., the lowest uncertainty) can be achieved with modern *wavemeters*, which we shall discuss in Sect. 4.4.2.

4.4.2 Today's Wavemeters

The different types of wavemeters for very accurate measurements of laser wavelengths are based on modifications of the Michelson interferometer [184], the Fizeau interferometer [185], or on a combination of several Fabry–Perot interferometers with different free spectral ranges [186–188]. The wavelength is measured either by monitoring the spatial distribution of the interference pattern with photodiode arrays, or by using traveling devices with electronic counting of the interference

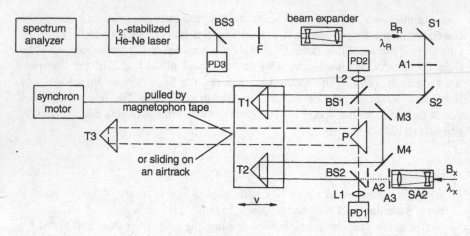

Figure 4.69 Traveling Michelson interferometer for accurate measurements of wavelengths of single-mode cw lasers

fringes. Nowadays several versions of wavemeters are commercially available which reach uncertainties of $\pm 0.2\,\text{pm}$ (accuracies $\nu/\delta\nu$ of about 10^{+7}). They can operate over a wide spectral range from $300\,\text{nm}$ to $5\,\mu\text{m}$.

a) The Michelson Wavemeter

Figure 4.69 illustrates the principle of a traveling-wave Michelson-type interferometer as used in our laboratory. Such a wavemeter was first demonstrated in a slightly different version by Hall and Lee [184] and by Kowalski et al. [190]. The beams B_R of a reference laser and B_x of a laser with unknown wavelength λ_x traverse the interferometer on identical paths, but in opposite directions. Both incoming beams are split into two partial beams by the beam splitters BS1 and BS2, respectively. One of the partial beams travels the constant path BS1–P–T3–P–BS2 for the reference beam, and in the opposite direction for the beam B_X. The second partial beam travels the variable path BS1–T1–M3–M4–T2–BS2 for B_R, and in the opposite direction for B_X. The moving corner-cube reflectors T1 and T2 are mounted on a carriage, which either travels with wheels on rods or slides on an airtrack.

The *corner-cube reflectors* guarantee that the incoming light beam is always reflected exactly parallel to its indicent direction, irrespective of slight misalignments or movements of the traveling reflector. The two partial beams (BS1–T1–M3–M4–T2–BS2 and BS1–P–T3–P–BS2) for the reference laser interfere at the detector PD1, and the two beams BS2–T2–M4–M3–T1–BS1 and BS2–P–T3–P–BS1 from the unknown laser interfere at the detector PD2. When the carriage is moving at a speed $v = \mathrm{d}x/\mathrm{d}t$ the phase difference $\delta(t)$ between the two interfering beams changes as

$$\delta(t) = 2\pi \frac{\Delta s}{\lambda} = 2\pi \cdot 4 \frac{\mathrm{d}x}{\mathrm{d}t} \frac{t}{\lambda} = 8\pi \frac{vt}{\lambda}\,, \qquad (4.107)$$

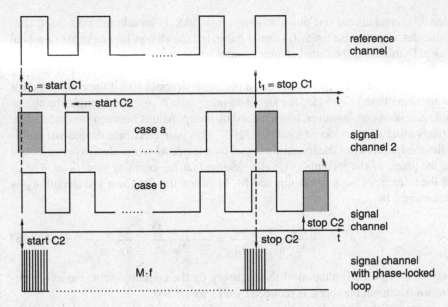

Figure 4.70 Signal sequences in the two detection channels of the traveling Michelson waveme-ter. The grey signal pulses are not counted

where the factor 4 stems from the fact that the optical path difference Δs has been doubled by introducing two corner-cube reflectors. The rates of interference maxima, which occur for $\delta = m2\pi$, are counted by PD2 for the unknown wavelength λ_X and by PD1 for the reference wavelength λ_R. The unknown wavelength λ_X can be obtained from the ratio of both counting rates if proper corrections are made for the dispersion $n(\lambda_R) - n(\lambda_X)$ of air. An electronic device produces a short voltage pulse each time the line-varying interference intensity passes through zero. These pulses are counted.

The signal lines to both counters are simultaneously opened at the time t_0 when the detector PD2 just delivers a trigger signal. Both counters are simultaneously stopped at the time t_1 when PD2 has reached the preset number N_0. From

$$\Delta t = t_1 - t_0 = N_0 \lambda_X / 4v = (N_R + \epsilon) \lambda_R / 4v ,$$

we obtain for the vacuum wavelength λ_X^0

$$\lambda_X^0 = \frac{N_R + \epsilon}{N_0} \lambda_R^0 \frac{n(\lambda_X, P, T)}{n(\lambda_R, P, T)} . \tag{4.108a}$$

The unknown fractional number $\epsilon < 2$ takes into account that the trigger signals from PD1, which define the start and stop times t_0 and t_1 (Fig. 4.69), may not exactly coincide with the pulse rise times in channel 2. The two worst cases are shown in Fig. 4.70. For case a, the trigger pulse at t_0 just misses the rise of the signal pulse, but the trigger at t_1 just coincides with the rise of a signal pulse. This means that the signal channel counts one pulse less than it should. In case b, the start pulse at t_0 coincides with the rise time of a signal pulse, but the stop pulse just misses a signal pulse. In this case, the signal channel counts one pulse more than it should.

For a maximum optical path difference $\Delta s = 4\,\mathrm{m}$, the number of counts for $\lambda = 500\,\mathrm{nm}$ is 8×10^6, which allows a precision of about 10^7, if the counting error is not larger than 1. Provided the signal-to-noise ratio is sufficiently high, the attainable precision can, however, be enhanced by interpolations between two successive counts using a *phase-locked loop* [191, 192]. This is an electronic device that multiplies the frequency of the incoming signal by a factor M while always being locked to the phase of the incoming signal. Assume that the counting rate $f_R = 4v/\lambda_R$ in the reference channel is multiplied by M. Then the unknown wavelength λ_X is determined by

$$\lambda_X^0 = \frac{MN_R + \epsilon}{MN_0}\lambda_R^0\frac{n_X}{n_R} = \frac{N_R + \varepsilon/M}{N_0}\lambda_R^0\frac{n_x}{n_R}. \tag{4.108b}$$

For $M = 100$ the limitation of the accuracy by the counting error due to the unknown fractional number ϵ is reduced by a factor of 100.

Instead of the phase-locked loop a coincidence curcuit may be employed. Here the signal paths to both counters are opened and closed at selected times t_0 and t_1, when both trigger signals from PD2 and PD1 coincide within a small time interval, say 10^{-8} s. Both techniques reduce the counting uncertainty to a value below 2×10^{-9}.

In general, the attainable accuracy, however, is lower because it is influenced by several sources of systematic errors. One is a misalignment of the interferometer, which causes both beams to travel slightly different path lengths. Another point that has to be considered is the curvature of the wavefronts in the diffraction-limited Gaussian beams (Sect. 5.3). This curvature can be reduced by expanding the beams through telescopes (Fig. 4.69). The uncertainty of the reference wavelength λ_R and the accuracy of measuring the refractive index $n(\lambda)$ of air are further error sources.

The maximum relative uncertainty of the absolute vacuum wavelength λ_X can be written as a sum of five terms:

$$\left|\frac{\Delta\lambda_X}{\lambda_X}\right| \leq \left|\frac{\Delta\lambda_R}{\lambda_R}\right| + \left|\frac{\epsilon}{MN_R}\right| + \left|\frac{\Delta r}{r}\right| + \left|\frac{\delta s}{\Delta s}\right| + \left|\frac{\delta\phi}{2\pi N_0}\right|, \tag{4.109}$$

where $r = n(\lambda_X)/n(\lambda_R)$ is the ratio of the refractive indices, δs is the difference of the travel paths for reference and signal beams, and $\delta\phi$ is the phase front variation in the detector plane. Let us briefly estimate the magnitude of the different terms in (4.109):

- The wavelength λ_R of the I_2-stabilized HeNe laser is known within an uncertainty $|\Delta\lambda_R/\lambda_R| < 10^{-10}$ [180]. Its frequency stability is better than $100\,\mathrm{kHz}$, i.e., $|\Delta v/v| < 2 \times 10^{-10}$. This means that the first term in (4.109) contributes at most 3×10^{-10} to the uncertainty of λ_X.
- With $\epsilon = 1.5$, $M = 100$, and $N_R = 8 \times 10^6$, the second term is about 2×10^{-9}.

- The index of refraction, $n(\lambda, p, T)$, depends on the wavelength λ, on the total air pressure, on the partial pressures of H_2O and CO_2, and on the temperature. If the total pressure is measured within 0.5 mbar, the temperature T within 0.1 K, and the relative humidity within 5%, the refractive index can be calculated from formulas given by Edlén [193] and Owens [194].

With the stated accuracies, the third term in (4.109) becomes

$$|\Delta r/r| \approx 1 \times 10^{-3} |n_0(\lambda_X) - n_0(\lambda_R)| , \qquad (4.110)$$

where n_0 is the refractive index for dry air under standard conditions ($T_0 = 15\,°C$, $p_0 = 1013\,hPa$). The contribution of the third term depends on the wavelength difference $\Delta\lambda = \lambda_R - \lambda_X$. For $\Delta\lambda = 1$ nm one obtains $|\Delta r/r| < 10^{-11}$, while for $\Delta\lambda = 200$ nm this term becomes, with $|\Delta r/r| \approx 5 \times 10^{-9}$, a serious limitation of the accuracy of $|\Delta\lambda_X/\lambda_X|$.

- The magnitude of the fourth term $|\delta s/\Delta s|$ depends on the effort put into the alignment of the two laser beams within the interferometer. If the two beams are tilted against each other by a small angle α, the two path lengths for λ_X and λ_R differ by

$$\delta s = \Delta s(\lambda_R) - \Delta s(\lambda_X) = \Delta s_R(1 - \cos\alpha) \approx (\alpha^2/2)\Delta s_R .$$

With $\alpha = 10^{-4}$ rad, the systematic relative error becomes

$$|\delta s/\Delta s| \approx 5 \times 10^{-9} .$$

It is therefore necessary to align both beams very carefully.

- With a surface quality of $\lambda/10$ for all mirrors and beam splitters, the distortions of the wavefront are already visible in the interference pattern. However, plane waves are focused onto the detector area and the phase of the detector signal is due to an average over the cross section of the enlarged beam ($\approx 1\,cm^2$). This averaging minimizes the effect of wavefront distortion on the accuracy of λ_X. If the modulation of the interference intensity (4.37) exceeds 90 %, this term may be neglected.

With careful alignment, good optical quality of all optical surfaces and accurate recording of p, T, and P_{H_2O}, the total uncertainty of λ_X can be pushed below 10^{-8}. This gives an absolute uncertainty $\Delta\nu_x \approx 3\,MHz$ of the optical frequency $\nu_x = 5 \times 10^{14}\,s^{-1}$ for a wavelength separation between λ_R and λ_x of $\Delta\lambda \approx 120$ nm. This has been proved by a comparison of independently measured wavelengths $\lambda_x = 514.5$ nm (I_2-stabilized argon laser) and $\lambda_R = 632.9$ nm (I_2-stabilized HeNe laser) [195].

When cw dye laser wavelengths are measured, another source of error arises. Due to air bubbles in the dye jet or dust particles within the resonator beam waist, the dye laser emission may be interrupted for a few microseconds. If this happens while counting the wavelength a few counts are missing. This can be avoided by using an additional phase-locked loop with a multiplication factor $M_x = 1$ in the counting channel of PD_x. If the time constant of the phase-locked loop is larger than $10 \, \mu s$, it continues to oscillate at the counting frequency during the few microseconds of dye laser beam interruptions.

There are several different designs of Michelson wavemeters that are commercially available and are described in [197–199].

b) Sigmameter

While the traveling Michelson is restricted to cw lasers, a motionless Michelson interferometer was designed by Jacquinot, et al. [200], which includes no moving parts and can be used for cw as well as for pulsed lasers. Figure 4.71 illustrates

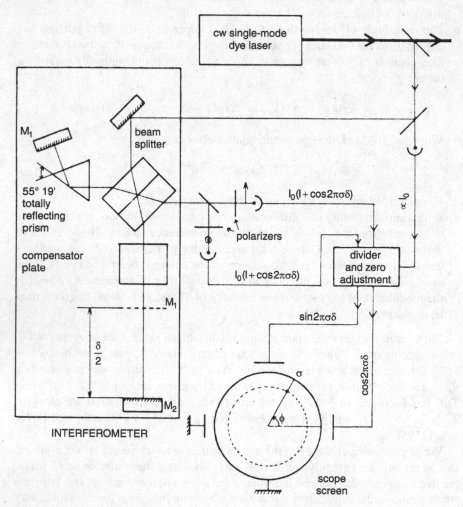

Figure 4.71 Sigmameter [200]

its operation. The basic element is a Michelson interferometer with a *fixed* path difference δ. The laser beam enters the interferometer polarized at 45° with respect to the plane of Fig. 4.71. When inserting a prism into one arm of the interferometer, where the beam is totally reflected at the prism base, a phase difference $\Delta\varphi$ is introduced between the two components polarized parallel and perpendicular to the totally reflecting surface. The value of $\Delta\varphi$ depends, according to Fresnel's formulas [129], on the incidence angle α and can be made $\pi/2$ for $\alpha = 55°19'$ and $n = 1.52$. The interference signal at the exit of the interferometer is recorded separately for the two polarizations and one obtains, because of the phase shifts $\pi/2$, $I_{\parallel} = I_0(1 + \cos 2\pi\delta/\lambda)$ and $I_{\perp} = I_0(1 + \sin 2\pi\delta/\lambda)$. From these signals it is possible to deduce the wave number $\sigma = 1/\lambda$ modulo $1/\delta$, since all wave numbers $\sigma_m = \sigma_0 + m/\delta$ ($m = 1, 2, 3, \ldots$) give the same interference signals. Using several interferometers of the same type with a common mirror M1 but different positions of M2, which have path differences in geometric ratios, such as 50 cm, 5 cm, 0.5 cm, and 0.05 cm, the wave number σ can be deduced unambiguously with an accuracy determined by the interferometer with the highest path difference. The actual path differences δ_i are calibrated with a reference line and are servo-locked to this line. The precision obtained with this instrument is about 5 MHz, which is comparable with that of the traveling Michelson interferometer. The measuring time, however, is much less since the different δ_i can be determined simultaneously. This instrument is more difficult to build but easier to handle. Since it measures wave numbers $\sigma = 1/\lambda$, the inventors called it a *sigmameter*.

c) Computer-Controlled Fabry–Perot Wavemeter

Another approach to accurate wavelength measurements of pulsed and cw lasers, which can be also applied to incoherent sources, relies on a combination of a small grating monochromator and three Fabry–Perot etalons [186–188]. The incoming laser beam is sent simultaneously through the monochromator and three temperature-stabilized Fabry–Perot interferometers with different free spectral ranges $\delta\nu_i$ (Fig. 4.72). In order to match the laser beam profile to the sensitive area of the linear diode arrays (25 mm × 50 μm), focusing with cylindrical lenses Z_i is utilized. The divergence of the beams in the plane of Fig. 4.72 is optimized by the spherical lenses L_i in such a way that the diode arrays detect 4–6 FPI fringes (Fig. 4.73). The linear arrays have to be properly aligned so that they coincide with a diameter through the center of the ring system. According to (4.72), the wavelength λ can be determined from the ring diameters D_p and the excess ϵ, provided the integer order m_0 is known, which means that λ must already be known at least within one-half of a free spectral range (Sect. 4.3).

The device is calibrated with different lines from a cw dye laser that are simultaneously measured with the traveling Michelson wavemeter (see above). This calibration allows:

- The unambiguous correlation between wavelength λ and the position of the illuminated diode of array 1 behind the monochromator with an accuracy of

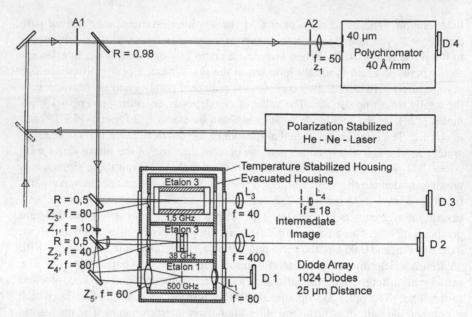

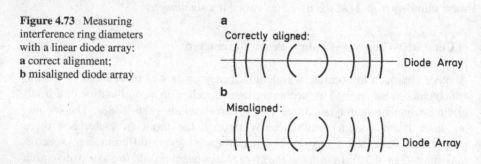

Figure 4.72 Wavemeter for pulsed and cw lasers, based on a combination of a small polychromator and three FPI with widely differing free spectral ranges [195]

Figure 4.73 Measuring
interference ring diameters
with a linear diode array:
a correct alignment;
b misaligned diode array

± 0.1 nm, which is sufficient to determine λ within 0.5 of the free spectral range
of etalon 1;

- The accurate determination of nd for all three FPI.

If the free spectral range $\delta\nu_1$ of the thin FPI is at least twice as large as the uncertainty $\Delta\nu$ of the monochromator measurement, the integer order m_0 of FPI1 can be unambiguously determined. The measurement of the ring diameters improves the accuracy by a factor of about 20. This is sufficient to determine the larger integer order m_0 of FPI2; from its ring diameters, λ can be measured with an accuracy 20 times higher than that from FPI1. The final wavelength determination uses the ring diameters of the large FPI3. Its accuracy reaches about 1 % of the free spectral range of FPI3.

Figure 4.74 Output signals at the polychromator and the three diode arrays of the FPI wavemeter, which had been illuminated by a cw HeNe laser oscillating on two axial modes (**a–d**). The lowest figure shows the ring intensity pattern of an excimer-pumped single-mode dye laser measured behind a FPI with 3.3 GHz free spectral range [187]

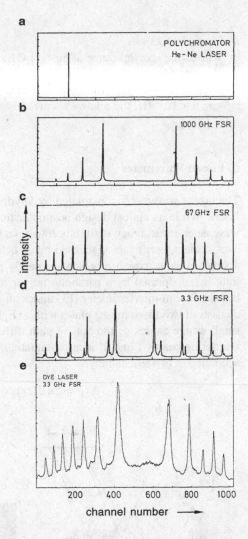

The whole measuring cycle is controlled by a computer. For pulsed lasers, one pulse (with an energy of $\geq 5\,\mu$J) is sufficient to initiate the device, while for cw lasers, a few microwatts input power are sufficient. The arrays are read out by the computer and the signals can be displayed on a screen. Such signals for the arrays D1–D4 are shown in Fig. 4.74 for a HeNe laser oscillating on two longitudinal modes and for a pulsed dye laser.

Since the optical distances $n_i d_i$ of the FPI depend critically on temperature and pressure, all FPI must be kept in a temperature-stabilized pressure-tight box. Furthermore, a stabilized HeNe laser can be used to control long-term drift of the FPI [195].

Example 4.23

With a free spectral range of $\delta\nu = 1\,\text{GHz}$, the uncertainty of calibration and of the determination of an unknown wavelength are both about $10\,\text{MHz}$. This gives an absolute uncertainty of less than $20\,\text{MHz}$. For the optical frequency $\nu = 6 \times 10^{14}\,\text{Hz}$, the relative accuracy is then $\Delta\nu/\nu \leq 3 \times 10^{-8}$.

d) Fizeau Wavemeter

The Fizeau wavemeter constructed by Snyder [201] can be used for pulsed and cw lasers. While its optical design is simpler than that of the sigmameter and the FPI wavemeter, its accuracy is slightly lower. Its basic principle is shown in Fig. 4.75b. The incident laser beam is focused by an achromatic microscope lens system onto a small pinhole, which represents a nearly pointlike light source. The divergent light is transformed by a parabolic mirror into an enlarged parallel beam, which hits the Fizeau interferometer (FI) under an incident angle α (Fig. 4.75a). The FI consists of two fused quartz plates with a slightly wedged air gap ($\phi \approx 1/20°$). For small wedge angles ϕ, the optical path difference Δs between the constructively interfering beams 1 and 1' is approximately equal to that of a plane-parallel plate according to (4.48a), namely

$$\Delta s_1 = 2nd(z_1)\cos\beta = m\lambda \ .$$

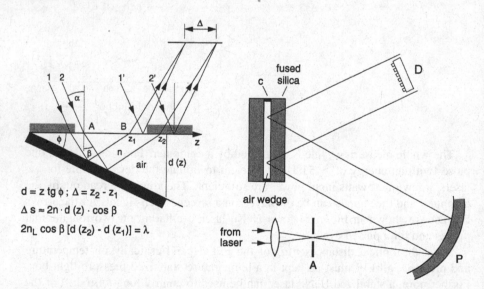

Figure 4.75 Fizeau wavemeter: **a** interference at a wedge (the wedge angle ϕ is greatly exaggerated); **b** schematic design; A, aperture as spatial filter; P, parabolic mirror; C, distance holder of cerodur; D, diode array

Figure 4.76 Densitometer trace of the fringe pattern in a Fizeau wavemeter [185]

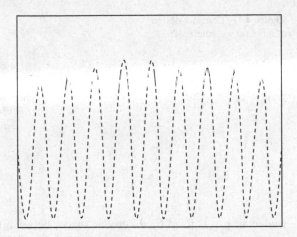

The path difference between the beams 2 and 2′, which belong to the next interference order, is $\Delta s_2 = (m + 1)\lambda$. The interference of the reflected light produces a pattern of parallel fringes (Fig. 4.76) with the separation

$$\Delta = z_2 - z_1 = \frac{d(z_2) - d(z_1)}{\tan\phi} = \frac{\lambda}{2n\tan\phi\cos\beta}, \qquad (4.111)$$

which depends on the wavelength λ, the wedge angle ϕ, the angle of incidence α, and the refractive index n of air.

Changing the wavelength λ causes a shift Δz of the fringe pattern and a slight change of the fringe separation Δ. For a change of λ by one free spectral range

$$\delta\lambda = \frac{\lambda^2}{2nd\cos\beta}. \qquad (4.112)$$

and Δz is equal to the fringe separation Δ. Therefore the two fringe patterns for λ and $\lambda + \delta\lambda$ look identical, apart from the slight change of Δ. It is therefore essential to know λ at least within $\pm\delta\lambda/2$. This is possible from a measurement of Δ. With a diode array of 1024 diodes, the fringe separation Δ can be obtained from a least-squares fit to the measured intensity distribution $I(z)$ with a relative accuracy of 10^{-4}, which yields an absolute value of λ within $\pm 10^{-4}\lambda$ [202].

With a value $d = 1\,\text{mm}$ of the air gap, the order of interference m is about 3000 at $\lambda = 500\,\text{nm}$. An accuracy of 10^{-4} is therefore sufficient for the unambiguous determination of m. Since the *position* of the interference fringes can be measured within 0.3 % of the fringe separation, the wavelength λ can be obtained within 0.3 % of a free spectral range, which gives the accuracy $\lambda/\Delta\lambda \approx 10^7$. The preliminary value of λ, deduced from the fringe separation Δ, and the final value, determined from the *fringe position*, are both obtained from the same FI after having calibrated the system with lines of known wavelengths.

The advantage of the Fizeau wavemeter is its compact design and its low price. A very elegant construction by Gardner [203, 204] is sketched in Fig. 4.77. The

Figure 4.77 Compact design
of a Fizeau wavemeter [203]

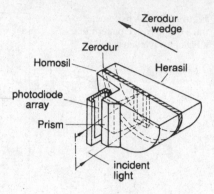

wedge air gap is fixed by a Zerodur spacer between the two interferometer plates
and forms a pressure tight volume. Variations of air pressure in the surroundings
therefore do not cause changes of n within the air gap. The reflected light is sent
to the diode array by a totally reflecting prism. The data are processed by a small
computer [205].

4.5   Detection of Light

For many applications in spectroscopy the sensitive detection of light and the accurate measurement of its intensity are of crucial importance for the successful
performance of an experiment. The selection of the proper detector for optimum
sensitivity and *accuracy* for the detection of radiation must take into account the
following characteristic properties, which may differ for the various detector types:

- The spectral relative response $R(\lambda)$ of the detector, which determines the wavelength range in which the detector can be used. The knowledge of $R(\lambda)$ is
  essential for the comparison of the true relative intensities $I(\lambda_1)$ and $I(\lambda_2)$ at
  different wavelengths.
- The absolute sensitivity $S(\lambda) = V_s/P$, which is defined as the ratio of output
  signal V_s to incident radiation power P. If the output is a voltage, as in photovoltaic devices or in thermocouples, the sensitivity is expressed in units of volts
  per watt. In the case of photocurrent devices, such as photomultipliers, $S(\lambda)$
  is given in amperes per watt. With the detector area A the sensitivity S can be
  expressed in terms of the irradiance I:

$$S(\lambda) = V_s/(AI) . \tag{4.113}$$

- The achievable signal-to-noise ratio V_s/V_n, which is, in principle, limited by
  the noise of the incident radiation. It may, in practice, be further reduced by
  inherent noise of the detector. The detector noise is often expressed by the *noise
  equivalent input power* (NEP), which means an incident radiation power that

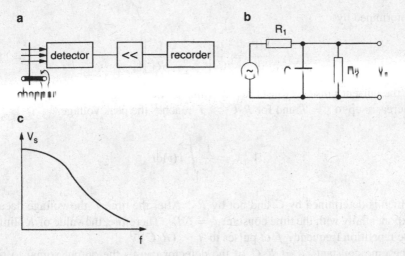

Figure 4.78 Typical detector: **a** schematic setup; **b** equivalent electrical circuit; **c** frequency response $V_s(f)$

generates the same output signal as the detector noise itself, thus yielding the signal-to-noise ratio $S/N = 1$. In infrared physics a figure of merit for the infrared detector is the detectivity

$$D = \frac{\sqrt{A\Delta f}}{P} \frac{V_s}{V_n} = \frac{\sqrt{A\Delta f}}{\text{NEP}} . \tag{4.114}$$

The specific detectivity D^* [cm s$^{-1/2}$ W^{-1}] gives the obtainable signalto-noise ratio V_s/V_n of a detector with the sensitive area $A = 1\,\text{cm}^2$ and the detector bandwidth $\Delta f = 1\,\text{Hz}$, at an incident radiation power of $P = 1\,\text{W}$. Because the noise equivalent input power is NEP $= P \cdot V_n/V_s$, the specific detectivity of a detector with the area $1\,\text{cm}^2$ and a bandwidth of $1\,\text{Hz}$ is $D^* = 1/\text{NEP}$.

- The maximum intensity range in which the detector response is linear. It means that the output signal V_s is proportional to the incident radiation power P. This point is particularly important for applications where a wide range of intensities is covered. Examples are output-power measurements of pulsed lasers, Raman spectroscopy, and spectroscopic investigations of line broadening, when the intensities in the line wings may be many orders of magnitude smaller than at the center.
- The time or frequency response of the detector, characterized by its time constant τ. Many detectors show a frequency response that can be described by the model of a capacitor, which is charged through a resistor R_1 and discharged through R_2 (Fig. 4.78b). When a very short light pulse falls onto the detector, its output pulse is smeared out. If the output is a current $i(t)$ that is proportional to the incident radiation power $P(t)$ (as, for example, in photomultipliers), the output capacitance C is charged by this current and shows a voltage rise and fall,

determined by

$$\frac{dV}{dt} = \frac{1}{C}\left[i(t) - \frac{V}{R_2}\right] .$$

(4.115)

If the current pulse $i(t)$ lasts for the time T, the voltage $V(t)$ at the capacitor increases up to $t = T$ and for $R_2 C \gg T$ reaches the peak voltage

$$V_{max} = \frac{1}{C}\int_0^T i(t)dt ,$$

which is determined by C and not by R_2! After the time T the voltage decays exponentially with the time constant $\tau = CR_2$. Therefore, the value of R_2 limits the repetition frequency f of pulses to $f < (R_2 C)^{-1}$.

The time constant $\tau_1 = R_1 C$ of the detector causes the output signal to rise slower than the incident input pulse and the time constant $\tau_2 = R_2 C$ causes a slower decay than that of the input pulse. The time constants can be determined by modulating the continuous input radiation at the frequency f. The output signal of such a device is characterized by (see Example 4.14)

$$V_s(f) = \frac{V_s(0)}{\sqrt{1 + (2\pi f \tau)^2}} ,$$

(4.116)

where $\tau = CR_1R_2/(R_1 + R_2) = \tau_1 \cdot \tau_2/(\tau_1 + \tau_2)$. At the modulation frequency $f = 1/(2\pi\tau)$, the output signal has decreased to $1/\sqrt{2}$ of its dc value. The knowledge of the detector time constant τ is essential for all applications where fast transient phenomena are to be monitored, such as atomic lifetimes or the time dependence of fast laser pulses (Vol. 2, Chap. 6).

• The price of a detector is another factor that cannot be ignored, since unfortunately it often restricts the optimum choice.

In this section we briefly discuss some detectors that are commonly used in laser spectroscopy. The different types can be divided into two categories, *thermal detectors* and *direct photodetectors*. In thermal detectors, the energy absorbed from the incident radiation raises the temperature and causes changes in the temperature-dependent properties of the detector, which can be monitored. Direct photodetectors are based either on the emission of photoelectrons from photocathodes, or on changes of the conductivity of semiconductors due to incident radiation, or on photovoltaic devices where a voltage is generated by the internal photoeffect. Whereas thermal detectors have a *wavelength-independent* sensitivity, photodetectors show a spectral response that depends on the work function of the emitting surface or on the band gap in semiconductors.

During recent years the development of image intensifiers, image converters, CCD cameras, and vidicon detectors has made impressive progress. At first pushed by military demands, these devices are now coming into use for light detection at

low levels, e.g., in Raman spectroscopy, or for monitoring the faint fluorescence
of spurious molecular constituents. Because of their increasing importance we
give a short survey of the principles of these devices and their application in laser
spectroscopy. In time-resolved spectroscopy, subnanosecond detection can now be
performed with fast phototubes in connection with transient digitizers, which resolve time intervals of less than 100 ps. Since such time-resolved experiments in
laser spectroscopy with streak cameras and correlation techniques are discussed in
Vol. 2, Chap. 6, we confine ourselves here to discussing only some of these modern
devices from the point of view of spectroscopic instrumentation. A more extensive treatment of the characteristics and the performance of various detectors can
be found in special monographs on detectors [206, 208, 211–217]. For reviews on
photodetection techniques relevant in laser physics, see also [218–221].

4.5.1 Thermal Detectors

Because of their wavelength-independent sensitivity, thermal detectors are useful for calibration purposes, e.g., for an absolute measurement of the radiation
power of cw lasers, or of the output energy of pulsed lasers. In the rugged form
of medium-sensitivity calibrated calorimeters, they are convenient devices for any
laser laboratory. With more sophisticated and delicate design, they have been developed as sensitive detectors for the whole spectral range, particularly for the infrared
region, where other sensitive detectors are less abundant than in the visible range.

For a simple estimate of the sensitivity and its dependence on the detector parameters, such as the heat capacitance and thermal losses, we shall consider the following model [222]. Assume that the fraction β of the incident radiation power P
is absorbed by a thermal detector with heat capacity H, which is connected to a heat
sink at constant temperature T_s (Fig. 4.79a). When G is the thermal conductivity of
the link between the detector and the heat sink, the temperature T of the detector
under illumination can be obtained from

$$\beta P = H \frac{dT}{dt} + G(T - T_s) . \tag{4.117}$$

If the time-independent radiation power P_0 is switched on at $t = 0$, the time-
dependent solution of (4.117) is

$$T = T_s + \frac{\beta P_0}{G}(1 - e^{-(G/H)t}) . \tag{4.118}$$

The temperature T rises from the initial value T_s at $t = 0$ to the temperature $T = T_s + \Delta T$ for $t = \infty$ with the time constant $\tau = H/G$. The temperature rise for
$t = \infty$

$$\Delta T = \frac{\beta P_0}{G} \tag{4.119}$$

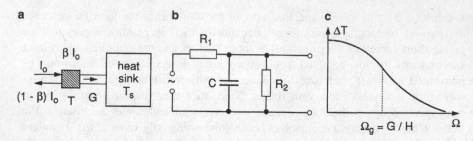

Figure 4.79 Model of a thermal detector: **a** schematic diagram; **b** equivalent electrical circuit; **c** frequency response $\Delta T(\Omega)$

is inversely proportional to the thermal losses G and does not depend on the heat capacity H, while the time constant of the rise $\tau = H/G$ depends on the ratio of both quantities. Small values of G make a thermal detector sensitive, but slow! It is therefore essential to realize small values of both quantities (H and G).

In general, P will be time dependent. When we assume the periodic function

$$P = P_0(1 + a\cos\Omega t)\,, \quad |a| \le 1\,, \tag{4.120}$$

we obtain, inserting (4.120) into (4.117), a detector temperature of

$$T(\Omega) = T_s + \Delta T(1 + \cos(\Omega t + \varphi))\,, \tag{4.121}$$

which depends on the modulation frequency Ω, and which shows a phase lag ϕ determined by

$$\tan\phi = \Omega H/G = \Omega\tau\,, \tag{4.122a}$$

and a modulation amplitude

$$\Delta T = \frac{a\beta P_0}{\sqrt{G^2 + \Omega^2 H^2}} = \frac{a\beta P_0}{G\sqrt{1 + \Omega^2\tau^2}}\,. \tag{4.122b}$$

At the frequency $\Omega_g = G/H = 1/\tau$, the amplitude ΔT decreases by a factor of $\sqrt{2}$ compared to its DC value.

Note The problem is equivalent to the analogous case of charging a capacitor ($C \leftrightarrow H$) through a resistor R_1 that discharges through R_2 ($R_2 \leftrightarrow 1/G$) (the charging current i corresponds to the radiation power P). The ratio $\tau = H/G$ ($H/G \leftrightarrow R_2C$) determines the time constant of the device (Fig. 4.79b).

We learn from (4.122b) that the sensitivity $S = \Delta T/P_0$ becomes large if G and H are made as small as possible. For modulation frequencies $\Omega > G/H$,

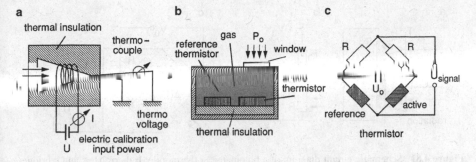

Figure 4.80 Calorimeter for measuring the output power of cw lasers or the output energy of pulsed lasers: **a** experimental design; **b** calorimeter with active irradiated thermistor and nonirradiated reference thermistor; **c** balanced bridge circuit

the amplitude ΔT will decrease approximately inversely to Ω. Since the time constant $\tau = H/G$ limits the frequency response of the detector, *a fast and sensitive detector should have a minimum heat capacity H*.

Since the specific heat decreases with decreasing temperature, thermal detectors with fast response but still high sensitivity, i.e. large value of G, should be operated at low temperatures.

For the calibration of the output power from cw lasers, the demand for high sensitivity is not as relevant since, in general, sufficiently large radiation power is available. Figure 4.80 depicts a simple home-made calorimeter and its circuit diagram. The radiation falls through a hole into a metal cone with a black inner surface. Because of the many reflections, the light has only a small chance of leaving the cone, ensuring that all light is absorbed. The absorbed power heats a thermocouple or a temperaturedependent resistor (thermistor) embedded in the cone. For calibration purposes, the cone can be heated by an electric wire. If the detector represents one part of a bridge (Fig. 4.80c) that is balanced for the electric input $W = UI$, but without incident radiation, the heating power has to be reduced by $\Delta W = P$ to maintain the balance with the incident radiation power P.

A system with higher accuracy uses the difference in output signals of two identical cones, where only one is irradiated (Fig. 4.80b).

For the measurement of output energies from pulsed lasers, the calorimeter should integrate the absorbed power at least over the pulse duration. From (4.117) we obtain

$$\int_0^{t_0} \beta P \, dt = H \Delta T + \int_0^{t_0} G(T - T_s) \, dt \; . \tag{4.123}$$

When the detector is thermally isolated, the heat conductivity G is small, therefore the second term may be completely neglected for sufficiently short pulse dura-

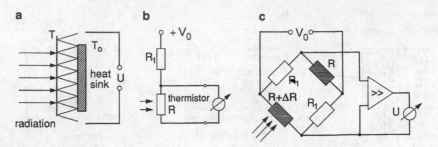

Figure 4.81 Schematic circuit diagram of a bolometer: **a** thermopile; **b** thermistor; and **c** bridge circuit with difference amplifier

tions t_0. The temperature rise

$$\Delta T = \frac{1}{H} \int\limits_0^{t_0} \beta P \, dt \, , \qquad (4.124)$$

is then directly proportional to the input energy. Instead of the cw electric input for calibration (Fig. 4.80a), now a charged capacitor C is discharged through the heating coil. If the discharge time is matched to the laser pulse time, the heat conduction is the same for both cases and does not enter into calibration. If the temperature rise caused by the discharge of the capacitor equals that caused by the laser pulse, the pulse energy is $\frac{1}{2}CU^2$.

For more sensitive detection of low incident powers, *bolometers* and *Golay cells* are used. A special design for a bolometer consists of N thermocouples in series, where one junction touches the backside of a thin electrically insulating foil that is exposed to the incident radiation (Fig. 4.81a). The other junction is in contact with a heat sink. The output voltage is

$$U = N \frac{dU}{dT} \Delta T \, ,$$

where dU/dT is the sensitivity of a single thermocouple.

Another version utilizes a thermistor that consists of a material with a large temperature coefficient $\alpha = (dR/dT)/R$ of the electrical resistance R. If a constant current i is fed through R (Fig. 4.81b), the incident power P that causes a temperature increase ΔT produces the voltage output signal

$$\Delta U = i \Delta R = i R \alpha \Delta T = \frac{V_0 R}{R + R_1} \alpha \Delta T \, , \qquad (4.125)$$

where ΔT is determined from (4.121) as $\Delta T = \beta P (G^2 + \Omega^2 H^2)^{-1/2}$. The response $\Delta U/P$ of the detector is therefore proportional to i, R, and α, and decreases with increasing H and G. At a constant supply voltage V_0, the current change Δi caused by the irradiation is, for $\Delta R \ll R + R_1$,

$$\Delta i = V_0 \left(\frac{1}{R_1 + R} - \frac{1}{R_1 + R + \Delta R} \right) \approx V_0 \frac{\Delta R}{(R_1 + R)^2} , \tag{4.121}$$

and can be generally neglected.

Since the input impedance of the following amplifier has to be larger than R, this puts an upper limit on R. Because any fluctuation of i causes a noise signal, the current i through the bolometer has to be extremely constant. This and the fact that the temperature rise due to Joule's heating should be small, limits the maximum current through the bolometer.

Equations (4.125 and 4.121) demonstrate again that small values of G and H are desirable. Even with perfect thermal isolation, heat radiation is still present and limits the lower value of G. At the temperature difference ΔT between a bolometer and its surroundings, the Stefan–Boltzmann law gives for the net radiation flux ΔP to the surroundings from the detector with the emitting area A^* and the emissivity $\epsilon \leq 1$

$$\Delta P = 4A\epsilon\sigma T^3 \Delta T , \tag{4.127}$$

where $\sigma = 5.77 \times 10^{-8} \, \text{W/m}^2 \, \text{K}^{-4}$ is the Stefan–Boltzmann constant. The minimum thermal conductivity is therefore

$$G_m = 4A\sigma\epsilon T^3 , \tag{4.128}$$

even for the ideal case where no other heat links to the surroundings exist. This limits the detection sensitivity to a minimum input radiation of about $10^{-10} \, \text{W}$ for detectors operating at room temperatures and with a bandwidth of 1 Hz. It is therefore advantageous to cool the bolometer, which furthermore decreases the heat capacity.

This cooling has the additional advantage that the slope of the function dR/dT becomes larger at low temperatures T. Two different materials can be utilized, as discussed below.

In semiconductors the electrical conductivity is proportional to the electron density n_e in the conduction band. With the band gap ΔE_G this density is, according to the Boltzmann relation

$$\frac{n_e(T)}{n_e(T + \Delta T)} = \exp\left(-\frac{\Delta E_G \Delta T}{2kT^2}\right) , \tag{4.129}$$

and is very sensitively dependent on temperature.

The quantity dR/dT becomes exceedingly large at the critical temperature T_c of superconducting materials. If the bolometer is always kept at this temperature T_c by a temperature control, the incident radiation power P can be very sensitively measured by the magnitude of the feedback control signal used to compensate for the absorbed radiation power [224–226].

Figure 4.82 Thermal excitation of electrons from donor levels into the conduction band

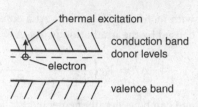

thermal excitation

conduction band
donor levels

electron

valence band

Example 4.24

With $\int P \, dt = 10^{-12}$ Ws, $\beta = 1$, $H = 10^{-11}$ Ws/K we obtain from (4.124): $\Delta T = 0.1$ K. With $\alpha = 10^{-4}$/K and $R = 10\,\Omega$, $R_1 = 10\,\Omega$, $V_0 = 1$ V, the current change is $\Delta i = 2.5 \times 10^{-6}$ A and the voltage change is $\Delta V = R\Delta i = 2.5 \times 10^{-5}$ V, which is readily detected.

Another material used for sensitive bolometers is a thin small disc of doped silicon, where the dopants are donor atoms with energy levels slightly below the conduction band (Fig. 4.82). A small temperature rise ΔT increases the fraction of ionized donors exponentially, thus producing free electrons in the conduction band. Such bolometers have to be operated at low temperatures in order to increase their sensitivity. The detectivity D^* (4.114) increases with falling temperature because the noise decreases (Fig. 4.83).

For thermal detectors the heat conduction G limits the sensitivity and the heat capacity H the frequency response. Since the specific heat decreases with the temperature, low temperatures improve both the sensitivity and the frequency response.

In Fig. 4.84 the whole setup for a bolometer operated at liquid helium temperatures is shown, including the liquid nitrogen and helium containers. Pumping the evaporating helium gas away drops the temperature below 1.5 K. The cold apertures

Figure 4.83 Specific detectivity D^* as a function of bolometer temperature

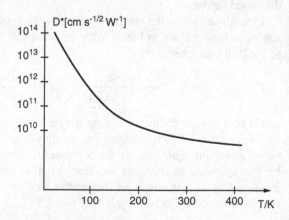

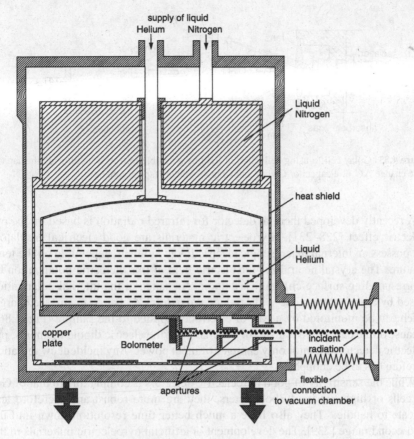

Figure 4.84 Bolometer with helium cryostat

in front of the bolometer disc stop thermal radiation from the walls of the vacuum vessel from reaching the detector. Using such a device radiation powers of less than 10^{-13} W can still be measured.

The Golay cell uses another method of thermal detection of radiation, namely the absorption of radiation in a closed gas capsule. According to the ideal gas law, the temperature rise ΔT causes the pressure rise $\Delta p = N(R/V)\Delta T$ (where N is the number of moles and R the gas constant), which expands a flexible membrane on which a mirror is mounted (Fig. 4.85a). The movement of the mirror is monitored by observing the deflection of a light beam from a light-emitting diode [227].

In modern devices the flexible membrane is part of a capacitor with the other plate fixed. The pressure rise causes a corresponding change of the capacitance, which can be converted to an AC voltage (Fig. 4.85b). This sensitive detector, which is essentially a *capacitor microphone*, is now widely used in photoacoustic spectroscopy (Vol. 2, Sect. 6.3) to detect the absorption spectrum of molecular gases by the pressure rise proportional to the absorption coefficient.

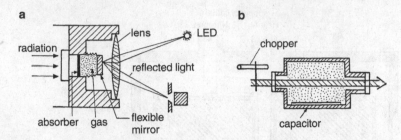

Figure 4.85 Golay cell: **a** using deflection of light by a flexible mirror; **b** monitoring the capacitance change ΔC of a capacitor C with a flexible membrane (spectraphone)

A recently developed thermal detector for infrared radiation is based on the pyroelectric effect [228–231]. Pyroelectric materials are good electrical insulators that possess an internal macroscopic electricdipole moment, depending on the temperature. The crystal neutralizes the electric field of this dielectric polarization by a corresponding surface-charge distribution. A change of the internal polarization caused by a temperature rise will produce a measurable change in surface charge, which can be monitored by a pair of electrodes applied to the sample (Fig. 4.86). Because of the capacitive transfer of the change of the electric dipole moments, pyroelectric detectors monitor only *changes* of input power. Any incident cw radiation therefore has to be chopped.

While the sensitivity of good pyroelectric detectors is comparable to that of Golay cells or high-sensitivity bolometers, they are more robust and therefore less delicate to handle. They also have a much better time resolution down into the nanosecond range [229]. The development of artificial pyroelectric materials in the form of thin films made of Gallium Nitride GaN or of Cesium Nitrate $CsNO_3$ has increased sensitivity and frequency response of pyroelectric detectors.

Figure 4.86 Pyroelectric detector

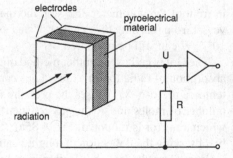

4.5.2   Photodiodes

Photodiodes are doped semiconductors that can be used as photovoltaic or photoconductive devices. When the n–n junction of the diode is irradiated the photovoltage U_{ph} is generated at the open output of the diode (Fig. 4.87a); within
a restricted range it is proportional to the absorbed radiation power. Diodes used as
photoconductive elements change their internal resistance upon irradiation and can
therefore be used as photoresistors in combination with an external voltage source
(Fig. 4.87b).

For their use as radiation detectors the spectral dependence of their absorption
coefficient is of fundamental importance. In an undoped semiconductor the absorption of one photon $h\nu$ causes an excitation of an electron from the valence band into
the conduction band (Fig. 4.88a). With the energy gap $\Delta E_g = E_c - E_v$ between
the valence and conduction band, only photons with $h\nu \geq \Delta E_g$ are absorbed. The
intrinsic absorption coefficient

$$\alpha_{\mathrm{intr}}(\nu) = \begin{cases} \alpha_0(h\nu - \Delta E_g)^{1/2}, & \text{for} \quad h\nu > \Delta E_g, \\ 0, & \text{for} \quad h\nu < \Delta E_g, \end{cases} \qquad (4.130)$$

is shown in Fig. 4.89 for different undoped materials. The quantity α_0 depends
on the material and is generally larger for semiconductors with direct transitions

Figure 4.87 Use of a photodiode: **a** as a photovoltaic
device; and **b** as a photoconductive resistor

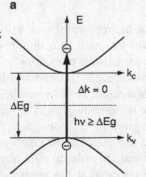

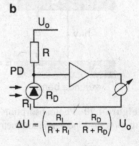

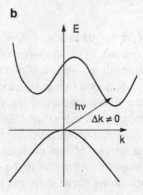

Figure 4.88 **a** Direct
band–band absorption in
an undoped semiconductor;
and **b** indirect transitions,
illustrated in a $E(k)$ band
diagram

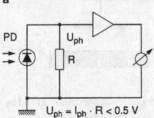

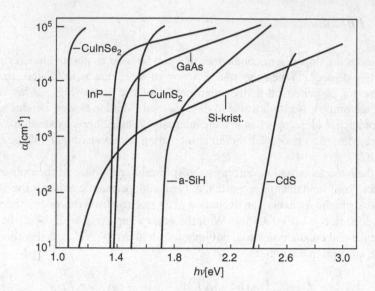

Figure 4.89 Spectral absorption $\alpha(v)$ of some semiconductors. Amorphous silicon a-SiH with indirect absorption transitions shows a smaller slope $d\alpha/dv$ of the curve $\alpha(v)$ while semiconductors with direct absorption (InP, GaAs or CuInSe$_2$) have a much steeper slope

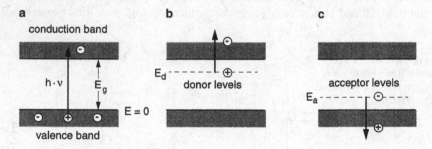

Figure 4.90 Photoabsorption in undoped semiconductors (**a**) and by donors (**b**) and acceptors (**c**) in n- or p-doped semiconductors

($\Delta k = 0$) as e.g. GaAs, than for indirect transitions with $\Delta k \neq 0$ (crystalline silicon). The steep rise of $\alpha(v)$ for $hv > E_g$ has only been observed for direct transitions, while it is much flatter for indirect transitions.

In doped semiconductors photon-induced electron transitions can occur between the donor levels and the conduction band, or between the valence band and the acceptor levels (Fig. 4.90). Since the energy gaps $\Delta E_d = E_c - E_d$ or $\Delta E_a = E_v - E_a$ are much smaller than the gap $E_c - E_v$, doped semiconductors absorb even at smaller photon energies hv and can therefore be employed for the detection of longer wavelengths in the midinfrared. In order to minimize thermal excitation of electrons, these detectors must be operated at low temperatures. For $\lambda \leq 10\,\mu\text{m}$

Figure 4.91 Detectivity $D^*(\lambda)$ of some photodetectors [218]

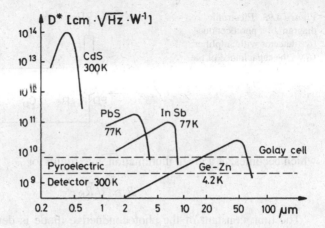

generally liquid-nitrogen cooling is sufficient, while for $\lambda > 10\,\mu m$ liquid-helium temperatures around 4–10 K are required.

Figure 4.91 plots the detectivity of commonly used photodetector materials with their spectral dependence, while Fig. 4.92 illustrates their useful spectral ranges and their dependence on the energy gap ΔE_g.

a) Photoconductive Diodes

When a photodiode is illuminated, its electrical resistance decreases from a "dark value" R_D to a value R_I under illumination. In the circuit shown in Fig. 4.87b, the change of the output voltage is given by

$$\Delta U = \left(\frac{R_D}{R_D + R} - \frac{R_I}{R_I + R} \right) U_0 = \frac{R(R_D - R_I)}{(R + R_D)(R + R_I)} U_0 , \qquad (4.131)$$

Figure 4.92 Energy gaps and useful spectral ranges of some semiconducting materials

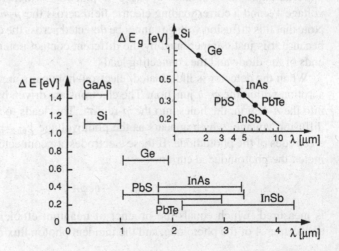

Figure 4.93 Electronic diagram of a photoconductive detector with amplifier; C_D is the capacitance of the photodiode and C_a is the capacitance of the amplifier

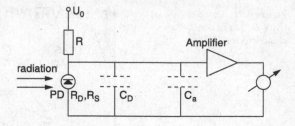

which becomes, at a given illumination, maximum for

$$R \approx \sqrt{R_D R_I} \,.$$

The time constant of the photoconductive diode is determined by $\tau \geq RC$, where $C = C_{PD} + C_a$ is the capacitance of the diode plus the input capacitance of the circuit. Its lower limit is set by the diffusion time of the electrons on their way from the p–n junction where they are generated to the electrodes. Detectors from PbS, for example, have typical time constants of 0.1–1 ms, while InSb detectors are much faster ($\tau \simeq 10^{-7}$–10^{-6} s). Although photoconductive detectors are generally more sensitive, photovoltaic detectors are better suited for the detection of fast signals.

b) Photovoltaic Detector

While photoconductors are passive elements that need an external power supply, photovoltaic diodes are active elements that generate their own photovoltage upon illumination, although they are often used with an external bias voltage. The principle of the photogenerated voltage is shown in Fig. 4.94.

In the nonilluminated diode, the diffusion of electrons from the n-region into the p-region (and the opposite diffusion of the holes) generates a space charge, with opposite signs on both sides of the p–n junction, which results in the diffusion voltage V_D and a corresponding electric field across the p–n junction (Fig. 4.94b). Note that this diffusion voltage cannot be detected across the electrodes of the diode, because it is just compensated by the different contact potentials between the two ends of the diode and the connecting leads.

When the detector is illuminated, electron–hole pairs are created by photon absorption within the p–n junction. The electrons are driven by the diffusion voltage into the n-region, the holes into the p-region. This leads to a *decrease* ΔV_D of the diffusion voltage, which appears as the photovoltage $V_{ph} = \Delta V_D$ across the open electrodes of the photodiode. If these electrodes are connected through an Ampére-meter, the photoinduced current

$$i_{ph} = -\eta e \phi A \,, \tag{4.132}$$

is measured, which equals the product of quantum efficiency η, the illuminated active area A of the photoiode, and the incident photon flux density $\phi = I / h\nu$.

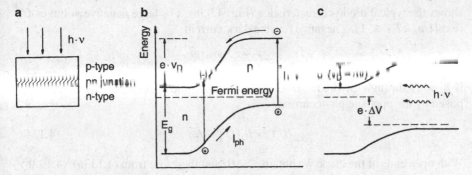

Figure 4.94 Photovoltaic diode: **a** schematic structure and **b** diffusion voltage and generation of an electron–hole pair by photon absorption within the p–n junction. **c** Reduction of the diffusion voltage V_D under illumination for an open circuit

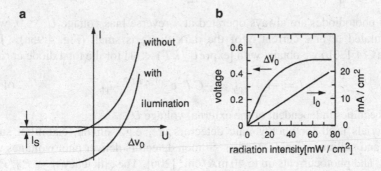

Figure 4.95 a Current–voltage characteristics of a dark and an illuminated diode; **b** diffusion voltage and photovoltage at the open ends and photocurrent in a shortened diode as a function of incident radiation power

The illuminated p–n photodetector can therefore be used either as a current generator or a voltage source, depending on the external resistor between the electrodes.

Note The photon-induced voltage $U_{ph} < \Delta E_g/e$ is always limited by the energy gap ΔE_g. The voltage U_{ph} across the open ends of the photodiode is reached even at relatively small photon fluxes, while the photocurrent is linear over a large range (Fig. 4.95b). When using photovoltaic detectors for measuring radiation power, the load resistor R_L must be sufficiently low to keep the output voltage $U_{ph} = i_{ph}R_L < U_s = \Delta E_g/e$ always below its saturation value U_s. Otherwise, the output signal is no longer proportional to the input power.

If an external voltage U is applied to the diode, the diode current without illumination

$$i_D(U) = C T^2 e^{-eV_D/kT}(e^{eU/kT} - 1), \tag{4.133a}$$

shows the typical diode characteristics (Fig. 4.95a). For large negative voltages U $(\exp(Ue/kT) \ll 1)$, a negative reverse dark current

$$i_s = -CT^2 e^{-eV_D/kT} \tag{4.133b}$$

is flowing through the diode. During illumination the dark current i_D is superimposed by the opposite photocurrent

$$i_{ill}(U) = i_D(U) - i_{ph} . \tag{4.134}$$

With open ends of the diode we obtain $i = 0$, and therefore from (4.133a), (4.133b) the photovoltage becomes

$$U_{ph}(i = 0) = \frac{kT}{e} \left[\ln \left(\frac{i_{ph}}{i_s} \right) + 1 \right] . \tag{4.135}$$

Fast photodiodes are always operated at a reverse bias voltage $U < 0$, where the saturated reverse current i_s of the dark diode is small (Fig. 4.95a). From (4.133a), (4.133b) we obtain, with $[\exp(eU/kT) \ll 1]$ for the total diode current,

$$i = -i_s - i_{ph} = -CT^2 e^{-eV_D/kT} - i_{ph} , \tag{4.136}$$

which becomes independent of the external voltage U.

Materials used for photovoltaic detectors are, e.g., silicon, cadmium sulfide (CdS), and gallium arsenide (GaAs). Silicon detectors deliver photovoltages up to 550 mV and photocurrents up to 40 mA/cm^2 [206]. The efficiency $\eta = P_{el}/P_{ph}$ of energy conversion reaches 10–14 %. New devices with a minimum number of crystal defects can even reach 20–30 %. Gallium arsenide (GaAs) detectors show larger photovoltages up to 1 V, but slightly lower photocurrents of about 20 mA/cm^2.

c) Fast Photodiodes

The photocurrent generates a signal voltage $V_s = U_{ph} = R_L i_{ph}$ across the load resistor R_L that is proportional to the absorbed radiation power over a large intensity range of several decades, as long as $V_s < \Delta E_g/e$ (Fig. 4.95b). From the circuit diagram in Fig. 4.96 with the capacitance C_s of the semiconductor and its series and parallel resistances R_s and R_p, one obtains for the upper frequency limit [232]

$$f_{max} = \frac{1}{2\pi C_s(R_s + R_L)(1 + R_s/R_p)} , \tag{4.137}$$

which reduces, for diodes with large R_p and small R_s, to

$$\boxed{f_{max} = \frac{1}{2\pi C_s R_L} .} \tag{4.138}$$

Figure 4.96 Equivalent circuit of a photodiode with internal capacity C_S, series internal resistor R_S, parallel internal resistor R_P, and external load resistor R_L

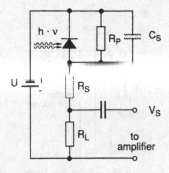

With small values of the resistor R_L, a high-frequency response can be achieved, which is limited only by the drift time of the carriers through the boundary layer of the p–n junction. This drift time can be reduced by an external bias voltage. Using diodes with large bias voltages and a 50-Ω load resistor matched to the connecting cable, rise times in the subnanosecond range can be obtained.

Example 4.25

$C_s = 10^{-11}\,\text{F}, \quad R_L = 50\,\Omega \quad \Rightarrow \quad f_{max} = 300\,\text{MHz}, \quad \tau = \dfrac{1}{2\pi f_{max}} \simeq 0.6\,\text{ns}.$

For photon energies $h\nu$ close to the band gap, the absorption coefficient decreases, see (4.130). The penetration depth of the radiation, and with it the volume from which carriers have to be collected, becomes large. This increases the collection time and makes the diode slow.

Definite collection volumes can be achieved in *PIN diodes*, where an undoped zone I of an intrinsic semiconductor separates the p- and n-regions (Fig. 4.97). Since no space charges exist in the intrinsic zone, the bias voltage applied to the

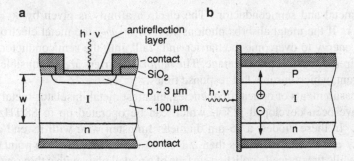

Figure 4.97 PIN photodiode with head-on (**a**) and side-on (**b**) illumination

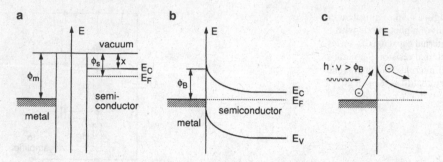

Figure 4.98 **a** Work functions ϕ_m of metal and ϕ_s of semiconductor and electron affinity χ. E_c is the energy at the bottom of the conduction band and E_F is the Fermi energy. **b** Schottky barrier at the contact layer between metal and n-type semiconductor. **c** Generation of a photocurrent

diode causes a constant electric field, which accelerates the carriers. The intrinsic region may be made quite wide, which results in a low capacitance of the p–n junction and provides the basis for a very fast and sensitive detector. The limit for the response time is, however, also set by the transit time $\tau = w/v_{th}$ of the carriers in the intrinsic region, which is determined by the width w and the thermal velocity v_{th} of the carriers. Silicon PIN diodes with a 700-µm wide zone I have response times of about 10 ns and a sensitivity maximum at $\lambda = 1.06\,\mu m$, while diodes with a 10-µm wide zone I reach 100 ps with a sensitivity maximum at a shorter wavelength–around $\lambda = 0.6\,\mu m$ [233]. Fast response combined with high sensitivity can be achieved when the incident radiation is focused from the side into the zone I (Fig. 4.97b). The only experimental disadvantage is the critical alignment necessary to hit the small active area.

Very fast response times can be reached by using the photoeffect at the metal–semiconductor boundary known as the Schottky barrier [234]. Because of the different work functions ϕ_m and ϕ_s of the metal and the semiconductor, electrons can tunnel from the material with low ϕ to that with high ϕ (Fig. 4.98). This causes a space-charge layer and a potential barrier

$$V_B = \phi_B/e\,, \quad \text{with } \phi_B = \phi_m - \chi\,, \tag{4.139}$$

between metal and semiconductor. The electron affinity is given by $\chi = \phi_s - (E_c - E_F)$. If the metal absorbs photons with $h\nu > \phi_B$, the metal electrons gain sufficient energy to overcome the barrier and "fall" into the semiconductor, which thus acquires a negative photovoltage. The *majority* carriers are responsible for the photocurrent, which ensures fast response times.

For measurements of optical frequencies, ultrafast metal–insulator-metal (MIM) diodes have been developed [235], which can be operated up to 88 THz ($\lambda = 3.39\,\mu m$). In these diodes, a 25-µm diameter tungsten wire with its end electrochemically etched to a point less than 200 nm in radius serves as the point contact element, while the optically polished surface of a nickel plate with a thin oxide layer forms the base element of the diode (Fig. 4.99).

Figure 4.99 Arrangement of a metal–insulator–metal (MIM) diode used for optical frequency mixing of laser frequencies

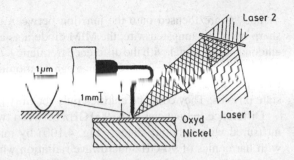

These MIM diodes can be used as mixing elements at optical frequencies. When illuminating the contact point with a focused CO_2 laser, a response time of 10^{-14} s or better has been demonstrated by the measurement of the 88-THz emission from the third harmonic of the CO_2 laser. If the beams of two lasers with the frequencies

Figure 4.100 Point-contact diode: **a** electron miocroscope picture **b** current-voltage characteristics [237]

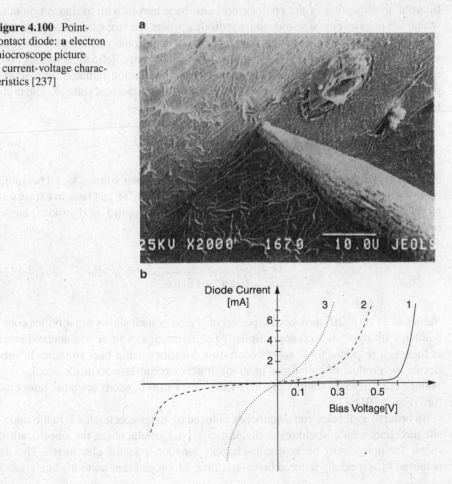

f_1 and f_2 are focused onto the junction between the nickel oxide layer and the sharp tip of the tungsten wire, the MIM diode acts as a rectifier and the wire as an antenna, and a signal with the difference frequency $f_1 - f_2$ is generated. Difference frequencies up into the terahertz range can be monitored [236] (see Sect. 6.6). The basic processes in these MIM diodes represent very interesting phenomena of solid-state physics. They could be clarified only recently [236].

Difference frequencies up to 900 GHz between two visible dye lasers have been measured with Schottky diodes (Fig. 4.100) by mixing the difference frequency with harmonics of 90-GHz microwave radiation which was also focused onto the diode [237]. Meanwhile, Schottky-barrier mixer diodes have been developed that cover the frequency range 1–10 THz [237].

d) Avalanche Diodes as Internal Amplifyers

Internal amplification of the photocurrent can be achieved with avalanche diodes, which are reverse-biased semiconductor diodes, where the free carriers acquire sufficient energy in the accelerating field to produce additional carriers on collisions with the lattice (Fig. 4.101). The multiplication factor M, defined as the average number of electron–hole pairs after avalanche multiplication initiated by a single photoproduced electron–hole pair, increases with the reverse-bias voltage. The multiplication factor

$$M = 1 / [1 - (V / V_{br})^n] \tag{4.140}$$

depends on the external bias voltage V and the breakdown voltage V_{br}. The value of n (2–6) depends on the material of the avalanche diode. M can be also expressed by the multiplication coefficient α for electrons and the length L of the space charge boundary:

$$M = \frac{1}{1 - \int_0^L \alpha(x)\,dx} . \tag{4.141}$$

Values of M up to 10^6 have been reported in silicon, which allows sensitivities comparable with those of a photomultiplier. The advantage of these avalanche diodes is their fast response time, which decreases with increasing bias voltage. In this device the product of gain times bandwidth may exceed 10^{12} Hz if the breakdown voltage is sufficiently high [208]. The value of M also depends upon the temperature (Fig. 4.101b).

In order to avoid electron avalanches induced by holes accelerated into the opposite direction, which would result in additional background noise, the amplification factor for holes must be kept considerably smaller than for electrons. This is achieved by a specially tailored layer structure, which yields a sawtooth-like graded

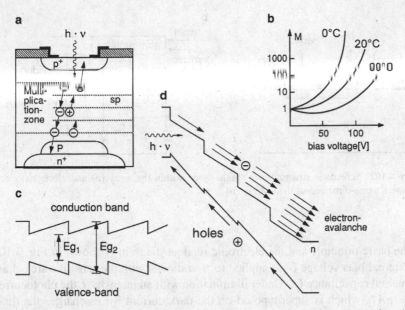

Figure 4.101 Avalanche diode: **a** schematic illustration of avalanche formation (n^+, p^+ are heavily doped layers); **b** amplification factor $M(V)$ as a function of the bias voltage V for a Si-avalanche diode; **c** spatial variation of band edges and bandgap without external field; and **d** within an external electric field

band-gap dependence $\Delta E_g(x)$ in the field x-direction (Fig. 4.101c,d). In an external field this structure results in an amplification factor M that is 50–100 times larger for electrons than for holes [238].

Such modern avalanche diodes may be regarded as the solid-state analog to photomultipliers (Sect. 4.5.5). Their advantages are a high quantum efficiency (up to 40 %) and a low supply voltage (10–100 V). Their disadvantage for fluorescence detection is the small active area compared to the much larger cathode area of photomultipliers [239–241].

Detailed data on avalanche photodiodes can be found on the homepage of Hamamatsu [242].

4.5.3 Photodiode Arrays

Many small photodiodes can be integrated on a single chip, forming a *photodiode array*. If all diodes are arranged in a line we have a onedimensional diode array, consisting of up to 2048 diodes. With a diode width $b = 15\,\mu\text{m}$ and a spacing of $d = 10\,\mu\text{m}$ between two diodes, the length L of an array of 1024 diodes becomes 25 mm with a height of about 40 μm [243].

a

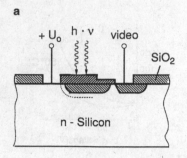

b

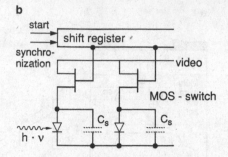

Figure 4.102 Schematic structure of a single diode within the array (**a**) and electronic circuit diagram of a one-dimensional diode array (**b**)

The basic principle and the electronic readout diagram is shown in Fig. 4.102. An external bias voltage U_0 is applied to p–n diodes with the sensitive area A and the internal capacitance C_s. Under illumination with an intensity I the photocurrent $i_{ph} = \eta A I$, which is superimposed on the dark current i_D, discharges the diode capacitance C_s during the illumination time ΔT by

$$\Delta Q = \int_t^{t+\Delta T} (i_D + \eta A I)\mathrm{d}t = C_s \Delta U \ . \tag{4.142}$$

Every photodiode is connected by a multiplexing MOS switch to a voltage line and is recharged to its original bias voltage U_0. The recharging pulse $\Delta U = \Delta Q/C_s$ is sent to a video line connected with all diodes. These pulses are, according to (4.142), a measure for the incident radiation energy $\int A I \, \mathrm{d}t$, if the dark current i_D is subtracted and the quantum efficiency η is known.

The maximum integration time ΔT is limited by the dark current i_D, which therefore also limits the attainable signal-to-noise ratio. At room temperature typical integration times are in the millisecond range. Cooling of the diode array by Peltier cooling down to $-40\,^\circ$C drastically reduces the dark current and allows integration times of 1–100 s. The minimum detectable incident radiation power is determined by the minimum voltage pulse ΔU that can be safely distinguished from noise pulses. The detection sensitivity therefore increases with decreasing temperature because of the possible increasing integration time. At room temperature typical sensitivity limits are about 500 photons per second and diode.

If such a linear diode array with N diodes and a length $L = N(b+d)$ is placed in the observation plane of a spectrograph (Fig. 4.1), the spectral interval

$$\delta\lambda = \frac{\mathrm{d}\lambda}{\mathrm{d}x} L \ ,$$

which can be detected simultaneously, depends on the linear dispersion $dx/d\lambda$ of the spectrograph. The smallest resolvable spectral interval

$$\delta\lambda = \frac{d\lambda}{dx} b ,$$

is limited by the width b of the diode. Such a system of spectrograph plus diode array is called an *optical multichannel analyzer* (OMA) or an *optical spectrum analyzer* (OSA) [243, 244].

Example 4.26
$b + d = 25\,\mu m, \quad L = 25\,mm, \quad d\lambda/dx = 5\,nm/mm$
$\Rightarrow \delta\lambda = 125\,nm, \quad \Delta\lambda = 0.125\,nm.$

The diodes can be also arranged in a two-dimensional array, which allows the detection of two-dimensional intensity distributions. This is, for instance, important for the observation of spatial distributions of light-emitting atoms in gas discharges or flames (Vol. 2, Sect. 15.4) or of the ring pattern behind a Fabry–Perot interferometer.

4.5.4 *Charge-Coupled Devices (CCDs)*

Photodiode arrays are now increasingly replaced by charge-coupled device (CCD) arrays, which consist of an array of small MOS junctions on a doped silicon substrate (Fig. 4.103) [245–250]. The incident photons generate electrons and holes in the n- or p-type silicon. The electrons or holes are collected and change the charge of the MOS capacitances. These changes of the charge can be shifted to the next MOS capacitance by applying a sequence of suitable voltage steps according to Fig. 4.103b. The charges are thus shifted from one diode to the next until they reach the last diode of a row, where they cause the voltage change ΔU, which is sent to a video line.

The quantum efficiency η of CCD arrays depends on the material used for the substrate, it reaches peak values over 90 %. The efficiency $\eta(\lambda)$ is generally larger than 20 % over the whole spectral range, which covers the region from 350–900 nm. Using fused quartz windows, even the UV and the IR from 200–1000 nm can be covered (Fig. 4.103c), and the efficiencies of most photocathodes are exceeded (Sect. 4.5.5). The spectral range of special CCDs ranges from 0.1–1000 nm. They can therefore be used in the VUV and X-ray regions, too. The highest sensitivity up to 90 % efficiency is achieved with backward-illuminated devices (Fig. 4.104). Table 4.2 compiles some relevant data for commercial CCD devices and Fig. 4.105

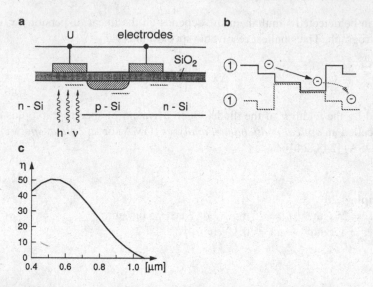

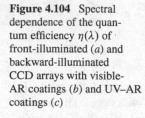

Figure 4.103 Principle of a CCD array: **a** alternately, a positive (*solid line*) and a negative (*dashed line*) voltage are applied to the electrodes. **b** This causes the charged carriers generated by photons to be shifted to the next diode. This shift occurs with the pulse frequency of the applied voltage. **c** Spectral sensitivity of CCD diodes

Figure 4.104 Spectral dependence of the quantum efficiency $\eta(\lambda)$ of front-illuminated (*a*) and backward-illuminated CCD arrays with visible-AR coatings (*b*) and UV–AR coatings (*c*)

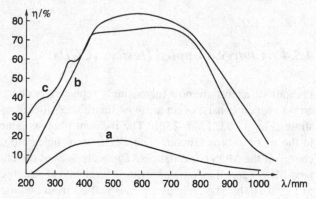

compares the spectral quantum efficiency of CCD detectors with those of the photographic plate and photomultiplier cathodes.

The dark current of cooled CCD arrays may be below 10^{-2} electrons per second and diode. The readout dark pulses are smaller than those of photodiode arrays. Therefore, the sensitivity is high and may exceed that of good photomultipliers. Particular advantages are their large dynamic range, which covers about five orders of magnitude, and their linearity.

The disadvantage is their small size compared to photographic plates. This restricts the spectral range that can be detected simultaneously. More information

Table 4.2 Characteristic data of CCD arrays

Active area [mm^2]	24.6 × 24.6
Pixel size [µm]	7.5 × 15 up to 24 × 24
Number of pixels	1024 ×· 1024 up to 4096 × 4096
Dynamic range [bits]	16
Readout noise at 50 kHz [electron charges]	4–6
Dark charge [electrons/(h pixel)]	< 1
Hold time at −120 °C [h]	> 10
Quantum efficiency peak	
Front illuminated	50 %
Backward illuminated	> 90 %
Spectral range [mm]	300–1100

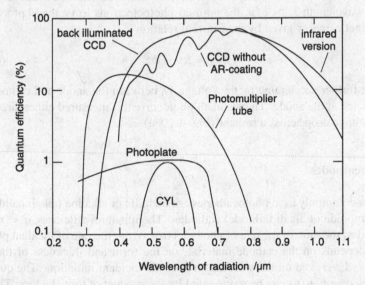

Figure 4.105 Comparison of the quantum efficiencies of CCD detectors, photoplates and photo-multipliers

about CCD detectors, which are becoming increasingly important in spectroscopy, can be found in [246, 250, 251].

4.5.5 Photoemissive Detectors

Photoemissive detectors, such as the photocell or the photomuliplier, are based on the external photoeffect. The photocathode of such a detector is covered with one or several layers of materials with a low work function ϕ (e.g., alkali metal com-

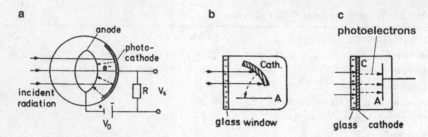

Figure 4.106 Photoemissive detector: **a** principle arrangement of a photocell; **b** opaque photocathode; and **c** semitransparent photocathode

pounds or semiconductor compounds). Under illumination with monochromatic light of wavelength $\lambda = c/\nu$, the emitted photoelectrons leave the photocathode with a kinetic energy given by the Einstein relation

$$E_{\text{kin}} = h\nu - \phi .$$ (4.143)

They are further accelerated by the voltage V_0 between the anode and cathode and are collected at the anode. The resultant photocurrent is measured either directly or by the voltage drop across a resistor (Fig. 4.106a).

a) Photocathodes

The most commonly used photocathodes are metallic or alkaline (alkali halides, alkali antimonide or alkali telluride) cathodes. The quantum efficiency $\eta = n_e/n_{\text{ph}}$ is defined as the ratio of the rate of photoelectrons n_e to the rate of incident photons n_{ph}. It depends on the cathode material, on the form and thickness of the photoemissive layer, and on the wavelength λ of the incident radiation. The quantum efficiency $\eta = n_a n_b n_c$ can be represented by the product of three factors. The first factor n_a gives the probability that an incident photon is actually absorbed. For materials with a large absorption coefficient, such as pure metals, the reflectivity R is high (e.g., for metallic surfaces $R \geq 0.8$–0.9 in the visible region), and the factor n_a cannot be larger than $(1 - R)$. For semitransparent photocathodes of thickness d, on the other hand, the absorption must be large enough to ensure that $\alpha d > 1$. The second factor n_b gives the probability that the absorbed photon really produces a photoelectron instead of heating the cathode material. Finally, the third factor n_c stands for the probability that this photoelectron reaches the surface and is emitted instead of being backscattered into the interior of the cathode.

Two types of photoelectron emitters are manufactured: opaque layers, where light is incident on the same side of the photocathode from which the photoelectrons are emitted (Fig. 4.106b); and semitransparent layers (Fig. 4.106c), where light enters at the opposite side to the photoelectron emission and is absorbed throughout the thickness d of the layer. Because of the two factors n_a and n_c, the quantum effi-

Figure 4.107 Spectral
sensitivity curves of some
commercial cathode types.
The *solid lines* give $S(\lambda)$
[mA/W]; *whereas the dashed*
curves given quantum efficiencies $\eta = n_e/n_{ph}$

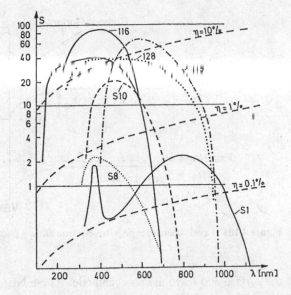

ciency of semitransparent cathodes and its spectral change are critically dependent
on the thickness d, and reach that of the reflection-mode cathode only if the value
of d is optimized.

Figure 4.107 shows the spectral sensitivity $S(\lambda)$ of some typical photocathodes,
scaled in milliamperes of photocurrent per watt incident radiation. For comparison,
the quantum efficiency curves for $\eta = 0.001, 0.01$ and 0.1 are also drawn (dashed
curves). Both quantities are related by

$$S = \frac{i}{P_{in}} = \frac{n_e e}{n_{ph} h\nu} \Rightarrow S = \frac{\eta e \lambda}{hc} \,. \tag{4.144}$$

For most emitters the threshold wavelength for photoemission is below $0.85\,\mu m$,
corresponding to a work function $\phi \geq 1.4\,eV$. An example for such a material
with $\phi \sim 1.4\,eV$ is a surface layer of NaKSb [252]. Only some complex cathodes
consisting of two or more separate layers have an extended sensitivity up to about
$\lambda \leq 1.2\,\mu m$. For instance, an InGaAs photocathode has an extended sensitivity in
the infrared, reaching up to 1700 nm. The spectral response of the most commonly
fabricated photocathodes is designated by a standard nomenclature, using the sym-
bols S1 to S20. Some newly developed types are labeled by special numbers, which
differ for the different manufacturers [253]. Examples are S1 $=$ Ag $-$ O $-$ Cs
(300–1200 nm) or S4 $=$ Sb $-$ Cs (300–650 nm).

Recently, a new type of photocathode has been developed that is based on photoconductive semiconductors whose surfaces have been treated to obtain a state of
negative electron affinity (NEA) (Fig. 4.108). In this state an electron at the bottom of the conduction band inside the semiconductor has a higher energy than the
zero energy of a free electron in vacuum [254]. When an electron is excited by
absorption of a photon into such an energy level within the bulk, it may travel to

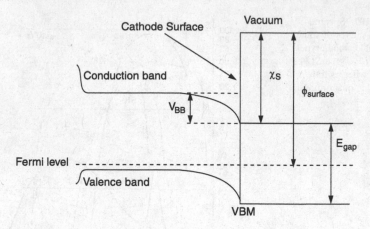

Figure 4.108 Level scheme for negative electron affinity photocathodes

the surface and leave the photocathode. These NEA cathodes have the advantage of a high sensitivity, which is fairly constant over an extended spectral range and even reaches into the infrared up to about $1.2\,\mu m$. Since these cathodes represent cold-electron emission devices, the dark current is very low. Until now, their main disadvantage has been the complicated fabrication procedure and the resulting high price.

Different devices of photoemissive detectors are of major importance in modern spectroscopy. These are the *the photomultiplier, the image intensifier, and the streak camera.*

b) Photomultipliers

Photomultipliers are a good choice for the detection of low light levels. They overcome some of the noise limitations by internal amplification of the photocurrent using secondary-electron emission from internal dynodes to mulitply the number of photoelectrons (Fig. 4.109).

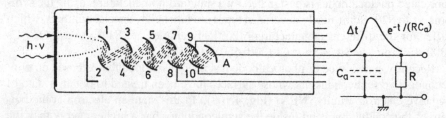

Figure 4.109 Photomultiplier with time-dependent output voltage pulse induced by an electron avalanche that was triggered by a delta-function light pulse

The photoelectrons emitted from the cathode are accelerated by a voltage of a few hundred volts and are focused onto the metal surface (e.g., Cu–Be) of the first "dynode" where each impinging electron releases, on the average, q secondary electrons. These electrons are further accelerated to a second dynode where each secondary electron again produces about q tertiary electrons, and so on. The amplification factor q depends on the accelerating voltage U, on the incidence angle α, and on the dynode material. Typical figures for $U = 200\,\text{V}$ are $q = 3\text{–}5$. A photomulitplier with ten dynodes therefore has a total current amplification of $G = q^{10} \sim 10^5\text{–}10^7$. Each photoelectron in a photomultiplier with N dynodes produces a charge avalanche at the anode of $Q = q^N e$ and a corresponding voltage pulse of

$$V = \frac{Q}{C_a} = \frac{q^N e}{C_a} = \frac{Ge}{C_a} , \qquad (4.145)$$

where C_a is the capacitance of the anode (including connections).

Example 4.27
$G = 2 \times 10^6$, $C_a = 30\,\text{pf} \Rightarrow V = 10.7\,\text{mV}$.

For cw operation the DC output voltage is given by $V = i_a \cdot R$, independent of the capacitance C_a.

For experiments demanding high time resolution, the rise time of this anode pulse should be as small as possible. Let us consider which effects may contribute to the anode pulse rise time, caused by the spread of transit times for the different electrons [255, 256]. Assume that a single photoelectron is emitted from the photocathode, and is accelerated to the first dynode. The initial velocities of the secondary electrons vary because these electrons are released at different depths of the dynode material and their initial energies, when leaving the dynode surface, are between 0 and 5 eV. The transit time between two parallel electrodes with distance d and potential difference V is obtained from $d = \frac{1}{2}at^2$ with $a = eV/(md)$, which gives

$$t = d\sqrt{\frac{2m}{eV}} , \qquad (4.146)$$

for electrons with mass m starting with zero initial energy. Electrons with the initial energy E_{kin} reach the next electrode earlier by the time difference

$$\Delta t_1 = \frac{d}{eV}\sqrt{2m E_{kin}} . \qquad (4.147)$$

Example 4.28
$E_{kin} = 0.5\,\text{eV}, \quad d = 1\,\text{cm}, \quad V = 250\,\text{V} \;\Rightarrow\; \Delta t_1 = 0.1\,\text{ns}.$

The electrons travel slightly different path lengths through the tube, which causes an additional time spread of

$$\Delta t_2 = \Delta d \sqrt{\frac{2m}{eV}}, \tag{4.148}$$

which is of the same magnitude as Δt_1. The rise time of an anode pulse started by a single photoelectron therefore decreases with increasing voltage proportional to $V^{-1/2}$. It depends on the geometry and form of the dynode structures.

When a short intense light pulse produces many photoelectrons simultaneously, the time spread is further increased by two phenomena:

- The initial velocities of the emitted photoelectrons differ, e.g., for a cesium antimonide S5 cathode between 0 and $2\,\text{eV}$. This spread depends on the wavelength of the incoming light [257]a.
- The time of flight between the cathode and the first dynode strongly depends on the locations of the spot on the cathode where the photoelectron is emitted. The resulting time spread may be larger than that from the other effects, but may be reduced by a focusing electrode between the cathode and the first dynode with careful optimization of its potential. Typical anode rise times of photomultipliers range from 0.5–20 ns. For specially designed tubes with optimized side-on geometry, where the curved opaque cathode is illuminated from the side of the tube, rise times of 0.4 ns have been achieved [257]b. Shorter rise times can be reached with channel plates and channeltrons [257]c.

Example 4.29
Photomultiplier type 1P28: $N = 9, q = 5.1$ at $V = 1250\,\text{V} \;\Rightarrow\; G = 2.5 \times 10^6$; anode capacitance and input capacitance of the amplifier $C_a = 15\,\text{pF}$. A single photoelectron produces an anode pulse of 27 mV with a rise time of 2 ns. With a resistor $R = 10^5\,\Omega$ at the PM exit, the trailing edge of the output pulse is $C_a = 1.5 \times 10^{-6}\,\text{s}$.

For low-level light detection, the question of noise mechanisms in photomultipliers is of fundamental importance [259]. There are three main sources of noise:

- Photomultiplier dark current;
- Noise of the incoming radiation;

- Shot noise and Johnson noise caused by fluctuations of the amplification and by noise of the load resistor.

We shall discuss these contributions separately:

- When a photomultiplier is operated in complete darkness, electrons are still emitted from the cathode. This dark current is mainly due to thermionic emission and is only partly caused by cosmic rays or by radioactive decay of spurious radioactive isotopes in the multiplier material. According to Richardson's law, the thermionic emission current

$$i = C_1 T^2 e^{-C_2\phi/T} , \qquad (4.149)$$

strongly depends on the cathode temperature T and on its work function ϕ.

If the spectral sensitvity extends into the infrared, the work function ϕ must be small, which increases the dark current. In order to decrease the dark current, the temperature T of the cathode must be reduced. For instance, cooling a cesium–antimony cathode from 20 °C to 0 °C reduces the dark current by a factor of about ten. The optimum operation temperature depends on the cathode type (because of ϕ). For S1 cathodes, e.g., those with a high infrared sensitivity and therefore a low work function ϕ, it is advantageous to cool the cathode down to liquid nitrogen temperatures. For other types with maximum sensitivity in the green, cooling below -40 °C gives no significant improvement because the thermionic part of the dark current has already dropped below other contributions, e.g., caused by high-energy β-particles from disintegration of ^{40}K nuclei in the window material. Excessive cooling can even cause undesirable effects, such as a reduction of the signal photocurrent or voltage drops across the cathode, because the electrical resistance of the cathode film increases with decreasing temperature [260].

For many spectroscopic applications only a small fraction of the cathode area is illuminated, e.g., for photomultipliers behind the exit slit of a monochromator. In such cases, the dark current can be futher reduced either by using photomulitpliers with a small effective cathode area or by placing small magnets around an extended cathode. The magnetic field defocuses electrons from the outer parts of the cathode area. These electrons cannot reach the first dynode and do not contribute to the dark current.

- The shot noise

$$\langle i_n \rangle_s = \sqrt{2e \cdot i \cdot \Delta f} \qquad (4.150a)$$

of the photocurrent [259] is amplified in a photomultiplier by the gain factor G. The root-mean-square (rms) noise voltage across the anode load resistor R is therefore

$$\langle V \rangle_s = GR\sqrt{2e\, i_c \Delta f} , \quad i_c : \text{ cathode current} ,$$

$$= R\sqrt{2e\, G i_a \Delta f} , \quad i_a = G i_c : \text{ anode current} , \qquad (4.150b)$$

if the gain factor G is assumed to be constant. However, generally G is not constant, but shows fluctuations due to random variations of the secondary-emission coefficient q, which is a small integer. This contributes to the total noise and multiplies the rms shot noise voltage by a factor $a > 1$, which depends on the mean value of q [261]. The shot noise at the anode is then:

$$\langle V_S \rangle = aR\sqrt{2eG\,i_a\Delta f} \; . \tag{4.150c}$$

- The Johnson noise of the load resistor R at the temperature T gives an rms-noise current

$$\langle i_n \rangle_J = \sqrt{4kT\Delta f/R} \tag{4.151a}$$

and a noise voltage

$$\langle V_n \rangle_J = R\langle i_n \rangle_J \; . $$

- From (4.150a)–(4.150c) we obtain with (4.151a) for the superposition $\langle V \rangle_{St+J} = \sqrt{\langle V \rangle_S^2 + \langle V_n \rangle_J^2}$ of shot noise and Johnson noise across the anode load resistor R at room temperature, where $4kT/e \approx 0.1\,\mathrm{V}$

$$\langle V \rangle_{J+s} = \sqrt{eR\Delta f(2RGa^2 i_a + 0.1)} \quad [\text{Volt}] \; . \tag{4.151}$$

For $GR\,i_a a^2 \gg 0.05\,\mathrm{V}$, the Johnson noise can be neglected. With the gain factor $G = 10^6$ and the load resistor of $R = 10^5\,\Omega$, this implies that the anode current i_a should be larger than 5×10^{-13} A. Since the anode dark current is already much larger than this limit, we see that *the Johnson noise does not contribute to the total noise of photomultipliers.*

The channel photomultiplier is a new photomultiplier design for low-level light detection. Here the photoelectrons released from the photocathode are not multiplied by a series of dynodes, but instead move from the cathode to the anode through a curved narrow semiconductive channel (Fig. 4.110). Each time a photoelectron hits the inner surface of the channel, it releases q secondary electrons, where the integer q depends on the voltage applied between the anode and the cathode. The curved geometry causes a grazing incidence of the electrons onto the surface, which enhances the secondary emission factor q. The total gain of these channel photomultipliers (CPM) can exceed $M = 10^8$ and is therefore generally higher than for PM with dynodes.

The main advantages of the CPM are its compact design, its greater dynamic range and its lower dark current (caused mainly by thermionic emission from the photocathode) which is smaller due to its reduced area. The noise caused by fluctuations in the multiplication factor is also smaller, due to the larger value of the secondary emission factor q.

A significant improvement of the signal-to-noise ratio in detection of low levels of radiation can be achieved with single-photon counting techniques, which enable spectroscopic investigations to be performed at incident radiation fluxes

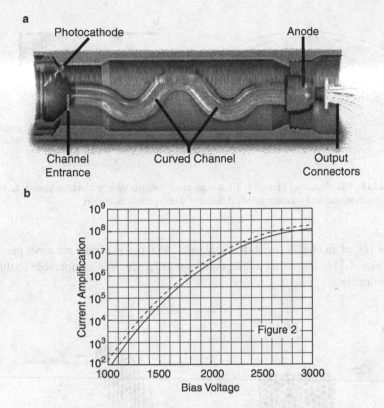

Figure 4.110 Channel photomultiplier. **a** Schematic design; **b** gain factor G as a function of the applied voltage between cathode and anode. [From Olympics Fluo View Resource Center]

down to 10^{-17} W. These techniques are discussed in Sect. 4.5.6. More details about photomultipliers and optimum conditions of performance can be found in excellent introductions issued by Hamamatsu, EMI or RCA [261, 262]. An extensive review of photoemissive detectors has been given by Zwicker [252]; see also the monographs [216, 217, 258, 263, 264].

c) Microchannel Plates

Photomultipliers are now often replaced by microchannel plates. They consist of a photocathode layer on a thin semiconductive glass plate (0.5–1.5 mm) that is perforated by millions of small holes with diameters in the range 10–25 μm (Fig. 4.111). The total area of the holes covers about 60 % of the glass plate area. The inner surface of the holes (channels) has a high secondary emission coefficient for electrons that enter the channels from the photocathode and are accelerated by a voltage applied between the two sides of the glass plate. The amplification factor

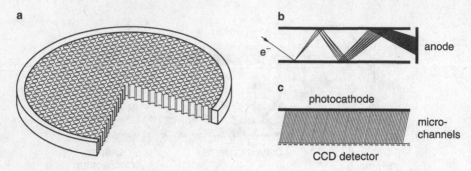

Figure 4.111   Microchannel plate (MCP): **a** schematic construction; **b** electron avalanche in one channel; **c** schematic arrangement of MCP detector with spatial resolution

is about 10^3 at an electric field of $500\,\text{V/mm}$. Placing two microchannel plates in series (Fig. 4.112) allows an amplification of 10^6, which is comparable to that of photomultipliers.

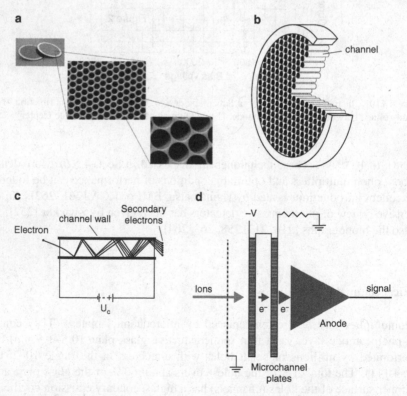

Figure 4.112   Microchannel plate: **a** microholes; **b** design of a microchannel plate; **c** principle of amplification; **d** two-stage microchannel plate

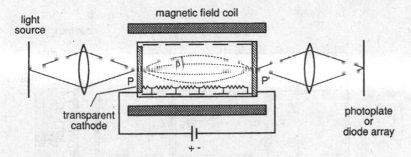

Figure 4.113 Single-stage image intensifier with magnetic focusing

The advantage of the microchannel plates is the short rise time (< 1 ns) of the electron avalanches generated by a single photon, the small size, and the possibility of spatial resolution [265].

d) Photoelectric Image Intensifiers

Image intensifiers consist of a photocathode, an electro-optical imaging device, and a fluorescence screen, where an intensified image of the irradiation pattern at the photocathode is reproduced by the accelerated photoelectrons. Either magnetic or electric fields can be used for imaging the cathode pattern onto the fluorescent screen. Instead of the intensified image being viewed on a phosphor screen, the electron image can be used in a camera tube to generate picture signals, which can be reproduced on the television screen and can be stored either photographically or on a recording medium [266–270].

For applications in spectroscopy, the following characteristic properties of image intensifiers are important:

- The intensity magnification factor M, which gives the ratio of output intensity to input intensity;
- The dark current of the system, which limits the minimum detectable input power;
- The spatial resolution of the device, which is generally given as the maximum number of parallel lines per millimeter of a pattern at the cathode which can still be resolved in the intensified output pattern;
- The time resolution of the system, which is essential for recording of fast transient input signals.

Figure 4.113 illustrates a simple, single-stage image intensifier with a magnetic field parallel to the accelerating electric field. All photoelectrons starting from the point P at the cathode follow helical paths around the magnetic field lines and are focused into P' at the phosphor screen after a few revolutions. The location of P' is, to a first approximation, independent of the direction β of the initial photoelec-

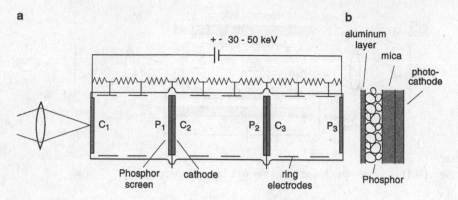

Figure 4.114 Cascade image intensifier: **a** schematic diagram with cathodes C_i, fluorescence screens P_i, and ring electrodes providing the acceleration voltage; **b** detail of phosphor–cathode sandwich structure

tron velocities. To get a rough idea about the possible magnification factor M, let us assume a quantum efficiency of 20 % for the photocathode and an accelerating potential of 10 kV. With an efficiency of 20 % for the conversion of electron energy to light energy in the phosphor screen, each electron produces about 1000 photons with $h\nu = 2$ eV. The amplification factor M giving the number of output photons per incoming photon is then $M = 200$. However, light from the phosphor is emitted into all directions and only a small fraction of it can be collected by an optical system. This reduces the total gain factor.

The collection efficiency can be enhanced when a thin mica window is used to support the phosphor screen and photographic contact prints of the image are made. Another way is the use of fiber-optic windows.

Larger gain factors can be achieved with cascade intensifier tubes (Fig. 4.114), where two or more stages of simple image intensifiers are coupled in series [268]. The critical components of this design are the phosphor–photocathode sandwich screens, which influence the sensitivity and the spatial resolution. Since light emitted from a spot around P on the phosphor should release photoelectrons from the opposite spot around P′ of the photocathode, the distance between P and P′ should be as small as possible in order to preserve the spatial resolution. Therefore, a thin layer of phosphor (a few microns) of very fine grain-size is deposited by electrophoresis on a mica sheet with a few microns thickness. An aluminum foil reflects the light from the phosphor back onto the photocathode (Fig. 4.114b) and prevents optical feedback to the preceding cathode.

The spatial resolution depends on the imaging quality, which is influenced by the thickness of the phosphor-screen–photocathode sandwiches, by the homogeneity of the magnetic field, and by the lateral velocity spread of the photoelectrons. Red-sensitive photocathodes generally have a lower spatial resolution since the initial velocities of the photoelectrons are larger. The resolution is highest at the center of

Table 4.3 Characteristic data of image intensifiers

Type	Useful diameter [mm]	Resolution [linepairs/mm]	Gain	Spectral range [nm]
⬚⬚⬚⬚⬚⬚	18	32	3×10^4	
RCA C33085DP	38	40	6×10^5	
EMI 9794	48	50	2×10^5	Depending on cathode type between 160 and 1000 nm
Hamamatsu				
V4435U	25	64	4×10^6	
I.I. with				
Multichannel plate	40	80	1×10^7	

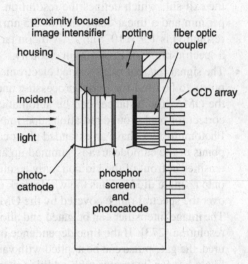

Figure 4.115 Modern version of a compact image intensifier

the screen and decreases toward the edges. Table 4.3 compiles some typical data of commercial three-stage image intensifiers [269]. In Fig. 4.115 a modern version of an image intensifier is shown. It consists of a photocathode, two short proximity-focused image intensifiers, and a fiber-optic coupler, which guides the intensified light generated at the exit of the second stage onto a CCD array.

Image intensifiers can be advantageously employed behind a spectrograph for the sensitive detection of extended spectral ranges [270]. Let us assume a linear dispersion of 1 mm/nm of a medium-sized spectrograph. An image intensifier with a useful cathode size of 30 mm and a spatial resolution of 30 lines/mm allows simultaneous detection of a spectral range of 30 nm with a spectral resolution of 3×10^{-2} nm. This sensitivity exceeds that of a photographic plate by many orders of magnitude. With cooled photocathodes, the thermal noise can be reduced to a level comparable with that of a photomultiplier, therefore incident radiation powers of a few photons can be detected. A combination of image intensifiers and vidicons or special diode arrays has been developed (optical multichannel analyzers, OMA) that has proved to be very useful for fast and sensitive measurements of extended spectral ranges, in particular for low-level incident radiation (Sect. 4.5.3).

Such intensified OMA systems are commercially available. Their advantages may be summarized as follows [271, 272]:

- The vidicon targets store optical signals and allow integration over an extended period, whereas photomultipliers respond only while the radiation falls on the cathode.
- All channels of the vidicon acquire optical signals simultaneously. Mounted behind a spectrometer, the OMA can measure an extended spectral range simultaneously, while the photomultiplier accepts only the radiation passing through the exit slit, which defines the resolution. With a spatial resolution of 30 lines per mm and a linear dispersion of $0.5\,\text{nm/mm}$ of the spectrometer, the spectral resolution is $1.7 \times 10^{-2}\,\text{nm}$. A vidicon target with a length of 16 mm can detect a spectral range of 8 nm simultaneously.
- The signal readout is performed electronically in digital form. This allows computers to be used for signal processing and data analyzing. The dark current of the OMA, for instance, can be automatically substracted, or the program can correct for background radiation superimposed on the signal radiation.
- Photomultipliers have an extended photocathode where the dark current from all points of the cathode area is summed up and adds to the signal. In the image intensifier in front of the vidicon, only a small spot of the photocathode is imaged onto a single diode. Thus the whole dark current from the cathode is distributed over the spectral range covered by the OMA.

  The image intensifier can be gated and allows detection of signals with high time resolution [273]. If the time dependence of a spectral distribution is to be measured, the gate pulse can be applied with variable delay and the whole system acts like a boxcar integrator with additional spectral display. The two-dimensional diode arrays also allow the time dependence of single pulses and their spectral distribution to be displayed, if the light entering the entrance slit of the spectrometer is swept (e.g., by a rotating mirror) parallel to the slit. The OMA or OSA systems therefore combine the advantages of high sensitivity, simultaneous detection of extended spectral ranges, and the capability of time resolution. These merits have led to their increased popularity in spectroscopy [271, 272].

4.5.6 Detection Techniques and Electronic Equipment

In addition to the radiation detectors, the detection technique and the optimum choice of electronic equipment are also essential factors for the success and the accuracy of spectroscopic measurements. This subsection is devoted to some modern detection techniques and electronic devices.

Figure 4.116 Schematic block-diagram of photon-counting electronics

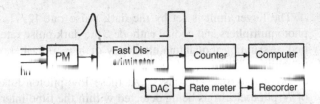

a) Photon Counting

At very low incident radiation powers it is advantageous to use the photomultiplier for counting single photoelectrons emitted at a rate n per second rather than to measure the photocurrent $i = n \cdot \Delta t \cdot e \cdot G / \Delta t$ averaged over a period Δt [274]. The electron avalanches arriving at the anode with the charge $Q = Ge$ generated by a single photoelectron produce voltage pulses $U = eG/C$ at the anode with the capacitance C. With $C = 1.5 \times 10^{-11}$ F, $G = 10^6 \rightarrow U = 10$ mV. These pulses with rise times of about 1 ns trigger a fast discriminator, which delivers a TTL-norm pulse of 5 V to a counter or to a digital–analog converter (DAC) driving a rate meter with variable time constant (Fig. 4.116) [275].

Compared with the conventional analog measurement of the anode current, the photon-counting technique has the following advantages:

- Fluctuations of the photomultiplier gain G, which contribute to the noise in analog measurements, see (4.151), are not significant here, since each photoelectron induces the same normalized pulse from the discriminator as long as the anode pulse exceeds the discriminator threshold.
- Dark curent generated by thermal electrons from the various dynodes can be suppressed by setting the discriminator threshold correctly. This discrimination is particularly effective in photomultipliers with a large conversion efficiency q at the first dynode, covered with a GaAsP layer.
- Leakage currents between the leads in the photomulitplier socket contribute to the noise in current measurements, but are not counted by the discriminator if it is correctly biased.
- High-energy β-particles from the disintegration of radioactive isotopes in the window material and cosmic ray particles cause a small, but nonnegligible, rate of electron bursts from the cathode with a charge $n \cdot e$ of each burst ($n \gg 1$). The resulting large anode pulses cause additional noise of the anode current. They can, however, be completely suppressed by a window discriminator used in photon counting.
- The digital form of the signal facilitates its further processing. The discriminator pulses can be directly fed into a computer that analyzes the data and may control the experiment [276].

The upper limit of the counting rate depends on the time resolution of the discriminator, which may be below 10 ns. This allows counting of randomly distributed pulse rates up to about 10 MHz without essential counting errors.

The lower limit is set by the dark pulse rate [277]. With selected low-noise photomultipliers and cooled cathodes, the dark pulse rate may be below 1 per second. Assuming a quantum efficiency of $\eta = 0.2$, it should therefore be possible to achieve, within a measuring time of 1 s, a signalto-noise ratio of unity even at a photon flux of 5 photons/s. At these low photon fluxes, the probability $p(N)$ of N photoelectrons being detected within the time interval Δt follows a Poisson distribution

$$p(N) = \frac{\overline{N}^{N} \mathrm{e}^{-\overline{N}}}{N!} , \tag{4.152}$$

where $\overline{N}$ is the average number of photoelectrons detected within a given time interval Δt [277]. If the probability that at least one photoelectron will be detected within Δt is 0.99, then $1 - p(0) = 0.99$ and

$$p(0) = \mathrm{e}^{-\bar{N}} = 0.01 , \tag{4.153}$$

which yields $\overline{N} \geq 4.6$. This means that we can expect a pulse during the observation time with 99 % certainty only if at least 20 photons fall onto the photocathode with a quantum efficiency of $\eta = 0.2$. For longer detection times, however, the detectable photoelectron rate may be even lower than the dark current rate if, for instance, lock-in detection is used. It is not the dark pulse rate N_{D} itself that limits the signal-to-noise ratio, but rather its fluctuations, which are proportional to $N_{\mathrm{D}}^{1/2}$.

Because of their low noise, channel photomultipliers or avalanche diodes are well suited to low-level photon counting.

b) Measurements of Fast Transient Events

Many spectroscopic investigations require the observation of fast transient events. Examples are lifetime measurements of excited atomic or molecular states, investigations of collisional relaxation, and studies of fast laser pulses (Vol. 2, Chap. 6). Another example is the transient response of molecules when the incident light frequency is switched into resonance with molecular eigenfrequencies (Vol. 2, Chap. 7). Several techniques are used to observe and to analyze such events and recently developed instruments help to optimize the measuring procedure. The combination of a CCD detector and a gated microchannel plate, which acts as an image intensifier with nanosecond resolution, allows the time-resolved sensitive detection of fast events. In addition, there are several devices that are particularly suited for the electronic handling of short pulses. We briefly present three examples of such equipment: the *boxcar integrator* with signal averaging, the *transient recorder*, and the *fast transient digitizer* with subnanosecond resolution.

The *boxcar integrator* measures the amplitudes and shapes of signals with a constant repetition rate integrated over a specific sampling interval Δt. It records these

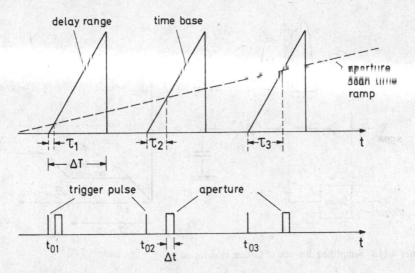

Figure 4.117 Principle of boxcar operation with synchronization of the repetitive signals. The time base determines the opening times of the gate with width Δt. The slow scan-time ramp shifts the delay times τ_i continuously over the signal–pulse time profile

signals repetitively over a selected number of pulses and computes the average value of those measurements. With a synchronized trigger signal it can be assured that one looks each time at the identical time interval of each sampled waveform. A delay circuit permits the sampled time interval Δt (called *aperture*) to be shifted to any portion of the waveform under investigation. Figure 4.117 illustrates a possible way to perform this sampling and averaging. The aperture delay is controlled by a ramp generator, which is synchronized to the signal repetition rate and which provides a sawtooth voltage at the signal repetition frequency. A slow aperture-scan ramp shifts the gating time interval Δt, where the signal is sampled at the time delay τ_i after the trigger pulse for a time interval Δt. Between two successive signals the gate time is shifted by an amount $\Delta \tau$, which depends on the slope of the ramp. This slope has to be sufficiently slow in order to permit a sufficient number of samples to be taken in each segment of the waveform. The output signal is then averaged over several scans of the time ramp by a signal averager [278]. This increases the signal-to-noise ratio and smooths the dc output, which follows the shape of the waveform under study.

The slow ramp is generally not a linearly increasing ramp as shown in Fig. 4.117, but rather a step function where the time duration of each step determines the number of samples taken at a given delay time τ. If the slow ramp is replaced by a constant selectable voltage, the system works as a gated integrator.

The integration of the input signal $U_s(t)$ over the sampling time interval Δt can be performed by charging a capacitance C through a resistor R (Fig. 4.118), which

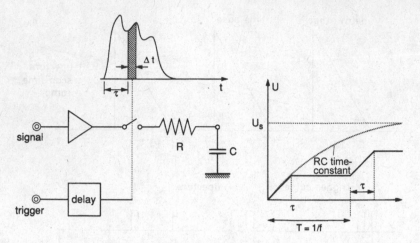

Figure 4.118 Simplified diagram of boxcar realization

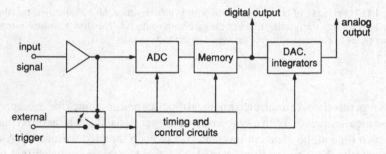

Figure 4.119 Block diagram of a transient recorder

gives a current $i(t) = U_s(t)/R$. The output is then

$$U(\tau) = \frac{1}{C} \int_{\tau}^{\tau+\Delta t} i(t)\mathrm{d}t = \frac{1}{RC} \int_{\tau}^{\tau+\Delta t} U_s(t)\mathrm{d}t \ . \tag{4.154}$$

For repetitive scans, the voltages $U(\tau)$ can be summed. Because of inevitable leakage currents, however, unwanted discharge of the capacitance occurs if the signal under study has a low duty factor and the time between successive samplings becomes large. This difficulty may be overcome by a digital output, consisting of a two-channel analog-todigital-to-analog converter. After a sampling switch opens, the acquired charge is digitized and loaded into a digital storage register. The digital register is then read by a digital-to-analog converter producing a dc voltage equal to the voltage $U(\tau) = Q(\tau)/C$ on the capacitor. This dc voltage is fed back to the integrator to maintain its output potential until the next sample is taken.

The boxcar integrator needs repetitive waveforms because it samples each time only a small time interval Δt of the input pulse and composes the whole period of the repetitive waveform by adding many sampling points with different delays. For many spectroscopic applications, however, only single-shot signals are available. Examples are shock-tube experiments or spectroscopic studies in laser-induced fusion. In such cases, the linear integrator is not useful and a *transient recorder* is a better choice. This instrument uses digital techniques to sample N preselected time intervals Δt_i which cover the total time $T = N\Delta t$ of an analog signal as it varies with time. The wave shape during the selected period of time is recorded and held in the instrument's memory until the operator instructs the instrument to make a new recording. The operation of a transient recorder is illustrated in Fig. 4.119 [279, 280]. A trigger, derived from the input signal or provided externally, initiates the sweep. The amplified input signal is converted at equidistant time intervals to its digital equivalent by an analog-to-digital converter and stored in a semiconductor memory in different channels. With 100 channels, for instance, a single-shot signal is recorded by 100 equidistant sampling intervals. The time resolution depends on the sweep time and is limited by the frequency response of the transient recorder. Sample intervals between 10 ns up to 20 s can be selected. This allows sweep times of 20 µs to 5 h for 2000 sampling points. With modern devices, sampling rates of up to 500 MHz are achievable.

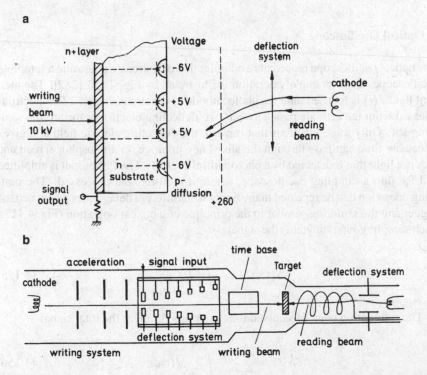

Figure 4.120 Fast transient digitizer: **a** silicon diode-array target; and **b** writing and reading gun [281]

Acquisition and analysis beyond 500 MHz has become possible by combining the features of a transient recorder with the fast response time of an electron beam that writes and stores information on a diode matrix target in a scan converter tube. Figure 4.120 illustrates the basic principle of the *transient digitizer* [281]. The diode array of about 640,000 diodes is scanned by the reading electron beam, which charges all reverse-biased p–n junctions until the diodes reach a saturation voltage. The writing electron beam impinges on the other side of the $10\,\mu$m thick target and creates electron–hole pairs, which diffuse to the anode and partially discharge it. When the reading beam hits a discharged diode, it becomes recharged, subsequently a current signal is generated at the target lead, which can be digitally processed.

The instrument can be used in a nonstoring mode where the operation is similar to that of a conventional television camera with a video signal, which can be monitored on a TV monitor. In the digital mode the target is scanned by the reading beam in discrete steps. The addresses of points on the target are transferred and stored in memory only when a trace has been written at those points on the target. This transient digitizer allows one to monitor fast transient signals with a time resolution of 100 ps and to process the data in digital form in a computer. It is, for instance, possible to obtain the frequency distribution of the studied signal from its time distribution by a Fourier transformation performed by the computer.

c) Optical Oscilloscope

The optical oscilloscope represents a combination of a streak camera and a sampling oscilloscope. Its principle of operation is illustrated by Fig. 4.121 [282]: The incident light $I(t)$ is focused onto the photocathode of the streak camera. The electrons released from the cathode pass between two deflecting electrodes toward the sampling slit. Only those electrons that traverse the deflecting electric field at a given selectable time can pass through the slit. They impinge on a phosphor screen and produce light that is detected by a photomultiplier (PM). The PM output is amplified and fed into a sampling oscilloscope, where it is stored and processed. The sampling operation can be repeated many times with different delay times t between the trigger and the sampling, similar to the principle of a boxcar operation (Fig. 4.117). Each sampling interval yields the signal

$$S(t, \Delta t) = \int\limits_{t}^{t+\Delta t} I(t)\mathrm{d}t \ . \tag{4.155}$$

The summation over all sampled time intervals Δt gives the total signal

$$S(t) = \sum_{n=1}^{N} \int\limits_{t=(n-1)\Delta t}^{t=n\cdot\Delta t} I(t)\mathrm{d}t \ , \tag{4.156}$$

which reflects the input time profile $I(t)$ of the incident light.

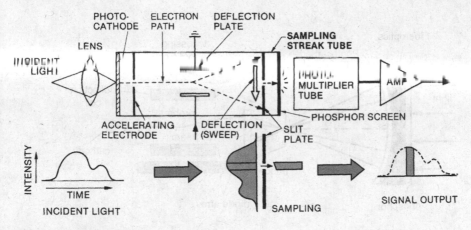

Figure 4.121 Optical oscilloscope [282]

The spectral response of the system depends on that of the first photocathode and reaches from 350 to 850 μm for the visible version and from 400 to 1550 μm for the extended infrared version. The time resolution is better than 10 ps and the sampling rate can be selected up to 2 MHz. The limitation is given by the time jitter, which was stated to be less than 20 ps.

d) Vidicon

In the Vidicon the electrons, emitted from the photocathode are accelerated by an electric field and are imaged onto a CCD array or a photodiode array. (Fig. 4.122). They produce electron–hole pairs. The voltage applied to the photodiodes drives the electrons to the positive electrode and the holes to the negative electrode resulting in a discharge of the diode capacitance. An electron beam imaged by an appropriate electron optics onto the different diodes of the array recharges the capacities up to their original voltage. This recharging current appears as voltage pulse on the common video line and gives the wanted signal.

Such photodiode arrays with the vidicon technique reach high sensitivities which are only limited by the quantum efficiency of the photocathode camparable to that of good photomultipliers. The advantage of these devices is the spatial resolution because each point on the photocathode is imaged onto a specific photodiode. If the light falling onto the photocathode has been dispersed by a spectrometer with a dispersion $d\lambda/dx$, the spatial resolution Δx gives a spectral resolution

$$\Delta\lambda = \frac{d\lambda}{dx}\,\Delta x\,.$$

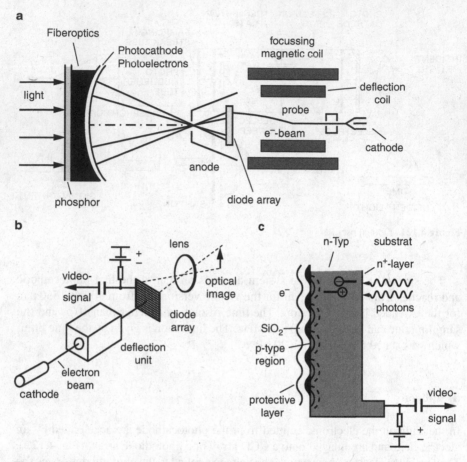

Figure 4.122 Basic principles of a vidicon. **a** Total setup, **b** details of detection system, **c** photodiode array

With a total legth L of the array the spectral range which can be simultaneously measured is

$$\delta\lambda = \frac{d\lambda}{dx} L \; .$$

These devices which store and analyse the information contained in the incident light simultaneously in many channels, are called *optical multichannel analysers* (OMA) or OSA (*optical spectrum analysers*). They measure two-dimensional images from weak extended light sources or cover simultaneously a spectral range $\delta\lambda$ [283]. The signals can be time integrated over a period of many seconds (for cooled devices even several hours) and are therefore superior to photomultipliers. They have found increasing importance in astronomy, where the faint images of distant galaxies are observes and their spectra are measured.

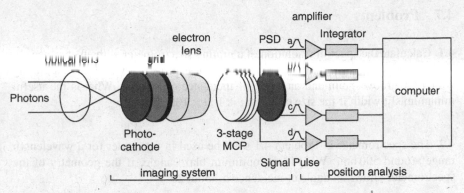

Figure 4.123 Schematic diagram of the PIAS as image intensifier with photon counting detection

e) PIAS System

PIAS is an acronym for *Photocounting Image Aquisation System*. Its basic principle is illustrated in Fig. 4.123. The photons from a light source fall onto the photocathode of an image intensifier where they release electrons which are amplified in a multichannel canal plate (MCP) and are imaged onto a position sensitive detector. A computer software analyses the data and gives the measured spectrum or the spatial variation of the intensity from extended sources [284, 285].

4.6 Conclusions

The aim of this chapter was to provide a general background in spectroscopic instrumentation, to summarize some basic ideas of spectroscopy, and to present some important relations between spectroscopic quantities. This background should be helpful in understanding the following chapters that deal with the main subject of this textbook: the applications of lasers to the solution of spectroscopic problems. Although until now we have only dealt with general spectroscopy, the examples given were selected with special emphasis on laser spectroscopy. This is especially true in Chap. 4, which is, of course, not a complete account of spectroscopic equipment, but is intended to give a survey on modern instrumentation used in laser spectroscopy.

There are several excellent and more detailed presentations of special instruments and spectroscopic techniques, such as spectrometers, interferometry, and Fourier spectroscopy. Besides the references given in the various sections, several series on optics [115], optical engineering [114], advanced optical techniques [284], and the monographs [117, 119, 283–288] may help to give more extensive information about special problems. Useful practical hints can be found in the handbooks [289, 290].

4.7   Problems

4.1 Calculate the spectral resolution of a grating spectrometer with an entrance slit width of $10\,\mu m$, focal lengths $f_1 = f_2 = 2\,m$ of the mirrors M_1 and M_2, a grating with 1800 grooves/mm and an angle of incidence $\alpha = 45°$. What is the useful minimum slit width if the size of grating is $100 \times 100\,mm^2$?

4.2 The spectrometer in Problem 4.1 shall be used in first order for a wavelength range around $500\,nm$. What is the optimum blaze angle, if the geometry of the spectrometer allows an angle of incidence α about $20°$?

4.3 Calculate the number of grooves/mm for a Littrow grating for a $25°$ incidence at $\lambda = 488\,nm$ (i.e., the first diffraction order is being reflected back into the incident beam at an angle $\alpha = 25°$ to the grating normal).

4.4 A prism can be used for expansion of a laser beam if the incident beam is nearly parallel to the prism surface. Calculate the angle of incidence α for which a HeNe laser beam ($\lambda = 632.8\,nm$) transmitted through a rectangular flint glass prism with $\epsilon = 60°$ is expanded tenfold.

4.5 Assume that a signal-to-noise ratio of 50 has been achieved in measuring the fringe pattern of a Michelson interferometer with one continuously moving mirror. Estimate the minimum path length ΔL that the mirror has to travel in order to reach an accuracy of $10^{-4}\,nm$ in the measurement of a laser wavelength at $\lambda = 600\,nm$.

4.6 The dielectric coatings of each plate of a Fabry–Perot interferometer have the following specifications: $R = 0.98$, $A = 0.3\,\%$. The flatness of the surfaces is $\lambda/100$ at $\lambda = 500\,nm$. Estimate the finesse, the maximum transmission, and the spectral resolution of the FPI for a plate separation of $5\,mm$.

4.7 A fluorescence spectrum shall be measured with a spectral resolution of $10^{-2}\,nm$. The experimentor decides to use a crossed arrangement of grating spectrometer (linear dispersion: $0.5\,nm/mm$) and FPI of Problem 4.6. Estimate the optimum combination of spectrometer slit width and FPI plate separation.

4.8 An interference filter shall be designed with peak transmission at $\lambda = 550\,nm$ and a bandwidth of $5\,nm$. Estimate the reflectivity R of the dielectric coatings and the thickness of the etalon, if no further transmission maximum is allowed between 350 and $750\,nm$.

4.9 A confocal FPI shall be used as optical spectrum analyzer, with a free spectral range of 3 GHz. Calculate the mirror separation d and the finesse that is necessary to resolve spectral features in the laser output within 10 MHz. What is the minimum reflectivity R of the mirrors, if the surface finesse is 500?

4.10 Calculate the transmission peaks of a Lyot filter with two plates ($d_1 = 1$ mm, $d_2 = 4$ mm) with $n = 1.40$ in the fast axis and $n = 1.45$ in the slow axis (a) as a function of λ for $\alpha = 45°$ in (4.97); and (b) as a function of α for a fixed wavelength λ. What is the contrast of the transmitted intensity $I(\alpha)$ for arbitrary values of λ if the absorption losses are 2 %?

4.11 Derive (4.116) for the equivalent electrical circuit of Fig. 4.79b.

4.12 A thermal detector has a heat capacity $H = 10^{-8}$ J/K and a thermal conductivity to a heat sink of $G = 10^{-9}$ W/K. What is the temperature increase ΔT for 10^{-9} W incident cw radiation if the efficiency $\beta = 0.8$? If the radiation is switched on at a time $t = 0$, how long does it take before the detector reaches a temperature increase $\Delta T(t) = 0.9 \Delta T_\infty$? What is the time constant of the detector and at which modulation frequency Ω of the incident radiation has the response decreased to 0.5 of its dc value?

4.13 A bolometer is operated at the temperature $T = 8$ K between superconducting and normal conducting states, where $R = 10^{-3} \Omega$. The heat capacity is $H = 10^{-8}$ J/K and the dc electrical current 1 mA. What is the change Δi of the heating current in order to keep the temperature constant when the bolometer is irradiated with 10^{-10} W?

4.14 The anode of a photomultiplier tube is connected by a resistor of $R = 1$ kΩ to ground. The stray capacitance is 10 pf, the current amplification 10^6, and the anode rise time 1.5 ns. What is the peak amplitude and the halfwidth of the anode output pulse produced by a single photoelectron? What is the dc output current produced by 10^{-12} W cw radiation at $\lambda = 500$ nm, if the quantum efficiency of the cathode is $\eta = 0.2$ and the anode resistor $R = 10^6 \Omega$? Estimate the necessary voltage amplification of a preamplifier (a) to produce 1 V pulses for single-photon counting; and (b) to read 1 V on a dc meter of the cw radiation?

4.15 A manufacturer of a two-stage optical image intensifier states that incident intensities of 10^{-17} W at $\lambda = 500$ nm can still be "seen" on the phosphor screen of the output state. Estimate the minimum intensity amplification, if the quantum efficiency of the cathodes and the conversion efficiency of the phosphor screens are both 0.2 and the collection efficiency of light emitted by the phosphor screens is 0.1. The human eye needs at least 20 photons/s to observe a signal.

4.16 Estimate the maximum output voltage of an open photovoltaic detector at room temperature under $10 \, \mu$W irradiation when the photocurrent of the shortened output is $50 \, \mu$A and the dark current is 50 nA.

Chapter 5
Lasers as Spectroscopic Light Sources

In this chapter we summarize basic laser concepts with regard to their applications in spectroscopy. A sound knowledge of laser physics with regard to passive and active optical cavities and their mode spectra, the realization of single-mode lasers, or techniques for frequency stabilization will help the reader to gain a deeper understanding of many subjects in laser spectroscopy and to achieve optimum performance of an experimental setup. Of particular interest for spectroscopists are the various types of tunable lasers, which are discussed in Sect. 5.7. Even in spectral ranges where no tunable lasers exist, optical frequency-doubling and mixing techniques may provide tunable coherent radiation sources, as outlined in Chap. 6.

5.1 Fundamentals of Lasers

This section gives a short introduction to the basic physics of lasers in a more intuitive than mathematical way. A more detailed treatment of laser physics and an extensive discussion of various types of lasers can be found in textbooks (see, for instance, [291–300]). For more advanced presentations based on the quantum-mechanical description of lasers, the reader is referred to [301–305].

5.1.1 Basic Elements of a Laser

A laser consists of essentially three components (Fig. 5.1a):

- The active medium, which amplifies an incident electromagnetic (EM) wave;
- The energy pump, which selectively pumps energy into the active medium to populate selected levels and to achieve population inversion;
- The optical resonator composed, for example, of two opposite mirrors, which stores part of the induced emission that is concentrated within a few resonator modes.

W. Demtröder, *Laser Spectroscopy 1*, DOI 10.1007/978-3-642-53859-9_5, © Springer-Verlag Berlin Heidelberg 2014

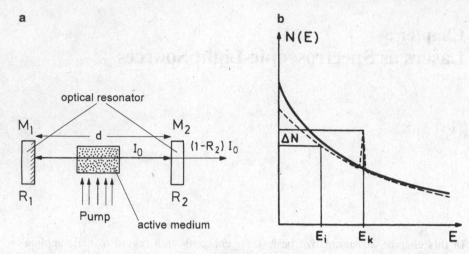

Figure 5.1 **a** Schematic setup of a laser; **b** population inversion (*dashed curve*), compared with a Boltzmann distribution at thermal equilibrium (*solid curve*)

The energy pump (e.g., flashlamps, gas discharges, or even other lasers) generates a population distribution $N(E)$ in the laser medium, which strongly deviates from the Boltzmann distribution (2.18) that exists for thermal equilibrium. At sufficiently large pump powers the population density $N(E_k)$ of the specific level E_k may exceed that of the lower level E_i (Fig. 5.1b).

For such a population inversion, the induced emission rate $N_k B_{ki} \rho(\nu)$ for the transition $E_k \to E_i$ exeeds the absorption rate $N_i B_{ik} \rho(\nu)$. An EM wave passing through this active medium is amplified instead of being attenuated according to (3.22).

The function of the optical resonator is the selective feedback of radiation emitted from the excited molecules of the active medium. Above a certain pump threshold this feedback converts the laser *amplifier* into a laser *oscillator*. When the resonator is able to store the EM energy of induced emission within a few resonator modes, the spectral energy density $\rho(\nu)$ may become very large. This enhances the induced emission into these modes since, according to (2.22), the induced emission rate already exceeds the spontaneous rate for $\rho(\nu) > h\nu$. In Sect. 5.1.3 we shall see that this concentration of induced emission into a small number of modes can be achieved with open resonators, which act as spatially selective and frequency-selective optical filters.

5.1.2 Threshold Condition

When a monochromatic EM wave with the frequency ν travels in the z-direction through a medium of molecules with energy levels E_i and E_k and $(E_k - E_i)/h = \nu$,

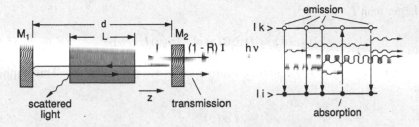

Figure 5.2 Gain and losses of an EM wave traveling back and forth along the resonator axis

the intensity $I(\nu, z)$ is, according to (3.23), given by

$$I(\nu, z) = I(\nu, 0)e^{-\alpha(\nu)z} , \tag{5.1}$$

where the frequency-dependent absorption coefficient

$$\alpha(\nu) = [N_i - (g_i/g_k)N_k]\sigma(\nu) , \tag{5.2}$$

is determined by the absorption cross section $\sigma(\nu)$ for the transition $(E_i \to E_k)$ and by the population densities N_i, N_k in the energy levels E_i, E_k with the statistical weights g_i, g_k, see (2.60). We infer from (5.2) that for $N_k > (g_k/g_i)N_i$, the absorption coefficient $\alpha(\nu)$ becomes negative and the incident wave is amplified instead of attenuated.

If the active medium is placed between two mirrors (Fig. 5.2), the wave is reflected back and forth, and traverses the amplifying medium many times, which increases the total amplification. With the length L of the active medium the total gain factor per single round-trip without losses is

$$G(\nu) = \frac{I(\nu, 2L)}{I(\nu, 0)} = e^{-2\alpha(\nu)L} . \tag{5.3}$$

A mirror with reflectivity R reflects only the fraction R of the incident intensity. The wave therefore suffers at each reflection a fractional reflection loss of $(1 - R)$. Furthermore, absorption in the windows of the cell containing the active medium, diffraction by apertures, and scattering due to dust particles in the beam path or due to imperfect surfaces introduce additional losses. When we summarize all these losses by a loss coefficient γ, which gives the fractional energy loss $\Delta W/W$ per round-trip time T, the intensity I decreases without an active medium per round-trip (if we assume the loss to be equally distributed along the resonator length d) as

$$I = I_0 e^{-\gamma} . \tag{5.4}$$

Including the amplification by the active medium with length L, we obtain for the intensity after a single round-trip through the resonator with length d, which may be larger than L:

$$I(v, 2d) = I(v, 0) \exp[-2\alpha(v)L - \gamma] \,. \tag{5.5}$$

The wave is amplified if the gain overcomes the losses per round-trip. This implies that

$$-2L\alpha(v) > \gamma \,. \tag{5.6}$$

With the absorption cross section $\sigma(v)$ from (5.2), this can be written as

$$2L \, \Delta N \, \sigma(v) > \gamma$$

which yields the *threshold condition* for the population difference

$$\boxed{\Delta N = N_k(g_i/g_k) - N_i > \Delta N_{\text{thr}} = \frac{\gamma}{2\sigma(v)L} \,.} \tag{5.7}$$

Example 5.1
$L = 10\,\text{cm}, \gamma = 10\,\%, \sigma = 10^{-12}\,\text{cm}^2 \rightarrow \Delta N = 5 \times 10^9/\text{cm}^3$. At a neon pressure of $0.2\,\text{mbar}$, ΔN corresponds to about 10^{-6} of the total density of neon atoms in a HeNe laser.

If the inverted population difference ΔN of the active medium is larger than ΔN_{thr}, a wave that is reflected back and forth between the mirrors will be amplified in spite of losses, therefore its intensity will increase.

The wave is initiated by spontaneous emission from the excited atoms in the active medium. Those spontaneously emitted photons that travel into the right direction (namely, parallel to the resonator axis) have the longest path through the active medium and therefore the greater chance of creating new photons by induced emission. Above the threshold they induce a photon avalanche, which grows until the depletion of the population inversion by stimulated emission just compensates the repopulation by the pump. Under steady-state conditions the inversion decreases to the threshold value ΔN_{thr}, the saturated net gain is zero, and the laser power limits itself to a finite value P_L. This laser power is determined by the pump power, the losses γ, and the gain coefficient $\alpha(v)$ (Sect. 5.7 and Chap. 7).

The frequency dependence of the gain coefficient $\alpha(v)$ is related to the line profile $g(v - v_0)$ of the amplifying transition. Without saturation effects (i.e., for small intensities), $\alpha(v)$ directly reflects this line shape, for homogeneous as well as for inhomogeneous profiles. According to (2.60) and (2.130) we obtain with the Einstein coefficienct B_{ik}

$$\alpha(v) = \Delta N \sigma_{ik}(v) = \Delta N(hv/c)B_{ik}g(v - v_0) \,, \tag{5.8}$$

which shows that the amplification is largest at the line center ν_0. For high intensities, saturation of the inversion occurs, which is different for homogeneous and for inhomogeneous line profiles (Vol. 2, Sects. 2.1 and 2.2).

The loss factor γ also depends on the frequency ν because the resonator losses are strongly dependent on ν. The frequency spectrum of the laser therefore depends on a number of parameters, which we discuss in more detail in Sect. 5.2.

5.1.3   Rate Equations

The photon number inside the laser cavity and the population densities of atomic or molecular levels under stationary conditions of a laser can readily be obtained from simple rate equations. **Note**, however, that this approach does *not* take into account coherence effects (Vol. 2, Chap. 7).

With the pump rate P (which equals the number of atoms that are pumped per second and per cm^3 into the upper laser level $|2\rangle$), the relaxation rates $R_i N_i$ (which equal the number of atoms that are removed per second and cm^3 from the level $|i\rangle$ by collision or spontaneous emission), and the spontaneous emission probability A_{21} per second, we obtain from (2.21) for equal statistical weights $g_1 = g_2$ the rate equations for the population densities N_i and the photon densities n (Fig. 5.3):

$$\frac{dN_1}{dt} = (N_2 - N_1)B_{21}nh\nu + N_2 A_{21} - N_1 R_1 , \qquad (5.9a)$$

$$\frac{dN_2}{dt} = P - (N_2 - N_1)B_{21}nh\nu - N_2 A_{21} - N_2 R_2 , \qquad (5.9b)$$

$$\frac{dn}{dt} = -\beta n + (N_2 - N_1)B_{21}nh\nu . \qquad (5.9c)$$

The loss coefficient β [s^{-1}] determines the loss rate of the photon density $n(t)$ stored inside the optical resonator. Without an active medium ($N_1 = N_2 = 0$), we obtain from (5.9c)

$$n(t) = n(0)e^{-\beta t} . \qquad (5.10)$$

Figure 5.3 Level diagram for pumping process P, relaxation rates $N_i R_i$, spontaneous and induced transitions in a four-level system

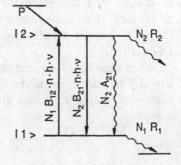

A comparison with the definition (5.4) of the dimensionless loss coefficient γ per round-trip yields for a resonator with length d and round-trip time $T = 2d/c$

$$\gamma = \beta T = \beta(2d/c) \ . \tag{5.11}$$

Under stationary conditions we have $dN_1/dt = dN_2/dt = dn/dt = 0$. Adding (5.9a and 5.9b) then yields

$$P = N_1 R_1 + N_2 R_2 \ , \tag{5.12}$$

which means that the pump rate P just compensates the loss rates $N_1 R_1 + N_2 R_2$ of the atoms in the two laser levels caused by relaxation processes into other levels. Further insight can be gained by adding (5.9b and 5.9c), which gives for stationary conditions

$$P = \beta n + N_2(A_{21} + R_2) \ . \tag{5.13}$$

In a continuous-wave (cw) laser the pump rate equals the sum of photon loss rate βn plus the total relaxation rate $N_2(A_{21} + R_2)$ of the upper laser level. A comparison of (5.12 and 5.13) shows that for a cw laser the relation holds

$$N_1 R_1 = \beta n + N_2 A_{21} \ . \tag{5.14}$$

Under stationary laser operation the relaxation rate $N_1 R_1$ of the lower laser level must always be larger than its feeding rate from the upper laser level!

The stationary inversion ΔN_{stat} can be obtained from the rate equation when multiplying (5.9a) by R_2, (5.9b) by R_1, and adding both equations. We find

$$\Delta N_{\text{stat}} = \frac{(R_1 - A_{21})P}{B_{12}nh\nu(R_1 + R_2) + A_{21}R_1 + R_1 R_2} \ . \tag{5.15}$$

This shows that a stationary inversion $\Delta N_{\text{stat}} > 0$ can only be maintained for $R_1 > A_{21}$. The relaxation probability R_1 of the lower laser level $|1\rangle$ must be larger than its refilling probability A_{21} by spontaneous transitions from the upper laser level $|2\rangle$. In fact, during the laser operation the induced emission mainly contributes to the population N_1 and therefore the more stringent condition $R_1 > A_{21} + B_{21}\rho$ must be satisfied. Continuous-wave lasers can therefore be realized on the transitions $|2\rangle \rightarrow |1\rangle$ only if the effective lifetime $\tau_{\text{eff}} = 1/R_1$ of level $|1\rangle$ is smaller than $(A_2 + B_{21}\rho)^{-1}$.

When starting a laser, the photon density n increases until the inversion density ΔN has decreased to the threshold density ΔN_{thr}. This can immediately be concluded from (5.9c), which gives for $dn/dt = 0$ and $d = L$

$$\Delta N = \frac{\beta}{B_{21}h\nu} = \frac{\gamma}{2L B_{21}h\nu/c} = \frac{\gamma}{2L\sigma} = \Delta N_{\text{thr}} \ , \tag{5.16}$$

where the relation (5.8) with

$$\int \alpha(\nu)d\nu = \Delta N \sigma_{12} = \Delta N (h\nu/c) B_{12} ,$$

has been used. Note, that $\int g(\nu - \nu_0)\, d\nu = 1$.

Example 5.2
With $N_2 = 10^{10}/\text{cm}^3$ and $(A_{21} + R_2) = 2 \times 10^7\,\text{s}^{-1}$, the total incoherent loss rate is $2 \times 10^{17}/\text{cm}^3 \cdot \text{s}$. In a HeNe laser discharge tube with $L = 10\,\text{cm}$ and 1 mm diameter, the active volume is about $0.075\,\text{cm}^3$. The total loss rate of the last two terms in (5.9c) then becomes $1.5 \times 10^{16}\,\text{s}^{-1}$.

For a laser output power of 3 mW at $\lambda = 633\,\text{nm}$, the rate of emitted photons is $\beta n = 10^{16}\,\text{s}^{-1}$. In this example the total pump rate has to be $P = (1.5 + 1) \times 10^{16}\text{s}^{-1} = 2.5 \times 10^{16}\,\text{s}^{-1}$, where the fluorescence emitted in all directions represents a larger loss than the mirror transmission.

5.2 Laser Resonators

In Sect. 2.1 it was shown that in a closed cavity a radiation field exists with a spectral energy density $\rho(\nu)$ that is determined by the temperature T of the cavity walls and by the eigenfrequencies of the cavity modes. In the optical region, where the wavelength λ is small compared with the dimension L of the cavity, we obtained the *Planck distribution* (2.13) at thermal equilibrium for $\rho(\nu)$. The number of modes per unit volume,

$$n(\nu)d\nu = 8\pi(\nu^2/c^3)d\nu ,$$

within the spectral interval $d\nu$ of a molecular transition turns out to be very large (Example 2.1a). When a radiation source is placed inside the cavity, its radiation energy will be distributed among all modes; the system will, after a short time, again reach thermal equilibrium at a correspondingly higher temperature. Because of the large number of modes in such a closed cavity, the mean number of photons per mode (which gives the ratio of induced to spontaneous emission rate in a mode) is very small in the optical region (Fig. 2.7). *Closed cavities with $L \gg \lambda$ are therefore not suitable as laser resonators.*

In order to achieve a concentration of the radiation energy into a small number of modes, the resonator should exhibit a strong feedback for these modes but large losses for all other modes. This would allow an intense radiation field to be built up in the modes with low losses but would prevent the system from reaching the oscillation threshold in the modes with high losses.

Assume that the kth resonator mode with the loss factor β_k contains the radiation energy W_k. The energy loss per second in this mode is then

$$\frac{dW_k}{dt} = -\beta_k W_k . \tag{5.17}$$

Under stationary conditions the energy in this mode will build up to a stationary value where the losses equal the energy input. If the energy input is switched off at $t = 0$, the energy W_k will decrease exponentially since integration of (5.17) yields

$$W_k(t) = W_k(0)e^{-\beta_k t} . \tag{5.18}$$

When we define the quality factor Q_k of the kth cavity mode as 2π times the ratio of energy stored in the mode to the energy loss per oscillation period $T = 1/\nu$

$$Q_k = -\frac{2\pi \nu W_k}{dW_k/dt} , \tag{5.19}$$

we can relate the loss factor β_k and the qualtiy factor Q_k by

$$Q_k = -2\pi \nu / \beta_k . \tag{5.20}$$

After the time $\tau = 1/\beta_k$, the energy stored in the mode has decreased to $1/e$ of its value at $t = 0$. This time can be regarded as the mean lifetime of a photon in this mode. If the cavity has large loss factors for most modes but a small β_k for a selected mode, the number of photons in this mode will be larger than in the other modes, even if at $t = 0$ the radiation energy in all modes was the same. If the unsaturated gain coefficient $\alpha(\nu)L$ of the active medium is larger than the loss factor $\gamma_k = \beta_k(2d/c)$ per round-trip but smaller than the losses of all other modes, the laser will oscillate only in this selected mode.

5.2.1 Open Optical Resonators

A resonator that concentrates the radiation energy of the active medium into a few modes can be realized with *open* cavities, which consist of two plane or curved mirrors aligned in such a way that light traveling along the resonator axis may be reflected back and forth between the mirrors. Such a ray traverses the active medium many times, resulting in a larger total gain. Other rays inclined against the resonator axis may leave the resonator after a few reflections before the intensity has reached a noticeable level (Fig. 5.4).

Besides these *walk-off losses*, *reflection losses* also cause a decrease of the energy stored in the resonator modes. With the reflectivities R_1 and R_2 of the resonator mirrors M_1 and M_2, the intensity of a wave in the passive resonator has decreased after a single round-trip to

$$I = R_1 R_2 I_0 = I_0 e^{-\gamma_R} , \tag{5.21}$$

Figure 5.4 Walk-off losses
of inclined rays and reflection
losses in an open resonator

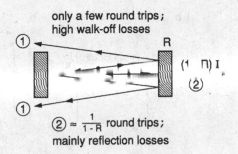

with $\gamma_R = -\ln(R_1 R_2)$. Since the round-trip time is $T = 2d/c$, the decay constant β in (5.18) due to reflection losses is $\beta_R = \gamma_R c/2d$. Therefore the mean lifetime of a photon in the resonator becomes without any additional losses

$$\tau = \frac{1}{\beta_R} = \frac{2d}{\gamma_R c} = -\frac{2d}{c \ln(R_1 R_2)}. \tag{5.22}$$

These open resonators are, in principle, the same as the Fabry–Perot interferometers discussed in Chap. 4; we shall see that several relations derived in Sect. 4.2 apply here. However, there is an essential difference with regard to the geometrical dimensions. While in a common FPI the distance between both mirrors is small compared with their diameter, the relation is generally reversed for laser resonators. The mirror diameter $2a$ is small compared with the mirror separation d. This implies that diffraction losses of the wave, which is reflected back and forth between the mirrrors, play a major role in laser resonators, while they can be completely neglected in the conventional FPI.

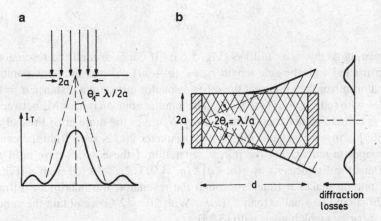

Figure 5.5 Equivalence of diffraction at an aperture **a** and at a mirror of equal size (**b**). The diffraction pattern of the transmitted light in (a) equals that of the reflected light in (b). The case $\theta_1 d = a \rightarrow N = 0.5$ is shown

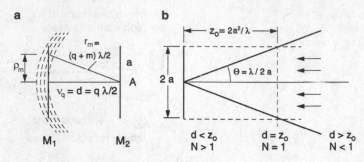

Figure 5.6 **a** Fresnel zones on mirror M_1, as seen from the center A of the other mirror M_2; **b** the three regions of d/a with the Fresnel number $N > 1$, $N = 1$, and $N < 1$

In order to estimate the magnitude of diffraction losses let us make use of a simple example. A plane wave incident onto a mirror with diameter $2a$ exhibits, after being reflected, a spatial intensity distribution that is determined by diffraction and that is completely equivalent to the intensity distribution of a plane wave passing through an aperture with diameter $2a$ (Fig. 5.5). The central diffraction maximum at $\theta = 0$ lies between the two first minima at $\theta_1 = \pm\lambda/2a$ (for circular apertures a factor 1.2 has to be included, see, e.g., [306]). About 16 % of the total intensity transmitted through the aperture is diffracted into higher orders with $|\theta| > \lambda/2a$. Because of diffraction the outer part of the reflected wave misses the second mirror M_2 and is therefore lost. This example demonstrates that the diffraction losses depend on the values of a, d, λ, and on the amplitude distribution $A(x, y)$ of the incident wave across the mirror surface. The influence of diffraction losses can be characterized by the dimensionless Fresnel number

$$N_F = \frac{a^2}{\lambda d} .$$ (5.23)

The meaning of this is as follows (Fig. 5.6a). If cones around the resonator axis are constructed with the side length $r_m = (q + m)\lambda/2$ and the apex point A on a resonator mirror they intersect the other resonator mirror at a distance $d = q\lambda/2$ in circles with radii $r_m = \frac{1}{2}(q+m)\cdot\lambda$. The annular zone on mirror M_1 between two circles is called Fresnel zone. The quantity N_F gives the number of Fresnel zones [306, 307] across a resonator mirror with diameter $2a$, as seen from the center A of the opposite mirror. For the mirror separation d these zones have radii $\rho_m = \sqrt{m\lambda d}$ and the distances $r_m = \frac{1}{2}(m+q)\lambda$ ($m = 0, 1, 2, \ldots \ll q$) from A (Fig. 5.6).

If a photon makes n transits through the resonator, the maximum diffraction angle 2θ should be smaller than $a/(nd)$. With $2\theta = \lambda/a$ we obtain the condition $\lambda/a < a/(n \cdot d)$ which gives with (5.23)

$$N_F > n .$$ (5.24)

This states that *the diffraction losses of a plane mirror resonator can be neglected if the Fresnel number N_F is larger than the number n of transits through the resonator.*

Example 5.3

a) A plane Fabry–Perot interferometer with $d = 1\,cm$, $a = 3\,cm$, $\lambda = 500\,nm$ has a Fresnel number $N = 1.8 \times 10^5$. The diffraction losses are completely negligible.

b) The resonator of a gas laser with plane mirrors at a distance $d = 50\,cm$, $a = 0.1\,cm$, $\lambda = 500\,nm$ has a Fresnel number $N = 4$. Since n should be about $n = 50$, $N_F \ll n$ and the diffraction losses are essential.

The fractional energy loss per transit due to diffraction of a plane wave reflected back and forth between the two plane mirrors is approximately given by

$$\gamma_D \sim \frac{1}{N}. \tag{5.25}$$

For our first example the diffraction losses of the plane FPI are about 5×10^{-6} and therefore completely negligible, whereas for the second example they reach 25 % and may already exceed the gain for many laser transitions. This means that a plane wave would not reach threshold in such a resonator. However, these high diffraction losses cause nonnegligible distortions of a plane wave and the amplitude $A(x, y)$ is no longer constant across the mirror surface (Sect. 5.2.2), but decreases towards the mirror edges. This decreases the diffraction losses, which become, for example, $\gamma_{Diffr} \leq 0.01$ for $N \geq 20$.

It can be shown [308] *that all resonators with plane mirrors that have the same Fresnel number also have the same diffraction losses, independent of the special choice of a, d, or λ.*

Resonators with curved mirrors may exhibit much lower diffraction losses than the plane mirror resonator because they can refocus the divergent diffracted waves of Fig. 5.5 (Sect. 5.2.5).

5.2.2 Spatial Field Distributions in Open Resonators

In Sect. 2.1 we have seen that any stationary field configuration in a *closed cavity* (called a *mode*) can be composed of plane waves. Because of diffraction, plane waves cannot give stationary fields in *open resonators*, since the diffraction losses depend on the coordinates (x, y) and increase from the z-axis of the resonator towards its edges. This implies that the distribution $A(x, y)$, which is independent of x and y for a plane wave, will be altered with each round-trip for a wave traveling back and forth between the mirrors of an open resonator until it approaches

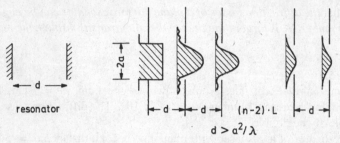

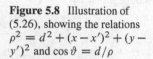

resonator |←d→|←d→| (n−2)·L |←d→|

$$d > a^2/\lambda$$

equivalent system of equidistant apertures

Figure 5.7 The diffraction of an incident plane wave at successive apertures separated by d is equivalent to the diffraction by successive reflections in a plane-mirror resonator with mirror separation d

Figure 5.8 Illustration of (5.26), showing the relations $\rho^2 = d^2 + (x − x')^2 + (y − y')^2$ and $\cos \vartheta = d/\rho$

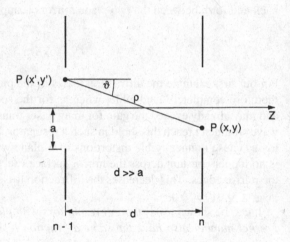

a stationary distribution. Such a stationary field configuration, called a *mode of the open resonator*, is reached when $A(x, y)$ no longer changes its form, although, of course, the losses result in a decrease of the total amplitude if they are not compensated by the gain of the active medium.

The mode configurations of open resonators can be obtained by an iterative procedure using the Kirchhoff–Fresnel diffraction theory [307]. Concerning the diffraction losses, the resonator with two plane square mirrors can be replaced by the equivalent arrangement of apertures with size $(2a)^2$ and a distance d between successive apertures (Fig. 5.7). When an incident plane wave is traveling into the z-direction, its amplitude distribution is successively altered by diffraction, from a constant amplitude to the final stationary distribution $A_n(x, y)$. The spatial distribution $A_n(x, y)$ in the plane of the nth aperture is determined by the distribution $A_{n-1}(x, y)$ across the previous aperture.

From Kirchhoff's diffraction theory we obtain (Fig. 5.8)

$$A_n(x, y) = -\frac{i}{\lambda} \iint A_{n-1}(x', y') \frac{1}{\rho} e^{-ik\rho} \cos \vartheta \, dx' dy' , \qquad (5.26)$$

A stationary field distribution is reached if

$$A_n(x, y) = C A_{n-1}(x, y) \quad \text{with} \quad C = e^{i\phi} \sqrt{1 - \gamma_D} . \qquad (5.27)$$

After the stationary state has been reached, the amplitude attenuation factor $|C|$ does not depend on x and y. The quantity γ_D represents the diffraction losses and ϕ the corresponding phase shift caused by diffraction.

Inserting (5.27) into (5.26) gives the following integral equation for the stationary field configuration

$$A(x, y) = -\frac{i}{\lambda}(1 - \gamma_D)^{-1/2} e^{-i\phi} \iint A(x', y') \frac{1}{\rho} e^{-ik\rho} \cos \vartheta \, dx' dy' . \qquad (5.28)$$

Because the arrangement of successive apertures is equivalent to the plane-mirror resonator, the solutions of this integral equation also represent the stationary modes of the open resonator. The diffraction-dependent phase shifts ϕ for the modes are determined by the condition of resonance. They are chosen in such a way that the diffracted wave reproduces itself after each round trip through the resonator.

The general integral equation (5.28) cannot be solved analytically, therefore one has to look for approximate methods. For two identical plane mirrors of quadratic shape $(2a)^2$, (5.28) can be solved numerically by splitting it into two one-dimensional equations, one for each coordinate x and y, if the Fresnel number $N = a^2/(d\lambda)$ is small compared with $(d/a)^2$, which means if $a \ll (d^3\lambda)^{1/4}$. The integral equation (5.28) can then be solved. The approximation implies $\rho \approx d$ in the denominator and $\cos \vartheta \approx 1$. In the phase term $\exp(-ik\rho)$, the distance ρ cannot be replaced by d, since the phase is sensitive even to small changes in the exponent. One can, however, for $x', x, y', y \ll d$, expand ρ into a power series

$$\rho = \sqrt{d^2 + (x' - x)^2 + (y' - y)^2} \approx d \left[1 + \frac{1}{2} \left(\frac{x' - x}{d} \right)^2 + \frac{1}{2} \left(\frac{y' - y}{d} \right)^2 \right] .$$
$$(5.29)$$

Inserting (5.29) into (5.28) allows the two-dimensional equation to be separated into two one-dimensional equations. Such numerical iterations for the "infinite strip" resonator have been performed by Fox and Li [309]. They showed that stationary field configurations do exist and computed the field distributions of these modes, their phase shifts, and their diffraction losses.

5.2.3 Confocal Resonators

The analysis has been extended by Boyd, Gordon, and Kogelnik to resonators with confocally-spaced spherical mirrors [310, 311] and later by others to general laser resonators [312–320]. For the symmetric confocal case (the two foci of the two mirrors with equal radii $R_1 = R_2 = R$ coincide, i.e., the mirror separation d is equal to the radius of curvature R).

For this case (5.28) can be separated into two one-dimensional homogeneous Fredholm equations that can be solved analytically [310, 314]. The solutions show that the stationary amplitude distributions for the confocal resonator can be represented by the product of Hermitian polynomials, a Gaussian function, and a phase factor:

$$A_{mn}(x, y, z) = C^* H_m(x^*) H_n(y^*) \exp(-r^2/w^2) \exp[-i\phi(z, r, R)] . \qquad (5.30)$$

Here, C^* is a normalization factor. The function H_m is the Hermitian polynomial of mth order. The last factor gives the phase $\phi(z_0, r)$ in the plane $z = z_0$ at a distance $r = (x^2 + y^2)^{1/2}$ from the resonator axis. The arguments x^* and y^* depend on the mirror separation d and are related to the coordinates x, y by $x^* = \sqrt{2}x/w$ and $y^* = \sqrt{2}y/w$, where

$$w^2(z) = \frac{\lambda d}{2\pi} \left[1 + (2z/d)^2\right] , \qquad (5.31)$$

is a measure of the radial intensity distribution. The coordinate z is measured from the center $z = 0$ of the confocal resonator.

From the definition of the Hermitian polynomials [321], one can see that the indices m and n give the number of nodes for the amplitude $A(x, y)$ in the x- (or the y-) direction. Figures 5.9 and 5.11 illustrate some of these "transverse electromagnetic standing waves," which are called TEM$_{m,n}$ modes. The diffraction effects do not essentially influence the transverse character of the waves. While Fig. 5.9a shows the one-dimensional amplitude distribution $A(x)$ for some modes, Fig. 5.9b depicts the two-dimensional field amplitude $A(x, y)$ in Cartesian coordinates and $A(r, \vartheta)$ in polar coordinates. Modes with $m = n = 0$ are called *fundamental modes* or *axial modes* (often zero-order transverse modes as well), while configurations with $m > 0$ or $n > 0$ are transverse modes of higher order. The intensity distribution of the fundamental mode $I_{00} \propto A_{00} A_{00}^*$ (Fig. 5.10) can be derived from (5.30). With $H_0(x^*) = H_0(y^*) = 1$ we obtain

$$I_{00}(x, y, z) = I_0 e^{-2r^2/w^2} . \qquad (5.32)$$

The fundamental modes have a Gaussian profile. For $r = w(z)$ the intensity decreases to $1/e^2$ of its maximum value $I_0 = C^{*2}$ on the axis ($r = 0$) the amplitude accordingly to $1/e$. The value $r = w(z)$ is called the *beam radius* or *mode radius*.

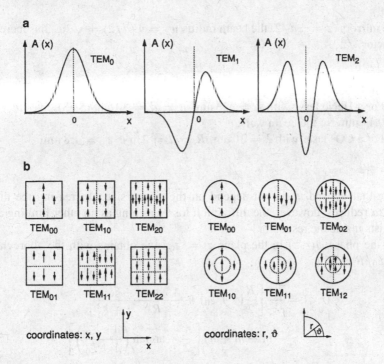

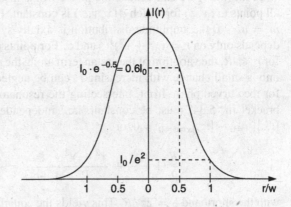

Figure 5.9 **a** Stationary one-dimensional amplitude distributions $A_m(x)$ in a confocal resonator; **b** two-dimensional presentation of linearly polarized resonator modes $\text{TEM}_{m,n}(x, y)$ for square and $\text{TEM}_{m,n}(r, \vartheta)$ for circular apertures

Figure 5.10 Radial intensity distribution of the fundamental TEM_{00} mode

The smallest beam radius w_0 within the confocal resonator is the *beam waist*, which is located at the center $z = 0$. From (5.31) we obtain with $d = R$

$$w_0 = (\lambda R/2\pi)^{1/2} . \tag{5.33}$$

On the mirrors ($z = \pm d/2$) the beam radius $w_s = w(d/2) = \sqrt{2}w_0$ has increased by a factor $\sqrt{2}$.

Example 5.4
a) For a HeNe laser with $\lambda = 633\,\text{nm}$, $R = d = 30\,\text{cm}$, (5.33) gives $w_0 = 0.17\,\text{mm}$ for the beam waist.
b) For a CO_2 laser with $\lambda = 10\,\mu\text{m}$, $R = d = 2\,\text{m}$ is $w_0 = 1.8\,\text{mm}$.

Note that w_0 and w do not depend on the mirror size. Increasing the mirror width $2a$ reduces, however, the diffraction losses as long as no other limiting aperture exists inside the resonator.

For the phase $\phi(r, z)$ in the plane $z = z_0$, one obtains with the abbreviation $\xi_0 = 2z_0/R$ [310]

$$\phi(r, z) = \frac{2\pi}{\lambda}\left[\frac{R}{2}(1 + \xi_0) + \frac{x^2 + y^2}{R}\frac{\xi_0}{1 + \xi_0^2}\right]$$
$$- (1 + m + n)\left[\frac{\pi}{2} - \arctan\left(\frac{1 - \xi_0}{1 + \xi_0}\right)\right]. \qquad (5.34)$$

Inside the resonator $0 < |\xi_0| < 1$, outside $|\xi_0| > 1$.

The equations (5.30) and (5.34) show that the field distributions $A_{mn}(x, y)$ and the form of the phase fronts depend on the location z_0 within the resonator.

From (5.34) we can deduce the phase fronts inside the confocal resonator, i.e., all points (x, y, z) for which $\phi(x, y, z)$ is constant. For the fundamental mode with $m = n = 0$ the amplitude distribution is axially symmetric and the phase $\phi(r, z)$ depends only on $r = (x^2 + y^2)^{1/2}$ and z. For points close to the resonator axis, i.e., for $r \ll R$, the variation of the arctan term along the phase front, where $z - z_0$ shows only a small change with increasing r, can be neglected. We obtain as a condition for the curved phase front, intersecting the resonator axis at $z = z_0$, that the first bracket in (5.34) must be constant, i.e., independent of x and y, which means: $[\ldots]_{x,y\neq0} - [\ldots]_{x=y=0} = 0$, or

$$\frac{R}{2}(1 + \xi) + \frac{x^2 + y^2}{R}\frac{\xi}{1 + \xi^2} = \frac{R}{2}(1 + \xi_0), \qquad (5.35)$$

with the shorthand $\xi = 2z/R$. This yields the equation

$$z_0 - z = \frac{x^2 + y^2}{R}\frac{\xi}{1 + \xi^2}, \qquad (5.36a)$$

which can be rearranged into the equation

$$x^2 + y^2 + (z - z_0)^2 = R'^2 \qquad (5.36b)$$

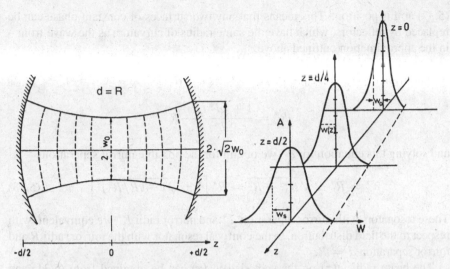

Figure 5.11 Phase fronts and intensity profiles of the fundamental TEM_{00} mode at several locations z in a confocal resonator with the mirrors at $z = \pm d/2$

of a spherical surface with the radius of curvature

$$R' \approx \left| \frac{1 + \xi_0^2}{2\xi_0} \right| R = \left[\frac{1}{4z_0} + \left(\frac{z_0}{R} \right)^2 \right] R . \tag{5.37}$$

The phase fronts of the fundamental modes inside a confocal resonator close to the resonator axis can be described as spherical surfaces with a z_0-dependent radius of curvature. For $z_0 = R/2 \to \xi_0 = 1 \Rightarrow R' = R$. This means that at the mirror surfaces of the confocal resonator close to the resonator axis the wavefronts are identical with the mirror surfaces. Due to diffraction this is not quite true at the mirror edges, (i.e., at larger distances r from the axis), where the approximation (5.35) is not correct.

At the center of the resonator $z = 0 \to \xi_0 = 0 \to R' = \infty$. The radius R' becomes infinite. **At the beam waist the constant phase surface becomes a plane $z = 0$.** This is illustrated by Fig. 5.11, which depicts the phase fronts and intensity profiles of the fundamental mode at different locations inside a confocal resonator.

5.2.4 General Spherical Resonators

It can be shown [291, 314] that in nonconfocal resonators with large Fresnel numbers N_F the field distribution of the fundamental mode can also be described by the Gaussian profile (5.32). The confocal resonator with $d' = R$ can be replaced by other mirror configurations without changing the field configurations if the radius R_i of each mirror at the position z_0 equals the radius R' of the wavefront in

(5.37) at this position. This means that any two surfaces of constant phase can be replaced by reflectors, which have the same radius of curvature as the wave front – in the approximation outlined above.

For symmetrical resonators with $R_1 = R_2 = R^*$ and the mirror separation d^*, we find from (5.37) with $z_0 = d^*/2 \rightarrow \xi_0 = d^*/R$

$$R^* = \frac{1 + (d^*/R)^2}{2d^*/R} R$$

and solving this equation for d^* we obtain for the possible mirror separations

$$d^* = R^* \pm \sqrt{R^{*2} - R^2} = R^* \left[1 \pm \sqrt{1 - (R/R^*)^2} \right] . \tag{5.38}$$

These resonators with mirror separation d^* and mirror radii R^* are equivalent, with respect to the field distribution, to the confocal resonator with the mirror radii R and mirror separation $d = R$.

The beam radii $w(z)$ on the spot size $w^2(z)$ can be obtained from (5.31) and (5.38). For the symmetric resonator with $R_1 = R_2 = R$ we get at the center ($z = 0$) and at the mirrors ($z = \pm d/2$)

$$w_0^2(z) = \left(\frac{d\lambda}{\pi} \right)^* \left[\frac{2R - d}{4d} \right]^{1/2} ; \quad w_1^2 = w_2^2 = \left(\frac{d\lambda}{\pi} \right) \left[\frac{R^2}{2dR - d^2} \right]^{1/2} . \tag{5.39a}$$

With the parameters

$$g = 1 - d/R$$

this can be written as

$$w_0^2(z = 0) = \frac{d\lambda}{\pi} \sqrt{\frac{1 + g}{4(1 - g)}} ; \quad w_1^2 = w_2^2 = \frac{d\lambda}{\pi} \sqrt{\frac{1}{1 - g^2}} . \tag{5.39b}$$

The mode waist $w_0^2(z = 0)$ is minimum for $g = 0$, i.e., $d = R$. The confocal resonator has the smallest beam waist. Also, the spot sizes $w_1^2 = w_2^2$ are minimum for $g = 0$. We therefore obtain the following result:

Of all symmetric resonators with a given mirror separation d the confocal resonator with $d = R$ has the smallest spot sizes at the mirrors and the smallest beam waist w_0.

5.2.5   Diffraction Losses of Open Resonators

The diffraction losses of a resonator depend on its Fresnel number $N_F = a^2/d\lambda$ (Sect. 5.2.1) and also on the field distribution $A(x, y, z = \pm d/2)$ at the mirror.

Figure 5.12 Diffraction losses of some modes in a confocal and in a plane-mirror resonator, plotted as a function of the Fresnel number N_F

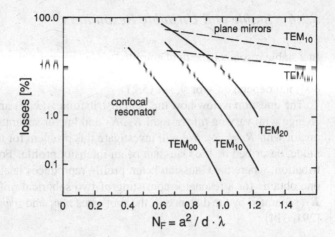

The fundamental mode, where the field energy is concentrated near the resonator axis, has the lowest diffraction losses, while the higher transverse modes, where the field amplitude has larger values toward the mirror edges, exhibit large diffraction losses. Using (5.31) with $z = d/2$ and (5.33) the Fresnel number $N_F = a^2/(d\lambda)$ can be expressed as

$$N_F = \frac{1}{\pi} \frac{\pi a^2}{\pi w_s^2} = \frac{1}{\pi} \frac{\text{effective resonator-mirror surface area}}{\text{confocal } TEM_{00} \text{ mode area on the mirror}}, \qquad (5.40)$$

which illustrates that the diffraction losses decrease with increasing N_F. Figure 5.12 presents the diffraction losses of a confocal resonator as a function of the Fresnel number N_F for the fundamental mode and some higherorder transverse modes. For comparison, the much higher diffraction losses of a plane-mirror resonator are also shown in order to illustrate the advantages of curved mirrors, which refocus the waves otherwise diverging by diffraction. From Fig. 5.12 it is obvious that higher-order transverse modes can be suppressed by choosing a resonator with a suitable Fresnel number, which may be realized, for instance, by a limiting aperture with the diameter $D < 2a$ inside the laser resonator. If the losses exceed the gain for these modes they do not reach threshold, and the laser oscillates only in the fundamental mode.

The confocal resonator with the smallest spot sizes at a given mirror separation d according to (5.39a), (5.39b) also has the lowest diffraction losses per round-trip, which can be approximated for circular mirrors and Fresnel numbers $N_F > 1$ by [291]

$$\gamma_D \sim 16\pi^2 N_F e^{-4\pi N_F}. \qquad (5.41)$$

5.2.6   Stable and Unstable Resonators

In a stable resonator the field amplitude $A(x, y)$ reproduces itself after each round-trip apart from a constant factor C, which represents the total diffraction losses but does not depend on x or y, see (5.27).

The question is now how the field distribution $A(x, y)$ and the diffraction losses change with varying mirror radii R_1, R_2 and mirror separation d for a general resonator with $R_1 \neq R_2$. We will investigate this problem for the fundamental TEM_{00} mode, described by the Gaussian beam intensity profile. For a stationary field distribution, where the Gaussian beam profile reproduces itself after each round-trip, one obtains for a resonator consisting of two spherical mirrors with the radii R_1, R_2, separated by the distance d, the spot sizes πw_1^2 and πw_2^2 on the mirror surfaces [291, 314]

$$\pi w_1^2 = \lambda d \left[\frac{g_2}{g_1(1 - g_1 g_2)} \right]^{1/2} , \quad \pi w_2^2 = \lambda d \left[\frac{g_1}{g_2(1 - g_1 g_2)} \right]^{1/2} , \qquad (5.42)$$

with the parameters g_i $(i = 1, 2)$

$$g_i = 1 - d/R_i . \qquad (5.43)$$

For $g_1 = g_2$ (confocal symmetric resonator), (5.42) simplifies to (5.39b). Equation (5.42) reveals that for $g_1 = 0$ the spot size πw_1^2 becomes ∞ at M_1 and $\pi w_2^2 = 0$ at M_2, while for $g_2 = 0$ the situation is reversed. For $g_1 g_2 = 1$ both spot sizes become infinite. This implies that the Gaussian beam diverges: the resonator becomes unstable. An exception is the confocal resonator with $g_1 = g_2 = 0$, which is "metastable", because it is only stable if both parameters g_i are exactly zero. For $g_1 g_2 > 1$ or $g_1 g_2 < 0$, the right-hand sides of (5.42) become imaginary, which means that the resonator is unstable. The condition for a stable resonator is therefore

$$\boxed{0 < g_1 g_2 < 1.} \qquad (5.44)$$

The beam waist w_0^2 of a confocal nonsymmetric resonator with $R_1 \neq R_2$ is no longer at the center of the resonator (as for symmetric resonators). Its distance from M_1 is

$$z_1(w_0) = \frac{d}{1 + (\lambda d/\pi w_1^2)^2} ; \quad z_2 = d - z_1 .$$

With the general stability parameter

$$G = 2g_1 g_2 - 1$$

we can distinguish stable resonators: $0 < |G| < 1$, unstable resonators: $|G| > 1$, metastable resonators: $|G| = 1$.

Table 5.1 Some commonly used optical resonators with their stability parameters $g_i = 1 - d/R_i$, and the resonator parameters $G = 2g_1 g_2 - 1$

Type of resonator	Mirror radii	Stability parameter	
Confocal	$R_1 + R_2 = 2d$	$g_1 + g_2 = 2g_1 g_2$	$|G| \leq 1$
Concentric	$R_1 + R_2 = d$	$g_1 g_2 = 1$	$G = 1$
Symmetric	$R_1 = R_2$	$g_1 = g_2 = g$	$|G| < 1$
Symmetric confocal	$R_1 = R_2 = d$	$g_1 = g_2 = 0$	$G = -1$
Symmetric concentric	$R_1 = R_2 = 1/2d$	$g_1 = g_2 = -1$	$G = 1$
Semiconfocal	$R_1 = 2d, R_2 = \infty$	$g_1 = 1, g_2 = 1/2$	$G = 0$
Plane	$R_1 = R_2 = \infty$	$g_1 = g_2 = +1$	$G = 1$

Example 5.5

a) $R_1 = 0.5\,\text{m}, d = 0.5\,\text{m}$. If the active medium close to M_1 with a diameter of $0.6\,\text{cm}$ needs to be completely filled with the TEM_{00} mode, the beam waist at M_1 should be $w_1 = 0.3\,\text{cm}$. With a Fresnel number $N_F = 3$ the diffraction losses are sufficiently small. The stability parameter for $\lambda = 1\,\mu\text{m}$

$$g_2 = \frac{w_1^2}{N_F\, 2d\lambda}$$

is then $g_2 = 3$. This gives for R_2: $g_2 = 1 - d/R_2 \Rightarrow R_2 = d/(1 - g_2) = -25\,\text{cm}$.

b) Confocal resonator with $d = 1\,\text{m}, \lambda = 500\,\text{nm}, R_1 = R_2 = 1\,\text{m} \Rightarrow w_1 = w_2 = 0.4\,\text{mm}$ at both mirrors.

If in a symmetric confocal resonator a plane mirror is placed at the beam waist (where the phase front is a plane), a semiconfocal resonator results (Fig. 5.14), with $R_1 = \infty, d = R_2/2, g_1 = 1, g_2 = 1/2, w_1^2 = \lambda d/\pi, w_2^2 = 2\lambda d/\pi$.

In Table 5.1 some resonators are compiled with their corresponding parameters g_i. Figure 5.13 displays the stability diagram in the g_1–g_2-plane. According to (5.44) the plane resonator ($R_1 = R_2 = \infty \Rightarrow g_1 = g_2 = 1$) is not stable, because the spot size of a Gaussian beam would increase after each round-trip. As was shown above, there are, however, other non-Gaussian field distributions, which form stable eigenmodes of a plane resonator, although their diffraction losses are much higher than those of resonators within the stability region. The symmetric confocal resonator with $g_1 = g_2 = 0$ might be called "metastable," since it is located between unstable regions in the stability diagram and even a slight deviation of g_1, g_2 into the direction $g_1 g_2 < 0$ makes the resonator unstable. For illustration, some commonly used resonators are depicted in Fig. 5.15.

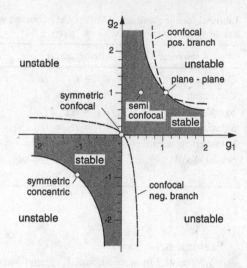

Figure 5.13 Stability diagram of optical resonators. The shaded areas represent stable resonators

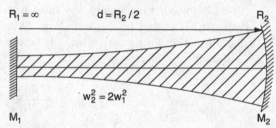

Figure 5.14 Semi-confocal resonator

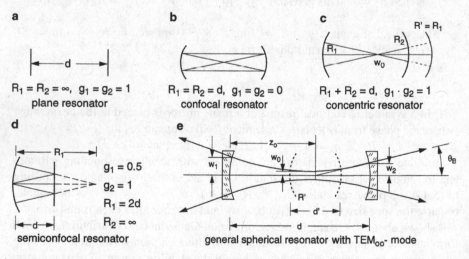

a

$R_1 = R_2 = \infty$, $g_1 = g_2 = 1$
plane resonator

b

$R_1 = R_2 = d$, $g_1 = g_2 = 0$
confocal resonator

c

$R_1 + R_2 = d$, $g_1 \cdot g_2 = 1$
concentric resonator

d

$g_1 = 0.5$
$g_2 = 1$
$R_1 = 2d$
$R_2 = \infty$
semiconfocal resonator

e

general spherical resonator with TEM$_{00}$- mode

Figure 5.15 Some examples of commonly used open resonators

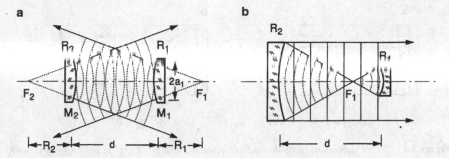

Figure 5.16 **a** Spherical waves in a symmetric unstable resonator emerging from the virtual focal points F_1 and F_2; **b** asymmetric unstable resonator with a real focal point between the two mirrors

For some laser media, in particular those with large gain, unstable resonators with $g_1 g_2 < 0$ may be more advantageous than stable ones for the following reason: in stable resonators the beam waist $w_0(z)$ of the fundamental mode is given by the mirror radii R_1, R_2 and the mirror separation d, see (5.33), and is generally small (Example 5.4). If the cross section of the active volume is larger than πw^2, only a fraction of all inverted atoms can contribute to the laser emission into the TEM_{00} mode, while in unstable resonators the beam fills the whole active medium. This allows extraction of the maximum output power. One has, however, to pay for this advantage by a large beam divergence.

Let us consider the simple example of a symmetric unstable resonator depicted in Fig. 5.16 formed by two mirrors with radii R_i separated by the distance d. Assume that a spherical wave with its center at F_1 is emerging from mirror M_1. The spherical wave geometrically reflected by M_2 has its center in F_2. If this wave after ideal reflection at M_1 is again a spherical wave with its center at F_1, the field configuration is stationary and the mirrors image the local point F_1 into F_2, and vice versa.

For the magnification of the beam diameter on the way from mirror M_1 to M_2 or from M_2 to M_1, we obtain from Fig. 5.16 the relations

$$M_{12} = \frac{d + R_1}{R_1}, \quad M_{21} = \frac{d + R_2}{R_2}. \tag{5.45}$$

We define the magnification factor $M = M_{12} M_{21}$ per round-trip as the ratio of the beam diameter after one round-trip to the initial one:

$$M = M_{12} M_{21} = \left(\frac{d + R_1}{R_1}\right)\left(\frac{d + R_2}{R_2}\right). \tag{5.46}$$

For $R_i > 0$ $(i = 1, 2)$ the virtual focal points are outside the resonator and the magnification factor becomes $M > 1$ (Fig. 5.16a).

In the resonator of Fig. 5.16a the waves are coupled out of both sides of the resonator. The resultant high resonator losses are generally not tolerable and for practical applications the resonator configurations of Fig. 5.16b and Fig. 5.17 consisting

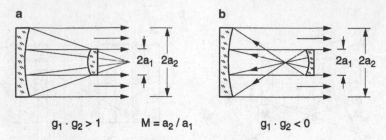

$g_1 \cdot g_2 > 1$ $M = a_2/a_1$ $g_1 \cdot g_2 < 0$

Figure 5.17 Two examples of unstable confocal resonators: **a** $g_1 \cdot g_2 > 1$; **b** $g_1 \cdot g_2 < 0$, with a definition of the magnification factor

of one large and one small mirror are better suited. Two types of nonsymmetric spherical unstable resonators are possible with $g_1 g_2 > 1 \Rightarrow G > 1$ (Fig. 5.17a) with the virtual beam waist outside the resonator and with $g_1 g_2 < 0 \Rightarrow G < -1$ (Fig. 5.17b) where the focus lies inside the resonator.

For these spherical resonators the magnification factor M can be expressed by the resonator parameter G [315]:

$$M_\pm = |G| \pm \sqrt{G^2 - 1} \, , \qquad (5.47a)$$

where the $+$ sign holds for $g_1 g_2 > 1$ and the $-$ sign for $g_1 g_2 < 0$.

If the intensity profile $I(x_1, y_1, z_0)$ in the plane $z = z_0$ of the outcoupling mirror does not change much over the mirror size, the fraction P_2/P_0 of the power P_0 incident on M_2 that is reflected back to M_1 equals the ratio of the areas

$$\frac{P_2}{P_0} = \frac{\pi w_2^2}{\pi w_1^2} = \frac{1}{M^2} \, . \qquad (5.47b)$$

The loss factor per round-trip is therefore

$$V = \frac{P_0 - P_2}{P_0} = 1 - \frac{1}{M^2} = \frac{M^2 - 1}{M^2} \, . \qquad (5.48)$$

Example 5.6
$R_1 = -0.5\,\mathrm{m}$, $R_2 = +2\,\mathrm{m}$, $d = 0.6\,\mathrm{m} \Rightarrow g_1 = 1 - d/R_1 = 2.4$; $g_2 = 1 - d/R_2 = 0.7$; $G = 2g_1 g_2 - 1 = 2.36$; $M_t = G + \sqrt{G^2 - 1} = 4.49$; $V = 1 - 1/M^2 = 0.95$. In these unstable resonators the losses per round trip are 95 %.

For the two unstable resonators of Fig. 5.17 the near-field pattern of the outcoupled wave is an annular ring (Fig. 5.18). The spatial farfield intensity distribution

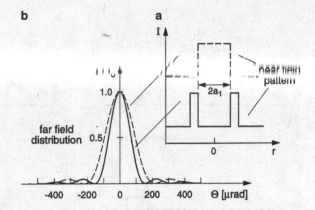

Figure 5.18 Diffraction pattern of the output intensity of a laser with an unstable resonator. **a** near field just at the output coupler and **b** far-field distribution for a resonator with $a = 0.66$ cm, $g_1 = 1.21$, $g_2 = 0.85$. The patterns obtained with a circular output mirror (*solid curve*) are compared with those of a circular aperture (*dashed curves*)

can be obtained as a numerical solution of the corresponding Kirchhoff–Fresnel integro-differential equation analog to (5.26). For illustration, the near-field and far-field patterns of an unstable resonator of the type shown in Fig. 5.17a is compared with the diffraction pattern of a circular aperture.

Note that the angular divergence of the central diffraction order in the far field is smaller for the annular-ring near-field distribution than that of a circular aperture with the same size as the small mirror of the unstable resonator. However, the higher diffraction orders are more intense, which means that the angular intensity distribution has broader wings.

In unstable resonators the laser beam is divergent and only a fraction of the divergent beam area may be reflected by the mirrors. The losses are therefore high and the effective number of round-trips is small. Unstable resonators are therefore suited only for lasers with a sufficiently large gain per round-trip [316–319].

In recent years, specially designed optics with slabs of cylindrical lenses have been used to make the divergent output beam more parallel, which allows one to focus the beam into a smaller spot size [320].

5.2.7 Ring Resonators

A ring resonator consists of at least three reflecting surfaces, which may be provided by mirrors or prisms. Four possible arrangements are illustrated in Fig. 5.19. Instead of the *standing* waves in a Fabry–Perot-type resonator, the ring resonator allows *traveling* waves, which may run clockwise or counter-clockwise through the resonator. With an "optical diode" inside the ring resonator unidirectional traveling waves can be enforced. Such an "optical diode" is a device that has low losses for light passing into one direction but sufficiently high losses to prevent laser oscillation for light traveling into the opposite direction. It consists of a Faraday rotator, which turns the plane of polarization by the angle $\pm\alpha$ (Fig. 5.20), a birefringent crystal, which also turns the plane of polarization by α, and elements with

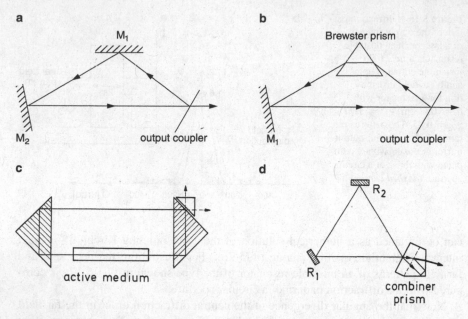

Figure 5.19 Four examples of possible ring resonators, using **a** three mirrors, **b** two mirrors and a Brewster prism, **c** total reflection: with corner-cube prism reflectors and frustrated total reflection for output coupling; **d** three-mirror arrangement with beam-combining prism

Figure 5.20 Optical diode consisting of a Faraday rotator, a birefringent crystal, and Brewster windows. Tilting of the polarization vector for the forward (**a**) and backward (**b**) directions

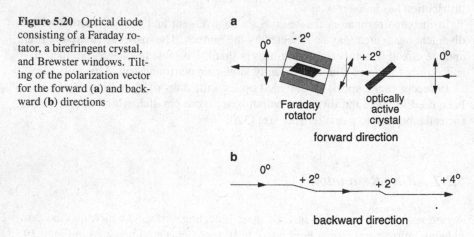

a polarization-dependent transmission, such as Brewster windows [322]. For the wanted direction the turning angles $-\alpha + \alpha = 0$ just cancel, and for the other direction they add to 2α, causing reflection losses at the Brewster windows. If these are larger than the gain this direction cannot reach the threshold.

The unidirectional ring laser has the advantage that spatial hole burning, which impedes single-mode oscillation of lasers (Sect. 5.3.3), can be avoided. In the case of homogeneous gain profiles, the ring laser can utilize the total population inversion within the active mode volume contrary to a standing-wave laser, where the inversion at the nodes of the standing wave cannot be utilized. One therefore expects larger output powers in single-mode operation than from standing-wave cavities at comparable pump powers

5.2.8   Frequency Spectrum of Passive Resonators

The stationary field configurations of open resonators, discussed in the previous sections, have an eigenfrequency spectrum that can be directly derived from the condition that the phase fronts at the reflectors have to be identical with the mirror surfaces. Because these stationary fields represent standing waves in the resonators, the mirror separation d must be an integer multiple of $\lambda/2$ and the phase factor in (5.30) becomes unity at the mirror surfaces. This implies that the phase ϕ has to be an integer multiple of π. Inserting the condition $\phi = q\pi$ into (5.34) gives the eigenfrequencies $\nu_r = c/\lambda_r$ of the confocal resonator with $R = d$, $\xi_0 = 1$, $x = y = 0$

$$\nu_r = \frac{c}{2d}\left[q + \frac{1}{2}(m+n+1)\right] . \tag{5.49}$$

The fundamental axial modes TEM_{00q} ($m = n = 0$) have the frequencies $\nu = (q + \frac{1}{2})c/2d$ and the frequency separation of adjacent axial modes is

$$\delta\nu = \frac{c}{2d} . \tag{5.50}$$

Equation (5.49) reveals that the frequency spectrum of the confocal resonator is degenerate because the transverse modes with $q = q_1$ and $m + n = 2p$ have the same frequency as the axial mode with $m = n = 0$ and $q = q_1 + p$. Between two axial modes there is always another transverse mode with $m + n + 1 =$ odd. The free spectral range of a *confocal resonator* is therefore

$$\delta\nu_{\text{confocal}} = \frac{c}{4d} . \tag{5.51}$$

If the mirror separation d deviates slightly from the radius of the mirror curvature R, the degeneracy is removed. We obtain from (5.34) with $\phi = q\pi$ and $\xi_0 = d/R \neq 1$ for a symmetric nonconfocal resonator with two equal mirror radii $R_1 = R_2 = R$

$$\nu_r = \frac{c}{2d}\left\{q + \frac{1}{2}(m+n+1)\left[1 + \frac{4}{\pi}\arctan\left(\frac{d-R}{d+R}\right)\right]\right\} . \tag{5.52}$$

Now the higher-order transverse modes are no longer degenerate with axial modes. The frequency separation depends on the ratio $(d - R)/(d + R)$. Figure 5.21 illustrates the frequency spectrum of the plane-mirror resonator, the confocal resonator

Figure 5.21 Degenerate mode frequency spectrum of a confocal resonator ($d = R$) (**a**), degeneracy lifting in a near-confocal resonator ($d = 1.1R$) (**b**), and the spectrum of fundamental modes in a plane-mirror resonator (**c**)

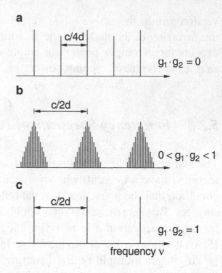

($R = d$), and of a nonconfocal resonator where d is slightly larger than R. Due to higher diffraction losses the amplitudes of the higher transverse modes decrease.

As has been shown in [311] the frequency spectrum of a general resonator with unequal mirror curvatures R_1 and R_2 can be represented by

$$v_r = \frac{c}{2d}\left[q + \frac{1}{\pi}(m + n + 1) \arccos \sqrt{g_1 g_2}\right], \qquad (5.53)$$

where $g_i = 1 - d/R_i$ ($i = 1, 2$) are the resonator parameters. The eigenfrequencies of the axial modes ($m = n = 0$) are no longer at $(c/2d)(q + \frac{1}{2})$, but are slightly shifted. The free spectral range, however, is again $\delta v = c/2d$.

Example 5.7

a) Consider a nonconfocal symmetric resonator: $R_1 = R_2 = 75$ cm, $d = 100$ cm. The free spectral range δv, which is the frequency separation of the adjacent axial modes q and $q + 1$, is $\delta v = (c/2d) = 150$ MHz. The frequency separation Δv between the $(q, 0, 0)$ mode and the $(q, 1, 0)$ mode is $\Delta v = 87$ MHz from (5.52).

b) Consider a confocal resonator: $R = d = 100$ cm. The frequency spectrum consists of equidistant frequencies with $\delta v = 75$ MHz. If, however, the higher-order transverse modes are suppressed, only axial modes oscillate with a frequency separation $\delta v = 150$ MHz.

Now we briefly discuss the spectral width Δv of the resonator resonances. The problem will be approached in two different ways.

Since the laser resonator is a Fabry–Perot interferometer, the spectral distribution of the transmitted intensity follows the Airy formula (4.61a)–(4.61d). According to (4.55b), the halfwidth $\Delta \nu_r$ of the resonances, expressed in terms of the free spectral range $\delta \nu$, is $\Delta \nu_r = \delta \nu / F^*$. If diffraction losses can be neglected, the finesse F^* is mainly determined by the reflectivity R of the mirrors, therefore the halfwidth of the resonance becomes

$$\Delta \nu = \frac{\delta \nu}{F^*} = \frac{c}{2d} \frac{1 - R}{\pi \sqrt{R}}. \tag{5.54}$$

Example 5.8
With the reflectivity $R = 0.98 \Rightarrow F^* = 150$. A resonator with $d = 1\,\mathrm{m}$ has the free spectral range $\delta \nu = 150\,\mathrm{MHz}$. The halfwidth of the resonator modes then becomes $\Delta \nu_r = 1\,\mathrm{MHz}$ if the mirrors are perfectly aligned and have nonabsorptive ideal surfaces.

Generally speaking, other losses such as diffraction, absorption, and scattering losses decrease the total finesse. Realistic values are $F^* = 50$–100, giving for Example 5.8 a resonance halfwidth of the passive resonator of about $2\,\mathrm{MHz}$.

The second approach for the estimate of the resonance width starts from the quality factor Q of the resonator. With total losses β per second, the energy W stored in a mode of a passive resonator decays exponentially according to (5.18). The Fourier transform of (5.18) yields the frequency spectrum of this mode, which gives a Lorentzian (Sect. 3.1) with the halfwidth $\Delta \nu_r = \beta / 2\pi$. With the mean lifetime $T = 1/\beta$ of a photon in the resonator mode, the frequency width can be written as

$$\Delta \nu_r = \frac{1}{2\pi T}. \tag{5.55}$$

If reflection losses give the main contribution to the loss factor, the photon lifetime is, with $R = \sqrt{R_1 R_2}$, see (5.22), $T = -d/(c \ln R)$. The width $\Delta \nu$ of the resonator mode becomes

$$\Delta \nu_r = \frac{c |\ln R|}{2\pi d} = \frac{\delta \nu (|\ln R|)}{\pi}, \tag{5.56}$$

which yields with $|\ln R| \approx 1 - R$ the same result as (5.54), apart from the factor $\sqrt{R} \approx 1$. The slight difference of the two results stems from the fact that in the second estimation we distributed the reflection losses uniformly over the resonator length.

5.3   Spectral Characteristics of Laser Emission

The frequency spectrum of a laser is determined by the spectral range of the active laser medium, i.e., its gain profile, and by the resonator modes falling within this spectral gain profile (Fig. 5.22). All resonator modes for which the gain exceeds the losses can participate in the laser oscillation. The active medium has two effects on the frequency distribution of the laser emission:

- Because of its index of refraction $n(\nu)$, it shifts the eigenfrequencies of the passive resonator (mode-pulling).
- Due to spectral gain saturation competition effects between different oscillating laser modes occur; they may influence the amplitudes and frequencies of the laser modes.

In this section we shall briefly discuss spectral characteristics of multimode laser emission and the effects that influence it.

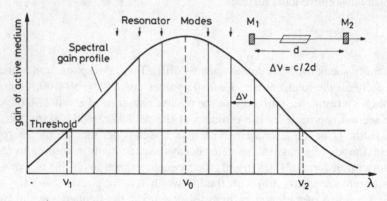

Figure 5.22   Gain profile of a laser transition with resonator eigenfrequencies of axial modes

5.3.1   Active Resonators and Laser Modes

Introducing the amplifying medium into the resonator changes the refractive index between the mirrors and with it the eigenfrequencies of the resonator. We obtain the frequencies of the *active resonator* by replacing the mirror separation d in (5.52) by

$$d^* = (d - L) + n(\nu)L = d + (n - 1)L , \qquad (5.57)$$

where $n(\nu)$ is the refractive index in the active medium with length L. The refractive index $n(\nu)$ depends on the frequency ν of the oscillating modes within the gain profile of a laser transition where anomalous dispersion is found. Let us at first consider how laser oscillation builds up in an active resonator.

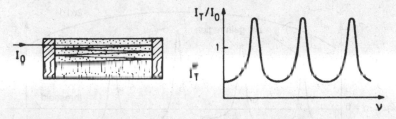

Figure 5.23 Transmission of an incident wave through an active resonator

If the pump power is increased continuously, the threshold is reached first at those frequencies that have a maximum net gain. According to (5.5) the net gain factor per round-trip

$$G(v, 2d) = \exp[-2\alpha(v)L - \gamma(v)], \tag{5.58}$$

is determined by the amplification factor $\exp[-2\alpha(v)L]$, which has the frequency dependence of the gain profile (5.8) and also by the loss factor $\exp(-2\beta d/c) = \exp[-\gamma(v)]$ per round-trip. While absorption or diffraction losses of the resonator do not strongly depend on the frequency within the gain profiles of a laser transition, the transmission losses exhibit a strong frequency dependence, which is closely connected to the eigenfrequency spectrum of the resonator. This can be illustrated as follows:

Assume that a wave with the spectral intensity distribution $I_0(v)$ traverses an interferometer with two mirrors, each having the reflectivity R and transmission factor T (Fig. 5.23). For the passive interferometer we obtain a frequency spectrum of the transmitted intensity according to (4.52a), (4.52b). With an amplifying medium inside the resonator, the incident wave experiences the amplification factor (5.58) per round-trip and we obtain, analogous to (4.65) by summation over all interfering amplitudes, the total transmitted intensity

$$I_T = I_0 \frac{T^2 G(v)}{[1 - G(v)]^2 + 4G(v)\sin^2(\phi/2)}. \tag{5.59}$$

The total amplification I_T/I_0 has maxima for $\phi = 2q\pi$, which corresponds to the condition (5.53) for the eigenfrequencies of the resonator with the modification (5.57). For $G(v) \to 1$, the total amplification I_T/I_0 becomes infinite for $\phi = 2q\pi$. This means that even an infinitesimally small input signal results in a finite output signal. Such an input is always provided, for instance, by the spontaneous emission of the excited atoms in the active medium. *For $G(v) = 1$ the laser amplifier converts to a laser oscillator.* This condition is equivalent to the threshold condition (5.7). Because of gain saturation (Sect. 5.3), the amplification remains finite and the total output power is determined by the pump power rather than by the gain.

According to (5.8) the gain factor $G_0(v) = \exp[-2\alpha(v)L]$ depends on the line profile $g(v - v_0)$ of the molecular transition $E_i \to E_k$. The threshold condition can

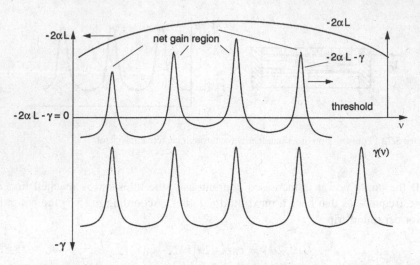

Figure 5.24 Reflection losses of a resonator (*lower curve*), gain curve $\alpha(\nu)$ (*upper curve*), and net gain $\Delta\alpha(\nu) = -2L\alpha(\nu) - \gamma(\nu)$ as difference between gain ($\alpha < 0$) and losses (*middle curve*). Only frequencies with $\Delta\alpha(\nu) > 0$ reach the oscillation threshold

be illustrated graphically by subtracting the frequency-dependent losses from the gain profile. Laser oscillation is possible at all frequencies ν_L where this subtraction gives a positive net gain (Fig. 5.24).

Example 5.9

a) In gas lasers, the gain profile is the Doppler-broadened profile of a molecular transition (Sect. 3.2) and therefore shows a Gaussian distribution with the Doppler width $\delta\omega_D$ (see Sect. 3.2),

$$\alpha(\omega) = \alpha(\omega_0) \exp\left(-\frac{\omega - \omega_0}{0.68\delta\omega_D}\right)^2 .$$

With $\alpha(\omega_0) = -0.01\,\text{cm}^{-1}$, $L = 10\,\text{cm}$, $\delta\omega_D = 1.3 \times 10^9\,\text{Hz} \cdot 2\pi$, and $\gamma = 0.03$, the gain profile extends over a frequency range of $\delta\omega = 2\pi \cdot 3\,\text{GHz}$ where $-2\alpha(\omega)L > 0.03$. In a resonator with $d = 50\,\text{cm}$, the mode spacing is 300 MHz and ten axial modes can oscillate.

b) Solid-state or liquid lasers generally exhibit broader gain profiles because of additional broadening mechanisms (Sect. 3.7). A dye laser has, for example, a gain profile with a width of about 10^{13} Hz. Therefore, in a resonator with $d = 50\,\text{cm}$ about 3×10^4 resonator modes fall within the gain profile.

The preceding example illustrates that the passive resonance halfwidth of typical resonators for gas lasers is very small compared with the linewidth of a laser transition, which is generally determined by the Doppler width. The active medium inside a resonator compensates the losses of the passive resonator resonances resulting in an exceedingly high quality factor Q. The linewidth of an oscillating laser mode should therefore be much smaller than the passive resonance width.

From (5.59) we obtain for the halfwidth $\Delta\nu$ of the resonances for an active resonator with a free spectral range $\delta\nu$ the expression

$$\Delta\nu_a = \delta\nu \frac{1 - G(\nu)}{2\pi\sqrt{G(\nu)}} = \delta\nu / F_a^* . \tag{5.60a}$$

The finesse

$$F_a^* = \frac{2\pi\sqrt{G(\nu)}}{1 - G(\nu)} \tag{5.60b}$$

of the active resonator approaches infinity for $G(\nu) \rightarrow 1$. Although the laser linewidth $\Delta\nu_L$ may become much smaller than the halfwidth of the passive resonator, it does not approach zero. This will be discussed in Sect. 5.6.

For frequencies between the resonator resonances, the losses are high and the threshold will not be reached. In the case of a Lorentzian resonance profile, for instance, the loss factor has increased to about ten times $\beta(\nu_0)$ at frequencies that are $3\Delta\nu_r$ away from the resonance center ν_0.

5.3.2   Gain Saturation

When the pump power of a laser is increased beyond its threshold value, laser oscillation will start at first at a frequency where the net gain, that is, the difference between total gain minus total losses, has a maximum. During the buildup time of the laser oscillation, the gain is larger than the losses and the stimulated wave inside the resonator is amplified during each round-trip until the radiation energy is sufficiently large to deplete the population inversion ΔN by stimulated emission down to the threshold value ΔN_{thr}. Under stationary conditions the increase of ΔN due to pumping is just compensated by its decrease due to stimulated emission. The gain factor of the active medium saturates from the unsaturated value $G_0(I = 0)$ at small intensities to the threshold value

$$G_{thr} = e^{-2L\alpha_{sat}(\nu)-\gamma} = 1 , \tag{5.61}$$

with $-2\alpha L - \gamma = 0$ where the gain just equals the total losses per round-trip. This gain saturation is different for homogeneous and for inhomogeneous line profiles of laser transitions (Sect. 3.6).

Figure 5.25 Saturation of
gain profiles: **a** for a homogeneous profile; **b** for an
inhomogeneous profile

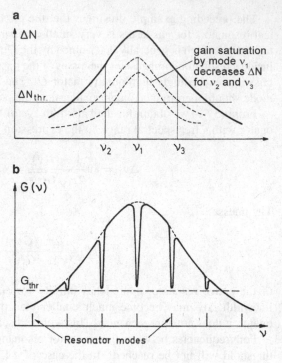

In the case of a homogeneous profile $g(\nu - \nu_0)$, all molecules in the upper level
can contribute to stimulated emission at the laser frequency ν_a with the probability
$B_{ik}\rho g(\nu_a - \nu_0)$, see (5.8). Although the laser may oscillate only with a single
frequency ν, the whole homogeneous gain profile $\alpha(\nu) = \Delta N\sigma(\nu)$ saturates until
the inverted population difference ΔN has decreased to the threshold value ΔN_{thr}
(Fig. 5.25a). The saturated amplification coefficient $\alpha_{sat}(\nu)$ at the intracavity laser
intensity I is, according to Sect. 3.6,

$$\alpha_s^{hom}(\nu) = \frac{\alpha_0(\nu)}{1 + S} = \frac{\alpha_0(\nu)}{1 + I/I_s}, \tag{5.62}$$

where $I = I_s$ is the intensity for which the saturation parameter $S = 1$, which
means that the induced transition rate equals the relaxation rate. For homogeneous
gain profiles, the saturation caused by one laser mode also diminishes the gain for
adjacent modes (mode competition).

In the case of inhomogeneous laser transitions, the whole line profile can be divided into homogeneously broadened subsections with the spectral width $\Delta\nu^{hom}$
(for example, the natural linewidth or the pressure- or power-broadened linewidth).
Only those molecules in the upper laser level that belong to the subgroup in the
spectral interval $\nu_L \pm \frac{1}{2}\Delta\nu^{hom}$, centered at the laser frequency ν_L, can contribute
to the amplification of the laser wave. A monochromatic wave therefore causes
selective saturation of this subgroup and burns a hole into the inhomogeneous distribution $\Delta N(\nu)$ (Fig. 5.25b). At the bottom of the hole, the inversion $\Delta N(\nu_L)$

has decreased to the threshold value ΔN_{thr}, but several homogeneous widths $\Delta \nu^{hom}$ away from ν_L, ΔN remains unsaturated. According to (3.68), the homogeneous width $\Delta \nu^{hom}$ of this hole increases with increasing saturating intensity as

$$\Delta \nu_s = \Delta \nu_0 \sqrt{1 + S} = \Delta \nu_0 \sqrt{1 + I/I_s} . \tag{3.63}$$

This implies that with increasing saturation *more* molecules from a larger spectral interval $\Delta \nu_s$ can contribute to the amplification. The gain factor decreases by the factor $1/(1 + S)$ because of a decrease of ΔN caused by saturation. It increases by the factor $(1 + S)^{1/2}$ because of the increased homogeneous width. The combination of both phenomena gives (Vol. 2, Sect. 2.2)

$$\alpha_s^{inh}(\nu) = \alpha_0(\nu) \frac{\sqrt{1 + S}}{1 + S} = \frac{\alpha_0(\nu)}{\sqrt{1 + I/I_s}} . \tag{5.64}$$

5.3.3   Spatial Hole Burning

A resonator mode represents a standing wave in the laser resonator with a z-dependent field amplitude $E(z)$, as illustrated in Fig. 5.26a. Since the saturation of the inversion ΔN, discussed in the previous section, depends on the intensity $I \propto |E|^2$, the inversion saturated by a single laser mode exhibits a spatial modulation $\Delta N(z)$, as sketched in Fig. 5.26c. Even for a completely homogeneous gain profile, there are always spatial regions of unsaturated inversion at the nodes of the standing wave $E_1(z)$. This may give sufficient gain for another laser mode $E_2(z)$ that is spatially shifted by $\lambda/4$ against $E_1(z)$, or even for a third mode with a shift of $\lambda/3$ of its amplitude maximum (Fig. 5.26b).

If the mirror separation d changes by only one wavelength (e.g., caused by acoustical vibrations of the mirrors), the maxima and nodes of the standing waves are shifted and the gain competition, governed by spatial hole burning, is altered. Therefore, every fluctuation of the laser wavelength caused by changes of the refractive index or the cavity length d results in a corresponding fluctuation of the coupling strength between the modes and changes the gain relations and the intensities of the simultaneously oscillating modes.

If the length L of the active medium is small compared to the resonator length (e.g., in cw dye lasers), it is possible to minimize the spatial hole-burning phenomenon by placing the active medium close to one cavity mirror (Fig. 5.26d). Consider two standing waves with the wavelengths λ_1 and λ_2. At a distance a from the end mirror, their maxima in the active medium are shifted by λ/p ($p = 2, 3, \ldots$). Since all standing waves must have nodes at the mirror surface, we obtain for two waves with the minimum possible wavelength difference $\Delta \lambda = \lambda_1 - \lambda_2$ the relation

$$m\lambda_1 = a = (m + 1/p)\lambda_2 , \tag{5.65}$$

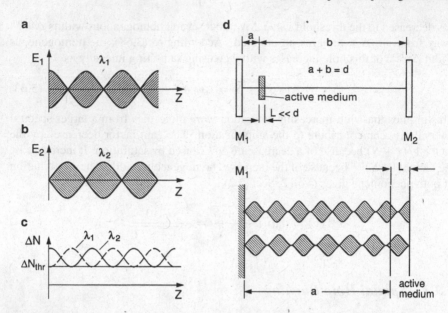

Figure 5.26 Spatial intensity distribution for two standing waves with slightly different wavelengths λ_1 and λ_2 (**a**), (**b**), and their corresponding saturation of the inversion $\Delta N(z)$ (**c**). Explanation of spatial hole-burning modes in the active medium **d** with a small length L, close to a resonator mirror M_1 ($a \ll b$)

or for their frequencies

$$\nu_1 = m\frac{c}{a}, \quad \nu_2 = \frac{c}{a}(m + 1/p) \;\Rightarrow\; \delta\nu_{sp} = \frac{c}{ap}. \tag{5.66}$$

In terms of the spacing $\delta\nu = c/2d$ of the longitudinal resonator modes, the spacing of the spatial hole-burning modes is

$$\delta\nu_{sp} = \frac{2d}{ap}\delta\nu. \tag{5.67}$$

Even when the net gain is sufficiently large to allow oscillation of, e.g., up to three spatially separated standing waves ($p = 1, 2, 3$), only one mode can oscillate if the spectral width of the homogeneous gain profile is smaller than $(2/3)(d/a)\delta\nu$ [323].

Example 5.10
$d = 100\,\text{cm}$, $L = 0.1\,\text{cm}$, $a = 5\,\text{cm}$, $p = 3$, $\delta\nu = 150\,\text{MHz}$, $\delta\nu_{sp} = 2000\,\text{MHz}$. Single-mode operation could be achieved if the spectral gain profile is smaller than 2000 MHz.

In gas lasers the effect of spatial hole burning is partly averaged out by diffusion of the excited molecules from nodes to maxima of a standing wave. It is, however, important in solid-state and in liquid lasers such as the ruby laser or the dye laser. Spatial hole burning can be completely avoided in unidirectional ring lasers (Sect. 5.2.7) where no standing waves exist. Waves propagating in one direction can saturate the entire spatially distributed inversion. This is the reason why ring lasers with sufficiently high pump powers have higher output powers than standing-wave lasers.

5.3.4   Multimode Lasers and Gain Competition

The different gain saturation of homogeneous and inhomogeneous transitions strongly affects the frequency spectrum of multimode lasers, as can be understood from the following arguments:

Let us first consider a laser transition with a purely *homogeneous* line profile. The resonator mode that is next to the center of the gain profile starts oscillating when the pump power exceeds the threshold. Since this mode experiences the largest net gain, its intensity grows faster than that of the other laser modes. This causes partial saturation of the whole gain profile (Fig. 5.25a), mainly by this strongest mode. This saturation, however, decreases the gain for the other weaker modes and their amplification will be slowed down, which further increases the differences in amplification and favors the strongest mode even more. This mode competition of different laser modes within a homogeneous gain profile will finally lead to a complete suppression of all but the strongest mode. Provided that no other mechanism disturbs the predominance of the strongest mode, this saturation coupling results in single-frequency oscillation of the laser, even if the homogeneous gain profile is broad enough to allow, in principle, simultaneous oscillation of several resonator modes [324].

In fact, such single-mode operation without further frequencyselecting elements in the laser resonator can be observed only in a few exceptional cases because there are several phenomena, such as spatial hole burning, frequency jitter, or time-dependent gain fluctuations, that interfere with the pure case of mode competition discussed above. These effects, which will be discussed below, prevent the unperturbed growth of one definite mode, introduce time-dependent coupling phenomena between the different modes, and cause in many cases a frequency spectrum of the laser which consists of a random superposition of many modes that fluctuate in time.

In the case of a purely *inhomogeneous* gain profile, the different laser modes do not share the same molecules for their amplification, and no mode competition occurs if the frequency spacing of the modes is larger than the saturation-broadened line profiles of the oscillating modes. Therefore all laser modes within that part of the gain profile, which is above the threshold, can oscillate simultaneously. The laser output consists of all axial and transverse modes for which the total losses are less than the gain (Fig. 5.27a).

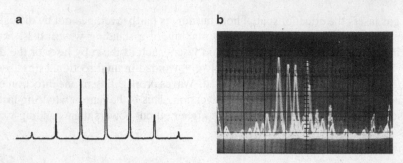

Figure 5.27 a Stable multimode operation of a HeNe laser (exposure time: 1 s); **b** two short-time exposures of the multimode spectrum of an argon laser superimposed on the same film to demonstrate the randomly fluctuating mode distribution

Real lasers do not represent these pure cases, but exhibit a gain profile that is a convolution of inhomogeneous and homogeneous broadening. It is the ratio of mode spacing $\delta\nu$ to the homogeneous width $\Delta\nu^{\text{hom}}$ that governs the strength of mode competition and that is crucial for the resulting single- or multi-mode operation. There is another reason why many lasers oscillate on many modes: if the gain exceeds the losses for higher transverse modes, mode competition between the modes TEM_{m_1,n_1} and TEM_{m_2,n_2} with $(m_1,n_1) \neq (m_2,n_2)$ is restricted because of their different spatial amplitude distributions. They gain their amplification from different regions of the active medium. This applies to laser types such as solid-state lasers (ruby or Nd:YAG lasers), flash-lamp-pumped dye lasers, or excimer lasers. In a nonconfocal resonator the frequencies of the transverse modes fill the gap between the TEM_{00} frequencies $\nu_a = (q + \frac{1}{2})c/(2nd)$ (Fig. 5.21). These transverse modes lead to a larger divergence of the laser beam, which is no longer a Gaussian-shaped beam.

The suppression of higher-order $\text{TEM}_{m,n}$ modes can be achieved by a proper choice of the resonator geometry, which has to be adapted to the cross section and the length L of the active medium (Sect. 5.4.2).

If only the axial modes TEM_{00} participate in the laser oscillation, the laser beam transmitted through the output mirrors has a Gaussian intensity profile (5.32), (5.42). It may still consist of many frequencies $\nu_a = qc/(2nd)$ within the spectral gain profile. **The spectral bandwidth of a multimode laser oscillating on an atomic or molecular transition is comparable to that of an incoherent source emitting on this transition!**

We illustrate this discussion by some examples:

Example 5.11
HeNe Laser at $\lambda = 632.8$ nm: The Doppler width of the Ne transition is about 1500 MHz, and the width of the gain profile above the threshold, which

depends on the pump power, may be 1200 MHz. With a resonator length of $d = 100$ cm, the spacing of the longitudinal modes is $\delta v = c/2d = 150$ MHz. If the higher transverse modes are suppressed by an aperture inside the resonator, seven to eight longitudinal modes reach the threshold. The homogeneous width Δv^{hom} is determined by several factors: the natural linewidth $\Delta v_n = 20$ MHz; a pressure broadening of about the same magnitude; and a power broadening, which depends on the laser intensity in the different modes. With $I/I_s = 10$, for example, we obtain with $\Delta v_0 = 30$ MHz a power-broadened linewidth of about 100 MHz, which is still smaller than the longitudinal modes spacing. The modes will therefore not compete strongly, and simultaneous oscillation of all longitudinal modes above threshold is possible. This is illustrated by Fig. 5.27a, which exhibits the spectrum of a HeNe laser with $d = 1$ m, monitored with a spectrum analyzer and integrated over a time interval of 1 s.

Example 5.12
Argon Laser: Because of the high temperature in the high-current discharge (about 10^3 A/cm^2), the Doppler width of the Ar$^+$ transitions is very large (about 8 to 10 GHz). The homogeneous width Δv^{hom} is also much larger than for the HeNe laser for two reasons: the long-range Coulomb interaction causes a large pressure broadening from electron–ion collisions and the high laser intensity (10–100 W) in a mode results in appreciable power broadening. Both effects generate a homogeneous linewidth that is large compared to the mode spacing $\delta v = 125$ MHz for a commonly used resonator length of $d = 120$ cm. The resulting mode competition in combination with the perturbations mentioned above cause the observed randomly fluctuating mode spectrum of the multimode argon laser. Figure 5.27b illustrates this by the superposition of two short-time exposures of the oscilloscope display of a spectrum analyzer taken at two different times.

Example 5.13
Dye Laser: The broad spectral gain profile of dye molecules in a liquid is predominantly homogeneously broadened (Sect. 3.7). About 10^5 modes of a laser resonator with $L = 75$ cm fall within a typical spectral width of 20 nm ($\hateq 2 \times 10^{13}$ Hz at $\lambda = 600$ nm). Without spectral hole burning and fluctuations of the optical length nd of the resonator, the laser would oscillate in a single mode at the center of the gain profile, despite the large number of

possible modes. However, fluctuations of the refractive index n in the dye liquid cause corresponding perturbations of the frequencies and the coupling of the laser modes, which results in a time-dependent multimode spectrum; the emission jumps in a random way between different mode frequencies. In the case of pulsed lasers, the timeaveraged spectrum of the dye laser emission fills more or less uniformly a broader spectral interval (about 1 nm) around the maximum of the gain profile. The spatial hole burning may result in oscillation of several groups of lines centered around the spatial hole-burning modes. In this case, the time-averaged frequency distribution generally does *not* result in a uniformly smoothed intensity profile $I(\lambda)$. In order to achieve tunable single-mode operation, extra wavelength-selective elements have to be inserted into the laser resonator (Sect. 5.4).

For spectroscopic applications of multimode lasers one has to keep in mind that the spectral interval Δv within the bandwidth of the laser is, in general, not uniformly filled. This means that, contrary to an incoherent source, the intensity $I(v)$ is not a smooth function within the laser bandwidth but exhibits holes. This is particularly true for multimode dye lasers with Fabry–Perot-type resonators where standing waves are present and spatial hole burning occurs (Sect. 5.3.4).

The spectral intensity distribution of the laser output is the superposition

$$I_{\mathrm{L}}(\omega,t) = \left| \sum_k A_k(t) \cos[\omega_k t + \phi_k(t)] \right|^2 , \qquad (5.68)$$

of the oscillating modes, where the phases $\phi_k(t)$ and the amplitudes $A_k(t)$ may randomly fluctuate in time because of mode competition and mode-pulling effects.

The time average of the spectral distribution of the output intensity

$$\langle I(\omega) \rangle = \frac{1}{T} \int_0^T \left| \sum_k A_k(t) \cos[\omega_k t + \phi_k(t)] \right|^2 \, dt , \qquad (5.69)$$

reflects the gain profile of the laser transition. The necessary averaging time T depends on the buildup time of the laser modes. It is determined by the unsaturated gain and the strength of the mode competition. In the case of gas lasers, the average spectral width $\langle \Delta v \rangle$ corresponds to the Doppler width of the laser transition. The coherence length of such a multimode laser is comparable to that of a conventional spectral lamp where a single line has been filtered out.

If such a multimode laser is used for spectroscopy and is scanned, for instance, with a grating or prism inside the laser resonator (Sect. 5.5), through the spectral range of interest, this nonuniform spectral structure $I_{\mathrm{L}}(0)$ may cause artificial structures in the measured spectrum. In order to avoid this problem and to obtain a smooth intensity profile $I_L(v)$, the length d of the laser resonator can be wobbled at the frequency $f > 1/\tau$, which should be larger than the inverse scanning time τ over a line in the investigated spectrum. This wobbling modulates all oscillating frequencies in the laser and results in a smoother time average, particularly, if $\tau > T$.

5.3.5   Mode Pulling

We now briefly discuss the frequency shift (called *mode pulling*) of the passive resonator frequencies by the presence of an active medium [325]. The phase shift for a stationary standing wave with frequency v_p and round-trip time T_p through a resonator with mirror separation d without an active medium is

$$\phi_p = 2\pi v_p T_p = 2\pi v_p 2d/c = m\pi \, , \tag{5.70}$$

where the integer m characterizes the oscillating resonator mode. On insertion of an active medium with refractive index $n(v)$, the frequency v_p changes to v_a in such a way that the phase shift per round-trip remains

$$\phi_a = 2\pi v_a T_a = 2\pi v_a n(v_a) 2d/c = m\pi \, . \tag{5.71}$$

This gives the condition

$$\frac{\partial \phi}{\partial v}(v_a - v_p) + [\phi_a(v_a) - \phi_p(v_a)] = 0 \, . \tag{5.72}$$

The index of refraction $n(v)$ is related to the absorption coefficient $\alpha(v)$ of a homogeneous absorption profile by the dispersion relation (3.24a, 3.24b)

$$n(v) = 1 + \frac{v_0 - v}{\Delta v_m} \frac{c}{2\pi v} \alpha(v) \, , \tag{5.73}$$

where $\Delta v_m = \gamma/2\pi$ is the linewidth of the amplifying transition in the active medium. In case of inversion ($\Delta N < 0$), $\alpha(v)$ becomes negative and $n(v) < 1$ for $v < v_0$, while $n(v) > 1$ for $v > v_0$ (Fig. 5.28). Under stationary conditions, the total gain per pass $\alpha(v)L$ saturates to the threshold value, which equals the total losses γ. These losses determine the resonance width $\Delta v_r = c\gamma/(4\pi d)$ of the cavity, see (5.54). We obtain from (5.70, 5.73) the final result for the frequency v_a of a laser mode for laser transitions with homogeneous line broadening Δv_m and center frequency v_0 in a resonator with mode-width Δv_r

$$v_a = \frac{v_r \Delta v_m + v_0 \Delta v_r}{\Delta v_m + \Delta v_r} \, . \tag{5.74}$$

The resonance width Δv_r of gas laser resonators is of the order of 1 MHz, while the homogeneous width of the amplifying medium is about 100 MHz. Therefore, when $\Delta v_r \ll \Delta v_m$, (5.74) reduces to

$$v_a = v_r + \frac{\Delta v_r}{\Delta v_m}(v_0 - v_r) \, . \tag{5.75}$$

Figure 5.28 Dispersion
curves for absorbing transitions ($\Delta N < 0$) and
amplifying transitions
($\Delta N > 0$) and phase shifts
$\Delta \phi$ per round-trip in the passive and active cavity

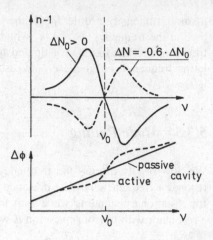

This demonstrates that the mode-pulling effect increases proportionally to the difference of cavity resonance frequency ν_r and central frequency ν_0 of the amplifying
medium. At the slopes of the gain profile, the laser frequency is pulled towards the
center.

5.4 Experimental Realization of Single-Mode Lasers

In the previous sections we have seen that without specific manipulation a laser
generally oscillates in many modes, for which the gain exceeds the total losses. In
order to select a single wanted mode, one has to suppress all others by increasing
their losses to such an amount that they do not reach the oscillation threshold. The
suppression of higher-order transverse TEM_{mn} modes demands actions other than
the selection of a single longitudinal mode out of many other TEM_{00} modes.

Many types of lasers, in particular, gaseous lasers, may reach oscillation threshold for several atomic or molecular transitions. The laser can then simultaneously
oscillate on these transitions [326]. In order to reach single-mode operation, one
has to first select a single transition.

5.4.1 Line Selection

In order to achieve single-line oscillation in laser media that exhibit gain for several transitions, wavelength-selecting elements inside or outside the laser resonator
can be used. If the different lines are widely separated in the spectrum, the selective reflectivity of the dielectric mirrors may already be sufficient to select a single
transition.

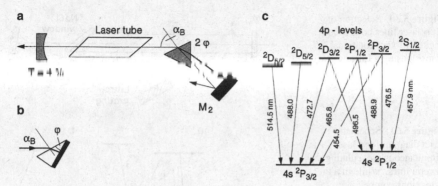

Figure 5.29 Line selection in an argon laser with a Brewster prism **a** or a Littrow prism reflector (**b**). Term diagram of laser transition in Ar^+ (**c**)

Example 5.14

The He-Ne laser can oscillate at $\lambda = 3.39\,\mu m$, $\lambda = 0.633\,\mu m$ and several lines around $\lambda = 1.15\,\mu m$.

The line at $\lambda = 3.39\,\mu m$ or at $\lambda = 0.633\,\mu m$ can be selected using special mirrors. The different lines around $1.15\,\mu m$ cannot be separated solely via the spectral reflectivity of the mirrors; other measures are required, as outlined below.

In the case of broadband reflectors or closely spaced lines, prisms, gratings, or Lyot filters are commonly utilized for wavelength selection. Figure 5.29 illustrates line selection by a prism in an argon laser. The different lines are refracted by the prism, and only the line that is vertically incident upon the end mirror is reflected back into itself and can reach the oscillation threshold, while all other lines are reflected out of the resonator. Turning the end reflector M_2 allows the desired line to be selected. To avoid reflection losses at the prism surfaces, a Brewster prism with $\tan \phi = 1/n$ is used, with the angle of incidence for both prism surfaces being Brewster's angle. The prism and the end mirror can be combined by coating the end face of a Brewster prism reflector (Fig. 5.29b). Such a device is called a *Littrow prism*.

Because most prism materials such as glass or quartz absorb in the infrared region, it is more convenient to use for infrared lasers a Littrow grating (Sect. 4.1) as wavelength selector in this wavelength range. Figure 5.30 illustrates the line selection in a CO_2 laser, which can oscillate on many rotational lines of a vibrational transition. Often the laser beam is expanded by a proper mirror configuration in order to cover a larger number of grating grooves, thus increasing the spectral resolution (Sect. 4.1). This has the further advantage that the power density is lower and damage of the grating is less likely.

Figure 5.30 Selection of CO_2 laser lines corresponding to different rotational transitions by a Littrow grating

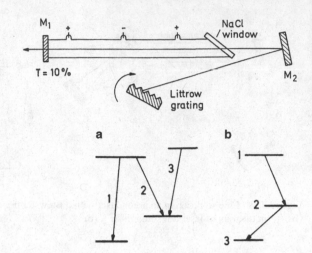

Figure 5.31 Schematic level diagram for a laser simultaneously oscillating on several lines. While in **a** the transitions compete with each other for gain, those in **b** enhance the gain for the other line

If some of the simultaneously oscillating laser transitions share a common upper or lower level, such as the lines 1, 2, and 3 in Fig. 5.29c and Fig. 5.31a, gain competition diminishes the output of each line. In this case, it is advantageous to use *intracavity* line selection in order to suppress all but one of the competing transitions. Sometimes, however, the laser may oscillate on cascade transitions (Fig. 5.31b). In such a case, the laser transition $1 \to 2$ increases the population of level 2 and therefore *enhances* the gain for the transition $2 \to 3$ [327]. Obviously, it is then more favorable to allow multiline oscillation and to select a single line by an external prism or grating. Using a special mounting design, it can be arranged so that no deflection of the output beam occurs when the multiline output is tuned from one line to the other [328].

For lasers with a broad continuous spectral gain profile, the preselecting elements inside the laser resonator restrict laser oscillation to a spectral interval, which is a fraction of the gain profile.

Some examples illustrate the situation (see also Sect. 5.7):

Example 5.15
HeNe Laser: The HeNe laser is probably the most thoroughly investigated gas laser [329]. From the level scheme (Fig. 5.32), which uses the Paschen notation [330], we see that two transitions around $\lambda = 3.39\,\mu m$ and the visible transitions at $\lambda = 0.6328\,\mu m$ share a common *upper* level. Suppression of the $3.39\,\mu m$ lines therefore enhances the output power at $0.6328\,\mu m$. The $1.15\,\mu m$ and the $0.6328\,\mu m$ lines, on the other hand, share a common *lower* level and also compete for gain, since both laser transitions increase the lower-level population and therefore decrease the inversion. If the 3.3903-μm transition is suppressed, e.g., by placing an absorbing CH_4 cell inside the resonator, the

Figure 5.32 Level diagram of the HeNe laser system in Paschen notation showing the most intense laser transitions

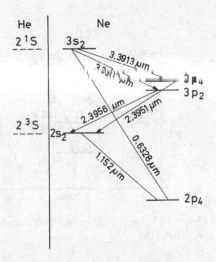

population of the upper $3s_2$ level increases, and a new line at $\lambda = 3.3913\,\mu m$ reaches the threshold.

This laser transition populates the $3p_4$ level and produces gain for another line at $\lambda = 2.3951\,\mu m$. This last line only oscillates together with the 3.3913-μm one, which acts as pumping source. This is an example of cascade transitions in laser media [327], as depicted in Fig. 5.31b.

The homogeneous width of the laser transitions is mainly determined by pressure and power broadening. At total pressures of above 5 mb and an intracavity power of 200 mW, the homogeneous linewidth for the transition $\lambda = 632.8\,nm$ is about 200 MHz, which is still small compared with the Doppler width $\Delta\nu_D = 1500\,MHz$. In single-mode operation, one can obtain about 20 % of the multimode power [331]. This roughly corresponds to the ratio $\Delta\nu_h/\Delta\nu_D$ of homogeneous to inhomogeneous linewidth above the threshold. The mode spacing $\delta\nu = \frac{1}{2}c/d$ equals the homogeneous linewidth for $d = d^* = \frac{1}{2}c/\Delta\nu_h$. For $d < d^*$, stable multimode oscillation is possible; for $d > d^*$, mode competition occurs.

Example 5.16

Argon Laser: The discharge of a cw argon laser exhibits gain for more than 15 different transitions. Figure 5.29c shows part of the energy level diagram, illustrating the coupling of different laser transitions. Since the lines at 514.5 nm, 488.0 nm, and 465.8 nm share the same lower level, suppression of the competing lines enhances the inversion and the output power of

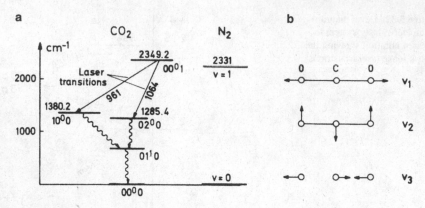

Figure 5.33 Level diagram and laser transitions in the CO_2 molecule **a** and normal vibrations (ν_1, ν_2, ν_3) **(b)**

the selected line. The mutual interaction of the various laser transitions has therefore been studied extensively [332, 333] in order to optimize the ouput power. Line selection is generally achieved with an internal Brewster prism (Fig. 5.29 and Fig. 5.41b). The homogeneous width $\Delta\nu_h$ is mainly caused by collision broadening due to electron–ion collisions and saturation broadening. Additional broadening and shifts of the ion lines result from ion drifts in the field of the discharge. At intracavity intensities of $350\,\mathrm{W/cm^2}$, which correspond to about 1 W output power, appreciable saturation broadening increases the homogeneous width, which may exceed 1000 MHz. This explains why the output at single-mode operation may reach 30 % of the multimode output on a single line [334].

Example 5.17
CO_2 Laser: A section of the level diagram is illustrated in Fig. 5.33. The vibrational levels (ν_1, ν_2^l, ν_3) are characterized by the number of quanta in the three normal vibrational modes. The upper index of the degenerate vibration ν_2 gives the quantum number of the corresponding vibrational angular momentum l which occurs when two degenerate bending vibrations ν_2 where the nuclei vibrate in orthogonal planes are superimposed [335]. Laser oscillation is achieved on many rotational lines within two vibrational transitions $(\nu_1, \nu_2^l, \nu_3) = 00^01 \rightarrow 10^00$ and $00^01 \rightarrow 02^00$ [336–338]. Without line selection, generally only the band around $961\,\mathrm{cm^{-1}}$ ($10.6\,\mu m$) appears because these transitions exhibit larger gain. The laser oscillation depletes the pop-

ulation of the $00^{0}1$ vibrational level and suppresses laser oscillation on the second transition, because of gain competition. With internal line selection (Fig. 8.30), many more lines can successively be optimized by tuning the wavelength-selecting grating. The output power of each line is then higher than that of the same line in multiline operation. Because of the small Doppler width (66 MHz), the free spectral range $\delta\nu = \frac{1}{2}c/d^{*}$ is already larger than the width of the gain profile for $d^{*} < 200$ cm. For such resonators, the mirror separation d has to be adjusted to tune the resonator eigenfrequency $\nu_{R} = \frac{1}{2}qc/d^{*}$ (where q is an integer) to the center of the gain profile. If the resonator parameters are properly chosen to suppress higher transverse modes, the CO_2 laser then oscillates on a single longitudinal mode.

5.4.2 Suppression of Transverse Modes

Let us first consider the selection of *transverse* modes. In Sect. 5.2.3 it was shown that the higher transverse TEM_{mnq} modes have radial field distributions that are less and less concentrated along the resonator axis with increasing transverse order n or m. This means that their diffraction losses are much higher than those of the fundamental modes TEM_{00q} (Fig. 5.12). The field distribution of the modes and therefore their diffraction losses depend on the resonator parameters such as the radii of curvature of the mirrors R_i, the mirror separation d, and, of course, the Fresnel number N_F (Sect. 5.2.1). Only those resonators that fulfill the stability condition [291, 314]

$$0 < g_1 g_2 < 1 \quad \text{or} \quad g_1 g_2 = 0 \quad \text{with} \quad g_i = (1 - d/R_i)$$

have finite spot sizes of the TEM_{00} field distributions inside the resonator (Sect. 5.2.6). The choice of proper resonator parameters therefore establishes the beam waist w of the fundamental TEM_{00q} mode and the radial extension of the higher-order TEM_{mn} modes. This, in turn, determines the diffraction losses of the modes.

In Fig. 5.34, the ratio γ_{10}/γ_{00} of the diffraction losses for the TEM_{10} and the TEM_{00} modes in a symmetric resonator with $g_1 = g_2 = g$ is plotted for different values of g as a function of the Fresnel number N_F. From this diagram one can obtain, for any given resonator, the diameter $2a$ of an aperture that suppresses the TEM_{10} mode but still has sufficiently small losses for the fundamental TEM_{00} mode with beam radius w. In gas lasers, the diameter $2a$ of the discharge tube generally forms the limiting aperture. One has to choose the resonator parameters in such a way that $a \simeq 3w/2$ because this assures that the fundamental mode nearly fills the whole active medium, but still suffers less than 1 % diffraction losses (Sect. 5.2.6).

Figure 5.34 Ratio γ_{10}/γ_{00} of diffraction losses for the TEM$_{10}$ and TEM$_{00}$ modes in symmetric resonators as a function of the Fresnel number N_F for different resonator parameters $g = 1 - d/R$

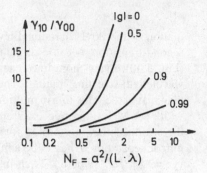

Because the frequency separation of the transverse modes is small and the TEM$_{10q}$ mode frequency is separated from the TEM$_{00q}$ frequency by less than the homogeneous width of the gain profile, the fundamental mode can partly saturate the inversion at the distance r_m from the axis, where the TEM$_{10q}$ mode has its field maximum. The resulting transverse mode competition (Fig. 5.35) reduces the gain for the higher transverse modes and may suppress their oscillation even if the unsaturated gain exceeds the losses. The restriction for the maximum-allowed aperture diameter is therefore less stringent. The resonator geometry of many commercial lasers has already been designed in such a way that "single-transverse-mode" operation is obtained. The laser can, however, still oscillate on several longitudinal modes, and for true single-mode operation, the next step is to suppress all but one of the longitudinal modes.

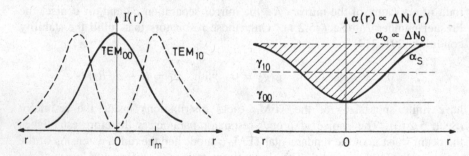

Figure 5.35 Transverse gain competition between the TEM$_{00}$ and TEM$_{10}$ modes

5.4.3 Selection of Single Longitudinal Modes

From the discussion in Sect. 5.3 it should have become clear that simultaneous oscillation on several longitudinal modes is possible when the inhomogeneous width $\Delta\nu_g$ of the gain profile exceeds the mode spacing $\frac{1}{2}c/d$ (Fig. 5.22). A simple way to

Figure 5.36 Single longitudinal mode operation by
reducing the cavity length d
to a value where the mode
spacing exceeds half of the
gain profile width above
threshold

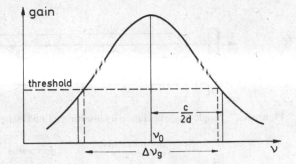

achieve single-mode operation is therefore the reduction of the resonator length $2d$
such that the width $\Delta\nu_g$ of the gain profile above threshold becomes smaller than
the free spectral range $\delta\nu = \frac{1}{2}c/d$ [339].

If the resonator frequency can be tuned to the center of the gain profile, single-
mode operation can be achieved even with the double length $2d$, because then the
two neighboring modes just fail to reach the threshold (Fig. 5.36). However, this
solution for the achievement of single-mode operation has several drawbacks. Since
the length L of the active medium cannot be larger than d ($L \leq d$), the threshold
can only be reached for transitions with a high gain. The output power, which is
proportional to the active mode volume, is also small in most cases. For single-
mode lasers with higher output powers, other methods are therefore preferable. We
distinguish between *external* and *internal* mode selection.

When the output of a multimode laser passes through an external spectral filter, such as an interferometer or a spectrometer, a single mode can be selected.
For perfect selection, however, high suppression of the unwanted modes and high
transmission of the wanted mode by the filter are required. This technique of external selection has the further disadvantage that only part of the total laser output
power can be used. *Internal* mode selection with spectral filters inside the laser resonator completely suppresses the unwanted modes even when without the selecting
element their gain exceeds their losses. Furthermore, the output power of a single-
mode laser is generally higher than the power in this mode at multimode oscillation
because the total inversion $V \cdot \Delta N$ in the active volume V is no longer shared by
many modes, as is the case for multimode operation with gain competition.

In single-mode operation with internal mode selection, we can expect output
powers that reach the fraction $\Delta\nu_{hom}/\Delta\nu_g$ of the multimode power, where $\Delta\nu_{hom}$ is
the homogeneous width within the inhomogeneous gain profile. This width $\Delta\nu_{hom}$
becomes even larger for single-mode operation because of power broadening by the
more intense mode. In an argon-ion laser, for example, one can obtain up to 30 %
of the multimode power in a single mode with internal mode selection.

This is the reason why virtually all single-mode lasers use *internal* mode selection. We now discuss some experimental possibilities that allow stable single-mode
operation of lasers with internal mode selection. As pointed out in the previous
section, all methods for achieving single-mode operation are based on mode sup-

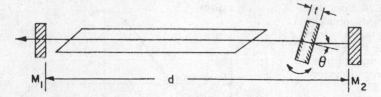

Figure 5.37 Single-mode operation by inserting a tilted etalon inside the laser resonator

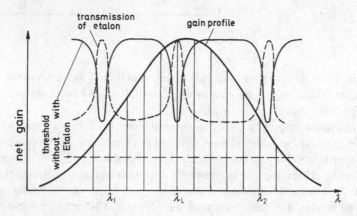

Figure 5.38 Gain profile, resonator modes, and transmission peaks of the intracavity etalon (*dashed curve*). Also shown are the threshold curves with and without etalon

pression by increasing the losses beyond the gain for all but the wanted mode. A possible realization of this idea is illustrated in Fig. 5.37, which shows longitudinal mode selection by a tilted plane-parallel etalon (thickness t and refractive index n) inside the laser resonator [340]. In Sect. 4.2.7, it was shown that such an etalon has transmission maxima at those wavelengths λ_m for which

$$m\lambda_m = 2nt \cos\theta , \tag{5.76}$$

for all other wavelengths the reflection losses should dominate the gain.

If the free spectral range of the etalon,

$$\delta\lambda = 2nt \cos\theta \left(\frac{1}{m} - \frac{1}{m+1}\right) = \frac{\lambda_m}{m+1} , \tag{5.77}$$

is larger than the spectral width $|\lambda_1 - \lambda_2|$ of the gain profile above the threshold, only a single mode can oscillate (Fig. 5.38). Since the wavelength λ is also determined by the resonator length d ($2d = q\lambda$), the tilting angle θ has to be adjusted so that

$$2nt \cos\theta/m = 2d/q \quad \text{(where q is an integer)}$$

$$\Rightarrow \cos\theta = \frac{m}{q} \cdot \frac{d}{n \cdot t} , \tag{5.78}$$

which means that the transmission peak of the etalon has to coincide with an eigen-resonance of the laser resonator.

Example 5.18
In the argon-ion laser the width of the gain profile is about 8 GHz. With a free spectral range of $\Delta \nu = c/(2nt) = 10\,\text{GHz}$ of the intracavity etalon, single-mode operation can be achieved. This implies with $n = 1.5$ a thickness $t = 1\,\text{cm}$.

The finesse F^* of the etalon has to be sufficiently high to ensure for the modes adjacent to the selected mode losses that overcome their gain (Fig. 5.38). Fortunately, in many cases their gain is already reduced by the oscillating mode due to gain competition. This allows the less stringent demand that the losses of the etalon must only exceed the saturated gain at a distance $\Delta \nu \geq \Delta \nu_{\text{hom}}$ away from the transmission peak.

Often a Michelson interferometer coupled by a beam splitter BS to the laser resonator is used for mode selection (Fig. 5.39). The free spectral range $\delta \nu = \frac{1}{2}c/(L_2 + L_3)$ of this *Fox–Smith cavity* [341] again has to be broader than the width of the gain profile. With a piezoelement PE, the mirror M_3 can be translated by a few microns to achieve resonance between the two coupled resonators. For the resonance condition

$$(L_1 + L_2)/q = (L_2 + L_3)/m = \lambda/2 \quad \text{(where } m \text{ and } q \text{ are integers)}, \quad (5.79)$$

the partial wave $M_1 \rightarrow$ BS, reflected by BS, and the partial wave $M_3 \rightarrow$ BS, transmitted through BS, interfere destructively. This means that for the resonance condition (5.79) the reflection losses by BS have a minimum (in the ideal case they are zero). For all other wavelengths, however, these losses are larger than the gain, They do not reach threshold and single-mode oscillation is achieved [342].

In a more detailed discussion the absorption losses A_{BS}^2 of the beam splitter BS cannot be neglected, since they cause the maximum reflectance R of the Fox–Smith cavity to be less than 1. Similar to the derivation of (4.80), the reflectance of the Fox–Smith selector, which acts as a wavelength-selecting laser reflector, can be calculated to be [343]

$$R = \frac{T_{\text{BS}}^2 R_2 (1 - A_{\text{BS}})^2}{1 - R_{\text{BS}}\sqrt{R_2 R_3} + 4R_{\text{BS}}\sqrt{R_2 R_3} \sin^2 \phi/2}. \quad (5.80)$$

Figure 5.39b exhibits the reflectance R_{max} for $\phi = 2m\pi$ and the additional losses of the laser resonator introduced by the Fox–Smith cavity as a function of the beam splitter reflectance R_{BS}. The finesse F^* of the selecting device is also plotted for $R_2 = R_3 = 0.99$ and $A_{\text{BS}} = 0.5\%$. The spectral width $\Delta \nu$ of the reflectivity

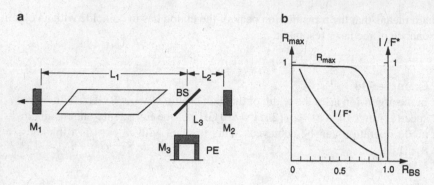

Figure 5.39 Mode selection with a Fox–Smith selector: **a** experimental setup; **b** maximum reflectivity and inverted finesse $1/F^*$ of the Michelson-type reflector as a function of the reflectivity R_{BS} of the beam splitter for $R_2 = R_3 = 0.99$ and $A_{BS} = 0.5\%$

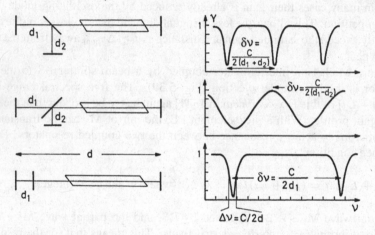

Figure 5.40 Some possible schemes of coupled resonators for longitudinal mode selection, with their frequency-dependent losses. For comparison the eigenresonances of the long laser cavity with a mode spacing $\Delta v = c/2d$ are indicated

maxima is determined by

$$\Delta v = \delta v/F^* = c/[2F^*(L_2 + L_3)] . \tag{5.81}$$

There are several other resonator-coupling schemes that can be utilized for mode selection. Figure 5.40 compares some of them, together with their frequency-selective losses [344].

In case of multiline lasers (e.g., argon or krypton lasers), line selection and mode selection can be simultaneously achieved by a combination of prism and Michelson interferometers. Figure 5.41 illustrates two possible realizations. The first replaces mirror M_2 in Fig. 5.39 by a Littrow prism reflector (Fig. 5.41a). In Fig. 5.41b, the front surface of the prism acts as beam splitter, and the two coated back surfaces replace the mirrors M_2 and M_3 in Fig. 5.39. The incident wave is split into the

Figure 5.41 a Simultaneous line selection and mode selection by a combination of prism selector and Michelson-type interferometer; **b** compact arrangement

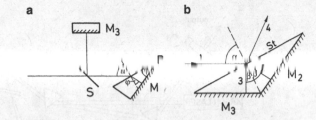

partial beams 4 and 2. After being reflected by M_2, beam 2 is again split into 3 and 1. Destructive interference between beams 4 and 3, after reflection from M_3, occurs if the optical path difference $\Delta s = 2n(S_2 + S_3) = m\lambda$. If both beams have equal amplitudes, no light is emitted in the direction of beam 4. This means that all the light is reflected back into the incident direction and the device acts as a wavelength-selective reflector, analogous to the Fox–Smith cavity [345]. Since the wavelength λ depends on the optical path length $n(L_2 + L_3)$, the prism has to be temperature stabilized to achieve wavelength-stable, single-mode operation. The whole prism is therefore embedded in a temperature-stabilized oven.

For lasers with a broad gain profile, one wavelength-selecting element alone may not be sufficient to achieve single-mode operation, therefore one has to use a proper combination of different dispersing elements. With preselectors, such as prisms, gratings, or Lyot filters, the spectral range of the effective gain profile is narrowed down to a width that is comparable to that of the Doppler width of fixed-frequency gas lasers. Figure 5.42 represents a possible scheme, that has been realized in practice. Two prisms are used as preselector to narrow the spectral width of a cw dye laser [346]; two etalons with different thicknesses t_1 and t_2 are used to achieve stable single-mode operation. Figure 5.42b illustrates the mode selection, depicting schematically the gain profile narrowed by the prisms and the spectral transmission curves of the two etalons. In the case of the dye laser with its homogeneous gain profile, not every resonator mode can oscillate, but only those that draw gain from the spatial hole-burning effect (Sect. 5.3.3). The "suppressed modes" at the bottom of Fig. 5.42 represent these spatial hole-burning modes that would simultaneously oscillate without the etalons. The transmission maxima of the two etalons have, of course, to be at the same wavelength λ_L. This can be achieved by choosing the correct tilting angles θ_1 and θ_2 such that

$$nt_1 \cos\theta_1 = m_1\lambda_L, \quad \text{and} \quad nt_2 \cos\theta_2 = m_2\lambda_L. \tag{5.82}$$

Example 5.19

The two prisms narrow the spectral width of the gain profile above threshold to about 100 GHz. If the free spectral range of the thin etalon 1 is 100 GHz ($\hat{=} \ \Delta\lambda \sim 1$ nm at $\lambda = 600$ nm) and that of the thick etalon 2 is 10 GHz, single-mode operation of the cw dye laser can be achieved. This demands $t_1 = 0.1$ cm and $t_2 = 1$ cm for $n = 1.5$.

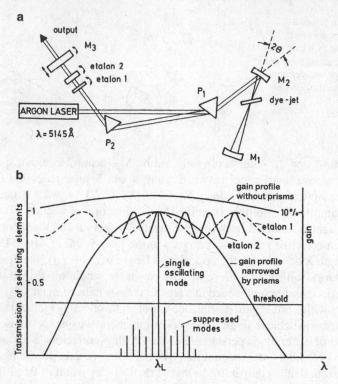

Figure 5.42 Mode selection in the case of broad gain profiles. The prisms narrow the net gain profile and two etalons enforce single-mode operation: **a** experimental realization for a jet stream cw dye laser; **b** schematic diagram of gain profile and transmission curves of the two etalons

Commercial cw dye laser systems (Sect. 5.5) generally use a different realization of single-mode operation (Fig. 5.43). The prisms are replaced by a birefringent filter, which is based on the combination of three Lyot filters (Sect. 4.2.11), and the thick etalon is substituted by a Fabry–Perot interferometer with the thickness t controllable by piezocylinders (Fig. 5.44). This is done because the walk-off losses of an etalon increase according to (4.64a), (4.64b) with the square of the tilting angle α and the etalon thickness t. They may become intolerably high if a large, uninterrupted tuning range shall be achieved by tilting of the etalon. Therefore the long intracavity FPI (Fig. 5.43) is kept at a fixed, small tilting angle while its transmission peak is tuned by changing the separation t between the reflecting surfaces.

In order to minimize the air gap between the reflecting surfaces of the FPI, the prism construction of Fig. 5.44b is often used, in which the small air gap is traversed by the laser beam at Brewster's angle to avoid reflection losses [347]. This design minimizes the influence of air pressure variations on the transmission peak wavelength λ_L.

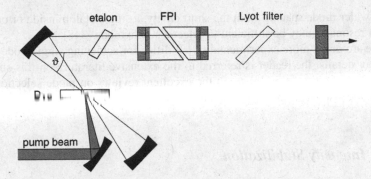

Figure 5.43 Mode selection in the cw dye laser with a folded cavity using a birefringent filter, a tilted etalon, and a prism FPI (Coherent model 599). The folding angle ϑ is chosen for optimum compensation of astigmatism introduced by the dye jet

Figure 5.44 Fabry–Perot interferometer tuned by a piezocylinder: **a** two plane-parallel plates with inner reflecting surfaces; **b** two Brewster prisms with the outer coated surfaces forming the FPI reflecting planes

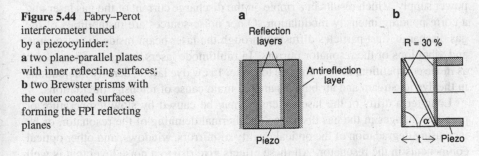

Figure 5.45 depicts the experimental arrangement for narrow-band operation of an excimer laser-pumped dye laser; the beam is expanded to fill the whole grating. Because of the higher spectral resolution of the grating (compared with a prism)

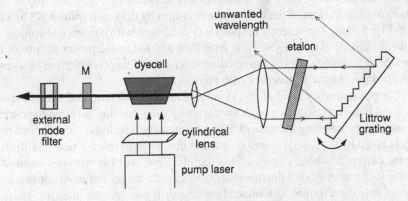

Figure 5.45 Short Hänsch-type dye laser cavity with Littrow grating and mode selection either with an internal etalon or an external FPI as "mode filter" [348]

and the wider mode spacing from the short cavity, a single etalon inside or outside the laser resonator may be sufficient to select a single mode [348].

There are many more experimental possibilities for achieving singlemode operation. For details, the reader is referred to the extensive literature on this subject, which can be found, for instance, in the excellent reviews on mode selection and single-mode lasers by Smith [344] or Goldsborough [349] and in [350, 351].

5.4.4   Intensity Stabilization

The intensity $I(t)$ of a cw laser is not completely constant, but shows periodic and random fluctuations and also, in general, long-term drifts. The reasons for these fluctuations are manifold and may, for example, be due to an insufficiently filtered power supply, which results in a ripple on the discharge current of the gas laser and a corresponding intensity modulation. Other noise sources are instabilities of the gas discharge, dust particles diffusing through the laser beam inside the resonator, and vibrations of the resonator mirrors. In multimode lasers, internal effects, such as mode competition, also contribute to noise. In cw dye lasers, density fluctuations in the dye jet stream and air bubbles are the main cause of intensity fluctuations.

Long-term drifts of the laser intensity may be caused by slow temperature or pressure changes in the gas discharge, by thermal detuning of the resonator, or by increasing degradation of the optical quality of mirrors, windows, and other optical components in the resonator. All these effects give rise to a noise level that is well above the theoretical lower limit set by the photon noise. Since these intensity fluctuations lower the signal-to-noise ratio, they may become very troublesome in many spectroscopic applications, therefore one should consider steps that reduce these fluctuations by stabilizing the laser intensity.

Of the various possible methods, we shall discuss two that are often used for intensity stabilization. They are schematically depicted in Fig. 5.46. In the first method, a small fraction of the output power is split by the beam splitter BS to a detector (Fig. 5.46a). The detector output V_D is compared with a reference voltage V_R and the difference $\Delta V = V_D - V_R$ is amplified and fed to the power supply of the laser, where it controls the discharge current. The servo loop is effective in a range where the laser intensity increases with increasing current.

The upper frequency limit of this stabilization loop is determined by the capacitances and inductances in the power supply and by the time lag between the current increase and the resulting increase of the laser intensity. The lower limit for this time delay is given by the time required by the gas discharge to reach a new equilibrium after the current has been changed. It is therefore not possible with this method to stabilize the system against fluctuations of the gas discharge. For most applications, however, this stabilization technique is sufficient; it provides an intensity stability where the fluctuations are less than 0.5 %.

To compensate fast intensity fluctuations, another technique, illustrated in Fig. 5.46b, is more suitable. The output from the laser is sent through a Pock-

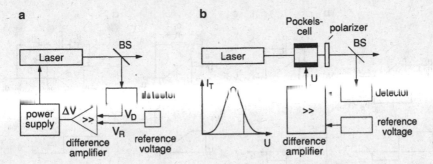

Figure 5.46 Intensity stabilization of lasers **a** by controlling the power supply, and **b** by controlling the transmission of a Pockels cell

els cell, which consists of an optically anisotropic crystal placed between two linear polarizers. An external voltage applied to the electrodes of the crystal causes optical birefringence, which rotates the polarization plane of the transmitted light and therefore changes the transmittance through the second polarizer. If part of the transmitted light is detected, the amplified detector signal can be used to control the voltage U at the Pockels cell. Any change of the transmitted intensity can be compensated by an opposite transmission change of the Pockels cell. This stabilization control works up to frequencies in the megahertz range if the feedback-control electronics are sufficiently fast. Its disadvantage is an intensity loss of 20 % to 50 % because one has to bias the Pockels cell to work on the slope of the transmission curve (Fig. 5.46b).

Figure 5.47 sketches how the electronic system of a feedback control can be designed to optimize the response over the whole frequency spectrum of the input signals. In principle, three operational amplifiers with different frequency responses are put in parallel. The first is a common *proportional* amplifier, with an upper frequency determined by the electronic time constant of the amplifier. The second is an integral amplifier with the output

$$U_{\text{out}} = \frac{1}{RC} \int_0^T U_{\text{in}}(t)\, dt\, .$$

This amplifier is necessary to bring the signal, which is proportional to the deviation of the intensity from its nominal value, really back to zero. This cannot be performed with a proportional amplifier. The third amplifier is a differentiating device that takes care of fast peaks in the perturbations. All three functions can be combined in a system called *PID control* [352, 353], which is widely used for intensity stabilization and wavelength stabilization of lasers.

For spectroscopic applications of dye lasers, where the dye laser has to be tuned through a large spectral range, the intensity change caused by the decreasing gain at both ends of the gain profile may be inconvenient. An elegant way to avoid this

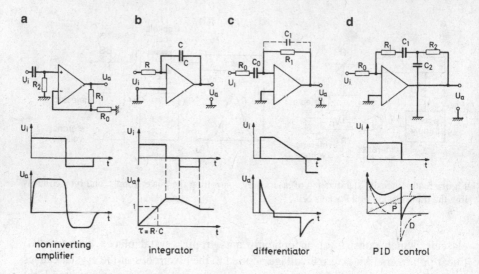

Figure 5.47 Schematic diagram of PID feedback control: **a** noninverting proportional amplifier; **b** integrator; **c** differentiating amplifier; **d** complete PID circuit that combines the functions (a–c)

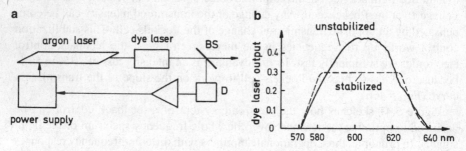

Figure 5.48 Intensity stabilization of a cw dye laser by control of the argon laser power: **a** experimental arrangement; **b** stabilized and unstabilized dye laser output $P(\lambda)$ when the dye laser is tuned across its spectral gain profile

change of $I_L(\lambda)$ with λ is to stabilize the dye laser output by controlling the argon laser power (Fig. 5.48). Since the servo control must not be too fast, the stabilization scheme of Fig. 5.48a can be employed. Figure 5.48b demonstrates how effectively this method works if one compares the stabilized with the unstabilized intensity profile $I(\lambda)$ of the dye laser.

5.4.5 Wavelength Stabilization

For many applications in high-resolution laser spectroscopy, it is essential that the laser wavelength stays as stable as possible at a preselected value λ_0. This means that the fluctuations $\Delta\lambda$ around λ_0 should be smaller than the molecular linewidths that are to be resolved. For such experiments only *single-mode* lasers can, in general, be used, because in most multimode lasers the momentary wavelengths fluctuate and only the time-averaged envelope of the spectral output profile is defined, as has been discussed in the previous sections. This stability of the wave length is important both for fixed-wavelength lasers, where the laser wavelength has to be kept at a time-independent value λ_0, as well as for tunable lasers, where the fluctuations $\Delta\lambda = |\lambda_L - \lambda_R(t)|$ around a controlled tunable wavelength $\lambda_R(t)$ have to be smaller than the resolvable spectral interval.

In this section we discuss some methods of wavelength stabilization with their advantages and drawbacks. Since the laser frequency $\nu = c/\lambda$ is directly related to the wavelength, one often speaks about *frequency* stabilization, although for most methods in the visible spectral region, it is not the frequency but the wavelength that is directly measured and compared with a reference standard. There are, however, new stabilization methods that rely directly on absolute frequency measurements (Vol. 2, Sect. 9.7).

In Sect. 5.3 we saw that the wavelength λ or the frequency ν of a longitudinal mode in the active resonator is determined by the mirror separation d and the refractive indices n_2 of the active medium with length L and n_1 outside the amplifying region. The resonance condition is

$$q\lambda = 2n_1(d - L) + 2n_2L . \qquad (5.83)$$

For simplicity, we shall assume that the active medium fills the whole region between the mirrors. Thus (5.83) reduces, with $L = d$ and $n_2 = n_1 = n$, to

$$q\lambda = 2nd , \quad \text{or} \quad \nu = qc/(2nd) . \qquad (5.84)$$

Any fluctuation of n or d causes a corresponding change of λ and ν. We obtain from (5.84)

$$\frac{\Delta\lambda}{\lambda} = \frac{\Delta d}{d} + \frac{\Delta n}{n} , \quad \text{or} \quad -\frac{\Delta\nu}{\nu} = \frac{\Delta d}{d} + \frac{\Delta n}{n} . \qquad (5.85)$$

Example 5.20
To illustrate the demands of frequency stabilization, let us assume that we want to keep the frequency $\nu = 6 \times 10^{14}$ Hz of an argon laser constant within 1 MHz. This means a relative stability of $\Delta\nu/\nu = 1.6 \times 10^{-9}$ and implies that the mirror separation of $d = 1$ m has to be kept constant within 1.6 nm!

From this example it is evident that the requirements for such stabilization are by no means trivial. Before we discuss possible experimental solutions, let us consider the causes of fluctuations or drifts in the resonator length d or the refractive index n.

Table 5.2 Linear thermal expansion coefficient of some relevant materials at room temperature $T = 20\,^{\circ}\mathrm{C}$

Material	$\alpha\ [10^{-6}\,\mathrm{K}^{-1}]$	Material	$\alpha\ [10^{-6}\,\mathrm{K}^{-1}]$
Aluminum	23	BeO	6
Brass	19	Invar	1.2
Steel	11–15	Soda-lime glass	5–8
Titanium	8.6	Pyrex glass	3
Tungsten	4.5	Fused quartz	0.4–0.5
Al_2O_3	5	Cerodur	< 0.1

If we could reduce or even eliminate these causes, we would already be well on the way to achieving a stable laser frequency. We shall distinguish between *long-term drifts* of d and n, which are mainly caused by temperature drifts or slow pressure changes, and *short-term fluctuations* caused, for example, by acoustic vibrations of mirrors, by acoustic pressure waves that modulate the refractive index, or by fluctuations of the discharge in gas lasers or of the jet flow in dye lasers.

To illustrate the influence of long-term drifts, let us make the following estimate. If α is the thermal expansion coefficient of the material (e.g., quartz or invar rods), which defines the mirror separation d, the relative change $\Delta d/d$ for a possible temperature change ΔT is, under the assumption of linear thermal expansion,

$$\Delta d/d = \alpha \Delta T\ . \tag{5.86}$$

Table 5.2 compiles the thermal expansion coefficients for some commonly used materials.

Example 5.21
For invar, with $\alpha = 1 \times 10^{-6}\,\mathrm{K}^{-1}$, we obtain from (5.86) for $\Delta T = 0.1\,\mathrm{K}$ a relative distance change of $\Delta d/d = 10^{-7}$, which gives for Example 5.20 a frequency drift of 60 MHz.

If the laser wave inside the cavity travels a path length $d - L$ through air at atmospheric pressure, any change Δp of the air pressure results in the change

$$\Delta s = (d - L)(n - 1)\Delta p/p\ , \quad \text{with} \quad \Delta p/p = \Delta n/(n - 1)\ , \tag{5.87}$$

of the optical path length between the resonator mirrors.

Example 5.22

With $n = 1.00027$ and $d - L = 0.2d$, which is typical for gas lasers, we obtain from (5.85) and (5.87) for pressure changes of $\Delta p = 3$ mbar (which can readily occur during one hour, particularly in air-conditioned rooms)

$$\Delta\lambda/\lambda = -\Delta v/v \approx (d - L)\Delta n/(nd) \geq 1.5 \times 10^{-7}.$$

For our example above, this means a frequency change of $\Delta v \geq 90$ MHz. In cw dye lasers, the length L of the active medium is negligible compared with the resonator length d, therefore we can take $d - L \simeq d$. This implies for the same pressure change a frequency drift that is five times larger than estimated above.

To keep these long-term drifts as small as possible, one has to choose distance holders for the resonator mirrors with a minimum thermal expansion coefficient α. A good choice is, for example, the recently developed cerodur–quartz composition with a temperature-dependent $\alpha(T)$ that can be made zero at room temperature [354]. Often massive granite blocks are used as support for the optical components; these have a large heat capacity with a time constant of several hours to smoothen temperature fluctuations. To minimize pressure changes, the whole resonator must be enclosed by a pressure-tight container, or the ratio $(d - L)/d$ must be chosen as small as possible. However, we shall see that such long-term drifts can be mostly compensated by electronic servo control if the laser wavelength is locked to a constant reference wavelength standard.

A more serious problem arises from the short-term fluctuations, since these may have a broad frequency spectrum, depending on their causes, and the frequency response of the electronic stabilization control must be adapted to this spectrum. The main contribution comes from acoustical vibrations of the resonator mirrors. The whole setup of a wavelengthstabilized laser should therefore be vibrationally isolated as much as possible. There are commercial optical tables with pneumatic damping, in their more sophisticated form even electronically controlled, which guarantee a stable setup for frequency-stabilized lasers. A homemade setup is considerably cheaper: Fig. 5.49 illustrates a possible table mount for the laser system as employed in our laboratory. The optical components are mounted on a heavy granite plate, which rests in a flat container filled with sand to damp the eigenresonances of the granite block. Styrofoam blocks and acoustic damping elements prevent room vibrations from being transferred to the system. The optical system is protected against direct sound waves through the air, air turbulence, and dust by a dust-free solid cover resting on the granite plate. A filtered laminar air flow from a flow box above the laser table avoids dust and air turbulence and increases the passive stability of the laser system considerably.

Figure 5.49 Experimental
realization of an acous-
tically isolated table for
a wavelength-stabilized laser
system

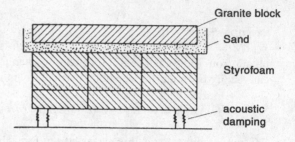

The high-frequency part of the noise spectrum is mainly caused by fast fluctuations of the refractive index in the discharge region of gas lasers or in the liquid
jet of cw dye lasers. These perturbations can only be reduced partly by choosing
optimum discharge conditions in gas lasers. In jet-stream dye lasers, density fluctuations in the free jet, caused by small air bubbles or by pressure fluctuations of the
jet pump and by surface waves along the jet surfaces, are the main causes of fast
laser frequency fluctuations. Careful fabrication of the jet nozzle and filtering of the
dye solution are essential to minimize these fluctuations.

All the perturbations discussed above cause fluctuations of the optical path length
inside the resonator that are typically in the nanometer range. In order to keep the
laser wavelength stable, these fluctuations must be compensated by corresponding
changes of the resonator length d. For such controlled and fast length changes in
the nanometer range, piezoceramic elements are mainly used [355, 356]. They consist of a piezoelectric material whose length in an external electric field changes
proportionally to the field strength. Either cylindrical plates are used, where the end
faces are covered by silver coatings that provide the electrodes or a hollow cylinder
is used, where the coatings cover the inner and outer wall surfaces (Fig. 5.50a). Typical parameters of such piezoelements are a few nanometers of length change per
volt. With stacks of many thin piezodisks, one reaches length changes of 100 nm/V.
When a resonator mirror is mounted on such a piezoelement (Fig. 5.50b,c), the resonator length can be controlled within a few microns by the voltage applied to the
electrodes of the piezoelement.

The frequency response of this length control is limited by the inertial mass of the
moving system consisting of the mirror and the piezoelement, and by the eigenresonances of this system. Using small mirror sizes and carefully selected piezos, one
may reach the 100 kHz range [357]. For the compensation of faster fluctuations, an
optical anisotropic crystal, such as potassium-dihydrogen-phosphate (KDP), can be
utilized inside the laser resonator. The optical axis of this crystal must be oriented
in such a way that a voltage applied to the crystal electrodes changes its refractive
index along the resonator axis without turning the plane of polarization. This allows
the optical path length nd, and therefore the laser wavelength, to be controlled with
a frequency response up into the megahertz range.

The wavelength stabilization system consists essentially of three elements
(Fig. 5.51):

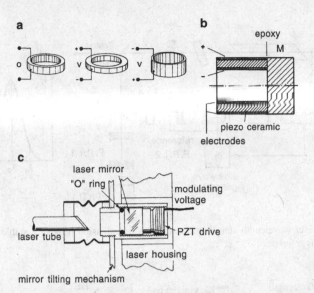

Figure 5.50 a Piezocylinders and their (exaggerated) change of length with applied voltage; **b** laser mirror epoxide on a piezocylinder; **c** mirror plus piezomount on a single-mode tunable argon laser

Figure 5.51 Schematic of laser wavelength stabilization

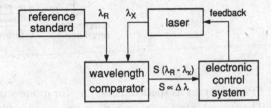

a) The wavelength reference standard with which the laser wavelength is compared. One may, for example, use the wavelength λ_R at the maximum or at the slope of the transmission peak of a Fabry–Perot interferometer that is maintained in a controlled environment (temperature and pressure stabilization). Alternately, the wavelength of an atomic or molecular transition may serve as reference. Sometimes another stabilized laser is used as a standard and the laser wavelength is locked to this standard wavelength.

b) The controlled system, which is in this case the resonator length nd defining the laser wavelength λ_L.

c) The electronic control system with the servo loop, which measures the deviation $\Delta \lambda = \lambda_L - \lambda_R$ of the laser wavelength λ_L from the reference value λ_R and which tries to bring $\Delta \lambda$ to zero as quickly as possible (Fig. 5.47).

A schematic diagram of a commonly used stabilization system is shown in Fig. 5.52. A few percent of the laser output are sent from the two beam splitters BS_1 and BS_2 into two interferometers. The first FPI1 is a scanning confocal

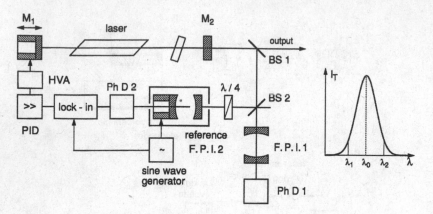

Figure 5.52 Laser wavelength stabilization onto the transmission peak of a stable Fabry–Perot interferometer as reference

Figure 5.53 Wavelength stabilization onto the slope of the transmission $T(\lambda)$ of a stable reference FPI

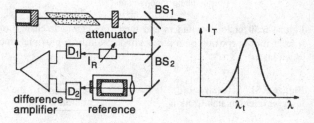

FPI and serves as spectrum analyzer for monitoring the mode spectrum of the laser. The second interferometer FPI2 is the wavelength reference and is therefore placed in a pressure-tight and temperature-controlled box to keep the optical path nd between the interferometer mirrors and with it the wavelength $\lambda_R = 2nd/m$ of the transmission peak as stable as possible (Sect. 4.2). One of the mirrors is mounted on a piezoelement. If a small ac voltage with the frequency f is fed to the piezo, the transmission peak of FPI2 is periodically shifted around the center wavelength λ_0, which we take as the required reference wavelength λ_R. If the laser wavelength λ_L is within the transmission range λ_1 to λ_2 in Fig. 5.52, the photodiode PD2 behind FPI2 delivers a dc signal that is modulated at the frequency f. The modulation amplitude depends on the slope of the transmission curve $dI_T/d\lambda$ of FPI2 and the phase is determined by the sign of $\lambda_L - \lambda_0$. Whenever the laser wavelength λ_L deviates from the reference wavelength λ_R, the photodiode delivers an ac amplitude that increases as the difference $\lambda_L - \lambda_R$ increases, as long as λ_L stays within the transmision range between λ_1 and λ_2. This signal is fed to a lock-in amplifier, where it is rectified, passes a PID control (Fig. 5.47), and a high-voltage amplifier (HVA). The output of the HVA is connected with the piezoelement of the laser mirror, which moves the resonator mirror M1 until the laser wavelength λ_L is brought back to the reference value λ_R.

Instead of using the maximum λ_0 of the transmission peak of $I_T(\lambda)$ as reference wavelength, one may also choose the wavelength λ_t at the turning point of $I_T(\lambda)$ where the slope $dI_T(\lambda)/d\lambda$ has its maximum (Fig. 5.53). This has the advantage that a modulation of the FPI transmission curve is not necessary and the lock-in amplifier can be dispensed with. The cw laser intensity $I_T(\lambda)$ transmitted through FPI2 is compared with a reference intensity I_R split by BS_2 from the same partial beam. The output signals S_1 and S_2 from the two photodiodes D1 and D2 are fed into a difference amplifier, which is adjusted so that its output voltage becomes zero for $\lambda_L = \lambda_t$. If the laser wavelength λ_L deviates from $\lambda_R = \lambda_t$, S_1 becomes smaller or larger, depending on the sign of $\lambda_L - \lambda_R$; the output of the difference amplifier is, for small differences $\lambda - \lambda_R$, proportional to the deviation. The output signal again passes a PID control and a high-voltage amplifier, and is fed into the piezoelement of the resonator mirror. The advantages of this difference method are the larger bandwidth of the difference amplifier (compared with a lock-in amplifier), and the simpler and less expensive composition of the whole electronic control system. Furthermore, the laser frequency does not need to be modulated which represents a big advantage for many spectroscopic applications [358]. Its drawback lies in the fact that different dc voltage drifts in the two branches of the difference amplifier result in a dc output, which shifts the zero adjustment and, with it, the reference wavelength λ_R. Such dc drifts are much more critical in dc amplifiers than in the ac-coupled devices used in the first method.

The stability of the laser wavelength can, of course, never exceed that of the reference wavelength. Generally it is worse because the control system is not ideal. Deviations $\Delta\lambda(t) = \lambda_L(t) - \lambda_R$ cannot be compensated immediately because the system has a finite frequency response and the inherent time constants always cause a phase lag between deviation and response.

Most methods for wavelength stabilization use a stable FPI as reference standard [359]. This has the advantage that the reference wavelength λ_0 or λ_t can be tuned by tuning the reference FPI. This means that the laser can be stabilized onto any desired wavelength within its gain profile. Because the signals from the photodiodes D1 and D2 in Fig. 5.53 have a sufficiently large amplitude, the signal-to-noise ratio is good, therefore the method is suitable for correcting short-term fluctuations of the laser wavelength.

For long-term stabilization, however, stabilization onto an external FPI has its drawbacks. In spite of temperature stabilization of the reference FPI, small drifts of the transmission peak cannot be eliminated completely. With a thermal expansion coefficient $\alpha = 10^{-6}$ of the distance holder for the FPI mirrors, even a temperature drift of 0.01 °C causes, according to (5.86), a relative frequency drift of 10^{-8}, which gives 6 MHz for a laser frequency of $\nu_L = 6 \times 10^{14}$ Hz. For this reason, an atomic or molecular laser transition is more suitable as a *long-term frequency standard*. A good reference wavelength should be reproducible and essentially independent of external perturbations, such as electric or magnetic fields and temperature or pressure changes. Therefore, transitions in atoms or molecules without permanent dipole moments, such as CH_4 or noble gas atoms, are best suited to serve as reference wavelength standards (Vol. 2, Chap. 9).

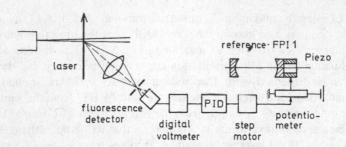

Figure 5.54 Long-term stabilization of the laser wavelength locked to a reference FPI that in turn is locked by a digital servo loop to a molecular transition

The accuracy with which the laser wavelength can be stabilized onto the center of such a transition depends on the linewidth of the transition and on the attainable signal-to-noise ratio of the stabilization signal. Doppler-free line profiles are therefore preferable. They can be obtained by some of the methods discussed in Vol. 2, Chaps. 2 and 4. In the case of small line intensities, however, the signal-to-noise ratio may be not good enough to achieve satisfactory stabilization. It is therefore advantageous to continue to lock the laser to the reference FPI, but to lock the FPI itself to the molecular line. In this double servo control system, the short-term fluctuations of λ_L are compensated by the fast servo loop with the FPI as reference, while the slow drifts of the FPI are stabilized by being locked to the molecular line.

Figure 5.54 illustrates a possible arrangement. The laser beam is crossed perpendicularly with a collimated molecular beam. The Doppler width of the absorption line is reduced by a factor depending on the collimation ratio (Vol. 2, Sect. 4.1). The intensity $I_F(\lambda_L)$ of the laser-excited fluorescence serves as a monitor for the deviation $\lambda_L - \lambda_c$ from the line center λ_c. The output signal of the fluorescence detector after amplification can be fed directly to the piezoelement of the laser resonator or to the reference FPI.

To decide whether λ_t drifts to lower or to higher wavelengths, one must either modulate the laser frequency or use a digital servo control, which shifts the laser frequency in small steps. A comparator compares whether the intensity has increased or decreased by the last step and activates accordingly a switch determining the direction of the next step. Since the drift of the reference FPI is slow, the second servo control can also be slow, and the fluorescence intensity can be integrated. This allows the laser to be stabilized for a whole day, even onto faint molecular lines where the detected fluorescence intensity is less than 100 photons per second [360].

Recently, cryogenic optical sapphire resonators with a very high finesse operating at $T = 4\,\mathrm{K}$ have proven to provide very stable reference standards [361]. They reach a relative frequency stability of 3×10^{-15} at an integration time of 20 s.

Since the accuracy of wavelength stabilization increases with decreasing molecular linewidth, spectroscopists have looked for particularly narrow lines that could be used for extremely well-stabilized lasers. It is very common to stabilize onto a hyperfine component of a visible transition in the I_2 molecule using Doppler-free saturated absorption inside [362] or outside [363] the laser resonator (Vol. 2, Sect. 2.3). The stabilization record was held for a long time by a HeNe laser at $\lambda = 3.39\,\mu m$ that was stabilized onto a Doppler-free infrared transition in CH_4 [364, 365].

Using the dispersion profiles of Doppler-free molecular lines in polarization spectroscopy (Vol. 2, Sect. 2.4), it is possible to stabilize a laser to the line center without frequency modulation. An interesting alternative for stabilizing a dye laser on atomic or molecular transitions is based on Doppler-free two-photon transitions (Vol. 2, Sect. 2.5) [368]. This method has the additional advantage that the lifetime of the upper state can be very long, and the natural linewidth may become extremely small. The narrow $1s - 2s$ two-photon transition in the hydrogen atom with a natural linewidth of 1.3 Hz provides the best known optical frequency reference to date [366].

Often the narrow Lamb dip at the center of the gain profile of a gas laser transition is utilized (Vol. 2, Sect. 2.2) to stabilize the laser frequency [369, 370]. However, due to collisional line shifts the frequency ν_0 of the line center slightly depends on the pressure in the laser tube and may therefore change in time when the pressure is changing (for instance, by He diffusion out of a HeNe laser tube).

By placing a thin Cs vapour cell inside the resonator of an external cavity diode laser, the laser can be readily stabilized onto the Lamb dip of the Cs resonance line [367].

A simple technique for wavelength stabilization uses the orthogonal polarization of two adjacent axial modes in a HeNe laser [371]. The two-mode output is split by a polarization beam splitter BS1 in the two orthogonally polarized modes, which are monitored by the photodetectors PD1 and PD2 (Fig. 5.55). The difference amplification delivers a signal that is used to heat the laser tube, which expands until the two modes have equal intensities (Fig. 5.55a). They are then kept at the frequencies $\nu_{\pm} = \nu_0 \pm \Delta\nu/2 = \nu_0 \pm c/(4nd)$. Only one of the modes is transmitted to the experiment.

Very high frequency stability can be achieved if the laser frequency is stabilized to the transition frequency of a single ion that is held in an ion trap under vacuum (see Vol. 2, Sect. 9.2) [382].

So far we have only considered the stability of the laser resonator itself. In the previous section we saw that wavelength-selecting elements inside the resonator are necessary for single-mode operation to be achieved, and that their stability and the influence of their thermal drifts on the laser wavelength must also be considered. We illustrate this with the example of single-mode selection by a tilted intracavity etalon. If the transmission peak of the etalon is shifted by more than one-half of the cavity mode spacing, the total gain becomes more favorable for the next cavity mode, and the laser wavelength will jump to the next mode. This implies that the optical pathlength of the etalon nt must be kept stable so that the peak transmission drifts by less than $c/4d$, which is about 50 MHz for an argon laser. One can use either an air-spaced etalon with distance holders with very small thermal expansion or a solid etalon in a temperature-stabilized oven. The air-spaced etalon is simpler

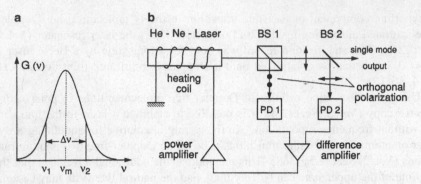

Figure 5.55 Schematic diagram of a polarization-stabilized HeNe laser: **a** symmetric cavity modes ν_1 and ν_2 within the gain profile; **b** experimental setup

but has the drawback that changes of the air pressure influence the transmission peak wavelength.

The actual stability obtained for a single-mode laser depends on the laser system, on the quality of the electronic servo loop, and on the design of the resonator and mirror mounts. With moderate efforts, a frequency stability of about 1 MHz can be achieved, while extreme precautions and sophisticated equipment allow a stability of better than 1 Hz to be achieved for some laser types [372].

A statement about the stability of the laser frequency depends on the averaging time and on the kind of perturbations. For short time periods the frequency stability is mainly determined by random fluctuations. The best way to describe short-term frequency fluctuations is the statistical root Allan variance. For longer time periods ($\Delta t \gg 1$ s), the frequency stability is limited by predictable and measurable fluctuations, such as thermal drifts and aging of materials. The stability against short-term fluctuations, of course, becomes better if the averaging time is increased, while long-term drifts increase with the sampling time. Figure 5.56 illustrates the stability of a single-mode argon laser, stabilized with the arrangement of Fig. 5.52. With more expenditure, a stability of better than 3 kHz has been achieved for this laser [373], with novel techniques even better than 1 Hz (Vol. 2, Sect. 9.7).

The residual frequency fluctuations of a stabilized laser can be represented in an *Allan plot*. The Allan variance [372, 374, 376]

$$\sigma = \frac{1}{\nu}\left(\sum_{i=1}^{N} \frac{\langle(\Delta\nu_i - \Delta\nu_{i-1})^2\rangle}{2(N-1)}\right)^{1/2} \tag{5.88}$$

is comparable to the relative standard deviation. It is determined by measuring at N times $t_i = t_0 + i\Delta t$ ($i = 0, 1, 2, 3 \ldots$) the relative frequency difference $\Delta\nu_i/\nu_R$ between two lasers stabilized onto the same reference frequency ν_R averaged over equal time intervals Δt. Figure 5.57 illustrates the Allan variance for different frequency reference devices: the He-Ne laser at $\lambda = 3.39\,\mu$m, locked to a vibration–rotation transition of the CH_4 molecule, the hydrogen maser at $\lambda =$

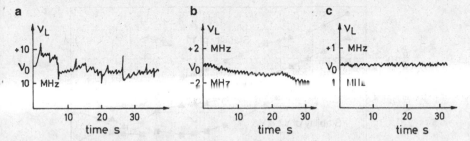

Figure 5.56 Frequency stability of a single-mode argon laser: **a** unstabilized; **b** stabilized with the arrangement of Fig. 5.52; **c** additional long-term stabilization onto a molecular transition. Note the different ordinate scales!

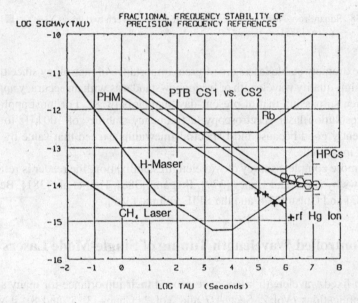

Figure 5.57 Allan variance obtained for different frequency-reference devices [374]

21 cm, two cesium clocks operated at the PTB (Physikalisch-Technische Bundesanstalt) in Braunschweig, Germany, the rubidium atomic clock, the clock based on the rf transition of the Hg^+-ion in a trap, and the pulsed hydrogen maser.

In Fig. 5.58 the Allan plot for the frequency stabilities of four Nd:YAG lasers stabilized onto a transition of the I_2 molecule are composed. The different lasers, called $Y_1 \ldots Y_4$, use different laser powers and beam diameters, which cause different saturations of the iodine transition.

The best frequency stability in the optical range can be achieved with the optical frequency-comb technique, which will be discussed in Vol. 2, Sect. 9.7 [377]. The relative frequency fluctuations go down to $\Delta v/v_0 < 10^{-15}$, which implies an absolute stability of about 0.5 Hz.

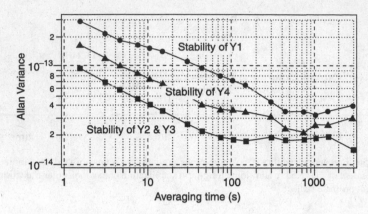

Figure 5.58 Square-root Allan variance of the beat notes between two lasers Y_2-Y_3 (■) Y_2-Y_4 (►) and Y_2-Y_1 (●) [375]

Such extremely stable lasers are of great importance in metrology since they can provide high-quality wavelength or frequency standards with an accuracy approaching or even surpassing that of present-day standards [378]. For most applications in high-resolution laser spectroscopy, a frequency stability of 100 kHz to 1 MHz is sufficiently good because most spectral linewidths exceed that value by several orders of magnitude.

For a more complete survey of wavelength stabilization, the reader is referred to the reviews by Baird and Hanes [379], Ikegami [380], Hall et al. [381], Bergquist et al. [383] and Ohtsu [384] and the SPIE volume [385].

5.5   Controlled Wavelength Tuning of Single-Mode Lasers

Although fixed-wavelength lasers have proved their importance for many spectroscopic applications (Vol. 2, Sect. 1.7 and Vol. 2, Chaps. 3, 5, and 8), it was the development of continuously tunable lasers that really revolutionized the whole field of spectroscopy. This is demonstrated by the avalanche of publications on tunable lasers and their applications (e.g., [386]). We shall therefore treat in this section some basic techniques for controlled tuning of single-mode lasers, while Sect. 5.7 gives a survey on tunable coherent sources developed in various spectral regions.

5.5.1   Continuous Tuning Techniques

Since the laser wavelength λ_L of a single-mode laser is determined by the optical path length nd between the resonator mirrors,

$$q\lambda = 2nd \,,$$

either the mirror separation d or the refractive index n can be continuously varied to obtain a corresponding tuning of λ_L. This can be achieved, for example, by a linear voltage ramp $U = U_0 + at$ applied to the piezoelement on which the resonator mirror is mounted, or by a continuous pressure variation in a tank containing the resonator or parts of it. However, as has been discussed in Sect. 5.4.3, most lasers need additional wavelength-selecting elements inside the laser resonator to ensure singlemode operation. When the resonator length is varied, the frequency v of the oscillating mode is tuned away from the transmission maximum of these elements (Fig. 5.38). During this tuning the neighboring resonator mode (which is not yet oscillating) approaches this transmission maximum and its losses may now become smaller than those of the oscillating mode. As soon as this mode reaches the threshold, it will start to oscillate and will suppress the former mode because of mode competition (Sect. 5.3). This means that the single-mode laser will jump back from the selected resonator mode to that which is next to the transmission peak of the wavelength-selecting element. Therefore the continuous tuning range is restricted to about half of the free spectral range $\delta v = \frac{1}{2}c/t$ of the intracavity selecting interferometer with thickness t, if no additional measures are taken. Similar but smaller mode hops $\Delta v = c/2d$ occur when the wavelength-selecting elements are continuously tuned but the resonator length d is kept constant.

Such discontinuous tuning of the laser wavelength will be sufficient if the mode hops $\delta v = \frac{1}{2}c/d$ are small compared with the spectral linewidths under investigation. As illustrated by Fig. 5.59a, which shows part of the neon spectrum excited in a HeNe gas discharge with a discontinuously tuned single-mode dye laser, the mode hops are barely seen and the spectral resolution is limited by the Doppler width of the neon lines. In sub-Doppler spectroscopy, however, the mode jumps appear as steps in the line profiles, as is depicted in Fig. 5.59b, where a single-mode argon laser is tuned with mode hops through some absorption lines of Na_2 molecules in a slightly collimated molecular beam where the Doppler width is reduced to about 200 MHz.

In order to enlarge the tuning range and to achieve truly continuous tuning, the transmission maxima of the wavelength selectors have to be tuned synchronously with the tuning of the resonator length. When a tilted etalon with the thickness t and refractive index n is employed, the transmission maximum λ_m that, according to (5.76), is given by

$$m\lambda_m = 2nt \cos\theta ,$$

can be continuously tuned by changing the tilting angle θ. In all practical cases, θ is very small, therefore we can use the approximation $\cos\theta \approx 1 - \frac{1}{2}\theta^2$. The wavelength shift $\Delta\lambda = \lambda_0 - \lambda$ is

$$\Delta\lambda = \frac{2nt}{m}(1 - \cos\theta) \approx \frac{1}{2}\lambda_0\theta^2 , \quad \lambda_0 = \lambda(\theta = 0) . \tag{5.89}$$

Equation (5.89) reveals that the wavelength shift $\Delta\lambda$ is proportional to θ^2 but is *independent of the thickness* t. Two etalons with different thicknesses t_1 and t_2 can

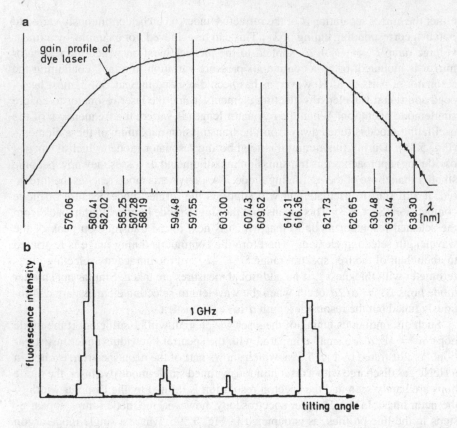

Figure 5.59 Discontinuous tuning of lasers: **a** part of the neon spectrum excited by a single-mode dye laser in a gas discharge with Doppler-limited resolution, which conceals the cavity mode hops of the laser; **b** excitation of Na_2 lines in a weakly collimated beam by a single-mode argon laser. In both cases the intracavity etalon was continuously tilted but the cavity length was kept constant

be mounted on the same tilting device, which may simply be a lever that is tilted by a micrometer screw driven by a small motor gearbox. The motor simultaneously drives a potentiometer, which provides a voltage proportional to the tilting angle θ. This voltage is electronically squared, amplified, and fed into the piezoelement of the resonator mirror. With properly adjusted amplification, one can achieve an exact sychronization of the resonator wavelength shift $\Delta\lambda_L = \lambda_L \Delta d/d$ with the shift $\Delta\lambda_l$ of the etalon transmission maximum. This can be readily realized with computer control.

Unfortunately, the reflection losses of an etalon increase with increasing tilting angle θ (Sect. 4.2 and [340, 387]). This is due to the finite beam radius w of the laser beam, which prevents a complete overlap of the partial beams reflected from the front and back surfaces of the etalon. These "walk-off losses" increases with the square of the tilting angle θ, see (4.64a), (4.64b) and Fig. 4.42.

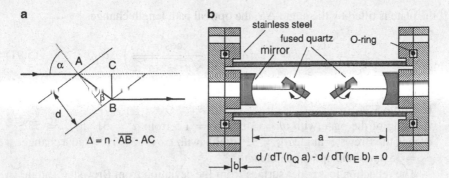

Figure 5.60   **a** Changing of resonator length by tilting of the Brewster plates inside the resonator; **b** temperature-compensated reference cavity with tiltable Brewster plates for wavelength tuning

---

Example 5.23
With $w = 1\,\mathrm{mm}$, $t = 1\,\mathrm{cm}$, $n = 1.5$, $R = 0.4$, we obtain for $\theta = 0.01$ ($\approx 0.6°$) transmission losses of 13 %. The frequency shift is, see (5.89): $\Delta\nu = \frac{1}{2}\nu_0\theta^2 \approx 30\,\mathrm{GHz}$. For a dye laser with the gain factor $G < 1.13$ the tuning range would therefore be smaller than $30\,\mathrm{GHz}$.

For wider tuning ranges interferometers with a variable air gap can be used at a fixed tilting angle θ (Fig. 5.44a). The thickness t of the interferometer and with it the transmitted wavelength $\lambda_m = 2nt\cos\theta/m$ can be tuned with a piezocylinder. This keeps the walk-off losses small. However, the extra two surfaces have to be antireflection-coated in order to minimize the reflection losses.

An elegant solution is shown in Fig. 5.44b, where the interferometer is formed by two prisms with coated backsides and inner Brewster surfaces. The air gap between these surfaces is very small in order to minimize shifts of the transmission peaks due to changes of air pressure.

The continuous change of the resonator length d is limited to about 5–10 μm if small piezocylinders are used (5–10 nm/V). A further drawback of piezoelectric tuning is the hysteresis of the expansion of the piezocylinder when tuning back and forth. Larger tuning ranges can be obtained by tilting a plane-parallel glass plate around the Brewster angle inside the laser resonator (Fig. 5.60). The additional optical path length through the plate with refractive index n at an incidence angle α is

$$s = (n\overline{AB} - \overline{AC}) = \frac{d}{\cos\beta}[n - \cos(\alpha - \beta)] = d\left[\sqrt{n^2 - \sin^2\alpha} - \cos\alpha\right].$$

$$(5.90)$$

If the plate is tilted by the angle $\Delta\alpha$, the optical path length changes by

$$\delta s = \frac{\mathrm{d}s}{\mathrm{d}\alpha}\Delta\alpha = d\sin\alpha\left(1 - \frac{\cos\alpha}{\sqrt{n^2 - \sin^2\alpha}}\right)\Delta\alpha \ . \tag{5.91}$$

Example 5.24
A tilting of the plate with $d = 3\,\mathrm{mm}$, $n = 1.5$ from $\alpha = 51°$ to $\alpha = 53°$ around the Brewster angle $\alpha_\mathrm{B} = 52°$ yields with $\Delta\alpha = 3\times10^{-2}$ rad a change $\delta s = 35\,\mu\mathrm{m}$ of the optical pathlength.

The reflection losses per surface from the deviation from Brewster's angle are less than $0.01\,\%$ and are therefore completely negligible.

If the free spectral range of the resonator is $\delta\nu$, the frequency-tuning range is

$$\Delta\nu = 2(\delta s/\lambda)\delta\nu \approx 116\,\delta\nu \quad\text{at}\quad \lambda = 600\,\mathrm{nm} \ . \tag{5.92}$$

With a piezocylinder with $\mathrm{d}s/\mathrm{d}V = 3\,\mathrm{nm/V}$ only a change of $\Delta\nu = 5\delta\nu$ can be realized at $V = 500\,\mathrm{V}$.

The Brewster plate can be tilted in a controllable way by a galvo-drive [388], where the tilting angle is determined by the strength of the magnetic field. In order to avoid a translational shift of the laser beam when tilting the plate, two plates with $\alpha = \pm\alpha_\beta$ can be used (Fig. 5.60b), which are tilted into opposite directions. This gives twice the frequency shift of (5.91). The frequency stability of the reference interferometer in Fig. 5.60b can be greatly improved by compensating for the thermal expansion of the quartz distance holder with the opposite expansion of the mirror holder. With the refractive indices n_Q of quartz and n_E of the mirror holder, the condition for exact compensation is:

$$\frac{\mathrm{d}}{\mathrm{d}T}(an_\mathrm{Q}) - \frac{\mathrm{d}}{\mathrm{d}T}(bn_\mathrm{E}) = 0 \ .$$

For illustration Fig. 5.61 shows a Doppler-free spectrum of naphthalene $C_{10}H_8$ recorded together with frequency markers from a stabilized etalon and an I_2-spectrum providing reference lines [392].

For many applications in high-resolution spectroscopy where the wavelength $\lambda(t)$ should be a linear function of the time t, it is desirable that the fluctuations of the laser wavelength λ_L around the programmed tunable value $\lambda(t)$ are kept as small as possible. This can be achieved by stabilizing λ_L to the reference wavelength λ_R of a stable external FPI (Sect. 5.4), while this reference wavelength λ_R is synchronously tuned with the wavelength-selecting elements of the laser resonator. The synchronization utilizes an electronic feedback system. A possible realization is shown in Fig. 5.62. A digital voltage ramp provided by a computer through

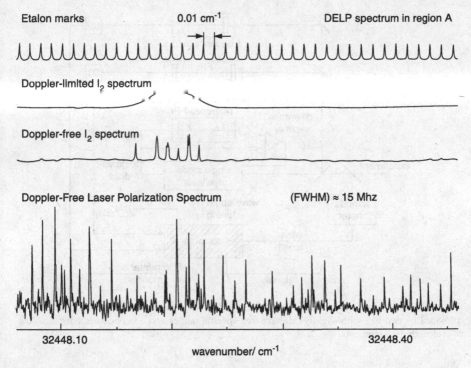

Figure 5.61 Frequency markers from an etalon with $n \cdot d = 50\,\text{cm}$, Doppler-limited and Doppler-free lines of I_2 as reference spectrum and a section of the Doppler-free spectrum of the naphthalene molecule, taken in a cell with about 5 mbar [392]

a digital–analog converter (DAC) activates the galvo-drive and results in a controlled tilting of the Brewster plates in a temperature-stabilized FPI. The laser wavelength is locked via a PID feedback control (Sect. 5.4.4) to the slope of the transmission peak of the reference FPI (Fig. 5.52). The output of the PID control is split into two parts: the low-frequency part of the feedback is applied to the galvo-plate in the laser resonator, while the high-frequency part is given to a piezoelement, which translates one of the resonator mirrors.

5.5.2   Wavelength Calibration

An essential goal of laser spectroscopy is the accurate determination of energy levels in atoms or molecules and their splittings due to external fields or internal couplings. This goal demands the precise knowledge of wavelengths and distances between spectral lines while the laser is scanned through the spectrum. There are several techniques for the solution of this problem: part of the laser beam is sent through a long FPI with mirror separation d, which is pressure-tight (or evacu-

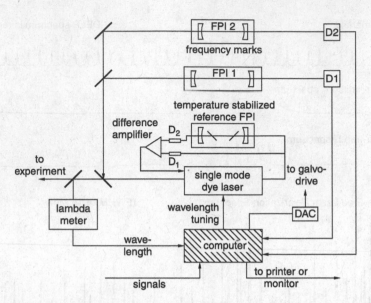

Figure 5.62 Schematic diagram of computer-controlled laser spectrometer with frequency marks provided by two FPI with slightly different free spectral ranges and a lambdameter for absolute wavelength measurement

ated) and temperature stabilized. The equidistant transmission peaks with distances $\delta\nu = \frac{1}{2}c/(nd)$ serve as frequency markers and are monitored simultaneously with the spectral lines (Fig. 5.62).

Most tunable lasers show an optical frequency $\nu(V)$ that deviates to a varying degree from the linear relation $\nu = \alpha V + b$ between laser frequency ν and input voltage V to the scan electronics. For a visible dye laser the deviations may reach 100 MHz over a 20-GHz scan. These deviations can be monitored and corrected for by comparing the measured frequency markers with the linear expression

$$\nu = \nu_0 + mc/(2nd) \quad (m = 0, 1, 2, \ldots).$$

For *absolute* wavelength measurements of spectral lines the laser is stabilized onto the center of the line and its wavelength λ is measured with one of the wavemeters described in Sect. 4.4. For Doppler-free lines (Vol. 2, Chaps. 2–6), one may reach absolute wavelength determinations with an uncertainty of smaller than $10^{-3}\,\text{cm}^{-1}$ ($\,\hat{=}\,20\,\text{pm}$ at $\lambda = 500\,\mu\text{m}$).

Often calibration spectra that are taken simultaneously with the unknown spectra are used. Examples are the I_2 spectrum, which has been published in the iodine atlas by Gerstenkorn and Luc [389] in the range of 14,800 to 20,000 cm^{-1} or with Doppler-free resolution by H. Kato [391]. Figure 5.61 illustrates this using the example of absorption lines of naphthalene molecules [392]. For wavelengths below 500 nm, thorium lines [390] measured in a hollow cathode by optogalvanic spectroscopy (Vol. 2, Sect. 1.5) or uranium lines [393] can be utilized.

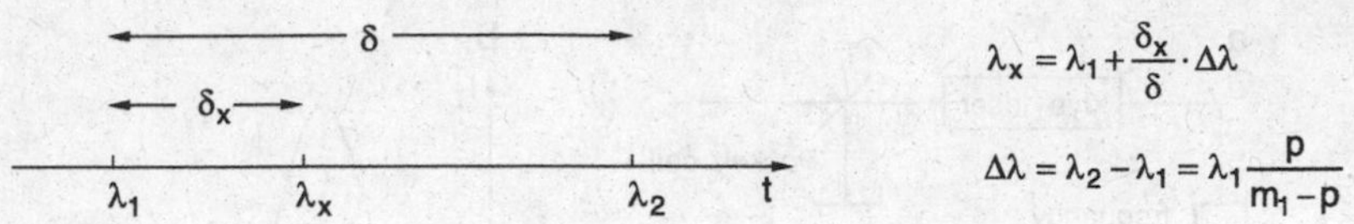

$$\lambda_x = \lambda_1 + \frac{\delta_x}{\delta} \cdot \Delta\lambda$$

$$\Delta\lambda = \lambda_2 - \lambda_1 = \lambda_1 \frac{p}{m_1 - p}$$

Figure 5.63 Scheme for wavelength determination according to (5.93c)

If no wavemeter is available, two FPIs with slightly different mirror separations d_1 and d_2 can be used for wavelength determination (Fig. 5.62b). Assume $d_1/d_2 = p/q$ equals the ratio of two rather large integers p and q with no common divisor and both interferometers have a transmission peak at λ_1:

$$\left.\begin{array}{l} m_1\lambda_1 = 2d_1 \\ m_2\lambda_1 = 2d_2 \end{array}\right\} \quad \text{with} \quad \frac{m_1}{m_2} = p/q \ . \tag{5.93a}$$

Let us assume that λ_1 is known from calibration with a spectral line. When the laser wavelength is tuned, the next coincidence appears at $\lambda_2 = \lambda_1 + \Delta\lambda$ where

$$(m_1 - p)\lambda_2 = 2d_1 \quad \text{and} \quad (m_2 - q)\lambda_2 = 2d_2 \ . \tag{5.93b}$$

From (5.93a, 5.93b) we obtain

$$\frac{\Delta\lambda}{\lambda_1} = \frac{p}{m_1 - p} = \frac{q}{m_2 - q} \Rightarrow \lambda_2 = \lambda_1 \frac{m_1}{m_1 - p} = \lambda_1 \frac{m_2}{m_2 - q} \ ,$$

where p and q are known integers that can be counted by the number of transmission maxima when λ is tuned from λ_1 to λ_2.

Between these two wavelengths λ_1 and λ_2 the maximum of a spectral line with the unknown wavelength λ_x may appear in a linear wavelength scan at the distance δ_x from the position of λ_1. Then we obtain from Fig. 5.63

$$\lambda_x = \lambda_1 + \frac{\delta_x}{\delta}\Delta\lambda = \lambda_1 \left(1 + \frac{\delta_x}{\delta}\frac{p}{m_1 - p}\right) \ . \tag{5.93c}$$

With the inputs for λ_1, p, q, d_1, and d_2 a computer can readily calculate λ_x from the measured value δ_x.

For a very precise measurement of small spectral invervals between lines a side-band technique is very useful. In this technique part of the laser beam is sent through a Pockels cell (Fig. 5.64), which modulates the transmitted intensity and generates sidebands at the frequencies $\nu_R = \nu_L \pm f$. When $\nu_R^+ = \nu_L + f$ is stabilized onto an external FPI, the laser frequency $\nu_L = \nu_R^+ - f$ can be continuously tuned by varying the modulation frequency f. This method does not need a tunable interferometer and its accuracy is only limited by the accuracy of measuring the modulation frequency f [394]a.

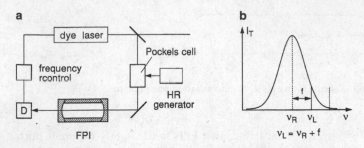

Figure 5.64 Optical sideband technique for precise tuning of the laser wavelength λ: **a** experimental setup; **b** stabilization of the sideband ν_R onto the transmission peak of the FPI

5.5.3   Frequency Offset Locking

This controllable shift of a laser frequency ν_L against a reference frequency ν_R can be also realized by electronic elements in the stabilization feedback circuit. This omits the Pockels cell of the previous method. A tunable laser is "frequency-offset locked" to a stable reference laser in such a way that the difference frequency $f = \nu_L - \nu_R$ can be controlled electronically. The experimental arrangement is shown in Fig. 5.65. The stable reference laser is a methane-stabilized He-Ne laser (see Vol. 2, Chap. 2) whose wavelength is stabilized on a rotational line in the vibrational band of the CH_4 molecule. The tunable frequency-offset laser is scanned through the spectral range of interest by stabilizing the difference frequency of the two lasers on a variable frequency, which is controlled by a frequency generator. This technique has been described by Hall [394b] and is used in many laboratories. More details will be discussed in Vol. 2, Chap. 2.

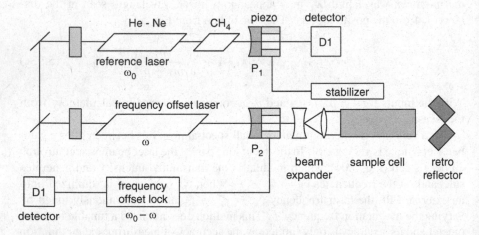

Figure 5.65 Schematic diagram of the frequency offset laser spectrometer

5.6 Linewidths of Single-Mode Lasers

In the previous sections we have seen that the frequency fluctuations of single-mode lasers caused by fluctuations of the product nd of the refractive index n and the by appropriate stabilization techniques. The output beam of such a single-mode laser can be regarded for most applications as a *monochromatic wave* with a radial Gaussian amplitude profile, see (5.32).

For some tasks in ultrahigh-resolution spectroscopy, the residual finite linewidth $\Delta \nu_L$, which may be small but nonzero, still plays an important role and must therefore be known. Furthermore, the question *why* there is an ultimate lower limit for the linewidth of a laser is of fundamental interest, since this leads to basic problems of the nature of electromagnetic waves. Any fluctuation of amplitude, phase, or frequency of our "monochromatic" wave results in a finite linewidth, as can be seen from a Fourier analysis of such a wave (see the analogous discussion in Sects. 3.1 and 3.2). Besides the "technical noise" caused by fluctuations of the product nd, there are essentially three noise sources of a *fundamental* nature, which cannot be eliminated, even by an ideal stabilization system. These noise sources are, to a different degree, responsible for the residual linewidth of a single-mode laser.

The first contribution to the noise results from the spontaneous emission of excited atoms in the upper laser level E_i. The total power P_{sp} of the fluorescence spontaneously emitted on the transition $E_i \to E_k$ is, according to Sect. 2.3, proportional to the population density N_i, the active mode volume V_m, and the transition probability A_{ik}, i.e.,

$$P_{sp} = N_i V_m A_{ik} . \tag{5.94}$$

This fluorescence is emitted into all modes of the EM field within the spectral width of the fluorescence line. According to Example 2.1 in Sect. 2.1, there are about 3×10^8 modes/cm³ within the Doppler-broadened linewidth $\Delta \nu_D = 10^9$ Hz at $\lambda = 500$ nm. The mean number of fluorescence photons per mode is therefore small.

Example 5.25
In a HeNe laser the stationary population density of the upper laser level is $N_i \simeq 10^{10}$ cm⁻³. With $A_{ik} = 10^8$ s⁻¹, the number of fluorescence photons per second is 10^{18} s⁻¹cm⁻³, which are emitted into 3×10^8 modes. Into each mode a photon flux $\phi = 3 \times 10^9$ photons/s is emitted, which corresponds to a mean photon density of $\langle n_{ph} \rangle = \phi/c \leq 10^{-1}$ in one mode. This has to be compared with 10^7 photons per mode due to induced emission inside the resonator at a laser output power of 1 mW through a mirror with $R = 0.99$.

When the laser reaches threshold, the number of photons in the laser mode increases rapidly by stimulated emission and the narrow laser line grows from

Figure 5.66 Linewidth
of a single-mode laser
just above threshold with
Doppler-broadened back-
ground due to spontaneous
emission. Note the logarithmic scale!

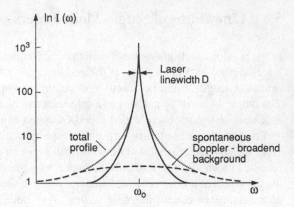

the weak but Doppler-broadened background radiation (Fig. 5.66). Far above the
threshold, the laser intensity is larger than this background by many orders of magnitude and we may therefore neglect the contribution of spontaneous emission to
the laser linewidth.

The second contribution to the noise resulting in line broadening is due to amplitude fluctuations caused by the statistical distribution of the number of photons
in the oscillating mode. At the laser output power P, the average number of photons that are transmitted per second through the output mirror is $\overline{n} = P/h\nu$. With
$P = 1\,\mathrm{mW}$ and $h\nu = 2\,\mathrm{eV}$ ($\hat{=} \lambda = 600\,\mathrm{nm}$), we obtain $n = 8 \times 10^{15}$. If the laser
operates far above threshold, the probability $p(n)$ that n photons are emitted per
second is given by the Poisson distribution [304, 305]

$$p(n) = \frac{\mathrm{e}^{-\overline{n}}(\overline{n}^n)}{n!} \ . \tag{5.95}$$

The average number $\overline{n}$ is mainly determined by the pump power P_p (Sect. 5.1.3).
If at a given value of P_p the number of photons increases because of an amplitude
fluctuation of the induced emission, saturation of the amplifying transition in the
active medium reduces the gain and decreases the field amplitude. Thus saturation
provides a self-stabilizing mechanism for amplitude fluctuations and keeps the laser
field amplitude at a value $E_\mathrm{s} \sim (\overline{n})^{1/2}$.

The main contribution to the residual laser linewidth comes from *phase fluctuations*. Each photon that is spontaneously emitted into the laser mode can be
amplified by induced emission; this amplified contribution is superimposed on the
oscillating wave. This does not essentially change the total amplitude of the wave
because these additional photons decrease the gain for the other photons (by gain
saturation) such that the average photon number $\overline{n}$ remains constant. However, the
phases of these spontaneously initiated photon avalanches show a random distribution, as does the phase of the total wave. There is no such stabilizing mechanism
for the total phase as there is for the amplitude. In a polar diagram, the total field
amplitude $E = A\mathrm{e}^{\mathrm{i}\varphi}$ can be described by a vector with the amplitude A, which is
restricted to a narrow range δA, and a phase angle φ that can vary from 0 to 2π

Figure 5.67 Polar diagram
of the amplitude vector $\mathbf{A}$
of a single-mode laser, for
illustration of phase diffusion

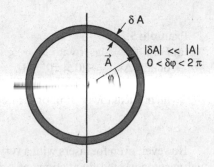

$|\delta A| \ll |A|$

$0 < \delta\varphi < 2\pi$

(Fig. 5.67). In the course of time, a *phase diffusion* φ occurs that can be described in a thermodynamic model by the diffusion coefficient D [395, 396].

For the spectral distribution of the laser emission in the ideal case in which all technical fluctuations of nd are totally eliminated, this model yields from a Fourier transform of the statistically varying phase the Lorentzian line profile

$$|E(\nu)|^2 = E_0^2 \frac{(D/2)^2}{(\nu - \nu_0)^2 + (D/2)^2} , \quad \text{with} \quad E_0 = E(\nu_0) , \tag{5.96}$$

with the center frequency ν_0, which may be compared with the Lorentzian line profile of a classical oscillator broadened by phase-perturbing collisions.

The full halfwidth $\Delta\nu = D$ of this intensity profile $I(\nu) \propto |E(\nu)|^2$ decreases with increasing output power because the contributions of the spontaneously initiated photon avalanches to the total amplitude and phase become less and less significant with increasing total amplitude.

Furthermore, the halfwidth $\Delta\nu_c$ of the resonator resonance must influence the laser linewidth, because it determines the spectral interval where the gain exceeds the losses. The smaller the value of $\Delta\nu_c$, the smaller is the fraction of spontaneously emitted photons (which are emitted within the full Doppler width) with frequencies within the interval $\Delta\nu_c$ that find enough gain to build up a photon avalanche. When all these factors are taken into account, one obtains for the theoretical lower limit $\Delta\nu_L = D$ for the laser linewidth the relation [397]

$$\Delta\nu_L = \frac{\pi h\nu_L(\Delta\nu_c)^2(N_{sp} + N_{th} + 1)}{2P_L} , \tag{5.97}$$

where N_{sp} is the number of photons spontaneously emitted per second into the oscillating laser mode, N_{th} is the number of photons in this mode due to the thermal radiation field, and P_L is the laser output power. At room temperature in the visible region, $N_{th} \ll 1$ (Fig. 2.7). With $N_{sp} = 1$ (at least one spontaneous photon starts the induced photon avalanche), we obtain from (5.97) the famous Schwalow–Townes relation [397]

$$\Delta\nu_L = \frac{\pi h\nu_L \Delta\nu_c^2}{P_L} . \tag{5.98}$$

Example 5.26

a) For a HeNe laser with $\nu_L = 5 \times 10^{14}$ Hz, $\Delta\nu_c = 1$ MHz, $P = 1$ mW, we obtain $\Delta\nu_L = 1.0 \times 10^{-3}$ Hz.

b) For an argon laser with $\nu_L = 6 \times 10^{14}$ Hz, $\Delta\nu_c = 3$ MHz, $P = 1$ W, the theoretical lower limit of the linewidth is $\Delta\nu_L = 1.1 \times 10^{-5}$ Hz.

However, even for lasers with a very sophisticated stabilization system, the residual uncompensated fluctuations of nd cause frequency fluctuations that are large compared with this theoretical lower limit. With moderate efforts, laser linewidths of $\Delta\nu_L = 10^4 - 10^6$ Hz have been realized for gas and dye lasers. With very great effort, laser linewidths of a few Hertz or even below 1 Hz [372, 398] can be achieved. However, several proposals have been made how the theoretical lower limit may be approached more closely [399, 400].

This linewidth should not be confused with the attainable frequency stability, which means the stability of the center frequency of the line profile. For dye lasers, stabilities of better than 1 Hz have been achieved, which means a relative stability $\Delta\nu/\nu \leq 10^{-15}$ [372]. For gas lasers, such as the stabilized HeNe laser or specially designed solid-state lasers, even values of $\Delta\nu/\nu \leq 10^{-16}$ are possible [401, 402].

5.7 Tunable Lasers

In this section we discuss experimental realizations of some tunable lasers, which are of particular relevance for spectroscopic applications. A variety of tuning methods have been developed for different spectral regions, which will be illustrated by several examples. While semiconductor lasers, color-center lasers, and vibronic solid-state lasers are the most widely used tunable *infrared* lasers to date, the dye laser in its various modifications and the titanium:sapphire laser are still by far the most important tunable lasers in the *visible* region. Great progress has recently been made in the development of new types of ultraviolet lasers as well as in the generation of coherent UV radiation by frequency-doubling or frequency-mixing techniques (Chap. 6). In particular, great experimental progress in optical parametric oscillators has been made; they are discussed in Sect. 6.7 in more detail. Meanwhile, the whole spectral range from the far infrared to the vacuum ultraviolet can be covered by a variety of tunable coherent sources. Of great importance for basic research on highly ionized atoms and for a variety of applications is the development of X-ray lasers, which is briefly discussed in Sect. 5.7.7.

This section can give only a brief survey of those tunable devices that have proved to be of particular importance for spectroscopic applications. For a more detailed discussion of the different techniques, the reader is referred to the literature cited in the corresponding subsections. A review of tunable lasers that covers the development up to 1974 has been given in [404], while more recent compilations can be found in [386, 405]. For a survey on infrared spectroscopy with tunable lasers see [406–408].

5.7.1 Basic Concepts

Tunable coherent light sources can be realized in different ways. One possibility, which has already been discussed in Sect. 5.5, relies on lasers with a *broad gain profile*. Wavelength-selecting elements inside the laser resonator restrict laser oscillation to a narrow spectral interval, and the laser wavelength may be continuously tuned across the gain profile by varying the transmission maxima of these elements. Dye lasers, color-center lasers, and excimer lasers are examples of this type of tunable device.

Another possibility of wavelength tuning is based on the shift of energy levels in the active medium by external perturbations, which cause a corresponding spectral shift of the gain profile and therefore of the laser wavelength. This level shift may be effected by an external magnetic field (spin-flip Raman laser and Zeeman-tuned gas laser) or by temperature or pressure changes (semiconductor laser).

A third possibility for generating coherent radiation with tunable wavelength uses the principle of optical frequency mixing, which is discussed in Chap. 6.

The experimental realization of these tunable coherent light sources is, of course, determined by the spectral range for which they are to be used. For the particular spectroscopic problem, one has to decide which of the possibilities summarized above represents the optimum choice. The experimental expenditure depends substantially on the desired tuning range, on the achievable output power, and, last but not least, on the realized spectral bandwidth $\Delta\nu$. Coherent light sources with bandwidths $\Delta\nu \simeq 1\,\text{MHz}$ to $30\,\text{GHz}$ (3×10^{-5}–$1\,\text{cm}^{-1}$), which can be continuously tuned over a larger range, are already commercially available. In the visible

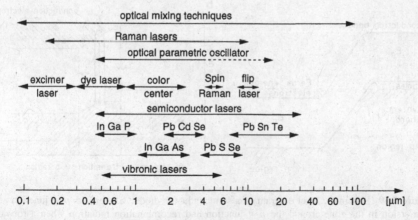

Figure 5.68 Spectral ranges of different tunable coherent sources

region, single-mode dye lasers are offered with a bandwidth down to about 1 MHz. These lasers are continuously tunable over a restricted tuning range of about 30 GHz ($1 \, cm^{-1}$). Computer control of the tuning elements allows a successive continuation of such ranges. In principle, "continuous" scanning of a single-mode laser over the whole gain profile of the laser medium, using automatic resetting of all tuning elements at definite points of a scan, is now possible. Examples are single-mode semiconductor lasers, dye lasers, or vibronic solid-state lasers.

We briefly discuss the most important tunable coherent sources, arranged according to their spectral region. Figure 5.68 illustrates the spectral ranges covered by the different devices.

5.7.2   Semiconductor-Diode Lasers

Many of the most widely used tunable coherent infrared sources use various semiconductor materials, either directly as the active laser medium (semiconductor lasers) or as the nonlinear mixing device (frequency-difference generation).

The basic principle of semiconductor lasers [409–413] may be summarized as follows. When an electric current is sent in the forward direction through a p–n semiconductor diode, the electrons and holes can recombine within the p–n junction and may emit the recombination energy in the form of electromagnetic radiation (Fig. 5.69). The linewidth of this spontaneous emission amounts to several cm^{-1}, and the wavelength is determined by the energy difference between the energy levels of electrons and holes, which is essentially determined by the band gap. The spectral range of spontaneous emission can therefore be varied within wide limits (about 0.4–40 μm) by the proper selection of the semiconductor material and its composition in binary compounds (Fig. 5.70).

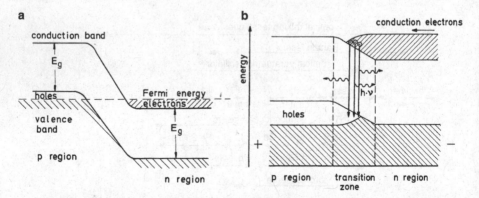

Figure 5.69 Schematic level diagram of a semiconductor diode: **a** unbiased p–n junction and **b** inversion in the zone around the p–n junction and recombination radiation when a forward voltage is applied

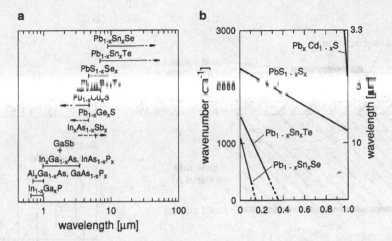

Figure 5.70   a Spectral ranges of laser emission for different semiconductor materials [411]; **b** dependence of the emission wave number on the composition x of $Pb_{1-x}Sn_x$Te, Se, or S–lead-salt lasers (courtesy of Spectra-Physics)

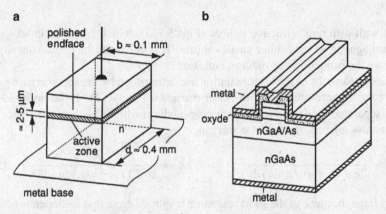

Figure 5.71   Schematic diagram of a diode laser: **a** geometrical structure; **b** concentration of the injection current in order to reach high current densities in the inversion zone

Above a certain threshold current, determined by the particular semiconductor diode, the radiation field in the junction becomes sufficiently intense to make the induced-emission rate exceed the spontaneous or radiationless recombination processes. The radiation can be amplified by multiple reflections from the plane end faces of the semiconducting medium and may become strong enough that induced emission occurs in the p–n junction before other relaxation processes deactivate the population inversion (Fig. 5.71a).

In order to increase the density of the electric current, one of the electrodes is formed as a small stripe (Fig. 5.71b). Continuous laser operation at room temperature has become possible with heterostructure lasers (Fig. 5.72), where both the electric current and the radiation are spatially confined by utilising a stack of thin

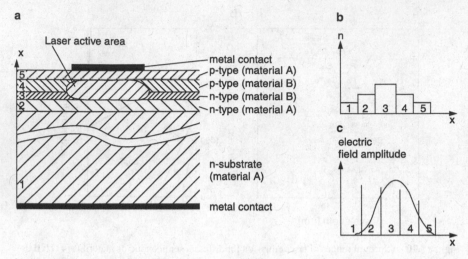

Figure 5.72 Heterostructure diode lasers. **a** Composition of p- and n-doped material with metal contacts; **b** refractive index profile; **c** laser field amplitude in the different layers

layers with different refractive indices (Fig. 5.72a), which cause an index-guided electromagnetic wave within a small volume. This enhances the photon density and therefore the probability of induced emission.

The wavelengths of the laser radiation are determined by the spectral gain profile and by the eigenresonances of the laser resonator (Sect. 5.3). If the polished end-faces (separated by d) of a semiconducting medium with refractive index n are used as resonator mirrors, the free spectral range

$$\delta\nu = \frac{c}{2nd\left(1 + (\nu/n)\mathrm{d}n/\mathrm{d}\nu\right)}, \quad \text{or} \quad \delta\lambda = \frac{\lambda^2}{2nd\left(1 - (\lambda/n)\mathrm{d}n/\mathrm{d}\lambda\right)}, \quad (5.99)$$

is very large, because of the short resonator length d. Note that $\delta\nu$ depends not only on d but also on the dispersion $\mathrm{d}n/\mathrm{d}\nu$ of the active medium.

Example 5.27
With $d = 0.5\,\mathrm{mm}$, $n = 2.5$ and $(\nu/n)\mathrm{d}n/\mathrm{d}\nu = 1.5$, the free spectral range is $\delta\nu = 48\,\mathrm{GHz} \stackrel{\wedge}{=} 1.6\,\mathrm{cm}^{-1}$, or $\delta\lambda = 0.16\,\mathrm{nm}$ at $\lambda = 1\,\mu\mathrm{m}$.

This illustrates that only a few axial resonator modes fit within the gain profile, which has a spectral width of several cm^{-1} (Fig. 5.73a).

For wavelength tuning, all those parameters that determine the energy gap between the upper and lower laser levels may be varied. A temperature change produced by an external cooling system or by a current change is most frequently utilized to generate a wavelength shift (Fig. 5.73b). Sometimes an external mag-

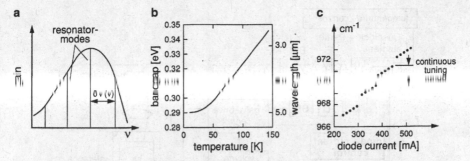

Figure 5.73 a Axial resonator modes within the spectral gain profile; **b** temperature tuning of the gain maximum; and **c** mode hops of a quasi-continuously tunable cw PbS$_n$Te diode laser in a helium cryostat. The points correspond to the transmission maxima of an external Ge etalon with a free spectral range of 1.955 GHz [408]

netic field or a mechanical pressure applied to the semiconductor is also employed for wavelength tuning. In general, however, no truly continuous tuning over the whole gain profile is possible. After a continuous tuning over about one wavenumber, mode hops occur because the resonator length is not altered synchronously with the maximum of the gain profile (Fig. 5.73c). In the case of temperature tuning this can be seen as follows:

The temperature difference ΔT changes the energy difference $E_g = E_1 - E_2$ between upper and lower levels in the conduction and valence band, and also the index of refraction by $\Delta n = (\mathrm{d}n/\mathrm{d}T)\Delta T$, and the length L of the cavity by $\Delta L = (\mathrm{d}L/\mathrm{d}T)\Delta T$.

The frequency $\nu_c = mc/(2nL)$ (m: integer) of a cavity mode is then shifted by

$$\Delta \nu_c = \frac{\partial \nu_c}{\partial n}\frac{\mathrm{d}n}{\mathrm{d}T}\Delta T + \frac{\partial \nu_c}{\partial L}\frac{\mathrm{d}L}{\mathrm{d}T}\Delta T = -\nu\left(\frac{1}{n}\frac{\mathrm{d}n}{\mathrm{d}T} + \frac{1}{L}\frac{\mathrm{d}L}{\mathrm{d}T}\right)\Delta T , \qquad (5.100)$$

while the maximum of the gain profile is shifted by

$$\Delta \nu_g = \frac{1}{h}\frac{\partial E_g}{\partial T}\Delta T . \qquad (5.101)$$

Although the first term in (5.100) is much larger than the second, the total shift $\Delta \nu_c/\Delta T$ amounts to only about 10–20 % of the shift $\Delta \nu_g/\Delta T$.

As soon as the maximum of the gain profile reaches the next resonator mode, the gain for this mode becomes larger than that of the oscillating one and the laser frequency jumps to this mode (Fig. 5.73c).

For a realization of continuous tuning over a wider range, it is therefore necessary to use external resonator mirrors with the distance d that can be independently controlled. Because of technical reasons this implies, however, a much larger distance d than the small length L of the diode and therefore a much smaller free spectral range. To achieve single-mode oscillation, additional wavelength-selecting

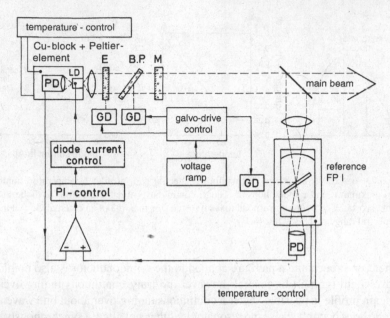

Figure 5.74 Tunable single-mode diode laser with external cavity. The etalon allows single-mode operation and the Brewster plate tunes the optical length of the cavity synchronized with etalon tilt and gain profile shift [416]

elements, such as optical reflection gratings or etalons, have to be inserted into the resonator. Furthermore, one end face of the semiconducting medium must be antireflection coated because the large reflection coefficient of the uncoated surfaces (with $n = 3.5$ the reflectivity becomes 0.3) causes large reflection losses. Such single-mode semiconductor lasers have been built [414–416].

An example is presented in Fig. 5.74. The etalon E enforces single-mode operation (see Sect. 5.4.3). The resonator length is varied by tilting a Brewster plate and the maximum of the gain profile is synchronously shifted through a change of the diode current. The laser wavelength is stabilized onto an external Fabry–Perot interferometer and can be controllably tuned by tilting a galvo-plate in this external cavity. Tuning ranges up to 100 GHz without mode hops have been achieved for a GaAlAs laser around 850 nm [416].

Another realization of tunable single-mode diode lasers uses a Littrow grating, which couples part of the laser output back into the gain medium (Fig. 5.75) [417]. When the grating with a groove spacing d_g is tilted by an angle $\Delta\alpha$, the wavelength shift is according to (4.21a)

$$\Delta\lambda = (2d_\mathrm{g})\cos\alpha \cdot \Delta\alpha . \tag{5.102}$$

Tilting of the grating is realized by mounting the grating on a lever of length L. If the tilting axis A in Fig. 5.75 is chosen correctly, the change $\Delta d_\mathrm{c} = L \cdot \cos\alpha \cdot \Delta\alpha$ of the cavity length d_c results in the same wavelength change $\Delta\lambda = (\Delta d_\mathrm{c}/d_\mathrm{c})\lambda$ of the

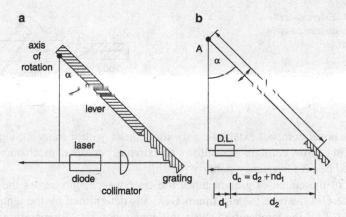

Figure 5.75 Continuously tunable diode laser with Littrow grating: **a** experimental setup, and **b** geometric condition for the location of the tilting axis for the grating. The rotation around point R_1 compensates only in first order, around R_2 in second order [418]

cavity modes, as given by (5.102). This gives the condition $d_c/L = \sin\alpha$, which shows that the tilting axis should be located at the crossing of the plane through the grating surface and the plane indicated by the dashed line that intersects the resonator axis at a distance $d_c = d_1 + n \cdot d_2$ from the grating, where n is the refractive index of the diode (Fig. 5.75b).

An improved version with a fixed Littman grating configuration and a tiltable end mirror (Fig. 5.76) allows a wider tuning range up to $500\,\mathrm{GHz}$, which is only limited by the maximum expansion of the piezo used for tilting the mirror lever

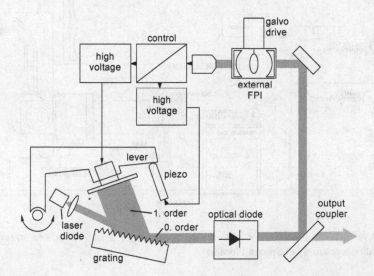

Figure 5.76 External-cavity widely tunable single-mode diode laser with Littman resonator

Figure 5.77 External-cavity
diode laser with transmission
Littrow grating

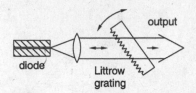

[418a]. A novel compact external-cavity diode laser with a transmission grating
(Fig. 5.77 in Littrow configuration allows an extremely compact mechanical design
with a good passive frequency stability [418b].

Tilting of the etalon or grating tunes the laser wavelength across the spectral
gain profile $G(\lambda)$, where the maximum $G(\lambda_m)$ is determined by the temperature.
A change ΔT of the temperature shifts this maximum λ_m. Temperature changes
are used for coarse tuning, whereas the mechanical tilting allows fine-tuning of the
single-mode laser.

A complete commercial diode laser spectrometer for convenient use in infrared
spectroscopy is depicted in Fig. 5.78.

Meanwhile tunable diode lasers in the visible region down to below 0.4 µm are
available [419].

Besides their applications as tunable light sources, diode lasers are more and
more used as pump lasers for tunable solid-state lasers and optical parametric amplifiers. Monolithic diode laser arrays can now deliver up to 100 W cw pump powers
[420].

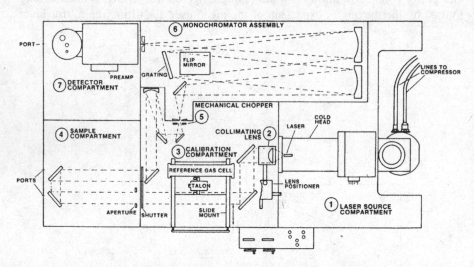

Figure 5.78 Schematic diagram of a diode laser spectrometer tunable from 3 to 200 µm with
different diodes (courtesy of Spectra-Physics)

5.7.3 Tunable Solid-State Lasers

The absorption and emission spectra of crystalline or amorphous solids can be varied within wide spectral ranges by doping them with atomic or molecular ions [411, 412]. The strong interaction of these ions with the host lattice causes broadenings and shifts of the ionic energy levels. The absorption spectrum shown in Fig. 5.79b for the example of alexandrite depends on the polarization direction of the pump light. Optical pumping of excited states generally leads to many overlapping fluorescence bands terminating on many higher "vibronic levels" in the electronic ground state, which rapidly relax by ion–phonon interaction back into the original ground state (Fig. 5.79a). These lasers are therefore often called *vibronic lasers*. If the fluorescence bands overlap sufficiently, the laser wavelength can be continously tuned over the corresponding spectral gain profile (Fig. 5.79c).

Vibronic solid-state laser materials are, e.g., alexandrite ($BeAl_2O_4$ with Cr^{3+} ions) titanium–sapphire ($Al_2O_3:Ti^+$) fluoride crystals doped with transition metal ions (e.g., $MgF_2:Co^{++}$ or $CsCaF_3:V^{2+}$) [405, 422–425].

The tuning range of vibronic solid-state lasers can be widely varied by a proper choice of the implanted ions and by selecting different hosts. This is illustrated in Fig. 5.80a, which shows the spectral ranges of laser-excited fluorescence of the same Cr^{3+} ion in different host materials [424] while Fig. 5.80b shows the tuning ranges of laser materials where different metal ions are doped in a MgF_2 crystal.

Table 5.3 compiles the operational modes and tuning ranges of different tunable vibronic lasers. A particularly efficient cw vibronic laser is the emerald laser

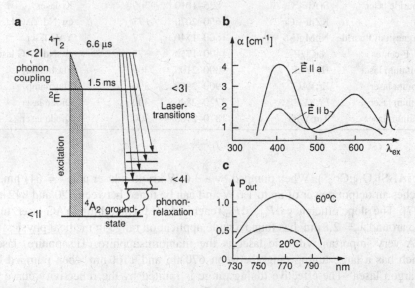

Figure 5.79 **a** Level scheme of a tunable "four-level solid-state vibronic laser"; **b** absorption spectrum for two different polarization directions of the pump laser; **c** output power $P_{out}(\lambda)$ for the example of the alexandrite laser

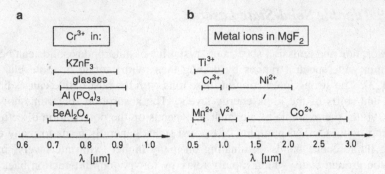

Figure 5.80 Spectral ranges of fluorescence for Cr^{3+} ions in different host materials (**a**) and different metal ions in MgF_2 (**b**)

Table 5.3 Characteristic data of some tunable solid-state lasers

Laser	Composition	Tuning range [nm]	Operation temperature [K]	Pump
Ti:sapphire	Al_2O_3:Ti^{3+}	670–1100	300	Ar laser
Alexandrite	$BeAl_2O_4$:Cr^{3+}	710–820	300–600	Flashlamp
		720–842	300	Kr laser
Emerald	$Be_3Al_2(SiO_3)_6$:Cr^{3+}	660–842	300	Kr^+ laser
Olivine	Mg_2SiO_4:Cr^{4+}	1160–1350	300	YAG laser
Flouride laser	$SrAlF_5$:Cr^{3+}	825–1010	300	Kr laser
	$KZnF_3$:Cr^{3+}	1650–2070	77	cw Nd:YAG laser
Magnesium fluoride	Ni:MgF_2	1600–1740	77	YAG laser
F_2^+ F-center	NaCl/F_2^+	1400–1750	77	cw Nd:YAG laser
Holmium laser	Ho:YLF	2000–2100	300	Flashlamp
Erbium laser	Er:YAG	2900–2950	300	Flashlamp
Erbium laser	Er:YLF	2720–2840	300	Diode laser
Thulium laser	Tm:YAG	1870–2160	300	Diode laser

($Be_3Al_2Si_6O_{18}$:Cr^{3+}). When pumped by a 3.6-W krypton laser at $\lambda_p = 641$ nm, it reaches an output power of up to 1.6 W and can be tuned between 720 and 842 nm [427]. The slope efficiency dP_{out}/dP_{in} reaches 64 %! The erbium:YAG laser, tunable around $\lambda = 2.8$ μm, has found a wide application range in medical physics.

A very important vibronic laser is the titanium:sapphire (Ti:sapphire) laser, which has a large tuning range between 670 nm and 1100 nm when pumped by an argon laser. The effective tuning range is limited by the reflectivity curve of the resonator mirrors, and for an optimum output power over the whole spectral range three different sets of mirrors are used. For spectral ranges with $\lambda > 700$ nm, the Ti:sapphire laser is superior to the dye laser (Sect. 5.7.4) because it has higher

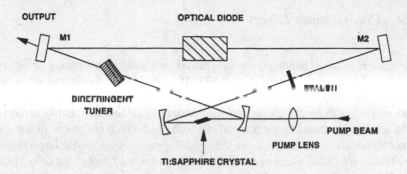

Figure 5.81 Experimental setup of a Ti:sapphire laser (courtesy of Schwartz Electro-Optics)

Figure 5.82 Tuning ranges
of some vibronic solid-state
lasers. *Black*: cw operation,
grey: pulsed operation

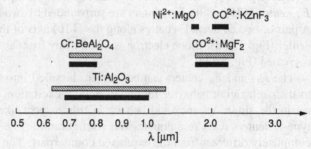

output power, better frequency stability and a smaller linewidth. The experimental
setup of a titanium-sapphire laser is depicted in Fig. 5.81.

The different vibronic solid-state lasers cover the red and near-infrared spectral
range from 0.65 to 2.5 µm (Fig. 5.82). Most of them can run at room temperature
in a pulsed mode, some of them also in cw operation.

The future importance of these lasers is derived from the fact that many of them
may be pumped by diode laser arrays. This has already been demonstrated for
Nd:YAG and alexandrite lasers, where very high total energy conversion efficiencies
were achieved. For the diode laser-pumped Nd:YAG laser, values of $\eta = 0.3$ for
the ratio of laser output power to electrical input power have been reported (30 %
plug-in efficiency) [428].

Intracavity frequency doubling of these lasers (Chap. 6) covers the visible and
near-ultraviolet range [429]. Although dye lasers are still the most important tunable lasers in the visible range, these compact and handy solid-state devices present
attractive alternatives and have started to replace dye lasers for many applications.

For more details about tunable solid-state lasers and their pumping by high-
power diode lasers, the reader is referred to [405, 430–433].

5.7.4   Color-Center Lasers

Color centers in alkali halide crystals are based on a halide ion vacancy in the crystal lattice of rock-salt structure (Fig. 5.83). If a single electron is trapped at such a vacancy, its energy levels result in new absorption lines in the visible spectrum, broadened to bands by the interaction with phonons. Since these visible absorption bands, which are caused by the trapped electrons and which are absent in the spectrum of the ideal crystal lattice, make the crystal appear colored, these imperfections in the lattice are called *F-centers* (from the German word "Farbe" for color) [434]. These F-centers have very small oscillator strengths for electronic transitions, therefore they are not suited as active laser materials.

If *one* of the six positive metal ions that immediately surround the vacancy is foreign (e.g., a Na^+ ion in a KCl crystal, Fig. 5.83b), the F-center is specified as an F_A-*center* [435], while F_B-*centers* are surrounded by *two* foreign ions (Fig. 5.83c). A pair of two adjacent F-centers along the (110) axis of the crystal is called an F_2-*center* (Fig. 5.83d). If one electron is taken away from an F_2-center, an F_2^+-*center* is created (Fig. 5.83e).

The F_A- and F_B-centers can be further classified into two categories according to their relaxation behavior following optical excitation. While centers of type I retain the single vacancy and behave in this respect like ordinary F-centers, the type-II centers relax to a double-well configuration (Fig. 5.84) with energy levels completely different from the unrelaxed counterpart. The oscillator strength for an electric-dipole transition between upper level $|k\rangle$ and lower level $|i\rangle$ in the relaxed double-well configuration is quite large. The relaxation times T_{R1} and T_{R2} for the transitions to the upper level $|k\rangle$ and from the lower level $|i\rangle$ back to the initial configuration are below 10^{-12} s. The lower level $|i\rangle$ is therefore nearly empty, which also allows sufficient inversion for cw laser operation. All these facts make the F_A- and F_B-type-II color centers – or, in shorthand, $F_A(II)$ and $F_B(II)$ – very suitable for tunable laser action [436–438].

The quantum efficiency η of $F_A(II)$-center luminescence decreases with increasing temperature. For a KCl:Li crystal, for example, η amounts to 40 % at liquid nitrogen temperatures (77 K) and approaches zero at room temperature (300 K). This implies that most color-center lasers must be operated at low temperatures,

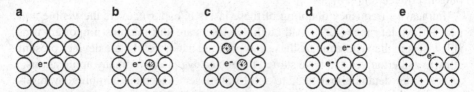

Figure 5.83   Color centers in alkali halides: **a** F-center; **b** F_A-center; **c** F_B-center; **d** F_2-center; and **e** F_2^+-center

Figure 5.84 Structural
change and level diagram of
optical pumping, relaxation,
and lasing of a $F_A(II)$-center

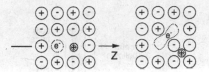

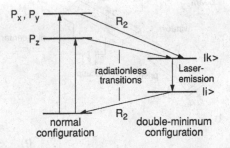

generally at 77 K. However, recently cw-operation has been observed at room temperature for diode-laser-pumped LiF:F_2-colour center lasers [438].

Two possible experimental arrangements of color-center lasers are shown
schematically in Fig. 5.85. The folded astigmatically compensated three-mirror
cavity design is identical to that of cw dye lasers of the Kogelnik type [439]
(Sect. 5.7.5). A collinear pump geometry allows optimum overlap between the
pump beam and the waist of the fundamental resonator mode in the crystal. The
mode-matching parameter (i.e., the ratio of pump-beam waist to resonator-mode
waist) can be chosen by appropriate mirror curvatures. The optical density of the
active medium, which depends on the preparation of the F_A centers [436], has to
be carefully adjusted to achieve optimum absorption of the pump wavelength. The
crystal is mounted on a cold finger cooled with liquid nitrogen in order to achieve
a high quantum efficiency η.

Coarse wavelength tuning can be accomplished by turning mirror M_3 of the
resonator with an intracavity dispersing sapphire Brewster prism. Because of the
homogeneous broadening of the gain profile, single-mode operation would be expected without any further selecting element (Sect. 5.3). This is, in fact, observed
except that neighboring spatial hole-burning modes appear, which are separated
from the main mode by

$$\Delta \nu = \frac{c}{4a} ,$$

where a is the distance between the end mirror M_1 and the crystal (Sect. 5.3). With
one Fabry–Perot etalon of 5-mm thickness and a reflectivity of 60–80 %, stable
single-mode operation without other spatial hole-burning modes can be achieved
[440]. With a careful design of the low-loss optical components inside the cavity
(made, e.g., of sapphire or of CaF_2), single-mode powers up to 75 % of the multimode output can be reached, since gain profile is homogeneous.

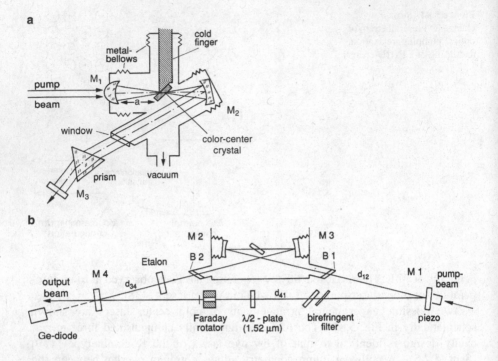

Figure 5.85 Two possible resonator designs for cw color-center lasers: **a** folded linear resonator with astigmatic compensation; and **b** ring resonator with optical diode for enforcing only one direction of the traveling laser wave and tuning elements (birefringent filter and etalon) [441]

Spatial hole burning can be avoided in the ring resonator (Fig. 5.85b). This facilitates stable single-mode operation and yields higher output powers. For example, a NaCl:OH color-center laser with a ring resonator yields 1.6 W output power at $\lambda = 1.55\,\mu m$ when pumped by 6 W of a cw YAG laser at $\lambda = 1.065\,\mu m$ [441].

When an $F_A(II)$- or F_2^+-color-center laser is pumped by a linearly polarized cw YAG laser, the output power degrades within a few minutes to a few percent of its initial value. The reason for this is as follows: many of the laser-active color centers possess a symmetry axis, for example, the (110) direction. Two-photon absorption of pump photons brings the system into an excited state of another configuration. Fluorescence releases the excited centers back into a ground state that, however, differs in its orientation from the absorbing state and therefore does not absorb the linearly polarized pump wave. This optical pumping process with changing orientation leads to a gradual bleaching of the original ground-state population, which could absorb the pump light. This orientation bleaching can be avoided when the crystal is irradiated during laser operation by the light of a mercury lamp or an argon laser, which "repumps" the centers with "wrong" orientation back into the initial ground state [437].

Figure 5.86 Spectral ranges of emission bands for different color-center crystals

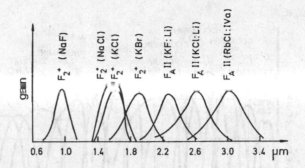

With different color-center crystals the total spectral range covered by existing color-center lasers extends from 0.65–3.4 µm. The luminescence bands of some color-center alkali halide crystals are exhibited in Fig. 5.86. Typical characteristics of some commonly used color-center lasers are compiled in Table 5.3 and are compared with some vibronic solid-state lasers. Recently room-temperature color-center lasers have been realized which are pumped by diode lasers [438].

The linewidth $\Delta\nu$ of a single-mode color-center laser is mainly determined by fluctuations of the optical path length in the cavity (Sect. 5.4). Besides the contribution $\Delta\nu_m$ caused by mechanical instabilities of the resonator, temperature fluctuations in the crystal, caused by pump power variations or by temperature variations of the cooling system, further increase the linewidth by adding contributions $\Delta\nu_p$ and $\Delta\nu_t$. Since all three contributions are independent, we obtain for the total frequency fluctuations

$$\Delta\nu = \sqrt{\Delta\nu_m^2 + \Delta\nu_p^2 + \Delta\nu_t^2}\,. \tag{5.103}$$

The linewidth of the unstabilized single-mode laser has been measured to be smaller than 260 kHz, which was the resolution limit of the measuring system [440]. An estimated value for the overall linewidth $\Delta\nu$ is 25 kHz [442]. This extremely small linewidth is ideally suited to perform high-resolution Doppler-free spectroscopy (Vol. 2, Chaps. 2–5).

More examples of color-center lasers in different spectral ranges are given in [443–445]. Good surveys on color-center lasers can be found in [437, 445] and, in particular, in [386], Vol. 2, Chap. 1. All these lasers, which provide tunable sources with narrow bandwidths, have serious competition from cw optical parametric oscillators (see Sect. 6.7), which are now available within the tuning range 0.4–4 µm.

5.7.5 Dye Lasers

Although tunable solid-state lasers and optical parametric oscillators are more and more competitive, dye lasers in their various modifications in the visible and UV

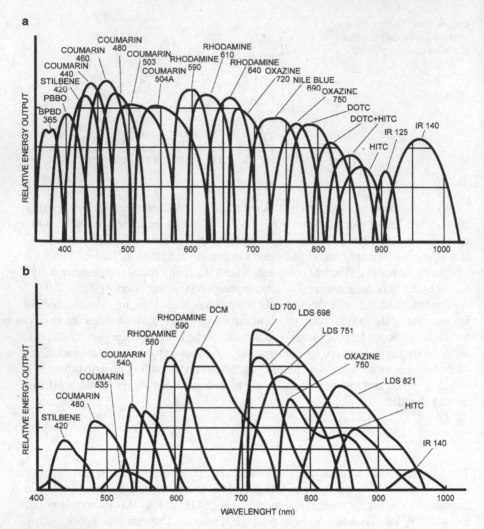

Figure 5.87 Spectral gain profiles of different laser dyes, illustrated by the output power of pulsed lasers (**a**) and cw dye lasers (**b**) (Lambda Physik and Spectra-Physics information sheets)

range are still the most widely used types of tunable lasers. Dye lasers were invented independently by P. Sorokin and F.P. Schäfer in 1966 [446]. Their active media are organic dye molecules solved in liquids. They display strong broadband fluorescence spectra under excitation by visible or UV light. With different dyes, the overall spectral range where cw or pulsed laser operation has been achieved extends from 300 nm to 1.2 μm (Fig. 5.87). Combined with frequency-doubling or mixing techniques (Chap. 6), the range of tunable devices where dye lasers are involved ranges from the VUV at 100 nm to the infrared at about 4 μm. In this section we briefly summarize the basic physical background and the most important exper-

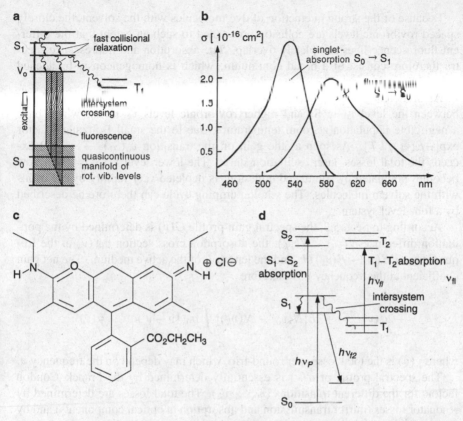

Figure 5.88 **a** Schematic energy level scheme and pumping cycle in dye molecules; **b** absorption and fluorescence spectrum of rhodamine 6G dissolved in ethanol; **c** structure of rhodamone 6G; **d** triplet absorption

imental realizations of dye lasers used in high-resolution spectroscopy. For a more extensive treatment the reader is referred to the laser literature [291, 298, 447, 448].

When dye molecules in a liquid solvent are irradiated with visible or ultra-violet light, higher vibrational levels of the first excited singlet state S_1 are populated by optical pumping from thermally populated rovibronic levels in the S_0 ground state (Fig. 5.88). Induced by collisions with solvent molecules, the excited dye molecules undergo very fast radiationless transitions into the lowest vibrational level v_0 of S_1 with relaxation times of 10^{-11} to 10^{-12} s. This level is depopulated either by spontaneous emission into the different rovibronic levels of S_0, or by radiationless transitions into a lower triplet state T_1 (intersystem crossing). Since the levels populated by optical pumping are generally above v_0 and since many fluorescence transitions terminate at higher rovibronic levels of S_0, the fluorescence spectrum of a dye molecule is redshifted against its absorption spectrum. This is shown in Fig. 5.88b for rhodamine 6G (Fig. 5.88c) the most widely used laser dye.

Because of the strong interaction of dye molecules with the solvent, the closely spaced rovibronic levels are collision broadened to such an extent that the different fluorescence lines completely overlap. The absorption and fluorescence spectra therefore consist of a broad continuum, which is homogeneously broadened (Sect. 3.3).

At sufficiently high pump intensity, population inversion may be achieved between the level v_0 in S_1 and higher rovibronic levels v_k in S_0, which have a negligible population at room temperature, due to the small Boltzmann factor $\exp[-E(v_k)/kT]$. As soon as the gain on the transition $v_0(S_1) \rightarrow v_k(S_0)$ exceeds the total losses, laser oscillation starts. The lower level $v_k(S_0)$, which now becomes populated by stimulated emission, is depleted very rapidly by collisions with the solvent molecules. The whole pumping cycle can therefore be described by a four-level system.

According to Sect. 5.2, the spectral gain profile $G(v)$ is determined by the population difference $N(v_0) - N(v_k)$, the absorption cross section $\sigma_{0k}(v)$ at the frequency $v = E(v_0) - E(v_k)/h$, and the length L of the active medium. The net gain coefficient at the frequency v is therefore

$$-2\alpha(v)L = +2L[N(v_0) - N(v_k)] \int \sigma_{0k}(v - v')\mathrm{d}v' - \gamma(v) ,$$

where $\gamma(v)$ is the total losses per round-trip, which may depend on the frequency v.

The spectral profile of $\sigma(v)$ is essentially determined by the Franck–Condon factors for the different transitions ($v_0 \rightarrow v_k$). The total losses are determined by resonator losses (mirror transmission and absorption in optical components) and by absorption losses in the active dye medium. The latter are mainly caused by two effects:

a) The intersystem crossing transitions $S_1 \rightarrow T_1$ not only diminish the population $N(v_0)$ and therefore the attainable inversion, but they also lead to an increased population $N(T_1)$ of the triplet state. The triplet absorption spectrum due to the transitions $T_1 \rightarrow T_m$ into higher triplet states T_m partly overlaps with the singlet fluorescence spectrum (Fig. 5.88d). This results in additional absorption losses $N(T_1)\alpha_T(v)L$ for the dye laser radiation. Because of the long lifetimes of molecules in this lowest triplet state, which can only relax into the S_0 ground state by slow phosphorescence or by collisional deactivation, the population density $N(T_1)$ may become undesirably large. One therefore has to take care that these triplet molecules are removed from the active zone as quickly as possible. This may be accomplished by mixing *triplet-quenching additives* to the dye solution. These are molecules that quench the triplet population effectively by spin-exchange collisions enhancing the intersystem crossing rate $T_1 \rightarrow S_0$. Examples are O_2 or cyclo-octotetraene (COT). Another solution of the triplet problem is *mechanical quenching*, used in cw dye lasers. This means that the triplet molecules are transported very rapidly through the active zone. The transit time should be much smaller than the triplet lifetime. This is achieved, e.g., by fast-flowing free jets, where the molecules pass the active zone in the focus of the pump laser in about 10^{-6} s.

b) For many dye molecules the absorption spectra $S_1 \rightarrow S_m$, corresponding to transitions from the optically pumped singlet state S_1 to still higher states S_m, partly overlap with the gain profile of the laser transition $S_1 \rightarrow S_0$. These inevitable losses often restrict the spectral range where the net gain is larger than the losses [447].

The essential characteristic of dye lasers is their broad homogeneous gain profile. Under ideal experimental conditions, homogeneous broadening allows *all* excited dye molecules to contribute to the gain at a single frequency. This implies that under single-mode operation the output power should not be much lower than the multimode power (Sect. 5.3), provided that the selecting intracavity elements do not introduce large additional losses.

The experimental realizations of dye lasers employ either flashlamps, pulsed lasers, or cw lasers as pumping sources. Recently, several experiments on pumping of dye molecules in the gas phase by high-energy electrons have been reported [449–451].

We now present the most important types of dye lasers in practical use for high-resolution spectroscopy.

a) Flashlamp-Pumped Dye Lasers

Flashlamp-pumped dye lasers [452, 453] have the advantage that they do not need expensive pump lasers. Figure 5.89 displays two commonly used pumping arrangements. The linear flashlamp, which is filled with xenon, is placed along one of the focal lines of a cylindric reflector with elliptical cross section. The liquid dye solution flowing through a glass tube in the second focal line is pumped by the focused light of the flashlamp. The useful maximum pumping time is again limited by the triplet conversion rate. By using additives as triplet quenchers, the triplet absorption is greatly reduced and long pulse emission has been obtained. Low-inductance pulsed power supplies have been designed to achieve short flashlamp pulses below 1 μs. A pulse-forming network of several capacitors is superior to the single energy storage capacitor because it matches the circuit impedance to that of the lamps, therefore a constant flashlight intensity over a period of 60–70 μs can be achieved [454]. With two linear flashlamps in a double-elliptical reflector, a reliable rhodamine 6G dye laser with 60-μs pulse duration, and a repetition rate up to 100 Hz, an *average* power of 4 W has been demonstrated. With the pumping geometry of Fig. 5.89b, which takes advantage of four linear flashlamps, a very high collection efficiency for the pump light is achieved. The light rays parallel to the plane of the figure are collected into an angle of about 85° by the rear reflector, the aplanatic lens directly in front of the flashlamp, the condenser lens, and the cylindrical mirrors. An average laser output power of 100 W is possible with this design [455].

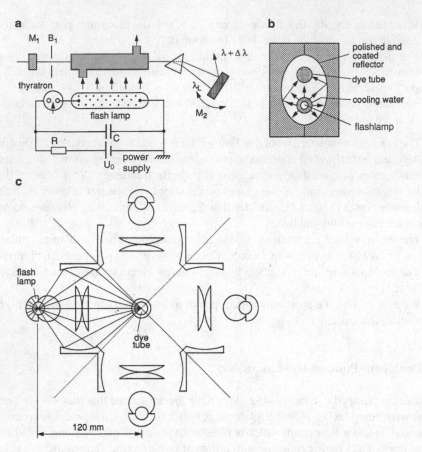

Figure 5.89 Two possible pumping designs for flashlamp-pumped dye lasers: **a** elliptical reflector geometry for pumping of a flowing dye solution by one linear xenon flashlamp; **b** side view showing the cylindrical mirror with elliptical cross-section with flashlamp and dye cell in the focal lines; **c** arrangement of four flashlamps for higher pump powers [455]

Similar to the laser-pumped dye lasers, reduction of the linewidth and wavelength tuning can be accomplished by prisms, gratings, interference filters [456], Lyot filters [457], and interferometers [458, 459].

One drawback of flashlamp-pumped dye lasers is the bad optical quality of the dye solution during the pumping process. Local variations of the refractive index due to schlieren in the flowing liquid, and temperature gradients due to the nonuniform absorption of the pump light deteriorate the optical homogeneity. The frequency jitter of narrow-band flashlamp-pumped dye lasers is therefore generally larger than the linewidth obtained in a single shot and they are mainly used in multimode operation. However, with three FPI inside the laser cavity, single-mode operation of a flashlamp-pumped dye laser has been reported [460]. The linewidth achieved was 4 MHz, stable to within 12 MHz. A better and more reliable solution for achieving single-mode operation is injection seeding. If a few milliwatts of narrow-band radiation from a single-mode cw dye laser is injected into the resonator of the flashlamp-pumped dye laser, the threshold is reached earlier for the injected wavelength than for the others. Due to the homogeneous gain profile, most of the induced emission power will then be concentrated at the injected wavelength [461].

A convenient tuning method of flashlamp-pumped dye lasers is based on intra-cavity electro-optically tunable Lyot filters (Sect. 4.2), which have the advantage that the laser wavelength can be tuned in a short time over a large spectral range [462, 463]. This is of particular importance for the spectroscopy of fast transient species, such as radicals formed in intermediate stages of chemical reactions. A single-element electro-optical birefringent filter can be used to tune a flashlamp-pumped dye laser across the entire dye emission band. With an electro-optically tunable Lyot filter (Sect. 4.2.11) in combination with a grating a spectral bandwidth of below 10^{-3} nm was achieved even without injection seeding [457].

b) Pulsed Laser-Pumped Dye Lasers

The first dye laser, developed independently by Schäfer [464] and Sorokin [465] in 1966, was pumped by a ruby laser. In the early days of dye laser development, giant-pulse ruby lasers, frequency-doubled Nd:glass lasers, and nitrogen lasers were the main pumping sources. All these lasers have sufficiently short pulse durations T_p, which are shorter than the intersystem crossing time constant $T_{IC}(S_1 \rightarrow T_1)$.

The short wavelength $\lambda = 337$ nm of the nitrogen laser permits pumping of dyes with fluorescence spectra from the near UV up to the near infrared. The high pump power available from this laser source allows sufficient inversion, even in dyes with lower quantum efficiency [466–470]. At present the most important dye laser pumps are the excimer laser [471, 472], the frequency-doubled or -tripled output of high-power Nd:YAG or Nd:glass lasers [473, 474], or copper-vapor lasers [475].

Various pumping geometries and resonator designs have been proposed or demonstrated [447]. In transverse pumping (Fig. 5.90), the pump laser beam is focused by a cylindrical lens into the dye cell. Since the absorption coefficient for the pump radiation is large, the pump beam is strongly attenuated and the maximum inversion in the dye cell is reached in a thin layer directly behind the entrance window along the focal line of the cylindrical lens. This geometrical restriction to a small gain zone gives rise to large diffraction losses and beam divergence. This divergent beam is converted by a telescope of two lenses into a parallel beam with enlarged diameter and is then reflected by a Littrow grating, which acts as wavelength selector (Hänsch-type arrangement) [467].

In longitudinal pumping schemes (Fig. 5.91), the pump beam enters the dye laser resonator at a small angle with respect to the resonator axis or collinear through one of the mirrors, which are transparent for the pump wavelength. This arrangement avoids the drawback of nonuniform pumping, present in the transverse pumping scheme. However, it needs a good beam quality of the pump laser and is there-

Figure 5.90 Hänsch-type dye laser with transverse pumping and beam expander [467]. The wavelength is tuned by turning the Littrow grating. Light with a different wavelength $\lambda_D + \Delta\lambda$ is diffracted out of the resonator

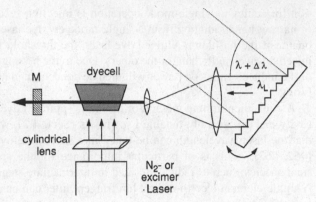

fore not suitable for excimer lasers as pump sources, but is used more and more frequently for pumping with frequency-doubled Nd:YAG lasers [473].

If wavelength selection is performed with a grating, it is preferable to expand the dye laser beam for two reasons.

a) The resolving power of a grating is proportional to the product Nm of the number N of illuminated grooves times the diffraction order m (Sect. 4.1). The more grooves that are hit by the laser beam, the better is the spectral resolution and the smaller is the resulting laser linewidth.

b) The power density without beam expansion might be high enough to damage the grating surface.

The enlargement of the beam can be accomplished either with a beam-expanding telescope (Hänsch-type laser [467, 468], Fig. 5.90) or by using grazing inci-

Figure 5.91 Possible resonator designs for longitudinal pumping of dye lasers [447]

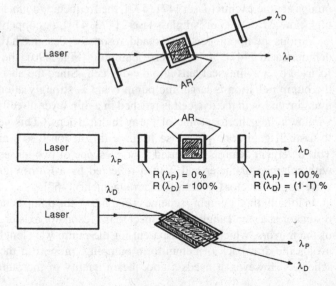

Figure 5.92 Short dye laser cavity with grazing incidence grating. Wavelength tuning is accomplished by turning the end mirror, which may also be replaced by a Littrow grating

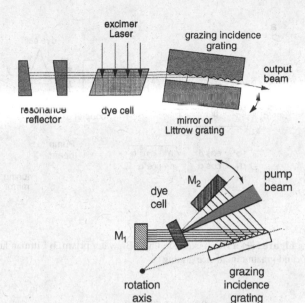

Figure 5.93 Littman laser with grazing incidence grating and Littrow grating using longitudinal pumping

dence under an angle of $\alpha \simeq 90°$ against the grating normal (Littman-type laser, Fig. 5.92). The latter arrangement [476] allows very short resonator lengths (below 10 cm). This has the advantage that even for short pump pulses, the induced dye laser photons can make several transits through the resonator during the pumping time. A further, very important advantage is the large spacing $\delta\nu = \frac{1}{2}c/d$ of the resonator modes, which allows single-mode operation with only one etalon or even without any etalon but with a fixed grating position and a turnable mirror M_2 (Fig. 5.93) [477, 478]. At the wavelength λ the first diffraction order is reflected from the grazing incidence grating ($\alpha \approx 88°$–$89°$) into the direction β determined by the grating equation (4.21)

$$\lambda = d(\sin\alpha + \sin\beta) \simeq d(1 + \sin\beta) .$$

For $d = 4 \times 10^{-5}$ cm (2500 lines/mm) and $\lambda = 400$ nm $\rightarrow \beta = 0°$, which means that the first diffraction order is reflected normal to the grating surface onto mirror M_2. With the arrangement in Fig. 5.93, a single-shot linewidth of less than 300 MHz and a time-averaged linewidth of 750 MHz have been achieved. Wavelength tuning is accomplished by tilting the mirror M_2.

For reliable single-mode operation of the Littman laser longitudinal pumping is better than transverse pumping, because the dye cell is shorter and inhomogenities of the refractive index caused by the pump process are less severe [479].

The reflectivity of the grating is very low at grazing incidence and the round-trip losses are therefore high. Using Brewster prisms for preexpansion of the laser beam (Fig. 5.94), the angle of incidence α at the grazing incidence grating can be

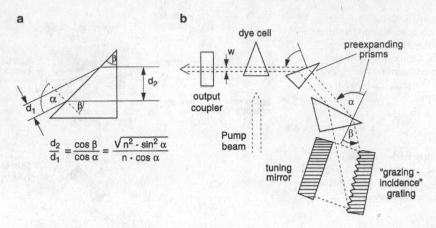

Figure 5.94 a Beam expansion by a Brewster prism; **b** Littman laser with beam-expanding prisms and grazing incidence grating

decreased from 89° to 85°–80° achieving the same total expansion factor. This reduces the reflection losses considerably [480, 481].

Example 5.28

Assume a reflectivity of $R(\alpha = 89°) = 0.05$ into the wanted first order at $\beta = 0°$. The attenuation factor per round-trip is then $(0.05)^2 \simeq 2.5 \times 10^{-3}$! The gain factor per round-trip must be larger than 4×10^2 in order to reach threshold. With preexpanding prisms and an angle $\alpha = 85°$, the reflectivity of the grating increases to $R(\alpha = 85°) = 0.25$, which yields the attenuation factor 0.06. Threshold is now reached if the gain factor exceeds 16.

In order to increase the laser power the output beam of the dye laser oscillator is sent through one or more amplifying dye cells, which are pumped by the same pump laser (Fig. 5.95).

A serious problem in all laser-pumped dye lasers is the spontaneous background, emitted from the pumped volume of the oscillator and the amplifier cells. This spontaneous emission is amplified when passing through the gain medium. It represents a perturbing, spectrally broad background of the narrow laser emission. This amplified spontaneous emission (ASE) can partly be suppressed by prisms and apertures between the different amplifying cells. An elegant solution is illustrated in Fig. 5.95. The end face of a prism expander serves as beam splitter. Part of the laser beam is refracted, expanded, and spectrally narrowed by the Littrow grating and an etalon [471] before it is sent back into the oscillator traversing the path 3–4–5–4–3. The spectral bandwidth of the oscillator is thus narrowed and only a small fraction of the ASE is coupled back into the oscillator. The partial beam 6 reflected

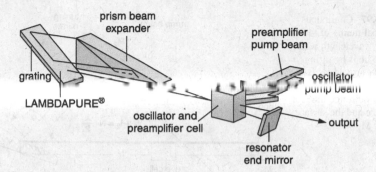

Figure 5.95 Oscillator and preamplifier of a laser-pumped dye laser with beam expander and grating. The same dye cell serves as gain medium for oscillator and amplifier [courtesy of Lambda Physik, Göttingen]

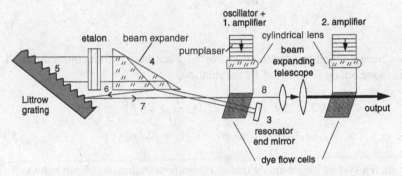

Figure 5.96 Excimer laser-pumped dye laser with oscillator and two amplifier stages. This design suppresses effectively the ASE (Lambda Physik FL 3002) (see text)

at the prism end face is sent to the same grating before it passes through another part of the first dye cell, where it is further amplified (path: 3–6–7–8). Again only a small fraction of the ASE can reach the narrow gain region along the focal lines of the cylindrical lenses used for pumping the amplifiers. The newly developed "-super pure" design shown in Fig. 5.96 further decreases the ASE by a factor of 10 compared to the former device [482].

For high-resolution spectroscopy the bandwidth of the dye laser should be as small as possible. With two etalons having different free spectral ranges, single-mode operation of the Hänsch-type laser (Fig. 5.90) can be achieved. For continuous tuning both etalons and the optical length of the laser resonator must be tuned synchronously. This can be realized with computer control (Sect. 5.4.5).

A simple mechanical solution for wavelength tuning of the dye laser in Fig. 5.92 without mode hops has been realized by Littman [478] for a short laser cavity (Fig. 5.97). If the turning axis of mirror M_2 coincides with the intersection of the two planes through mirror M_2 and the grating surface, the two conditions for the resonance wavelength (cavity length $l_1 + l_2 = N \cdot \lambda/2$ and the diffracted light must

Figure 5.97 Continuous mechanical tuning of the dye laser wavelength without mode hops by tilting mirror M_2 around an axis through the intersection of two planes through the grating surface and the surface of mirror M_2

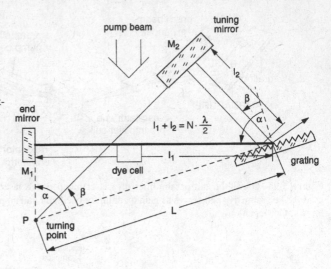

always have vertical incidence on mirror M_2) can be simultaneously fulfilled. In this case we obtain from Fig. 5.97 the relations:

$$N\lambda = 2(l_1 + l_2) = 2L(\sin\alpha + \sin\beta) , \quad \text{and}$$

$$\lambda = d(\sin\alpha + \sin\beta) \implies L = Nd/2 . \tag{5.104}$$

With such a system single-mode operation without etalons has been achieved. The wave number $\bar{\nu} = 1/\lambda$ could be tuned over a range of $100\,\text{cm}^{-1}$ without mode hops.

The spectral bandwidth of a single-mode pulsed laser with pulse duration ΔT is, in principle, limited by the Fourier limit, that is,

$$\Delta\nu = a/\Delta T , \tag{5.105}$$

where the constant $a \simeq 1$ depends on the time profile $I(t)$ of the laser pulse. This limit is, however, generally not reached because the center frequency ν_0 of the laser pulse shows a jitter from pulse to pulse, due to fluctuations and thermal instabilities. This is demonstrated by Fig. 5.98 where the spectral profile of a Littman-type single-mode pulsed laser was measured with a Fabry–Perot wavemeter for a single shot and compared with the average over 500 shots. A very stable resonator design and, in particular, temperature stabilization of the dye liquid, which is heated by absorption of the pump laser, decreases both the jitter and the drift of the laser wavelength.

A more reliable technique for achieving really Fourier-limited pulses is based on the amplification of a cw single-mode laser in several pulsed amplifier cells. The expenditure for this setup is, however, much larger because one needs a cw dye laser with a cw pump laser and a pulsed pump laser for the amplifier cells. Since

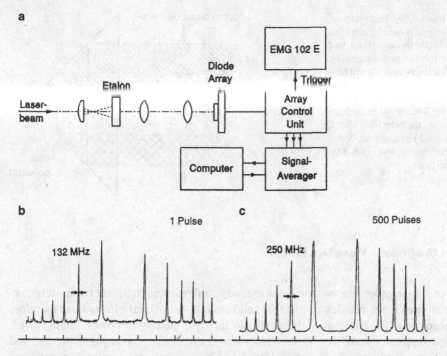

Figure 5.98 Linewidth of a single-mode pulsed laser measured with a Fabry–Perot wavemeter: **a** experimental setup; **b** single shot; and **c** signal averaged over 500 pulses

the Fourier limit $\Delta \nu = 1/\Delta T$ decreases with increasing pulse width ΔT, copper-vapor lasers with $\Delta T = 50$ ns are optimum for achieving spectrally narrow and frequency-stable pulses. A further advantage of copper-vapor lasers is their high repetition frequency up to $f = 20$ kHz.

In order to maintain the good beam quality of the cw dye laser during its amplification by transversely pumped amplifier cells, the spatial distribution of the inversion density in these cells should be as uniform as possible. Special designs (Fig. 5.99) of prismatic cells, where the pump beam traverses the dye several times after being reflected from the prism end faces, considerably improves the quality of the amplified laser beam profile.

Example 5.29
When the output of a stable cw dye laser ($\Delta \nu \simeq 1$ MHz) is amplified in three amplifier cells, pumped by a copper-vapor laser with a Gaussian time profile $I(t)$ with the halfwidth Δt, Fourier-limited pulses with $\Delta \nu \simeq 40$ MHz and peak powers of 500 kW can be generated. These pulses are wavelength tunable with the wavelength of the cw dye laser.

Figure 5.99 Transversely pumped prismatic amplifier cell (Berthune cell) for more uniform isotropic pumping. The laser beam should have a diameter about four times larger than the bore for the dye. The partial beam 1 traverses the bore from above, beam 2 from behind, beam 4 from below, and beam 3 from the front

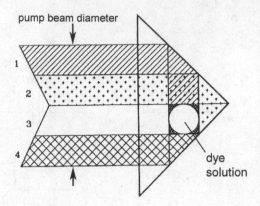

c) Continuous-Wave Dye Lasers

For sub-Doppler spectroscopy, single-mode cw dye lasers represent the most important laser types besides cw tunable solid-state lasers. Great efforts have therefore been undertaken in many laboratories to increase the output power, tuning range, and frequency stability. Various resonator configurations, pump geometries, and designs of the dye flow system have been successfully tried to realize optimum dye-laser performance. In this section we can only present some examples of the numerous arrangements used in high-resolution spectroscopy.

Figure 5.100 illustrates three possible resonator configurations. The pump beam from an argon or krypton laser enters the resonator either collinearly through the semitransparent mirror M_1 and is focused by L_1 into the dye (Fig. 5.100a), or the pump beam and dye laser beam are separated by a prism (Fig. 5.100b). In both arrangements the dye laser wavelength can be tuned by tilting the flat end mirror M_2. In another commonly used arrangement (Fig. 5.100c), the pump beam is focused by the spherical mirror M_p into the dye jet and crosses the dye medium under a small angle against the resonator axis.

In all these configurations the active zone consists of the focal spot of the pump laser within the dye solution streaming in a laminar free jet of about 0.5–1-mm thickness, which is formed through a carefully designed polished nozzle. At flow velocities of 10 m/s the time of flight for the dye molecules through the focus of the pump laser (about 10 μm) is about 10^{-6} s. During this short period the intersystem crossing rate cannot build up a large triplet concentration, and the triplet losses are therefore small.

For free-running dye jets the viscosity of the liquid solvent must be sufficiently large to ensure the laminar flow necessary for high optical quality of the gain zone. Most jet-stream dye lasers use ethylene glycol or propylene glycol as solvents. Since these alcohols decrease the quantum efficiency of several dyes and also do not have optimum thermal properties, the use of water-based dye solutions with viscosity-raising additives can improve the power efficiency and frequency stability

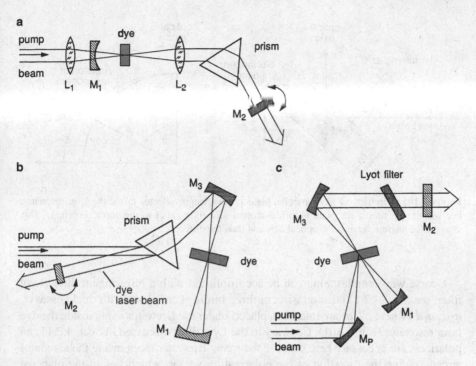

Figure 5.100 Three possible standing-wave resonator configurations used for cw dye lasers: **a** collinear pumping geometry; **b** folded astigmatically compensated resonator of the Kogelnik type [439] with a Brewster prism for separation of pump beam and dye-laser beam; and **c** the pump beam is focused by an extra pump mirror into the dye jet and is tilted against the resonator axis

of jet-stream cw dye lasers [483]. Output powers of more than 30 W have been reported for cw dye lasers [484].

In order to achieve a symmetric beam waist profile of the dye laser mode in the active medium, the astigmatism produced by the spherical folding mirror M_3 in the folded cavity design has to be compensated by the plane-parallel liquid slab of the dye jet, which is tilted under the Brewster angle against the resonator axis [439]. The folding angle for optimum compensation depends on the optical thickness of the jet and on the curvature of the folding mirror.

The threshold pump power depends on the size of the pump focus and on the resonator losses, and varies between 1 mW and several watts. The size of the pump focus should be adapted to the beam waist in the dye laser resonator (mode matching). If it is too small, less dye molecules are pumped and the maximum output power is smaller. If it is too large, the inversion for transverse modes exceeds threshold and the dye laser oscillates on several transverse modes. Under optimum conditions, pump efficiencies (dye laser output/pump power input) up to $\eta = 35\,\%$ have been achieved, yielding dye output powers of 2.8 W for only 8 W pump power.

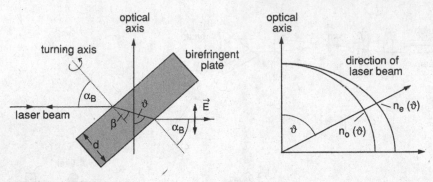

Figure 5.101 Birefringent plane-parallel plate as wavelength selector inside the laser resonator. For wavelength tuning the plate is turned around an axis parallel to the surface normal. This changes the angle ϑ against the optical axis and thus the difference $n_e(\vartheta) - n_o(\vartheta)$

Coarse wavelength tuning can be accomplished with a birefringent filter (Lyot filter, see Sect. 4.2.11) that consists of three birefringent plates with thicknesses d, $q_1 d$, $q_2 d$ (where q_1, q_2 are integers), placed under the Brewster angle inside the dye laser resonator (Fig. 5.101). Contrary to the Lyot filter discussed in Sect. 4.2.11, no polarizers are necessary here because the many Brewster faces inside the resonator already define the direction of the polarization vector, which lies in the plane of Fig. 5.101.

When the beam passes through the birefringent plate with thickness d under the angle β against the plate-normal, a phase difference $\Delta\varphi = (2\pi/\lambda) \cdot (n_e - n_o) \Delta s$ with $\Delta s = d/\cos\beta$ develops between the ordinary and the extraordinary waves. Only those wavelengths λ_m can reach oscillation threshold for which this phase difference is $2m\pi$ ($m = 1, 2, 3, \ldots$). In this case, the plane of polarization of the incident wave has been turned by $m\pi$ and the transmitted wave is again linearly polarized in the same direction as the incident wave. For all other wavelengths the transmitted wave is elliptically polarized and suffers reflection losses at the Brewster end faces. The transmission curve $T(\lambda)$ of a three-stage birefringent filter is depicted in Fig. 5.102 for a fixed angle ϑ. The laser will oscillate on the transmission maximum that is closest to the gain maximum of the dye medium [485, 486]. Turning the Lyot filter around the axis in Fig. 5.101 will shift all these maxima.

For single-mode operation additional wavelength-selecting elements have to be inserted into the resonator (Sect. 5.4.3). In most designs two FPI etalons with different free spectral ranges are employed [487, 488]. Continuous tuning of the single-mode laser demands synchronous control of the cavity length and the transmission maxima of all selecting elements (Sect. 5.5). Figures 5.103a and b show two commercial versions of a single-mode cw dye laser. The optical path length of the cavity can be conveniently tuned by turning a tilted plane-parallel glass plate inside the resonator (galvo-plate). If the tilting range is restricted to a small interval around the Brewster angle, the reflection losses remain negligible (see Sect. 5.5.1).

Figure 5.102 Transmission
$T(\lambda)$ of a birefringent filter
with three Brewster plates
of KDP, with plate thickness
$d_1 = 0.34\,\text{mm}$, $d_2 = 4d_1$,
$d_3 = 16d_1$ [485]

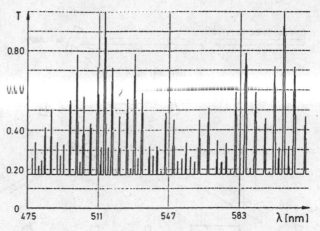

The scanning etalon can be realized by the piezo-tuned prism FPI etalon in
Fig. 5.44 with a free spectral range of about 10 GHz. It can be locked to the oscillating cavity eigenfrequency by a servo loop: if the transmission maximum ν_T of
the FPI is slightly modulated by an ac voltage fed to the piezoelement, the laser intensity will show this modulation with a phase depending on the difference $\nu_c - \nu_T$
between the cavity resonance ν_c and the transmission peak ν_T. This phase-sensitive
error signal can be used to keep the difference $\nu_c - \nu_T$ always zero. If only the prism
FPI is tuned synchronously with the cavity length, tuning ranges of about 30 GHz
($\cong 1\,\text{cm}^{-1}$) can be covered without mode hops. For larger tuning ranges the second thin etalon and the Lyot filter must also be tuned synchronously. This demands
a more sophisticated servo system, which can, however, be provided by computer
control.

A disadvantage of cw dye lasers with standing-wave cavities is spatial hole burning (Sect. 5.3.3), which impedes single-mode operation and prevents all of the
molecules within the pump region from contributing to laser emission. This effect can be avoided in ring resonators, where the laser wave propagates in only one
direction (Sect. 5.2.7). Ring lasers therefore show, in principle, higher output powers and more stable single-mode operation [489]. However, their design and their
alignment are more critical than for standing-wave resonators.

In order to avoid laser waves propagating in both directions through the ring
resonator, losses must be higher for one direction than for the other. This can be
achieved with an optical diode [322]. This diode essentially consists of a birefringent crystal and a Faraday rotator (Fig. 5.20), which turns the bifringent rotation
back to the input polarization for the wave incident in one direction but increases
the rotation for the other direction.

The specific characteristics of a cw ring dye laser regarding output power and
linewidth have been studied in [489]. A theoretical treatment of mode selection in
Fabry–Perot-type and in ring resonators can be found in [490]. Because of the many
optical elements in the ring resonator, the losses are generally slightly higher than

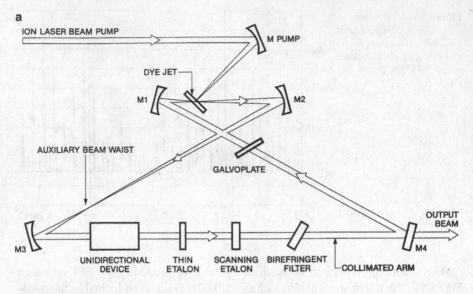

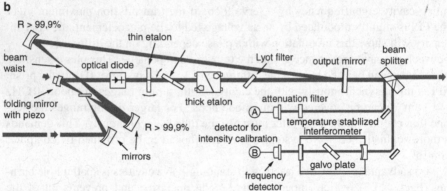

Figure 5.103 a Commercial version of a single-mode cw ring dye laser (Spectra-Physics), **b** single-mode tunable cw ring dye laser (Coherent model CR699-21)

in standing-wave resonators. This causes a higher threshold. Since more molecules contribute to the gain, the slope efficiency $\eta_{al} = dP_{out}/dP_{in}$ is, however, higher. At higher input powers well above threshold, the output power of ring lasers is therefore higher (Fig. 5.104).

The characteristic data of different dye laser types are compiled in Table 5.4 for "typical" operation conditions in order to give a survey on typical orders of magnitude for these figures. The tuning ranges depend not only on the dyes but also on the pump lasers. They are slightly different for pulsed lasers pumped by excimer lasers from that of cw lasers pumped by argon or krypton lasers. Meanwhile, frequency-doubled Nd:YAG lasers are used more and more frequently as pump sources for

Figure 5.104 Comparison of output powers of ring lasers (*full circles* and *squares*) and standing wave lasers (*open circles* and *crosses*) for two different laser dyes

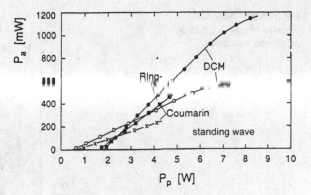

Table 5.4 Characteristic parameters of some dye lasers pumped by different sources

Pump	Tuning range [nm]	Pulse width [ns]	Peak power [W]	Pulse energy [mJ]	Repetition rate [s⁻¹]	Average output [W]
Excimer laser	370–985	10–200	$\leq 10^7$	≤ 300	20–200	0.1–30
N_2 laser	370–1020	1–10	$< 10^5$	< 1	$< 10^3$	0.01–0.1
Flashlamp	300–800	300–10^4	10^2–10^5	< 5000	1–100	0.1–200
Ar^+ laser	350–900	cw	cw	–	cw	0.1–10
Kr^+ laser	400–1100	cw	cw	–		0.1–5
Nd:YAG laser	400–920	10–20	10^5–10^7	10–100	10–30	0.1–5
$\lambda/2$: 530 nm						
$\lambda/3$: 355 nm						
Copper-vapor laser	530–890	30–50	$\simeq 10^4$–5	≈ 1	$\simeq 10^4$	≤ 10

dye lasers. Many data on dye laser wavelengths, tuning ranges and possible pump lasers can be found in [296].

5.7.6 Excimer Lasers

Excimers (that is, excited dimers) are molecules that are bound in excited states but are unstable in their electronic ground states. Examples are diatomic molecules composed of closed-shell atoms with $^1S_0$ ground states, such as the rare gases, which form stable *excited dimers* He_2^*, Ar_2^*, etc., but have a mainly repulsive potential in the ground state with a very shallow van der Waals minimum (Fig. 5.105). The well depth ϵ of this minimum is small compared to the thermal energy kT at room temperature, which prevents the stable formation of ground-state molecules. Mixed excimers such as KF or XeNa can be formed from combinations of closed-shell/open-shell atoms (for example, combination of atomic states $^1S+^2S$, $^1S+^2P$, $^1S + ^3P$, etc.), which lead to repulsive ground-state potentials [491, 492].

Figure 5.105 Schematic potential energy diagram of an exciter molecule

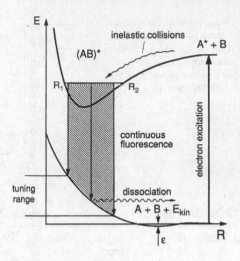

These excimers are ideal candidates for forming the active medium of tunable lasers since inversion between the pumped upper bound state and the dissociating lower state is automatically maintained because the lower state dissociates very rapidly ($\simeq 10^{-12}$–10^{-13} s) and the frequently occurring bottleneck caused by a small depletion rate of the lower laser level is prevented. The output power of excimer lasers mainly depends on the excitation rate of the *upper* state.

The tunability range depends on the slope of the repulsive potential and on the internuclear distances R_1 and R_2 of the classical turning points in the excited vibrational levels. The spectral gain profile is determined by the Franck–Condon factors for bound–free transitions. The corresponding intensity distribution $I(\omega)$ of the fluorescence from the upper vibrational levels shows a modulatory structure (see Fig. 2.21) reflecting the R dependence $|\psi_{\text{vib}}(R)|^2$ of the vibrational wave function in these levels [493].

The gain of the active medium at the frequency $\omega = (E_k - E_i)/\hbar$ is, according to (5.2), given by

$$\alpha(\omega) = [N_i - (g_i/g_k)N_k]\sigma(\omega) , \qquad (5.106)$$

where the absorption cross section $\sigma(\omega)$ is related to the spontaneous transition probability $A_{ki} = 1/\tau_k$ [491] by

$$\int_{\omega_1}^{\omega_2} \sigma(\omega)\mathrm{d}\omega = (\lambda/2)^2 A_{ki} = \frac{(\lambda/2)^2}{\tau_k} . \qquad (5.107)$$

Because of the broad spectral range $\Delta\omega = \omega_1 - \omega_2$, the cross section $\sigma(\omega)$ may be very small in spite of the large overall transition probability indicated by the short upper-state lifetime τ_k. Consequently, a high population density N_k is necessary to achieve sufficient gain. Since the pumping rate R_p has to compete with the

Table 5.5 Characteristic data of some excimer lasers. (Pulse width: 10–200 μs; repetition frequency: 1–200 s^{-1}, depending on the model; output beam divergence: 2 × 4 mrad; jitter of the pulse energy: 3–10; time jitter: 1–10 μs, depending on the model)

Laser medium	F_2	ArF	KrCl	KrF	XeCl	XeF
Wavelength [nm]	157	193	222	248	308	357
Pulse energy [mJ]	15	≤ 500	≤ 60	≤ 1000	≤ 600	500
Pulse repetition rate [Hz]	10	20	20	≤ 300	≤ 300	≤ 300

spontaneous transition rate, which is proportional to the third power of the transition frequency ω, the pumping power $R_p \hbar \omega$ at laser threshold scales at least as the fourth power of the lasing frequency. *Short-wavelength lasers therefore require high pumping powers* [494, 495].

Pumping sources are provided by high-voltage, high-current electron beam sources, such as the FEBETRON [496] or by fast transverse discharges [497]. The primary step is the excitation of atoms by electron impact. Since the excitation of the upper excimer states needs collisions between these excited atoms and ground-state atoms (remember that there are no ground-state excimer molecules), high atom densities are required to form a sufficient number N^* of excimers in the upper state. A typical gas mixture of a XeCl laser is: Xe: 40 mbar, HCl: 5 mbar, He: 2000–4000 mbar. These high pressures impede a uniform discharge along the whole active zone in the channel. Preionization by fast electrons or by ultraviolet radiation is required to achieve a large and uniform density of excimers, and specially formed electrodes are used [498]. Fast switches, such as magnetically confined thyratrons have been developed, and the inductances of the discharge circuits must be matched to the discharge time [499].

Up to now the rare-gas halide excimers, such as KrF, ArF, or XeCl, form the active medium of the most advanced UV excimer lasers. Similar to the nitrogen laser, these rare-gas halide lasers can be pumped by fast transverse discharges, and lasers of this type are the most common commercial excimer lasers (Table 5.5).

Inversion is reached by a sufficiently fast and large population increase of the upper laser level. This is achieved through a chain of different collision processes that are still not been completely understood for all excimer lasers. As an example of the complexity of these processes, some possible paths to inversion in XeCl excimer lasers, which use a mixture of Xe, HCl, and He or Ne as gas filling, are given by

$$\mathrm{Xe} + \mathrm{e}^- \begin{cases} \to \mathrm{Xe}^* + \mathrm{e}^- \;, \\ \to \mathrm{Xe}^+ + 2\mathrm{e}^- \;, \end{cases}$$
$$\mathrm{Xe}^* + \mathrm{Cl}_2 \to \mathrm{XeCl}^* + \mathrm{Cl} \;,$$
$$\mathrm{Xe}^* + \mathrm{HCl} \to \mathrm{XeCl}^* + \mathrm{H} \;,$$
$$\mathrm{Xe}^+ + \mathrm{Cl}^- + \mathrm{M} \to \mathrm{XeCl}^* + \mathrm{M} \;. \tag{5.108}$$

All these formation processes of XeCl* occur very rapidly on a time scale of 10^{-8}–10^{-9} s and have to compete with quenching processes such as

$$XeCl^* + He \rightarrow Xe + Cl + He \ ,$$

which diminish the inversion.

The pulse width of most excimer lasers lies within 5–20 ns. Recently, long-pulse XeCl lasers have been developed, which have pulse widths of $T > 300$ ns [500]. They allow amplification of single-mode cw dye lasers with Fourier-limited bandwidths of $\Delta \nu < 2$ MHz at peak powers of $P > 10$ kW. Because of the large volume of the gain medium, unstable resonators are often used to match the mode volume to the gain volume (see Sect. 5.2.6).

More details on experimental designs and on the physics of excimer lasers can be found in [492, 500–502].

5.7.7 Free-Electron Lasers

In recent years a completely novel concept of a tunable laser has been developed that does not use atoms or molecules as an active medium, but rather "free" elec-

Figure 5.106 a Schematic arrangement of a free-electron laser; **b** radiation of a dipole at rest ($v = 0$) and a moving dipole with $v \simeq c$; **c** phase-matching condition

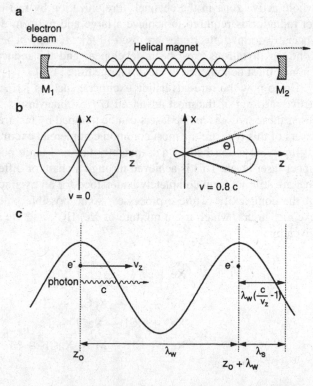

trons in a specially designed magnetic field. It converts part of the electron kinetic energy into electromagnetic radiation as stimulated synchrotron radiation. The first free-electron laser (FEL) was realized by Madey and coworkers [503]. A schematic diagram of the FEL is shown in Fig. 5.106. The high-energy relativistic electrons from an accelerator pass along a static, spatially periodic magnetic field B, which can be realized, for example, by a periodic arrangement of magnets with alternating directions of the magnetic field perpendicular to the electron beam propagation (Fig. 5.107) or by a doubly-wound helical superconducting magnet (wiggler) providing a circularly polarized B field.

The basic physics of the FEL and the process in which FEL radiation originates can be understood in a classical model, following the representation in [504]. Because of the Lorentz force, the electrons passing through the wiggler undergo periodic oscillations, resulting in the emission of radiation. For an electron oscillating in the x-direction around a point at rest, the angular distribution of such a dipole radiation is $I(\theta) = I_0 \cdot \sin^2 \theta$ (Fig. 5.106b). In contrast, for the relativistic electron with the velocity $v \simeq c$, it is sharply peaked in the forward direction (Fig. 5.106b) within a cone of solid angle $\theta \simeq (1 - v^2/c^2)^{1/2}$. For electrons of energy $E = 100\,\text{MeV}$, for instance, θ is about 2 mrad. This relativistic dipole radiation is the analog to the spontaneous emission in conventional lasers and can be used to initiate induced emission in the FEL.

The wavelength λ of the emitted light is determined by the wiggler period Λ_w and the following *phase-matching* condition: assume the oscillating electron at the position z_0 in the wiggler emits radiation of all wavelengths. However, the light moves faster than the electron (velocity v_z) in the z-direction. After one wiggler period at $z_1 = z_0 + \Lambda_\text{w}$, there will be a time lag

$$\Delta t = \Lambda_\text{w} \left(\frac{1}{v_z} - \frac{1}{c} \right) ,$$

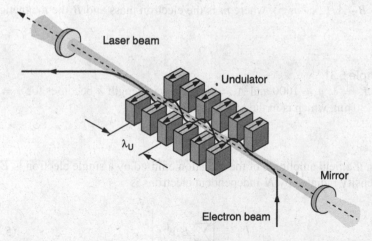

Figure 5.107 Principle of the free-electron laser [Institute of Nuclear Physics, Darmstadt]

between the electron and the light emitted at z_0. The light emitted by the electron in z_1 will therefore not be in phase with the light emitted in z_0 unless the time difference $\Delta t = n \cdot T = n \cdot \lambda/c$ is an integer multiple of the light period T. Phase matching can be therefore only be achieved for certain wavelengths

$$\lambda_n = \frac{\Delta L}{n} = \frac{\Lambda_{\mathrm{w}}}{n} \left(\frac{c}{v_z} - 1 \right) \quad (n = 1, 2, 3, \ldots). \tag{5.109}$$

Only for these wavelengths λ_n are the contributions emitted by the electron at different locations in phase and therefore interfere constructively. The lowest harmonic λ_1 ($n = 1$) of the emitted light has therefore the wavelength $\lambda_1 = \Lambda_{\mathrm{w}}(c/v_z - 1)$ and can be tuned with the velocity v_z of the electron.

Example 5.30

With $\Lambda_{\mathrm{w}} = 3\,\mathrm{cm}$, $E_{\mathrm{el}} = 10\,\mathrm{MeV} \to v_z \simeq 0.999c$, we obtain $\lambda = 40\,\mu\mathrm{m}$ for $n = 1$ and $\lambda = 13\,\mu\mathrm{m}$ for $n = 3$, which lies in the mid-infrared. For $E_{\mathrm{el}} = 100\,\mathrm{MeV} \Rightarrow v_z = (1 - 1.25 \times 10^{-5})c$ and the phase-matching wavelength has decreased to $\lambda_1 = 1.25 \times 10^{-5} \Lambda_{\mathrm{w}} = 375\,\mathrm{nm}$, which is in the UV range.

Since the electrons move with nearly the velocity of light, they have to be treated relativistically. A rigorous relativistic treatment gives instead of (5.109) the correct relation

$$\lambda_n = \frac{\Lambda_{\mathrm{w}}}{2n\gamma^2} \left(1 + K^2 \right) \tag{5.110}$$

with the relativistic factor $\gamma = (1 - v^2/c^2)^{-1/2}$ and the magnetic field parameter $K = \mathrm{e} \cdot B \cdot \Lambda/(2\pi \cdot mc^2)$, where m is the electron mass and B the magnetic field strength.

Example 5.31

With $K = 1$, $\gamma = 1000$ and $\Lambda = 3\,\mathrm{cm}$ the wavelength λ becomes for $n = 1$ $\lambda_1 = 30\,\mathrm{nm}$, which is in the soft X-ray region.

When the field amplitude of the radiation emitted by a single electron is E_j, the total intensity radiated by N independent electrons is

$$I_{\mathrm{tot}} = \left| \sum_{j=1}^{N} E_j \mathrm{e}^{\mathrm{i}\varphi_j} \right|^2, \tag{5.111a}$$

where the phases φ_j of the different contributions may be randomly distributed.

If somehow all electrons emit with the same phase, the total intensity for the case of equal amplitudes $E_j = E_0$ becomes

$$I_{\text{tot}} = \left| \sum_{j=1}^{N} E_j \right|^2 \propto |N E_0|^2 \propto N^2 I_{\text{el}} , \tag{5.111b}$$

when $I_{\text{el}} \propto E_0^2$ is the intensity emitted by a single electron. This coherent emission with equal phases therefore yields N times the intensity of the incoherent emission with random phases. It is realized in the FEL.

In order to understand how this can be achieved, we first consider a laser beam with the correct wavelength λ_{m} that passes along the axis of the wiggler. Electrons that move at the critical velocity $v_c = c \Lambda_w / (\Lambda_w + m \lambda_m)$ are in phase with the laser wave and can be induced to emit a photon that amplifies the laser wave (stimulated Compton scattering). The electron loses the emitted radiation energy and becomes slower. All electrons that are a little bit faster than v_c can lose energy by adding radiation to the laser wave without coming out of phase as long as they are not slower than v_c. On the other hand, electrons that are slower than v_c can *absorb* photons, which makes them faster until they reach the velocity v_c.

This means that the faster electrons contribute to the amplification of the incident laser wave, whereas the slower electrons attenuate it. This stimulated emission of the faster electrons and the absorption of photons by the slower electrons leads to a velocity bunching of the electrons toward the critical velocity v_c and enhances the coherent superposition of their contributions to the radiation field. The energy pumped by the electrons into the radiation field comes from their kinetic energy and has to be replaced by acceleration in RF cavities, if the same electrons in storage rings are to be used for multiple traversions through the wiggler.

This free-electron radiation amplifier can be converted into a laser by providing reflecting mirrors for optical feedback. Such FELs are now in operation at several places in the world. Their advantages are their tunability over a large spectral range from millimeter waves into the VUV region by changing the electron energy. Their potential high output power represents a further plus for FELs. Their definitive disadvantage is the large experimental expenditure that demands, besides a delicate wiggler structure, a high-energy accelerator or a storage ring.

At present FELs with output powers of several kilowatts in the infrared and several watts in the visible have been realized. The Stanford FEL reaches, for example, 130 kW at 3.4 μm, whereas from a cooperation between TRW and Stanford University, peak powers of 1.2 MW at $\lambda = 500$ mm were reported. During recent years a large FEL called *FLASH* has been build at DESY, where part of the linear accelerator TESLA is used for the FEL. At electron energies between 0.37 and 1.25 GeV the spectral range covers the soft X-ray region between 4.2 and 45 nm. The parameters of this FEL are compiled in Table 5.6.

As shown in Fig. 5.108 a beam switch for the high energy electrons allows the operation of two FELs.

Table 5.6 Relevant operation parameters of the FEL Flash at DESY, Hamburg [http://flash.desy. de/]

Parameter	Value
Wavelength range	4.2–45 nm
Average single pulse energy	10–500 μJ
Pulse duration (FWHM)	< 50–200 fs
Peak power (from av.)	1–3 GW
Average power (example from 5000 pulses/sec)	up to 600 mW
Spectral width (FWHM)	0.7–2 %
Photons per pulse	10^{11}–10^{13}
Average brilliance	10^{17}–10^{21} photons/s/mrad²/mm²/0.1 % bw
Peak brilliance	10^{29}–10^{31} photons/s/mrad²/mm²/0.1 % bw

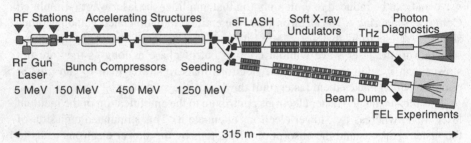

Figure 5.108 Schematic experimental setup for the FEL's *Flash 1* and *Flash 2* [http://flash2.desy. de/]

5.7.8   X-Ray Lasers

For many problems in atomic, molecular, and solid-state physics intense sources of tunable X-rays are required. Examples are inner-shell excitation of atoms and molecules or spectroscopy of multiply charged ions. Until now, these demands could only partly be met by X-ray tubes or by synchrotron radiation. The development of lasers in the spectral range below 100 nm is therefore of great interest. Besides the free electron laser, which represents the most powerful but expensive X-ray laser, there are other possibilities which can realize much less expensive table top lasers in the X-ray region. They are based on different excitation mechanisms:

a) capillary discharges: A high electric voltage capacitor is discharged through a few centimetre long capillary filled with a gas at low pressure. This gives a short ($< 10^{-6}$ s) high current electrical pulse which ionizes the gas atoms and excites the multiply charged ions A^{n+}. The recombination radiation from the process $A^{n+} + e^- \rightarrow A^{(n-1)+}$ can cover the far UV to the X-ray region. Using argon as

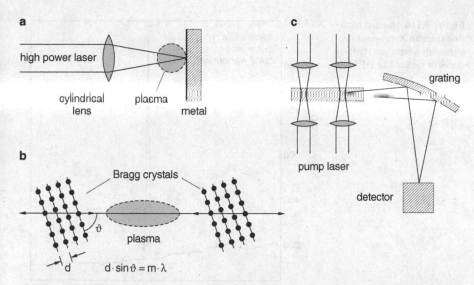

Figure 5.109 Experimental setup for realizing X-ray lasers: **a** production of high-temperature plasma; **b** X-ray resonator using Bragg reflection by crystals; **c** measurement of single-pass gain and line narrowing

the gas in the capillary radiation at around $\lambda = 47$ nm can be observed from the recombination of Ar^{8+}.

b) Plasma obtained from laser irradiation of solid surfaces. Here a high power laser pulse evaporates part of the solid material and produces a plasma which is further excited by a second laser pulse generating highly excited multiply charged ions.

c) Gas breakdown by focussed high power laser pulses. Here a high temperature plasma is formed by focussing a powerful laser pulse into a gas (gas breakdown). While the first part of the pulse ionizes the gas, where multiple charged ions are produced, the rest of the pulse energy excites these ions into high lying states.

The difficulties of experimental realization of X-ray lasers are the following:

According to (2.22), the spontaneous transition probability A_i scales with the third power ν^3 of the emitted frequency. The energy losses of the upper-state by fluorescence are therefore proportional to $A_i h\nu \propto \nu^4$! This means that high pumping powers are required to achieve inversion. Therefore only pulsed operation has a chance to be realized where ultrashort laser pulses with high peak powers are used as pumping sources. Possible candidates that can serve as active media for X-ray lasers are highly excited multiply charged ions. They can be produced in a laser-induced high-temperature plasma (Fig. 5.109) or in a capillary plasma discharge. If the pump laser beam is focused by a cylindrical lens onto the target, a high-temperature plasma is produced along the focal line. The q-fold ionized species

Figure 5.110 An emission
line from an X-ray laser on
a transition between Rydberg
states of nickel-like Pd^{18+}
[508]

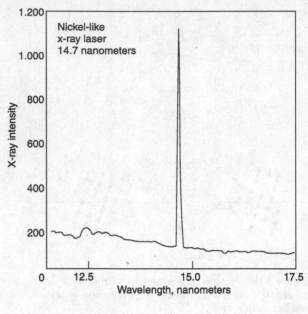

Figure 5.111 Level scheme
for inversion by ion–electron
recombination

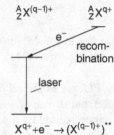

with nuclear charge $Z \cdot e$ in the plasma plume recombines with electrons to form
Rydberg states of ions with electron charge $Q_{el} = -(Z - q + 1)$. In favorable cases
these high Rydberg levels are more strongly populated than lower states of this ion
and inversion is achieved (Fig. 5.111). The conditions for achieving inversion and
thus amplification of X-ray radiation can only be maintained for very short times
(on the order of picoseconds).

An example of X-ray amplification in nickel-like palladium Pd^{18+} is shown in
Fig. 5.110, where a terawatt laser pulse created a hot plasma from a palladium
surface. By recombining electrons with highly charged palladium ions, inversion
between two Rydberg states of Pd^{18+} could be achieved, resulting in an intense
laser line at $\lambda = 14.7$ nm [508, 509].

An efficient way to generate inversion is to use double pulses [510], where the
first pulse heats and explodes a thin metal foil, producing a hot plasma. The second
pulse further ionizes the plasma, generating highly charged ions, which can recombine with electrons creating inversion between two Rydberg levels (Fig. 5.111).

Figure 5.112 Double pulses for creating and further ionizing the plasma generated by the first pulse [510]

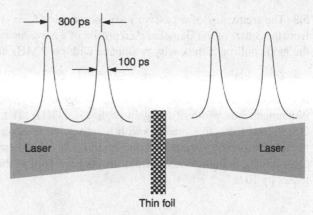

In order to improve the efficiency of X-ray lasers below 20 nm, a grazing incidence pumping scheme has been proposed that should allow inversion to be achieved with pump pulse energies of below 150 mJ [510]. A preformed plasma is first produced in a flat target by a laser pulse in order to generate the optimum gain region. Then a second short pulse (1 ps, $\lambda = 800$ nm) is released at a grazing angle to strongly heat this gain region, producing efficient on-axis X-ray lasing.

Such soft X-ray lasers have already been realized [511–514]. The shortest wavelength reported to date is 6 nm [513]. Resonators for X-ray lasers can be composed of Bragg reflectors, which consist of suitable crystals that can be tilted to fulfill the Bragg condition $2d \cdot \sin \vartheta = m \cdot \lambda$ for constructive interference between the partial waves reflected by the crystal planes with distance d (Fig. 5.109b).

Another way to realize coherent X-ray radiation is based on the generation of high harmonics of high-power femtosecond laser pulses (see Vol. 2, Chap. 13). More detailed information on this interesting subject can be found in [512–519].

The historical development of X-ray lasers can be found in [522].

5.8 Problems

5.1 Calculate the necessary threshold inversion of a gas laser transition at $\lambda = 500$ nm with the transition probability $A_{ik} = 5 \times 10^7 \, \text{s}^{-1}$ and a homogeneous linewidth $\Delta\nu_{\text{hom}} = 20$ MHz. The active length is $L = 20$ cm and the resonator losses per round-trip are 5 %.

5.2 A laser medium has a Doppler-broadened gain profile of halfwidth 2 GHz and central wavelength $\lambda = 633$ nm. The homogeneous width is 50 MHz, and the transition probability $A_{ik} = 1 \times 10^8 \, \text{s}^{-1}$. Assume that one of the resonator modes ($L = 40$ cm) coincides with the center frequency ν_0 of the gain profile. What is the threshold inversion for the central mode, and at which inversion does oscillation start on the two adjacent longitudinal modes if the resonator losses are 10 %?

5.3 The frequency of a passive resonator mode ($L = 15$ cm) lies $0.5\Delta\nu_D$ away from the center of the Gaussian gain profile of a gas laser at $\lambda = 632.8$ nm. Estimate the mode pulling if the cavity resonance width is 2 MHz and $\Delta\nu_D = 1$ GHz.

5.4 Assume a laser transition with a homogeneous width of 100 MHz, while the inhomogeneous width of the gain profile is 1 GHz. The resonator length is $d = 200$ cm and the active medium with length $L \ll d$ is placed 20 cm from one end mirror. Estimate the spacing of the spatial hole-burning modes. How many modes can oscillate simultaneously if the unsaturated gain at the line center exceeds the losses by 10 %?

5.5 Estimate the optimum transmission of the laser output mirror if the unsaturated gain per round trip is 2 and the internal resonator losses are 10 %.

5.6 The output beam from an HeNe laser with a confocal resonator ($R = L = 30$ cm) is focused by a lens of $f = 30$ cm, 50 cm away from the output mirror. Calculate the location of the focus, the Rayleigh length, and the beam waist in the focal plane.

5.7 A nearly parallel Gaussian beam with $\lambda = 500$ nm is expanded by a telescope with two lenses of focal lengths $f_1 = 1$ cm and $f_2 = 10$ cm. The spot size at the entrance lens is $w = 1$ mm. An aperture in the common focal plane of the two lenses acts as a spatial filter to improve the quality of the wavefront in the expanded beam (why?). What is the diameter of this aperture, if 95 % of the intensity is transmitted?

5.8 A HeNe laser with an unsaturated gain of $G_0(\nu_0) = 1.3$ per round trip at the center of the Gaussian gain profile with halfwidth 1.5 GHz has a resonator length of $d = 50$ cm and total losses of 4 %. Single-mode operation at ν_0 is achieved with a coated tilted etalon inside the resonator. Design the optimum combination of etalon thickness and finesse.

5.9 An argon laser oscillating at $\lambda = 488$ nm with resonator length $d = 100$ cm and two mirrors with radius $R_1 = \infty$ and $R_2 = 400$ cm has an intracavity circular aperture close to the spherical mirror to prevent oscillation on transversal modes. Estimate the maximum diameter of the aperture that introduces losses $\gamma_{\text{diffr}} < 1$ % for the TEM$_{00}$ mode, but prevents oscillation of higher transverse modes, which without the aperture have a net gain of 10 %.

5.10 A single-mode HeNe laser with resonator length $L = 15\,\text{cm}$ is tuned by moving a resonator mirror mounted on a piezo. Estimate the maximum tuning range before a mode hop will occur, assuming an unsaturated gain of $10\,\%$ at the line center and resonator losses of $3\,\%$. What voltage has to be applied to the piezo (expansion $1\,\text{nm}/\text{V}$) for this tuning range?

5.11 Estimate the frequency drift of a laser oscillating at $\lambda = 500\,\text{nm}$ because of thermal expansion of the resonator at a temperature drift of $1\,^\circ\text{C/h}$, when the resonator mirrors are mounted on distance-holder rods (a) made of invar and (b) made of fused quartz.

5.12 Mode selection in an argon laser is often accomplished with an intra-cavity etalon. What is the frequency drift of the transmission maximum

a) for a solid fused quartz etalon with thickness $d = 1\,\text{cm}$ due to a temperature change of $2\,^\circ\text{C}$?
b) For an air-space etalon with $d = 1\,\text{cm}$ due to an air pressure change of $4\,\text{mb}$?
c) Estimate the average time between two mode hopes (cavity length $L = 100\,\text{cm}$) for a temperature drift of $1\,^\circ\text{C/h}$ or a pressure drift of $2\,\text{mbar/h}$.

5.13 Assume that the output power of a laser shows random fluctuations of about $5\,\%$. Intensity stabilization is accomplished by a Pockels cell with a half-wave voltage of $600\,\text{V}$. Estimate the ac output voltage of the amplifier driving the Pockels cell that is necessary to stabilize the transmitted intensity if the Pockels cell is operated around the maximum slope of the transmission curve.

5.14 A single-mode laser is frequency stabilized onto the slope of the transmission maximum of an external reference Fabry–Perot interferometer made of invar with a free spectral range of $8\,\text{GHz}$. Estimate the frequency stability of the laser

a) against temperature drifts, if the FPI is temperature stabilized within $0.01\,^\circ\text{C}$,
b) against acoustic vibrations of the mirror distance d in the FPI with amplitudes of $1\,\text{nm}$.
c) Assume that the *intensity* fluctuations are compensated to $1\,\%$ by a difference amplifier. Which *frequency* fluctuations are still caused by the residual intensity fluctuations, if a FPI with a free spectral range of $10\,\text{GHz}$ and a finesse of 50 is used for frequency stabilization at the slope of the FPI transmission peak?

Chapter 6
Nonlinear Optics

When an electromagnetic wave interacts with atoms the electrons perform oscillations around their equilibrium position which results in the emission of radiation. If the intensity of the wave is sufficiently small the amplitude of these oscillations is small and the restoring force

$$F_r = -k \cdot r \qquad (6.1)$$

is proportional to the displacement r of the electron from its equilibrium position $r = 0$ (Hooke's Law). This is the regime of *linear Optics*.

For higher intensities, however, as can be reached with lasers, the amplitude becomes so large that the restoring force is no longer linearly dependent on the displacement and higher order terms have to be included in (6.1), which has to be replaced by the sum

$$F_r = -\sum k_i r^i . \qquad (6.2)$$

Since the force on an electron with charge $-e$ in an electric field E is $F = -e \cdot E$ and the oscillating charge equals an induced dipole moment, this leads for high intensities to a nonlinear dependence between electric field amplitude and induced dipole moment. The radiation emitted by the oscillating electrons contains besides the fundamental frequency ω of the driving field also higher frequencies $n \cdot \omega$ (*higher harmonics generation*).

This harmonics generation is generally realized by focussing a laser beam into optical crystals with special symmetries. Also gases or metal vapours have been used although the efficiency of harmonics generation is here lower because of the much lower atomic densities. The generation of higher harmonics or of sum-and difference frequencies, (if two lasers with different frequencies are focussed into the optical crystal) has considerably widened the frequency range for coherent radiation from the extreme UV to the far infrared region.

We will now study this nonlinear behaviour and the applications of nonlinear optics in more detail. Several examples shall illustrate the subject [523–534].

W. Demtröder, *Laser Spectroscopy 1*, DOI 10.1007/978-3-642-53859-9_6,
© Springer-Verlag Berlin Heidelberg 2014

6.1   Mathematical Description

The dielectric polarization of a medium

$$P = \sum p_i \, ,$$

which is the sum of all induced dipole moments $p = e \cdot r$ can be written as the expansion in powers of the applied field

$$P = \epsilon_0(\tilde{\chi}^{(1)} E + \tilde{\chi}^{(2)} E^2 + \tilde{\chi}^{(3)} E^3 + \dots) \, , \tag{6.3}$$

where $\tilde{\chi}^{(k)}$ is the kth-order susceptibility tensor of rank $k + 1$.

Example 6.1

Consider, for example, the EM wave

$$E = E_1 \cos(\omega_1 t - k_1 z) + E_2 \cos(\omega_2 t - k_2 z) \, , \tag{6.4}$$

composed of two components incident on the nonlinear medium. The induced polarization at a fixed position (say, $z = 0$) in the crystal is generated by the combined action of both components. The linear term in (6.3) describes the Rayleigh scattering. The quadratic term $\chi^{(2)} E^2$ gives the contributions

$$
\begin{aligned}
P^{(2)} &= \epsilon_0 \tilde{\chi}^{(2)} E^2 (z = 0) \\
&= \epsilon_0 \tilde{\chi}^{(2)} \left(E_1^2 \cos^2 \omega_1 t + E_2^2 \cos^2 \omega_2 t + 2 E_1 E_2 \cos \omega_1 t \cdot \cos \omega_2 t \right) \\
&= \epsilon_0 \tilde{\chi}^{(2)} \left\{ \frac{1}{2} (E_1^2 + E_2^2) + \frac{1}{2} E_1^2 \cos 2\omega_1 t \right. \\
&\quad \left. + \frac{1}{2} E_2^2 \cos 2\omega_2 t + E_1 \cdot E_2 [\cos(\omega_1 + \omega_2)t + \cos(\omega_1 - \omega_2)t] \right\} \, ,
\end{aligned}
\tag{6.5}
$$

where the trigonometric relations $\cos^2 x = \frac{1}{2}(1 + \cos 2x)$ and $\cos x \cdot \cos y = \frac{1}{2}(\cos(x + y) + \cos(x - y))$ have been used. The summands of this sum represent dc polarization, ac components at the second harmonics $2\omega_1$, $2\omega_2$, and components at the sum or difference frequencies $\omega_1 \pm \omega_2$.

Note The direction of the polarization vector P may be different from those of E_1 and E_2. The components χ_{ijk} are generally complex and the phase of the polarization differs from that of the driving fields.

Taking into account that the field amplitudes E_1, E_2 are vectors and that the second-order susceptibility $\tilde{\chi}^{(2)}$ is a tensor of rank 3 with components χ_{ijk} depending on the symmetry properties of the nonlinear crystal [529], we can write (6.3) in the explicit form

$$P_i^{(2)} = \epsilon_0 \left(\sum_{k=1}^{3} \chi_{ik}^{(1)} E_k + \sum_{j,k=1}^{3} \chi_{ijk}^{(2)} E_j E_k \right) \quad (1 \triangleq x, 2 \triangleq y, 3 \triangleq z), \quad (6.6)$$

if we restrict the expansion to the linear and quadratic terms.

Here P_i $(i = x, y, z)$ gives the ith component of the dielectric polarization $P = \{P_x, P_y, P_z\}$.

The components P_i $(i = x, y, z)$ of the induced polarization are determined by the polarization characteristics of the incident wave (i.e., which of the components E_x, E_y, E_z are nonzero), and by the components of the susceptibility tensor, which in turn depend on the symmetries of the nonlinear medium.

Let us first discuss the linear part of (6.6), which can be written as

$$\begin{pmatrix} P_x^{(1)} \\ P_y^{(1)} \\ P_z^{(1)} \end{pmatrix} = \epsilon_0 \begin{pmatrix} \chi_{xx} & \chi_{xy} & \chi_{xz} \\ \chi_{yx} & \chi_{yy} & \chi_{yz} \\ \chi_{zx} & \chi_{zy} & \chi_{zz} \end{pmatrix} \begin{pmatrix} E_x \\ E_y \\ E_z \end{pmatrix} . \quad (6.7a)$$

One can always choose a coordinate system (ξ, η, ς) in which the tensor $\chi^{(1)}$ becomes diagonal (principal axis transformation). If we align the crystal in such a way that the (ξ, η, ς)-axes coincide with the (x, y, z)-axes, (6.7a) simplifies in the principal axes system to:

$$\begin{pmatrix} P_x^{(1)} \\ P_y^{(1)} \\ P_z^{(1)} \end{pmatrix} = \epsilon_0 \begin{pmatrix} \chi_1 & 0 & 0 \\ 0 & \chi_2 & 0 \\ 0 & 0 & \chi_3 \end{pmatrix} \begin{pmatrix} E_x \\ E_y \\ E_z \end{pmatrix} . \quad (6.7b)$$

This shows that generally P and E are no longer parallel because the χ_i may be different. Using the relation $\varepsilon_i = 1 + \chi_i$ we can replace the susceptibility χ by the relative dielectric constant ϵ, which is related to the refractive index n through $\epsilon = n^2$. In nonlinear optical crystals there are generally three different refractive indices n_1, n_2 and n_3 along the three principal axes. We call the corresponding refractive indices the principal indices. This can be visualized by plotting vectors with length $n = \epsilon^{1/2}$ in all directions in a principal coordinate system (n_1, n_2, n_3) from its origin. The endpoints of these vectors form an ellipsoid, called the *index ellipsoid*, which can be described by the equation

$$\frac{n_x^2}{n_1^2} + \frac{n_y^2}{n_2^2} + \frac{n_z^2}{n_3^2} = 1 . \quad (6.8)$$

For uniaxial crystals, two of the n $(n_1 = n_2)$ are equal and the index ellipsoid has rotational symmetry around the principal axis, called the optical axis of the

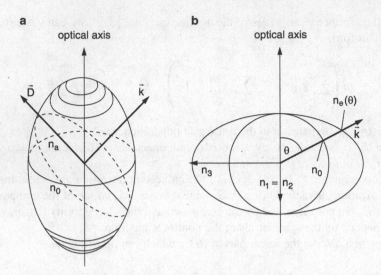

Figure 6.1 a Index ellipsoid for uniaxial birefringent optical crystals. **b** Cutting through the ellipsoid in a plane that contains the optical axis and the propagation direction k

uniaxial crystal, which we choose to be the z-axis of our laboratory coordinate system (Fig. 6.1a).

An incident wave $E = E_0 e^{i(\omega t - kr)}$ with a small amplitude $E = \{E_x, E_y, E_z\}$ generates a polarization $P = \{n_1^{1/2} E_x, n_1^{1/2} E_y, n_3^{1/2} E_z\}$ in the optical material.

If the wave vector k forms an angle $\theta \neq 0$ or $90°$ with the optical axis, the wave in the crystal splits into an ordinary beam (refractive index $n_1 = n_2 = n_0$) where the phase velocity is independent of θ, and an extraordinary wave (refractive index n_e) where n_e and therefore the phase velocity does depend on the direction θ (Fig. 6.1b).

In such birefringent crystals, the direction k of the wave propagation and the direction of the Poynting vector $S = c\,\epsilon_0 (E \times B)$, which is the direction of energy flow, do not coincide (Fig. 6.2). Only in the directions parallel or perpendicular to the optical axis the two vectors point into the same direction.

Now we turn to the second term in (6.6) with the nonlinear susceptibility tensor $\chi^{(2)}$. We assume that the incident wave contains only two frequencies ω_1 and ω_2. With $\omega = (\omega_1 \pm \omega_2)$ we have the detailed description

$$
\begin{pmatrix} P_x^{(2)}(\omega) \\ P_y^{(2)}(\omega) \\ P_z^{(2)}(\omega) \end{pmatrix} = \epsilon_0 \begin{pmatrix} \chi_{xxx}^{(2)} & \chi_{xxyz}^{(2)} & \cdots & \chi_{xzz}^{(2)} \\ \chi_{yxx}^{(2)} & \chi_{yxy}^{(2)} & \cdots & \chi_{yzz}^{(2)} \\ \chi_{zxx}^{(2)} & \chi_{zxy}^{(2)} & \cdots & \chi_{zzz}^{(2)} \end{pmatrix} \begin{pmatrix} E_x(\omega_1) \cdot E_x(\omega_2) \\ E_x(\omega_1) \cdot E_y(\omega_2) \\ E_x(\omega_1) \cdot E_z(\omega_2) \\ E_y(\omega_1) \cdot E_x(\omega_2) \\ E_y(\omega_1) \cdot E_y(\omega_2) \\ \vdots \quad \vdots \\ E_z(\omega_1) \cdot E_z(\omega_2) \end{pmatrix} . \tag{6.9}
$$

Figure 6.2 Directions of electric field E, polarization P, magnetic field B, wave propagation k, and energy flow S in a birefringent crystal

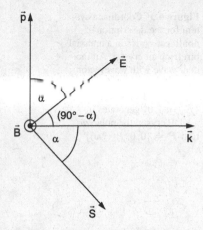

Equation (6.6) demonstrates that the components of the induced polarization P are determined by the tensor components χ_{ijk} and the components of the incident fields. Since the sequence $E_j E_k$ produces the same polarization as $E_k E_j$, we obtain

$$\chi_{ijk} = \chi_{ikj} \ .$$

This reduces the 27 components of the susceptibility tensor $\tilde{\chi}^{(2)}$ to 18 independent components.

In isotropic media the reflection of all vectors at the origin should not change the nonlinear susceptibility. This yields $\chi_{ijk} = -\chi_{ijk}$, which can be only fulfilled by $\chi_{ijk} \equiv 0$. **In all media with an inversion center the second-order susceptibility tensor vanishes! This means, for instance, that optical frequency doubling in gases is not possible.**

In order to reduce the number of indices in the formulas, the components χ_{ijk} are often written in the reduced *Voigt notation*. For the first index the convention $x = 1$, $y = 2$, $z = 3$ is used, whereas the second and third indices are combined as follows: $xx = 1$, $yy = 2$, $zz = 3$, $yz = zy = 4$, $xz = zx = 5$, $xy = yx = 6$. The coefficients in this Voigt notation are named d_{im}. Equation (6.6) can then be written as:

$$\begin{pmatrix} P_1^{(2)} \\ P_2^{(2)} \\ P_3^{(2)} \end{pmatrix} = \varepsilon_0 \begin{pmatrix} d_{11} & d_{12} & d_{13} & d_{14} & d_{15} & d_{16} \\ d_{21} & d_{22} & d_{23} & d_{24} & d_{25} & d_{26} \\ d_{31} & d_{32} & d_{33} & d_{34} & d_{35} & d_{36} \end{pmatrix} \begin{pmatrix} E_1^2 \\ E_2^2 \\ E_3^2 \\ 2E_2 E_3 \\ 2E_1 E_3 \\ 2E_1 E_2 \end{pmatrix} . \qquad (6.10)$$

Figure 6.3 Coordinate system for the description of nonlinear optics in a uniaxial birefringent crystal. An incident wave with wavevector k_λ and $k = (k_x, k_y, 0)$ electric field vector $E = \{E_x, E_y, 0\}$ generates in a KDP crystal the polarization $P = \{0, 0, P_z(2\omega)\}$

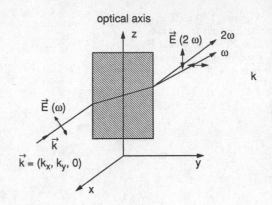

Example 6.2

In potassium dihydrogen phosphate (KDP) the only nonvanishing components of the susceptibility tensor are

$$\chi^{(2)}_{xyz} = d_{14} = \chi^{(2)}_{yxz} = d_{25} \quad \text{and} \quad \chi^{(2)}_{zxy} = d_{36} \ .$$

The components of the induced polarization are therefore with $d_{25} = d_{14}$

$$P_x = 2\epsilon_0 d_{14} E_y E_z \ , \quad P_y = 2\epsilon_0 d_{14} E_x E_z \ , \quad P_z = 2\epsilon_0 d_{36} E_x E_y \ .$$

Suppose there is only one incident wave traveling in a direction k with the polarization vector E normal to the optical axis of a uniaxial birefringent crystal, which we choose to be the z-axis (Fig. 6.3). In this case, $E_z = 0$ and the only nonvanishing component of $P(2\omega)$,

$$P_z(2\omega) = 2\epsilon_0 d_{36} E_x(\omega) E_y(\omega) \ ,$$

is perpendicular to the polarization plane of the incident wave.

Example 6.3

We will consider another example, the GaAs crystal with T_d symmetry, where the d_{ij} tensor is

$$d_{ij} = \begin{pmatrix} 0 & 0 & 0 & d_{14} & 0 & 0 \\ 0 & 0 & 0 & 0 & d_{14} & 0 \\ 0 & 0 & 0 & 0 & 0 & d_{14} \end{pmatrix} \ .$$

Figure 6.4 Second harmonic generation in a GaAs crystal, where the propagation of fundamental and SH waves are perpendicular to each other

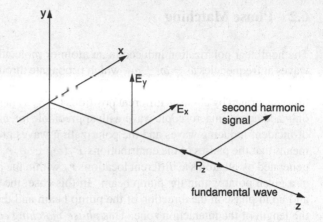

According to (6.10), this gives the polarization components

$$P_x = 2d_{14}E_yE_z$$
$$P_y = 2d_{14}E_zE_x$$
$$P_z = 2d_{14}E_xE_y \, .$$

For a fundamental wave $(E_x, E_y, 0)$ traveling in z-direction, the only component $\neq 0$ is P_z. Since the propagation of the second harmonic is perpendicular to P, this shows that the second harmonic signal will be always perpendicular to the propagation of the fundamental wave (Fig. 6.4) and an efficient generation of second harmonic waves is not possible. This material is therefore not suited for second harmonic generation.

In gases the susceptibility has maxima for resonance frequencies ω_0 of atoms or molecules. One obtains

$$\chi_{res} = \frac{A}{\omega - \omega_0 + i\gamma/2}$$

where A is proportional to the transition matrix element and γ is the full linewidth of the transition. Choosing the fundamental frequency close to a resonance frequency therefore enhances the efficiency of harmonic generation. The disadvantage is that also the absorption of the fundamental wave increases.

6.2   Phase Matching

The nonlinear polarization induced in an atom or molecule acts as a source of new waves at frequencies $\omega = \omega_1 \pm \omega_2$, which propagate through the nonlinear medium with the phase velocity $v_{\mathrm{ph}} = \omega/k = c/n(\omega)$. However, the microscopic contributions generated by atoms at different positions (x, y, z) in the nonlinear medium can only add up to a macroscopic wave with appreciable intensity if the phase velocities of incident inducing waves and the polarization waves are properly matched. This means that the phases of the contributions $\boldsymbol{P}_i(\omega_1 \pm \omega_2, \boldsymbol{r}_i)$ to the polarization wave generated by all atoms at different locations $\boldsymbol{r}_i$ within the pump beam must be equal at a given point within the pump beam. In this case, the amplitudes $\boldsymbol{E}_i(\omega_1 \pm \omega_2)$ add up in phase in the direction of the pump beam and the intensity increases with the length of the interaction zone. This *phase-matching condition* can be written as

$$\boldsymbol{k}(\omega_1 \pm \omega_2) = \boldsymbol{k}(\omega_1) \pm \boldsymbol{k}(\omega_2) \,, \tag{6.11}$$

which may be interpreted as *momentum conservation for the three photons participating in the mixing process.*

The phase-matching condition (6.11) is illustrated by Fig. 6.5. If the angles between the three wave vectors are too large, the overlap region between focused beams becomes too small and the efficiency of the sum- or difference-frequency generation decreases. Maximum overlap is achieved for collinear propagation of all three waves. In this case, $\boldsymbol{k}_1 \| \boldsymbol{k}_2 \| \boldsymbol{k}_3$ and we obtain with $c/n = \omega/k$ and $\omega_3 = \omega_1 \pm \omega_2$ the condition

$$n_3\omega_3 = n_1\omega_1 \pm n_2\omega_2 \quad \Rightarrow \quad n_3 = n_1 = n_2 \,, \tag{6.12}$$

for the refractive indices n_1, n_2, and n_3.

This condition can be fulfilled in unaxial birefringent crystals that have two different refractive indices n_{o} and n_{e} for the ordinary and the extraordinary waves. The ordinary wave is polarized in the x–y-plane perpendicular to the optical axis, while the extraordinary wave has its $\boldsymbol{E}$-vector in a plane defined by the optical axis and

Figure 6.5 Phase-matching condition as momentum conservation for **a** noncollinear and **b** collinear propagation of the three waves

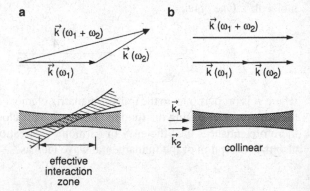

Figure 6.6 **a** Index ellipsoid and refractive indices n_o and n_e for two directions of the electric vector of the wave in a plane perpendicular to the wave propagation $\boldsymbol{k}$, **b** Dependence of n_o and n_e on the angle θ between the wave vector $\boldsymbol{k}$ and the optical axis of a uniaxial positive birefringent crystal

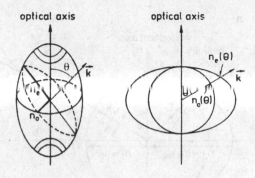

the incident beam. While the ordinary index n_o does not depend on the propagation direction, the extraordinary index n_e depends on the directions of both $\boldsymbol{E}$ and $\boldsymbol{k}$. The refractive indices n_o, n_e and their dependence on the propagation direction in uniaxial birefringent crystals can be illustrated by the index ellipsoid (6.8). If we specify a propagation direction $\boldsymbol{k}$, we can illustrate the refractive indices n_o and n_e experienced by the EM wave $\boldsymbol{E} = \boldsymbol{E}_0 \cos(\omega t - \boldsymbol{k} \cdot \boldsymbol{r})$ in the following way (Fig. 6.6a): consider a plane through the center of the index ellipsoid with its normal in the direction of $\boldsymbol{k}$. The intersection of this plane with the ellipsoid forms an ellipse. The principal axes of this ellipse give the ordinary and extraordinary indices of refraction n_o and n_e, respectively. These principal axes are plotted in Fig. 6.6b as a function of the angle θ between the optical axis and the wave vector $\boldsymbol{k}$. If the angle θ between $\boldsymbol{k}$ and the optical axis (which is assumed to coincide with the z-axis) is varied, n_o remains constant, while the extraordinary index $n_e(\theta)$ changes according to

$$\frac{1}{n_e^2(\theta)} = \frac{\cos^2 \theta}{n_o^2} + \frac{\sin^2 \theta}{n_e^2(\theta = \pi/2)} . \tag{6.13}$$

The uniaxial crystal is called *positively* birefringent if $n_e \geq n_o$ and *negatively* birefringent if $n_e \leq n_o$ (Fig. 6.7). It is possible to find nonlinear birefringent crystals where the phase-matching condition (6.12) for collinear phase matching can be fulfilled if one of the three waves at ω_1, ω_2, and $\omega_1 \pm \omega_2$ propagates as an extraordinary wave and the others as ordinary waves through the crystal in a direction θ specified by (6.13) [530].

One distinguishes between **type-I and type-II phase-matching** depending on which of the three waves with ω_1, ω_2, $\omega_3 = \omega_1 \pm \omega_2$ propagates as an ordinary or as an extraordinary wave. Type I corresponds to $(1 \to e, 2 \to e, 3 \to o)$ in positive uniaxial crystals and to $(1 \to o, 2 \to o, 3 \to e)$ in negative uniaxial crystals, whereas type II is characterized by $(1 \to o, 2 \to e, 3 \to o)$ for positive and $(1 \to e, 2 \to o, 3 \to e)$ for negative uniaxial crystals [533]. Let us now illustrate these general considerations with some specific examples.

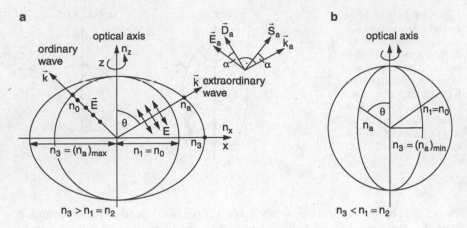

Figure 6.7   Index ellipsoid for **a** positive **b** negative birefingent uniaxial optical crystals

6.3   Second-Harmonic Generation

For the case $\omega_1 = \omega_2 = \omega$, the phase-matching condition (6.11) for second-harmonic generation (SHG) becomes

$$k(2\omega) = 2k(\omega) \quad \Rightarrow \quad v_{\text{ph}}(2\omega) = v_{\text{ph}}(\omega) \,, \tag{6.14}$$

which implies that *the phase velocities of the incident and SH wave must be equal.* This can be achieved in a negative birefringent uniaxial crystal (Fig. 6.8) in a certain direction θ_p against the optical axis if in this direction the extraordinary refractive index $n_e(2\omega)$ for the SH wave equals the ordinary index $n_o(\omega)$ for the fundamental wave. When the incident wave propagates as an ordinary wave in this direction θ_p through the crystal, the local contributions of $P(2\omega, r)$ can all add up in phase and a macroscopic SH wave at the frequency 2ω will develop as an extraordinary wave. The polarization direction of this SH wave is orthogonal to that of the fundamental wave. In uniaxial positive birefringent crystals, the phase-matching condition can be fulfilled for type-I phase matching when the fundamental wave at ω travels as an extraordinary wave through the crystal and the second harmonic at 2ω travels as an ordinary wave.

In favorable cases phase-matching is achieved for $\theta = 90°$. This has the advantage that both the fundamental and the SH beams travel collinearly through the crystal, whereas for $\theta \neq 90°$ the power flow direction of the extraordinary wave differs from the propagation direction k_e. This results in a decrease of the overlap region between both beams.

Let us estimate how a possible slight phase mismatch $\Delta n = n(\omega) - n(2\omega)$ affects the intensity of the SH wave. The nonlinear polarization $P(2\omega)$ generated at the position r by the driving field $E_0 \cos[\omega t - k(\omega) \cdot r]$ can be deduced from

Figure 6.8 Index matching
for SHG in a uniaxial negatively birefringent crystal

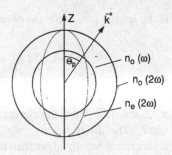

$n_0(\omega)$

$n_0(2\omega)$

$n_e(2\omega)$

(6.5) as

$$P(2\omega) = \tfrac{1}{2}\epsilon_0\chi_{\text{eff}}^{(2)}E_0^2(\omega)[1 + \cos(2\omega t)] . \tag{6.15}$$

This nonlinear polarization generates a wave

$$P(2\omega, r) = E_0(2\omega) \cdot \cos(2\omega t - k(2\omega) \cdot r) ,$$

with amplitude $E(2\omega)$, which travels with the phase velocity $v(2\omega) = 2\omega/k(2\omega)$
through the crystal. The effective nonlinear coefficient $\chi_{\text{eff}}^{(2)}$ depends on the nonlinear crystal and on the propagation direction.

Assume that the pump wave propagates in the z-direction. Over the path length z
a phase difference

$$\Delta\varphi = \Delta k \cdot z = [2k(\omega) - k(2\omega)] \cdot z , \tag{6.16}$$

between the fundamental wave at ω and the second-harmonic wave at 2ω has developed. If the field amplitude $E(2\omega)$ always remains small compared to $E(\omega)$
(low conversion efficiency), we may neglect the decrease of $E(\omega)$ with increasing z. Therefore we obtain the total amplitude of the SH wave summed over the
path length $z = 0$ to $z = L$ through the nonlinear crystal by integration over the microscopic contribution $dE(2\omega, z)$ generated by $P(2\omega, z)$. From (6.15), one obtains
with $\Delta k = |2k(\omega) - k(2\omega)|$ and $dE(2\omega)/dz = [2\omega/\epsilon_0 nc]P(2\omega)$ [533]

$$E(2\omega, L) = \int_{z=0}^{L} \chi_{\text{eff}}^{(2)}(\omega/nc)E_0^2(\omega)\cos(\Delta k z)dz$$

$$= \chi_{\text{eff}}^{(2)}(\omega/nc)E_0^2(\omega)\frac{\sin \Delta k L}{\Delta k} . \tag{6.17a}$$

The intensity $I = (nc\epsilon_0/2n)|E(2\omega)|^2$ of the SH wave is then

$$I(2\omega, L) = I^2(\omega)\frac{2\omega^2|\chi_{\text{eff}}^{(2)}|^2L^2}{n^3c^3\epsilon_0}\frac{\sin^2(\Delta k L)}{(\Delta k L)^2} . \tag{6.17b}$$

If the length L exceeds the coherence length

$$L_{\text{coh}} = \frac{\pi}{2\Delta k} = \frac{\lambda}{4(n_{2\omega} - n_\omega)} , \qquad (6.18)$$

the fundamental wave (λ) and the SH wave $(\lambda/2)$ have a phase difference $\Delta\varphi >$ $\pi/2$, and destructive interference begins, which diminishes the amplitude of the SH wave. The difference $n_{2\omega} - n_\omega$ should therefore be sufficiently small to provide a coherence length larger than the crystal length L.

According to the definition at the end of Sect. 6.2, type-I phase matching is achieved in uniaxial negatively birefringent crystals when $n_e(2\omega, \theta) = n_o(\omega)$. The polarizations of the fundamental wave and the SH wave are then orthogonal. From (6.13) and the condition $n_e(2\omega, \theta) = n_o(\omega)$, we obtain the phase-matching angle θ as

$$\sin^2\theta = \frac{v_o^2(\omega) - v_o^2(2\omega)}{v_e^2(2\omega, \pi/2) - v_o^2(2\omega)} . \qquad (6.19a)$$

For type-II phase matching the polarization of the fundamental wave does not fall into the plane defined by the optical axis and the k-vector. It therefore has one component in the plane, which travels with $v = c/n_o$, and another component with $v = c/n_e$ perpendicular to the plane. The phasematching condition now becomes

$$n_e(2\omega, \theta) = \tfrac{1}{2}\left[n_e(\omega, \theta) + n_o(\omega)\right] . \qquad (6.19b)$$

The choice of the nonlinear medium depends on the wavelength of the pump laser and on its tuning range (Table 6.1). For SHG of lasers around $\lambda = 1\,\mu\text{m}$, $90°$ phase matching can be achieved with LiNbO_3 crystals, while for SHG of dye lasers around $\lambda = 0.5\text{–}0.6\,\mu\text{m}$, KDP crystals or ADA can be used. Figure 6.9 illustrates the dispersion curves $n_o(\lambda)$ and $n_e(\lambda)$ of ordinary and extraordinary waves in KDP and LiNbO_3, which show that $90°$ phase matching can be achieved in LiNbO_3 for $\lambda_p = 1.06\,\mu\text{m}$ and in KDP for $\lambda_p \simeq 515\,\text{nm}$ [530].

Since the intensity $I(2\omega)$ of the SH wave is proportional to the square of the pump intensity $I(\omega)$, most of the work on SHG has been performed with pulsed lasers, which offer high peak powers.

Focusing of the pump wave into the nonlinear medium increases the power density and therefore enhances the SHG efficiency. However, the resulting divergence of the focused beam decreases the coherence length because the wave vectors k_p are spread out over an interval Δk_p, which depends on the divergence angle. The partial compensation of both effects leads to an optimum focal length of the focusing lens, which depends on the angular dispersion $dn_e/d\theta$ of the refractive index n_e and on the spectral bandwidth $\Delta\omega_p$ of the pump radiation [539].

If the wavelength λ_p of the pump laser is tuned, phase matching can be maintained either by turning the crystal orientation θ against the pump beam propagation k_p (angle tuning) or by temperature control (temperature tuning), which relies

Table 6.1 Characteristic data of nonlinear crystals used for frequency doubling or sum-frequency generation [537, 538]

Material	Transparency range [nm]	Spectral range of phase matching of type I or II	Damage threshold [GW/cm^2]	Relative doubling effi- ciency	Reference
ADP	220–2000	500–1100	0.5	1.2	[295]
KD*P	200–2500	517–1500 (I)	8.4	1.0	[556]
		732–1500 (II)		8.4	
Urea	210–1400	473–1400 (I)	1.5	6.1	[567]
BBO	197–3500	410–3500 (I)	9.9	26.0	[540–546]
		750–1500 (II)			
LiJO$_3$	300–5500	570–5500 (I)	0.06	50.0	[550, 568]
KTP	350–4500	1000–2500 (II)	1.0	215.0	[566]
LiNbO$_3$	400–5000	800–5000 (II)	0.05	105.0	[556]
LiB$_3$O$_5$	160–2600	550–2600	18.9	3	[557]
CdGeAs$_2$	1–20 μm	2–15 μm	0.04	9	[572]
AgGaSe$_2$	3–15 μm	3.1–12.8 μm	0.03	6	
Te	3.8–32 μm		0.045	270	[556]

Table 6.2 Abbreviations for some commonly used nonlinear crystals

ADP	=	Ammonium dihydrogen phosphate	NH$_4$H$_2$PO$_4$
KDP	=	Potassium dihydrogen phosphate	KH$_2$PO$_4$
KD*P	=	Potassium dideuterium phosphate	KD$_2$PO$_4$
KTP	=	Potassium titanyl phosphate	KTiOPO$_4$
KNbO$_3$	=	Potassium niobate	KNbO$_3$
LBO	=	Lithium triborate	LiB$_3$O$_5$
LiIO$_3$	=	Lithium iodate	LiIO$_3$
LiNbO$_3$	=	Lithium niobate	LiNbO$_3$
BBO	=	Beta-barium borate	β-BaB$_2$O$_4$

on the temperature dependence $\Delta n(T, \lambda) = n_o(T, \lambda) - n_e(T, \lambda/2)$. The tuning range $2\omega \pm \Delta_2\omega$ of the SH wave depends on that of the pump wave ($\omega \pm \Delta_1\omega$) and on the range where phase matching can be maintained. Generally, $\Delta_2\omega < 2\Delta_1\omega$ because of the limited phase-matching range. With frequency-doubled pulsed dye lasers and different dyes the whole tuning range between $\lambda = 195$–500 nm can be completely covered. The strong optical absorption of most nonlinear crystals below 220 nm causes a low damage threshold, and the shortest wavelength achieved by SHG is, at present, $\lambda = 200$ nm [534, 539–543].

Example 6.4

The refractive indices $n_o(\lambda)$ and $n_e(\lambda)$ of ADP (ammonium dihydrogen phosphate) for $\theta = 90°$ are plotted in Fig. 6.10, together with the phasematching

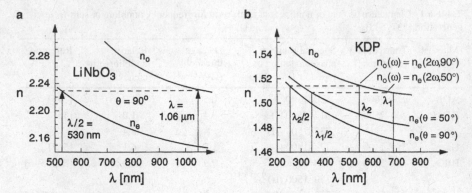

Figure 6.9 Refractive indices $n_o(\lambda)$ and $n_e(\lambda)$: **a** for $\theta = 90°$ in LiNbO$_3$ [533] and **b** for $\theta = 50°$ and $90°$ KDP [529]. Collinear phase matching can be achieved in LiNbO$_3$ for $\theta = 90°$ and $\lambda = 1.06\,\mu$m (Nd$^+$ laser) and in KDP for $\theta = 50°$ at $\lambda = 694\,$nm (ruby laser) or for $\theta = 90°$ at $\lambda = 515\,$nm (argon laser)

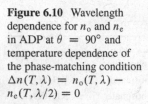

Figure 6.10 Wavelength dependence for n_o and n_e in ADP at $\theta = 90°$ and temperature dependence of the phase-matching condition $\Delta n(T,\lambda) = n_o(T,\lambda) - n_e(T,\lambda/2) = 0$

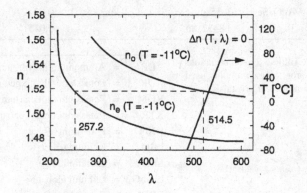

curve: $\Delta(T,\lambda) = n_o(T,\lambda) - n_e(T,\lambda/2) = 0$. This plot shows that at $T = -11\,°$C, the phase-matching condition $\Delta(T,\lambda) = 0$ is fulfilled for $\lambda = 514.5\,$mm, and thus $90°$ phase matching for SHG of the powerful green argon laser line at $\lambda = 514.5\,$nm is possible.

Limitations of the SH output power generated by pulsed lasers are mainly set by the damage threshold of available nonlinear crystals. Very promising new crystals are the negative uniaxial BBO (beta-barium borate) β-BaB$_2$O$_4$ [541–545] and lithium borate LiBO, which have high damage thresholds and which allow SHG from 205 nm to above 3000 nm.

Figure 6.11 External ring resonator for efficient optical frequency doubling. The mirrors M$_2$ and M$_4$ are highly reflective, while mirror M$_1$ transmits the fundamental wave and mirror M$_3$ transmits the second harmonic wave

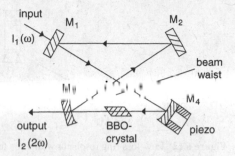

Example 6.5

The five nonvanishing nonlinear coefficients of BBO are d_{11}, d_{22}, d_{31}, d_{13}, d_{14}, where the largest coefficient d_{11} is about 6 times larger than d_{36} of KDP. The transmission range of BBO is 195–3500 nm. It has a low temperature dependence of its birefringence and a high optical homogeneity. Its damage threshold is about $10\,\mathrm{GW/cm^2}$.

Type-I phase matching is possible in the range 410–3500 nm, type-II phase-matching in the range 750–1500 nm.

The effective nonlinear coefficient for type-I phase-matching is

$$d_{\mathrm{eff}} = d_{31} \sin \theta + (d_{11} \cos 3\phi - d_{22} \sin 3\phi) \cos \phi \,,$$

where θ and ϕ are the polar angles between the $\boldsymbol{k}$-vector of the incident wave and the $z(= c)$-axis and the $x(= a)$-axis of the crystal, respectively. For $\phi = 0$ d_{eff} becomes maximum.

With cw dye lasers in the visible (output power ≤ 1 W), generally UV powers of only a few milliwatts are achieved by frequency doubling. The doubling efficiency $\eta = I(2\omega)/I(\omega)$ can be greatly enhanced when the doubling crystal is placed inside the laser cavity where the power of the fundamental wave is much higher [548–552]. The auxiliary beam waist in a ring laser resonator is the best location for placing the crystal (Fig. 5.103). With an intracavity LiIO$_3$ crystal, for example, UV output powers in the range 20–50 mW have been achieved at $\lambda/2 = 300$ nm [550].

If the dye laser must be used for visible as well as for UV spectroscopy, a daily change of the configuration is troublesome, therefore it is advantageous to apply an extra external ring resonator for frequency doubling [553–555]. This resonator must, of course, always be kept in resonance with the dye laser wavelength λ_L and therefore must be stabilized by a feedback control to the wavelength λ_L when the dye laser is tuned.

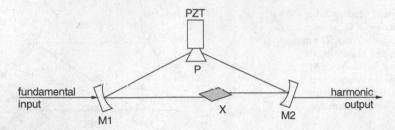

Figure 6.12 Low-loss ring resonator with wide tuning range for optical frequency doubling with astigmatic compensation [533]

One example is illustrated in Fig. 6.11. In order to avoid feedback into the laser, ring resonators are used and the crystal is placed under the Brewster angle in the beam waist of the resonator. Since the enhancement factor for $I(\omega)$ depends on the resonator lasers, the mirrors should be highly reflective for the fundamental wave, but the output mirror should have a high transmission for the second-harmonic wave. An elegant solution is shown in Fig. 6.12, where only two mirrors and a Brewster prism form the ring resonator. The resonator length can be conveniently tuned by shifting the prism with a piezo-translating device in the z-direction.

Many more examples of external and intracavity frequency doubling with different nonlinear crystals [556] can be found in the literature [558–560]. Table 6.1 compiles some optical properties of commonly utilized nonlinear crystals.

6.4 Quasi Phase Matching

Recently, optical frequency doubling devices have been developed that consist of many thin slices of a crystal with periodically varying directions of their optical axes. This can be achieved by producing many thin electrodes with lithographic techniques on the two side faces of the crystal and then placing the crystal at higher temperatures in a spatially periodic electric field. This results in a corresponding anisotropy of the charge distribution (induced electric dipole moments), which determines the optical axis of the crystal (Fig. 6.13a). If there is a phase mismatch

$$\Delta k = \frac{2\pi}{\lambda}[n(2\omega - n(\omega))] \,, \tag{6.20a}$$

the phases of fundamental and second-harmonic waves differ by π after the coherence length

$$L_c = \frac{\pi}{k(2\omega) - 2k(\omega)} = \frac{\lambda}{2[n(2\omega) - n(\omega)]} \,. \tag{6.20b}$$

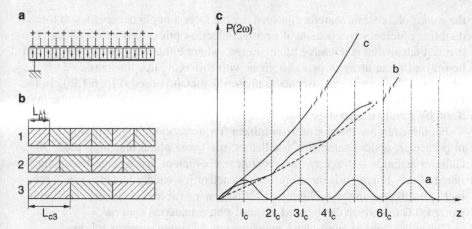

Figure 6.13 Quasi phase matching: **a** periodic poling of crystal orientation; **b** array of crystals with different period lengths for choosing the optimum doubling efficiency for a given wavelength; **c** second-harmonic output power as a function of total length $L_i = n \cdot L_c$ for one crystal with slight phase mismatch (curve *a*), for a periodically poled crystal (curve *b*), and for a single crystal with ideal phase matching

A nonlinear crystal with a length $L \gg L_c$ shows the output power $P(2\omega)$ of the second-harmonic wave as a function of the propagation length z depicted by curve *a* in Fig. 6.13c. After one coherence length the power decreases again because of destructive interference between the second-harmonic and the out-of-phase fundamental wave.

If, however, the crystal has length $L = L_c$ followed by a second crystal with $L = 2L_c$ but opposite orientation of its optical axis, then the phase mismatch is reversed and the phase difference decreases from π to $-\pi$. Now the next layer follows with the orientation of the first one and the phase difference again increases from $-\pi$ to $+\pi$, and so on. This yields the output power of the second harmonic as shown in Fig. 6.13c, curve *b*.

For comparison, the curve *c* of a perfectly phase-matched long crystal is shown in Fig. 6.13c. This demonstrates that the quasi-phase-matching device gives a lower output power than the perfectly matched crystal, but a much larger power than for a single crystal in the case of slight phase mismatches. The advantage of this quasi-phase-matching is the possible larger spectral range of the fundamental wave, which can be frequency-doubled.

For frequency doubling of tunable lasers, it is difficult to maintain perfect phase matching for all wavelengths; therefore phase mismatches cannot be avoided. Furthermore, for angle tuning of the crystal, noncollinear propagation of the fundamental and the second-harmonic wave occurs. This limits the effective interaction length and therefore the doubling efficiency. With correctly designed quasi-phase-matched devices, collinear noncritical phase matching can be realized, which allows long interaction lengths. Furthermore, fundamental and second-harmonic waves can have the same polarization; therefore one can use the largest nonlinear coefficient for the doubling efficiency by choosing the correct electro-optic poling of the slices. The greatest advantage is the large tuning range, where either temperature tuning can be utilized or an array of periodic slices with different slice thicknesses $L = L_c$ adapted to the wavelength-dependent phase mismatch is used (Fig. 6.13b). In the latter case the different devices, all on the same chip, can be shifted into the laser beam by a translational stage.

For these reasons many modern nonlinear frequency-doubling or mixing devices, in particular, optical parametric oscillators, use quasi phase matching [562, 563]. Gallium arsenide has a very high nonlinear coefficient and a wide transparency range of 0.7–17 µm. It is therefore very attractive for widely tunable optical parametric oscillators in the mid-infrared. It is now possible to fabricate orientation-patterned GaAs which can be used as quasi-phase-matched material.

The advantages of quasi-phase-matching can be summarized as follows:

a) Unlike birefringent crystals, where the propagation direction and polarization of the fundamental wave are severely constrained, both of these parameters can be chosen to maximize the effective nonlinear coefficient d_{eff}.
b) The Poynting vector has the same direction for fundamental and harmonic waves. There is no walk-off as in birefringent crystals for $\theta \neq 90$.
c) Any wavelength within the transparency range of the material can be phase-matched, whereas in birefringent crystals only a narrow wavelength range can be phase-matched for a given direction with respect to the optical axis.

6.5 Sum-Frequency Generation

In the case of laser-pumped dye lasers, it is often more advantageous to generate tunable UV radiation by optical mixing of the pump laser and the tunable dye laser outputs rather than by frequency doubling of the dye laser. Since the intensity $I(\omega_1 + \omega_2)$ is proportional to the product $I(\omega_1)I \cdot (\omega_2)$, the larger intensity $I(\omega_1)$ of the pump laser allows enhanced UV intensity $I(\omega_1 + \omega_2)$. Furthermore, it is often possible to choose the frequencies ω_1 and ω_2 in such a way that 90° phase matching can be achieved. The range $(\omega_1 + \omega_2)$ that can be covered by sum-frequency generation is generally wider than that accessible to SHG. Radiation at wavelengths too short to be produced by frequency doubling can be generated by the mixing of two different frequencies ω_1 and ω_2. This is illustrated by Fig. 6.14, which depicts possible wavelength combinations λ_1 and λ_2 that allow 90° phase-matched sum-frequency mixing in KDP and ADP at room temperature or along the b-axis of biaxial KB5 crystals [564].

Some examples are given to demonstrate experimental realizations of the sum-frequency mixing technique [565–576].

Figure 6.14 Possible combinations of wavelength pairs (λ_1, λ_2) that allow 90° phase-matched sum-frequency generation in ADP, KDP, and KB5 [565, 571]

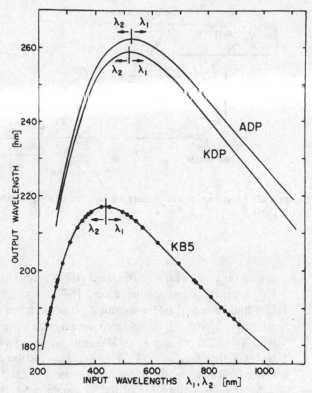

Example 6.6

a) The output of a cw rhodamine 6G dye laser pumped with 15 W on all lines of an argon laser is mixed with a selected line of the same argon laser (Fig. 6.15). The superimposed beams are focused into the temperature-stabilized KDP crystal. Tuning is accomplished by simultaneously tuning the dye laser wavelength and the orientation of the KDP crystal. The entire wavelength range from 257 to 320 nm can be covered by using different argon lines with a single Rhodamine 6G dye laser without changing dyes [565].

b) The generation of intense tunable radiation in the range 240–250 nm has been demonstrated by mixing in a temperature-tuned 90° phase-matched ADP crystal the second harmonic of a ruby laser with the output of an infrared dye laser pumped by the ruby laser's fundamental output [564].

c) UV radiation tunable between 208 and 259 nm has been generated efficiently by mixing the fundamental output of a Nd:YAG laser and the output of a frequency-doubled dye laser. Wavelengths down to 202 nm

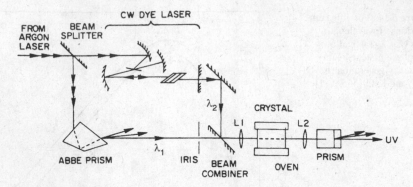

Figure 6.15 Experimental arrangement for sum-frequency generation of cw radiation in a KDP crystal [565]

can be obtained with a refrigerated ADP crystal because ADP is particularly sensitive to temperature tuning [572].

d) In lithium borate (LBO) noncritical phase-matched sum-frequency generation at $\theta = 90°$ can be achieved over a wide wavelength range. Starting with $\lambda_1 < 220$ nm and $\lambda_2 \geq 1064$ nm, sum-frequency radiation down to wavelengths of $\lambda_3 = (1/\lambda_1 + 1/\lambda_2)^{-1} = 160$ nm can be generated. The lower limit is set by the transmission cutoff of LBO [574].

e) After frequency doubling of the Ti:sapphire wavelength 920–960 nm in a LBO crystal, and sum-frequency mixing of the fundamental ω with the second harmonic 2ω in another 90 %-phase-matched LBO, the third harmonic 3ω could be obtained with an overall efficiency of 35 %, tunable between 307–320 nm [575].

A novel device for efficiently generating intense radiation at wavelengths around 202 nm is shown in Fig. 6.16. A laser diode-pumped Nd:YVO$_4$ laser is frequency doubled and delivers intense radiation at $\lambda = 532$ nm, which is again frequency doubled to $\lambda = 266$ nm in a BBO crystal inside a ring resonator. The output from this resonator is superimposed in a third enhancement cavity with the output from a diode laser at $\lambda = 850$ nm to generate radiation at $\lambda = 202$ nm by sum-frequency mixing. This 202-nm radiation is polarized perpendicularly to that at the two other waves and can be therefore efficiently coupled out of the cavity by a Brewster plate [576].

The lower-wavelength limit for nonlinear processes in crystals (SHG or sum-frequency mixing) is generally given by the absorption (transmission cutoff) of the crystals.

For shorter wavelengths sumfrequency mixing or higher-harmonic generation in homogeneous mixtures of rare gases and metal vapors can be achieved. Because

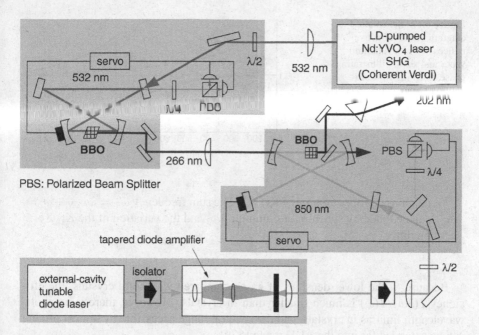

PBS: Polarized Beam Splitter

Figure 6.16 Sum-frequency generation in an enhancement cavity down to $\lambda = 202$ nm [576]

in centro-symmetric media the second-order susceptibility must vanish, SHG is not posssible, but all third-order processes can be utilized for the generation of tunable ultraviolet radiation. Phase matching is achieved by a proper density ratio of raregas atoms to metal atoms. Several examples illustrate the method.

Example 6.7

a) Third-harmonic generation of Nd:YAG laser lines around $\lambda = 1.05\,\mu m$ can be achieved in mixtures of xenon and rubidium vapor in a heat pipe. Figure 6.17 is a schematic diagram for the refractive indices $n(\lambda)$ for Xe and rubidium vapor. Choosing the proper density ratio $N(\text{Xe})/N(\text{Rb})$, phase matching is obtained for $n(\omega) = n(3\omega)$, where the refractive index $n = n(\text{Xe}) + n(\text{Rb})$ is determined by the rubidium and Xe densities. Figure 6.17 illustrates that this method utilizes the compensation of the normal dispersion in Xe by the anomalous dispersion for rubidium [577].

b) A second example is the generation of tunable VUV ratiation between 110 and 130 nm by phase-matched sum-frequency generation in a xenon–krypton mixture [578]. This range covers the Lyman-α line of hydrogen and is therefore particularly important for many experiments in plasma diagnostics and in fundamental physics. A frequency-doubled dye laser at $\omega_{\text{UV}} = 2\omega_1$ and a second tunable dye laser at ω_2 are focused into a cell that

Figure 6.17 Schematic diagram of the refractive indices $n(\lambda)$ for rubidium vapor and xenon, illustrating phase matching for third-harmonic generation

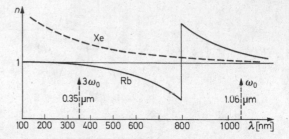

contains a proper mixture of Kr/Xe. The sum frequency $\omega_3 = 2\omega_{UV} + \omega_2$ can be tuned by synchronous tuning of ω_2 and the variation of the Kr/Xe mixture.

Because of the lower densities of gases compared with solid crystals, the efficiency $I(3\omega)/I(\omega)$ is much smaller than in crystals. However, there is no short-wavelength limit as in crystals, and the spectral range accessible by optical mixing can be extended far into the VUV range [579].

The efficiency may be greatly increased by resonance enhancement if, for example, a resonant two-photon transition $2\hbar\omega_1 = E_1 \to E_k$ can be utilized as a first step of the sum-frequency generation $\omega = 2\omega_1 + \omega_2$. This is demonstrated by an early experiment shown in Fig. 6.18. The orthogonally polarized outputs from two N_2 laser-pumped dye lasers are spatially overlapped in a Glan–Thompson prism. The collinear beams of frequencies ω_1 and ω_2 are then focused into a heat pipe containing the atomic metal vapor. One laser is fixed at half the frequency of an appropriate two-photon transition and the other is tuned. For a tuning range of the dye laser between 700 and 400 nm achievable with different dyes, tunable VUV

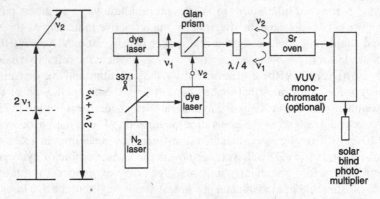

Figure 6.18 Generation of tunable VUV radiation by resonant sum-frequency mixing in metal vapors: **a** level scheme; **b** experimental arrangement

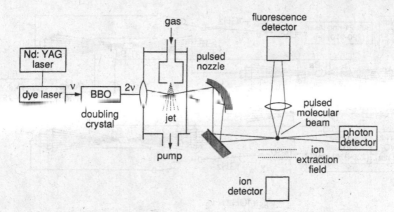

Figure 6.19 Generation of VUV radiation by resonant frequency mixing in a jet [586]

radiation at the frequencies $\omega = 2\omega_1 + \omega_2$ is generated, which can be tuned over a large range. Third-harmonic generation can be eliminated in this experiment by using circularly polarized ω_1 and ω_2 radiation, since the angular momentum will not be conserved for frequency tripling in an isotropic medium under these conditions. The sum frequency $\omega = 2\omega_1 + \omega_2$ corresponds to an energy level beyond the ionization limit [580–585].

Windows cannot be used for wavelengths below 120 nm because all materials absorb the radiation, therefore apertures and differential pumping is needed. An elegant solution is the VUV generation in pulsed laser jets (Fig. 6.19), where the density of wanted molecules within the focus of the incident lasers can be made large without having too much absorption for the generated VUV radiation because the molecular density is restricted to the small path length across the molecular jet close to the nozzle [586, 587]. The output of a tunable dye laser is frequency doubled in a BBO crystal. Its UV radiation is then focused into the gas jet where frequency tripling occurs. The VUV radiation is now collimated by a parabolic mirror and imaged into a second molecular beam within the same vacuum chamber, where the experiment is performed.

An intense coherent tunable Fourier-transform-limited narrow-band all-solid-state vacuum-ultraviolet (VUV) laser system has been developed by Merkt and coworkers [573]. Its bandwidth is less than 100 MHz and the tuning range covers a wide spectral interval around $120,000\,\text{cm}^{-1}$ (15 eV). At a repetition rate of 20 Hz the output reaches 10^8 photons per pulse, which corresponds to an energy of 0.25 nJ per pulse, a peak power of 25 mW for a pulse length of 10 ns, and an average power of 5 nW. For these short VUV wavelengths of around $\lambda = 80\,\text{nm}$ this is remarkable and is sufficient for many experiments in the VUV.

Its principle is illustrated in Fig. 6.20: The setup consists of two cw Ti:sapphire near-infrared single-mode ring lasers with wavenumbers ν_1 and ν_2. The output radiation of these lasers is amplified by nanosecond pump laser pulses, resulting in amplified Fourier-limited pulses in the near IR. Tunable VUV radiation

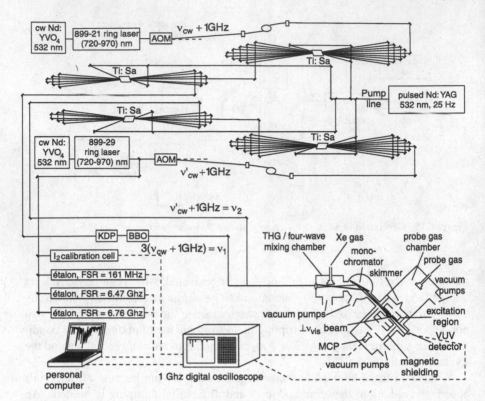

Figure 6.20 Narrow-band VUV laser source. The *upper part* displays the generation of amplified NIR pulses from two cw ring Ti:sapphire lasers with pulse amplification in a multipass amplifier arrangement. The *middle part* shows the KDP and BBO crystals for sum-frequency generation. The VUV radiation is generated in a Xe jet shown in the *lower part* [589]

with wavenumbers $\nu_{VUV} = 2(\nu_3) + \nu_2$ was produced by resonance-enhanced sum-frequency mixing in a supersonic jet of xenon, using the two-photon resonance $(5p)^6 S_0 \to (5p)^5 6p(1/2)\,(J = 0)$ at $2\nu_3 = 80,119\,\text{cm}^{-1}$. The tripled wavenumber $\nu_3 = 3\nu_1$ was produced by generating the third harmonics of ν_1 in successive KDP and BBO crystals. While the wavenumber ν_3 was fixed, the infrared wavenumber ν_2 could be tuned between $12,000–13,900\,\text{cm}^{-1}$, and therefore the VUV wavenumber could be tuned over $1900\,\text{cm}^{-1}$.

Although second harmonic generation is not possible in centro-symmetric media, such as gases, the symmetry is broken at the plane surface of solid materials. Therefore efficient SHG has been found at the boundary between a solid and a gas or vapour. Molecules in a thin layer at solid state surfaces can be studied with SHG or sum frequency mixing, even if the have centro-symmetry. This surface enhanced spectroscopy has been already successfully used for Raman spectroscopy (see Vol. 2, Chap. 3) [591].

More information on the generation of VUV radiation by nonlinear mixing techniques can be found in [574–590].

6.6 Difference-Frequency Spectrometer

While generation of *sum frequencies* yields tunable ultraviolet radiation by mixing the output from two lasers in the visible range, the phase-matched generation of *difference* frequencies allows one to construct tunable coherent *infrared* sources. One early example is the difference-frequency spectrometer of Pine [591], which has proved to be very useful for highresolution infrared spectroscopy.

Two collinear cw beams from a stable single-mode argon laser and a tunable single-mode dye laser are mixed in a $LiNbO_3$ crystal (Fig. 6.21). For 90° phase matching of collinear beams, the phase-matching condition

$$k(\omega_1 - \omega_2) = k(\omega_1) - k(\omega_2) , \tag{6.21a}$$

can be written as $|k(\omega_1 - \omega_2)| = |k(\omega_1)| - |k(\omega_2)|$, which gives for the refractive index $n = c(k/\omega)$ the relation

$$n(\omega_1 - \omega_2) = \frac{\omega_1 n(\omega_1) - \omega_2 n(\omega_2)}{\omega_1 - \omega_2} . \tag{6.21b}$$

The whole spectral range from 2.2 to 4.2 µm can be continuously covered by tuning the dye laser and the phase-matching temperature of the $LiNbO_3$ crystal ($-0.12\,°C/cm^{-1}$). The infrared power is, according to (6.6), (6.17b), proportional to the product of the incident laser powers and to the square of the coherence length. For typical operating powers of 100 mW (argon laser) and 10 mW (dye laser), a few microwatts of infrared radiation is obtained. This is 10^4 to 10^5 times higher than the noise equivalent input power of standard IR detectors.

The spectral linewidth of the infrared radiation is determined by that of the two pump lasers. With frequency stabilization of the pump lasers, a linewidth of a few megahertz has been reached for the difference-frequency spectrometer. In

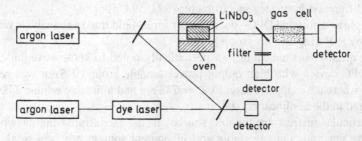

Figure 6.21 Difference-frequency spectrometer [592]

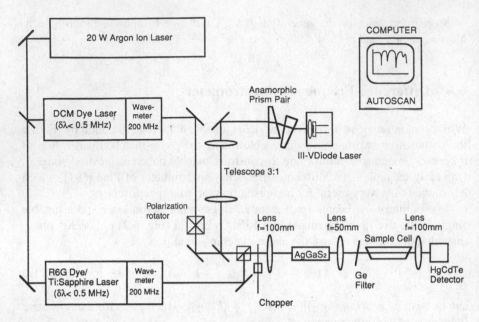

Figure 6.22 Difference-frequency spectrometer based on mixing a cw Ti:sapphire ring laser with a single-frequency III–V diode laser in the nonlinear crystal AgGaS$_2$ [593]

combination with a multiplexing scheme devised for calibration, monitoring, drift compensation, and absolute stabilization of the difference spectrometer, a continuous scan of 7.5 cm^{-1} has been achieved with a reproducibility of better than 10 MHz [592].

A very large tuning range has been achieved with a cw laser spectrometer based on difference-frequency generation in AgGaS$_2$ crystals. By mixing the output of two single-mode tunable dye lasers, infrared powers up to 250 µW have been generated in the spectral range 4–9 µm (Fig. 6.22) [594]. Widely tunable diode lasers in the near infrared, mixed with the output of a fixed frequency high power Nd:YAG laser produces a highly efficient difference frequency spectrometer (Fig. 6.23). Even more promising is the difference-frequency generation of two tunable diode lasers (Fig. 6.23), which allows the construction of a very compact and much cheaper difference-frequency spectrometer [593, 594, 596].

A simple and portable DFG-spectrometer for in-field trace gas analysis was constructed by P. Hering and his group [597].

Using quasi-phase matching in a periodically poled LiNbO$_3$ waveguide structure, a DFG-device with high output power tunable around 1.5 µm was reported in [598], where a Ti:sapphire laser at $\lambda = 748$ nm and a tunable erbium fiber laser were mixed in the nonlinear crystal.

Of particular interest are tunable sources in the far infrared region where no microwave generators are available and incoherent sources are very weak. With selected crystals such as proustite (Ag$_3$AsS$_3$), LiNbO$_3$, or GaAs, phase matching

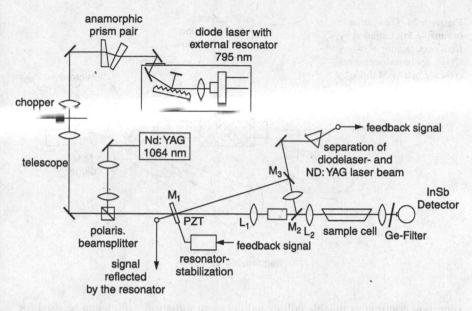

Figure 6.23   Difference-frequency spectrometer with diode lasers [569]

for difference-frequency generation can be achieved for the middle infrared using CO_2 lasers and spin-flip Raman lasers. The search for new nonlinear materials will certainly enhance the spectroscopic capabilities in the whole infrared region [599].

A very useful frequency-mixing device is the MIM diode (Sect. 4.5.2), which allows the realization of continuously tunable FIR radiation covering the difference-frequency range from the microwave region (GHz) to the submillimeter range (THz) [600–602]. It consists of a specially shaped tungsten wire with a very sharp tip that is pressed against a nickel surface covered with a thin layer of nickel oxide (Fig. 4.99). If the beams of two lasers with freqencies ν_1 and ν_2 are focused onto the contact point (Fig. 6.24), frequency mixing due to the nonlinear response of the diode occurs. The tungsten wire acts as an antenna that radiates waves at the difference frequency $(\nu_1 - \nu_2)$ into a narrow solid angle corresponding to the antenna lobe. These waves are collimated by a parabolic mirror with a focus at the position of the diode.

Using CO_2 lasers with different isotope mixtures, laser oscillation on several hundred lines within the spectral range between 9 and 10 µm can be achieved. This laser oscillation can be fine-tuned over the pressure-broadened gain profiles. Therefore their difference frequencies cover the whole FIR region with only small gaps. These gaps can be closed when the radiation of a tunable microwave generator is additionally focused onto the MIM mixing diode. The waves at frequencies

$$\nu = \nu_1 - \nu_2 \pm \nu_{MW}$$

Figure 6.24 Generation of tunable FIR radiation by frequency mixing of two CO_2 laser beams with a microwave in a MIM diode

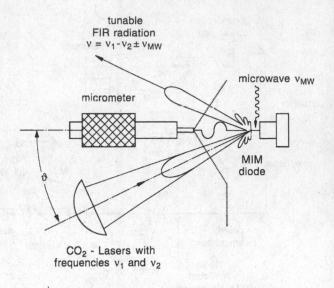

represent continuous tunable collimated coherent radiation, which can be used for absorption spectroscopy in the far infrared [602, 603]. An interesting technique for the difference frequency generation in the far infrared region (terahertz frequencies) was reported by Belkin et al. [604]. They used the active region of a quantum cascade laser inside a resonator as highly nonlinear medium with a giant second order nonlinear susceptibility. When the two input beams in the mid infrared at $\lambda_1 = 8.9\,\mu m$ and $\lambda_2 = 10.5\,\mu m$ were mixed in the cascade laser a difference frequency output at $\lambda = 60\,\mu m$ was obtained with output powers of $7\,\mu W$ at a temperature of 80 K and $0.3\,\mu W$ at room temperature.

6.7 Optical Parametric Oscillators

The optical parametric oscillator (OPO) [604–606, 608–611] is based on the parametric interaction of a strong pump wave $E_p \cos(\omega_p t - \mathbf{k}_p \cdot \mathbf{r})$ with molecules in a crystal that have a sufficiently large nonlinear susceptibility. This interaction can be described as an inelastic scattering of a pump photon $\hbar\omega_p$ by a molecule where the pump photon is absorbed and two new photons $\hbar\omega_s$ and $\hbar\omega_i$ are generated. Because of energy conservation, the frequencies ω_i and ω_s are related to the pump frequency ω_p by

$$\omega_p = \omega_i + \omega_s \,. \tag{6.22}$$

Analogous to the sum-frequency generation, the parametrically generated photons ω_i and ω_s can add up to a macroscopic wave if the phase-matching condition

$$\mathbf{k}_p = \mathbf{k}_i + \mathbf{k}_s \tag{6.23}$$

is fulfilled, which may be regarded as the conservation of momentum for the three photons involved in the parametric process. Simply stated, parametric generation splits a pump photon into two photons that satisfy conservation of energy and momentum at every point in the nonlinear crystal. For a given wave vector k_p of the pump wave, the phase-matching condition (6.23) selects, out of the infinite number of possible combinations $\omega_1 + \omega_2$ allowed by (1) ω_2, a single pair (ω_i, k_i) and (ω_s, k_s) that is determined by the orientation of the nonlinear crystal with respect to k_p. The two resulting macroscopic waves $E_s \cos(\omega_s t - k_s \cdot r)$ and $E_i \cos(\omega_i t - k_i \cdot r)$ are called the *signal* wave and *idler* wave. The most efficient generation is achieved for collinear phase matching where $k_p \| k_i \| k_s$. For this case, the relation (6.12) between the refractive indices gives

$$n_p \omega_p = n_s \omega_s + n_i \omega_i . \tag{6.24}$$

If the pump is an extraordinary wave, collinear phase matching can be achieved for some angle θ against the optical axis, if $n_p(\theta)$, defined by (6.13), lies between $n_o(\omega_p)$ and $n_e(\omega_p)$.

The gain of the signal and idler waves depends on the pump intensity and on the effective nonlinear suceptibility. Analogous to the sum- or difference-frequency generation, one can define a parametric gain coefficient per unit pathlength $\Gamma = I_s/I_p$ or I_i/I_p

$$\Gamma = \frac{\omega_i \omega_s |d|^2 |E_p|^2}{n_i n_s c^2} = \frac{2\omega_i \omega_s |d|^2 I_p}{n_i n_s n_p \epsilon_0 c^3} , \tag{6.25}$$

which is proportional to the pump intensity I_p and the square of the effective nonlinear susceptibility $|d| = \chi_{eff}^{(2)}$. For $\omega_i = \omega_s$, (6.25) becomes identical with the gain coefficient for SHG in (6.17b).

If the nonlinear crystal that is pumped by the incident wave E_p is placed inside a resonator, oscillation on the idler or signal frequencies can start when the gain exceeds the total losses. The optical cavity may be resonant for both the idler and signal waves (doubly-resonant oscillator) or for only one of the waves (singly-resonant oscillator) [608]. Often, the cavity is also resonant for the pump wave in order to increase I_p and thus the gain coefficient Γ.

Figure 6.25 shows schematically the experimental arrangement of a collinear optical parametric oscillator. Due to the much higher gain, pulsed operation is generally preferred where the pump is a Q-switched laser source. The threshold of a doubly-resonant oscillator occurs when the gain equals the product of the signal and idler losses. If the resonator mirrors have high reflectivities for both the signal and idler waves, the losses are small, and even cw parametric oscillators can reach threshold [612]. For singly-resonant cavities, however, the losses for the nonresonant waves are high and the threshold increases.

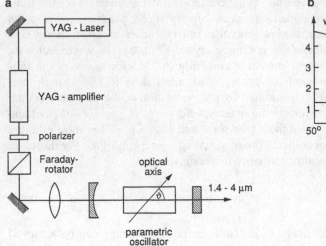

Figure 6.25 Optical parametric oscillator: **a** schematic diagram of experimental arrangement; **b** pairs of wavelengths (λ_1, λ_2) for idler and signal wave for collinear phase matching in LiNbO$_3$ as a function of angle θ [606]

Example 6.8
For a 5-cm long 90° phase-matched LiNbO$_3$ crystal pumped at $\lambda_p = 0.532\,\mu$m, threshold is at 38-mW pump power for the doubly-resonant cavity with 2 % losses at ω_i and ω_s. For the singly-resonant cavity, threshold increases by a factor of 100 to 3.8 W [609].

Tuning of the OPO can be accomplished either by crystal rotation or by controlling the crystal temperature. The tuning range of a LiNbO$_3$ OPO, pumped by various frequency-doubled wavelengths of a Q-switched Nd:YAG laser, extends from 0.55 to about 4 μm. Turning the crystal orientation by only 4° covers a tuning range between 1.4 and 4.4 μm (Fig. 6.25b). Figure 6.26 shows temperature tuning curves for idler and signal waves generated in LiNbO$_3$ by different pump wavelengths. Angle tuning has the advantage of faster tuning rates than in the case of temperature tuning.

Previously, one of the drawbacks of the OPO was the relatively low damage threshold of available nonlinear crystals. The growth of advanced materials with high damage thresholds, large nonlinear coefficients, and broad transparency spectral ranges has greatly aided the development of widely tunable and stable OPOs [610]. Examples are BBO (β-barium borate) and lithium borate (LBO) [611]. For illustration of the wide tuning range, Fig. 6.27 displays wavelength tuning of the BBO OPO for different pump wavelengths.

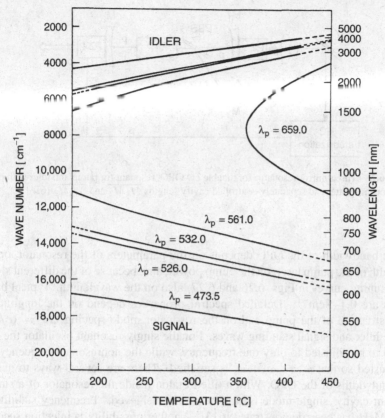

Figure 6.26 Temperature tuning curves of signal and idler wavelengths for a LiNbO$_3$ optical parametric oscillator pumped by different pump wavelengths [608]

Figure 6.27 Wavelengths of signal and idler waves in BBO as a function of the phase-matching angle ϑ for different pump wavelengths λ_p [611]

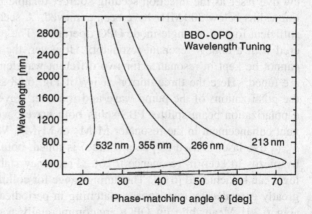

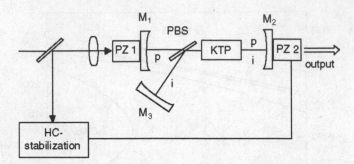

Figure 6.28 Three-mirror resonator for tunable cw OPO, resonant for pump and idler with polarization beam splitter and separately controlled cavity lengths $\overline{M_1 M_2}$ and $\overline{M_3 M_2}$ [616]

The bandwidth of the OPO depends on the parameters of the resonator, on the linewidth of the pump laser, on the pump power, and, because of the different slopes of the tuning curves in Figs. 6.26 and 6.27, also on the wavelength. Typical bandwidths are 0.1–$5\,\mathrm{cm}^{-1}$. Detailed spectral properties depend on the longitudinal mode structure of the pump and on the resonator mode spacing $\Delta \nu = (c/2L)$ for the idler and signal standing waves. For the singly-resonant oscillator the cavity has to be adjusted to only one frequency, while the nonresonant frequency can be adjusted so that $\omega_\mathrm{p} = \omega_\mathrm{i} + \omega_\mathrm{s}$ is satisfied. There are several ways to narrow the bandwidths of the OPO. With a tilted etalon inside the resonator of a singly-resonant cavity, single-mode operation can be achieved. Frequency stability of a few MHz has been demonstrated [613]. Another possibility is injection seeding. Stable single-mode operation was, for example, obtained by injecting the beam of a single-mode Nd:YAG pumplaser into the OPO cavity [614]. Using a single mode cw dye laser as the injection seeding source, tunable pulsed OPO-radiation with linewidths below 500 MHz have been achieved. A seed power of $0.3\,\mathrm{mW}$(!) was sufficient for stable single-mode OPO operation. The pump threshold can be lowered with a doubly-resonant resonator. However, the simple cavity of Fig. 6.25 cannot be kept in resonance for two different wavelengths, if these wavelengths are tuned. Here the three-mirror cavity of Fig. 6.28 solves this problem. Since the polarizations of the pump wave and the idler wave are generally orthogonal, a polarization beam splitter PBS splits both waves, which now experience resonant enhancement in the resonator $M_1 M_2$ or $M_3 M_2$. When the pump wavelength λ_p (a dye laser is used as pump source) is tuned, both cavities can be controlled by piezos to keep in resonance [616]. Frequency stabilities of below the $1\,\mathrm{kHz}$ level can be achieved [617]. The tuning range for collinear phase matching can be greatly extended by quasi phase matching in periodically poled $\mathrm{LiNbO_3}$ (PPLN) (Fig. 6.29). Meanwhile, cw OPOs are commercially available [618].

Impressive progress has been achieved with femtosecond optical parametric amplifiers, which can be used as ultrashort pulse generators with wavelengths tunable over a wide spectral range. They will be discussed in Vol. 2, Chap. 6.

Figure 6.29 High-power
cw OPO with periodically
poled LiNbO$_3$ crystal with
temperature control in a ring
cavity [618]

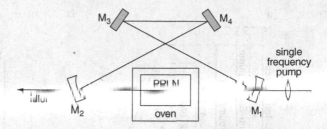

An interesting realization of a widely tunable OPO is a fiber OPO pumped by a fiber laser. One example is the sub-1 μm operation of a fiber OPO in a fiber ring resonator, pumped by an all-fiber master oscillator plus power amplifier based on a photonic crystal fiber as gain medium [592]. A conversion efficiency of 8:6 % from the pump at 1079 nm to the anti-Stokes signal at 715 nm. This means a frequency shift of 142 THz between pump and anti-Stokes signal.

Summary: Optical parametric oscillators are coherent devices similar to lasers. There are, however, important differences. While lasers can be pumped by incoherent sources, OPOs require coherent pump sources. Often diode laser-pumped solid state lasers are used. While in lasers coherent amplification can last until the inversion in the active medium has fallen below threshold, in OPO's the time dependence of the coherent output is directly coupled to that of the pump laser. Since the pump photon is split into signal and idler photon with $\omega_p = \omega_s + \omega_i$, the energy of the output equals that of the input i.e. there is no energy, i.e. heat deposited in the active crystal. The spectral tuning range is by far wider than for tunable lasers. Most OPOs operate in the near infrared but can be tuned from the visible region to the far infrared.

A good survey on different aspects of OPOs can be found in [615].

6.8   Tunable Raman Lasers

The tunable "Raman laser" may be regarded as a parametric oscillator based on stimulated Raman scattering. Since stimulated Raman scattering is discussed in more detail in Vol. 2, Sect. 3.3, we here summarize only very briefly the basic concept of these devices.

The ordinary Raman effect can be described as an inelastic scattering of pump photons $\hbar\omega_p$ by molecules in the energy level E_i. The energy loss $\hbar(\omega_p - \omega_s)$ of the scattered *Stokes photons* $\hbar\omega_s$ is converted into excitation energy (vibrational, rotational, or electronic energy) of the molecules

$$\hbar\omega_p + M(E_i) \to M^*(E_f) + \hbar\omega_s , \qquad (6.26)$$

where $E_f - E_i = \hbar(\omega_p - \omega_s)$. For the vibrational Raman effect this process can be interpreted as parametric splitting of the pump photon $\hbar\omega_p$ into a Stokes photon

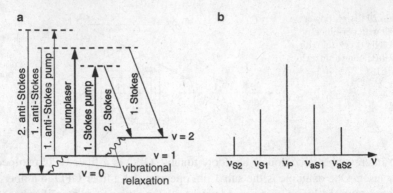

Figure 6.30 a Term diagram of Raman processes with several Stokes and anti-Stokes lines at frequencies $\nu = \nu_p \pm m\nu_v$; **b** spectral distribution of Raman lines and their overtones

$\hbar\omega_s$ and an optical phonon $\hbar\omega_v$ representing the molecular vibrations (Fig. 6.30a). The contributions $\hbar\omega_s$ from all molecules in the interaction region can add up to macroscopic waves when the phase-matching condition

$$\boldsymbol{k}_p = \boldsymbol{k}_s + \boldsymbol{k}_v ,$$

is fulfilled for the pump wave, the Stokes wave, and the phonon wave. In this case, a strong Stokes wave $E_s \cos(\omega_s t - \boldsymbol{k}_s \cdot \boldsymbol{r})$ develops with a gain that depends on the pump intensity and on the Raman scattering cross section. If the active medium is placed in a resonator, oscillation arises on the Stokes component as soon as the gain exceeds the total losses. Such a device is called a *Raman oscillator* or *Raman laser*, although, strictly speaking, it is not a laser but a parametric oscillator.

Those molecules that are initially in *excited* vibrational levels can give rise to superelastic scattering of *anti-Stokes radiation*, which has gained energy $(\hbar\omega_s - \hbar\omega_p) = (E_i - E_f)$ from the deactivation of vibrational energy.

The Stokes and the anti-Stokes radiation have a constant frequency shift against the pump radiation, which depends on the vibrational eigenfrequencies ω_v of the molecules in the active medium.

$$\omega_s = \omega_p - \omega_v , \quad \omega_{as} = \omega_p + \omega_n \dots .$$

If the Stokes or anti-Stokes wave becomes sufficiently strong, it can again produce another Stokes or anti-Stokes wave at $\omega_s^{(2)} = \omega_s^{(1)} - \omega_v = \omega_p - 2\omega_v$ and $\omega_{as}^{(2)} = \omega_p + 2\omega_v$. Therefore, several Stokes and anti-Stokes waves are generated at frequencies $\omega_s^{(n)} = \omega_p - n\omega_v$: $\omega_{as}^{(n)} = \omega_p + n\omega_v$ $(n = 1, 2, 3, \dots)$ (Fig. 6.30b). Tunable lasers as pumping sources therefore allow one to transfer the tunability range $(\omega_p \pm \Delta\omega)$ into other spectral regions $(\omega_p \pm \Delta\omega \pm n\omega_v)$.

The experimental realization uses a high-pressure cell filled with a molecular gas (H_2, N_2, CO, etc.) at pressures of up to 100 bar. The pump laser is either focussed into the gas cell with a lens of long focal length or a waveguide structure

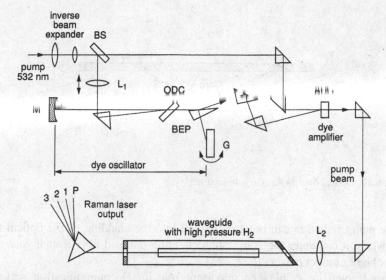

Figure 6.31 Infrared Raman waveguide laser in compressed hydrogen gas H_2, pumped by a tunable dye laser. The frequency-doubled output beam of a Nd:YAG laser is split by BS in order to pump a dye laser oscillator and amplifier. The dye laser oscillator is composed of mirror M, grating G, and beam-expanding prism BEP. The different Stokes lines are separated by the prism P (ODC: oscillator dye cell) [620]

is used (Fig. 6.31) where the pump laser beam is totally reflected at the walls of the waveguide, thus increasing the pathlength in the gain medium.

Stimulated Raman scattering (SRS) of dye laser radiation in hydrogen gas can cover the whole spectrum between 185 and 880 nm without any gaps, using three different laser dyes and frequency doubling the dye laser radiation [619]. A broadly tunable IR waveguide Raman laser pumped by a dye laser can cover the infrared region from 0.7 to 7 μm without gaps, using SRS up to the third Stokes order ($\omega_s = \omega_p - 3\omega_v$) in compressed hydrogen gas. Energy conversion efficiencies of several percent are possible and output powers in excess of 80 kW for the third Stokes component ($\omega_p - 3\omega_v$) have been achieved [620].

Instead of a high-pressure gas cell, solid bulk crystals can also be used as Raman gain medium. Because of their high density, the gain per cm is much higher and shorter pathlengths can be sufficient to obtain a high conversion efficiency. This can be further enhanced if the crystal is placed inside the pump laser resonator where the pump power is much higher.

If the gain medium is an optical fiber a long pathlength can be realized and the threshold is therefore low, which means that a low-power pump laser can be used. Since the most of the pump power is confined inside the core of the fiber by total reflection at the boundary between cladding and core (Fig. 6.32), the pump intensity inside the core is high. Even cw operation of Raman lasers has been demonstrated with silicon as the gain medium [621].

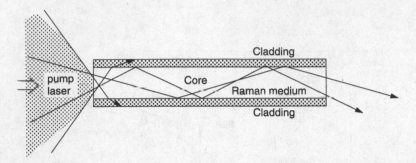

Figure 6.32 Optical fiber as Raman gain medium

The pump radiation can be also coupled into the cladding of the optical fiber, from where it can enter the core. Such cladding-pumped fiber Raman lasers can deliver higher output powers [622, 623].

Fiber Raman lasers play an important role in telecommunication networks, where optical fibers are used as pump sources for the signal wave [625].

For infrared spectroscopy, Raman lasers pumped by the numerous intense lines of CO_2, CO, HF, or DF lasers may be advantageous. Besides the vibrational Raman scattering, the rotational Raman effect can be utilized, although the gain is much lower than for vibrational Raman scattering, due to the smaller scattering cross section. For instance, H_2 and D_2 Raman lasers excited with a CO_2 laser can produce many Raman lines in the spectral range from 900 to $400\,\mathrm{cm}^{-1}$, while liquid N_2 and O_2 Raman lasers pumped with an HF laser cover a quasi-continuous tuning range between 1000 and $2000\,\mathrm{cm}^{-1}$. With high-pressure gas lasers as pumping sources, the small gaps between the many rotational–vibrational lines can be closed by pressure broadening (Sect. 3.3) and a true continuous tuning range of IR Raman lasers in the far infrared region becomes possible. Recently, a cw tunable Raman oscillator has been realized that utilizes as active medium a 650-m long single-mode silica fiber pumped by a 5-W cw Nd:YAG laser. The first Stokes radiation is tunable from 1.08 to 1.13 µm, the second Stokes from 1.15 to 1.175 µm [627]. With stimulated Raman scattering up to the seventh anti-Stokes order, efficient tunable radiation down to 193 nm was achieved when an excimer-laser pumped dye laser tunable around 440 nm was used [628].

A more detailed presentation of IR Raman lasers may be found in the review by Grasiuk et al. [629] and in [630–633].

Chapter 7
Optics of Gaussian Beams

In most textbooks on optics generally plane waves and their transformation by optical elements are treated. Since the fundamental modes in the laser output have a Gaussian beam profile (see Sect. 5.2), some knowledge about imaging and focusing of Gaussian beams are important for proper applications of laser beams. Although a nearly parallel laser beam is in many aspects similar to a plane wave, it shows several features that are different but that are important when laser beams are imaged by optical elements.

In this chapter we will discuss the basic characteristics of Gaussian beams and their transformation by optical elements such as lenses, mirrors, prisms and optical gratings. The following presentation follows that of the recommendable review by Kogelnik and Li [314].

7.1 Basic Characteristics of Gaussian Beams

A laser beam traveling into the z-direction can be represented by the field amplitude

$$E = A(x, y, z)e^{-i(\omega t - kz)} \quad \text{with } k = \frac{\omega}{c} . \tag{7.1}$$

While $A(x, y, z)$ is constant for a plane wave, it is a slowly varying complex function for a Gaussian beam. Since every wave obeys the general wave equation

$$\Delta E + k^2 E = 0 , \tag{7.2}$$

we can obtain the amplitude $A(x, y, z)$ of our particular laser wave by inserting (7.1) into (7.2). We assume the trial solution

$$A = C \cdot e^{-i[\varphi(z)+(k/2q)r^2]} , \tag{7.3}$$

where $r^2 = x^2 + y^2$, and $\varphi(z)$ represents a complex phase shift. In order to understand the physical meaning of the complex parameter $q(z)$, we express it in terms

W. Demtröder, *Laser Spectroscopy 1*, DOI 10.1007/978-3-642-53859-9_7,
© Springer-Verlag Berlin Heidelberg 2014

of two real parameters $w(z)$ and $R(z)$

$$\frac{1}{q} = \frac{1}{R} - i\frac{\lambda}{\pi w^2} . \tag{7.4}$$

With (7.4) we obtain from (7.3) the amplitude $A(x, y, z)$ in terms of R, w, and φ

$$A = C \cdot \exp\left(-\frac{r^2}{w^2}\right) \exp\left[-i\frac{kr^2}{2R(z)} - i\varphi(z)\right] \tag{7.5}$$

with

$$C = \frac{w_0}{w(z)} \quad w_0 = w(z = 0) .$$

This illustrates that $R(z)$ represents the radius of curvature of the wavefronts intersecting the axis at z (Fig. 7.1), and $w(z)$ gives the distance $r = (x^2 + y^2)^{1/2}$ from the axis where the amplitude has decreased to $1/e$ and thus the intensity has decreased to $1/e^2$ of its value on the axis (Sect. 5.2.3 and Fig. 5.11). Inserting (7.5) into (7.2) and comparing terms of equal power in r yields the relations

$$\frac{dq}{dz} = 1 , \quad \text{and} \quad \frac{d\varphi}{dz} = -i/q , \tag{7.6}$$

which can be integrated and gives, with $R(z = 0) = \infty$ from (7.4)

$$q(z) = q_0 + z = i\frac{\pi w_0^2}{\lambda} + z , \tag{7.7a}$$

where $q_0 = q(z = 0)$ and $w_0 = w(z = 0)$ (Fig. 7.1) and when we measure z from the beam waist at $z = 0$.

From (7.7a) we obtain:

$$\frac{1}{q(z)} = \frac{1}{q_0 + z} = \frac{1}{z + i\pi w_0^2/\lambda} . \tag{7.7b}$$

Multiplying nominator and denominator with $z - i\pi w_0^2/\lambda$ yields

$$\frac{1}{q(z)} = \frac{z}{z^2 + (\pi w_0^2/\lambda)^2} - i\frac{\lambda}{\pi w_0^2 \left(1 + (\lambda z/\pi w_0^2)^2\right)}$$

$$= \frac{1}{R} - i\frac{\lambda}{\pi w^2} \tag{7.7c}$$

where the last line equals (7.4).

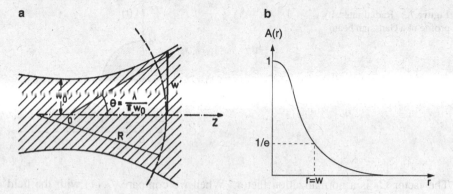

Figure 7.1 **a** Gaussian beam with beam waist w_0 and phase-front curvature $R(z)$; **b** radial dependence of the amplitude $A(r)$ with $r = (x^2 + y^2)^{1/2}$ [314]

This gives for the beam waist $w(z)$ and the radius of curvature $R(z)$ the relations:

$$w^2(z) = w_0^2 \left[1 + \left(\frac{\lambda z}{\pi w_0^2} \right)^2 \right] , \tag{7.8}$$

$$R(z) = z \left[1 + \left(\frac{\pi w_0^2}{\lambda z} \right)^2 \right] . \tag{7.9}$$

Integration of the phase relation (7.6)

$$\frac{d\varphi}{dz} = -i/q = -\frac{i}{z + i\pi w_0^2/\lambda} ,$$

yields the z-dependent phase factor

$$i\varphi(z) = \ln \sqrt{1 + (\lambda z/\pi w_0^2)} - i \arctan(\lambda z/\pi w_0^2) . \tag{7.10}$$

The second term in (7.10) is called the Gouy-phase. It describes the fact, that a Gaussian beam acquires an additional phase (besides the normal Phase $\exp(ikz)$) when it passes through a focus (beam waist w_0). It can be written as $\Phi_G = \arctan(z/z_R)$ where $z_R = \pi w_0^2/\lambda$ is the Rayleigh length (see below (7.26)).

Having found the relations between φ, R, and w, we can finally express the Gaussian beam (7.1) by the real beam parameters R and w. From (7.10) and (7.5), we get

$$E = C_1 \frac{w_0}{w} e^{(-r^2/w^2)} e^{[ik(z-r^2/2R)-i\phi]} e^{-i\omega t} . \tag{7.11}$$

The first exponential factor gives the radial Gaussian distribution, the second the phase, which depends on z and r. We have used the abbreviation

$$\phi = \arctan(\lambda z/\pi w_0^2) .$$

Figure 7.2 Radial intensity
profile of a Gaussian beam

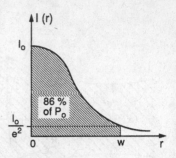

The factor C_1 is a normalization factor. When we compare (7.11) with the field
distribution (5.30) of the fundamental mode in a laser resonator, we see that both
formulas are identical for $m = n = 0$.

The radial intensity distribution (Fig. 7.2) is

$$I(r,z) = \frac{c\epsilon_0}{2}|E|^2 = C_2 \frac{w_0^2}{w^2} \exp\left(-\frac{2r^2}{w^2}\right) . \qquad (7.12)$$

The normalization factor C_2 allows

$$\int\limits_{r=0}^{\infty} 2\pi r I(r)\mathrm{d}r = P_0 \qquad (7.13)$$

to be normalized, which yields $C_2 = (2/\pi w_0^2)P_0$, where P_0 is the total power in
the beam. This yields

$$I(r,z) = \frac{2P_0}{\pi w^2} \exp\left(-\frac{2r^2}{w(z)^2}\right) . \qquad (7.14)$$

The peak intensity I_p is reached for $r = 0$ and we obtain from (7.14) $I_p = 2P_0/(\pi w^2)$. Since the average intensity is $I = P_0/(\pi w^2)$ the **peak intensity
in a Gaussian beam is just twice the average intensity**.

When the Gaussian beam is sent through an aperture with diameter $2a$, the fraction

$$\frac{P_t}{P_i} = \frac{2}{\pi w^2} \int\limits_{r=0}^{a} 2r\pi e^{-2r^2/w^2}\mathrm{d}r = 1 - e^{-2a^2/w^2} , \qquad (7.15)$$

of the incident power is transmitted through the aperture. Figure 7.3 illustrates this
fraction as a function of a/w. For $a = (3/2)w$ 99 % of the incident power is
transmitted, and for $a = 2w$ more than 99.9 % of the incident power is transmitted.
In this case diffraction losses are therefore negligible.

Figure 7.3 Fraction P_t/P_i of the incident power P_i of a Gaussian beam transmitted through an aperture with radius a

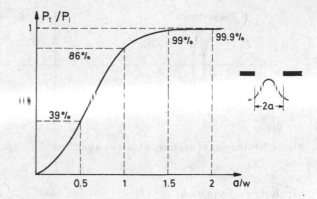

7.2 Imaging of Gaussian Beams by Lenses

A Gaussian beam can be imaged by lenses or mirrors, and the imaging equations are similar to those of spherical waves. When a Gaussian beam passes through a focusing thin lens with focal length f, the spot size w_s is the same on both sides of the lens (Fig. 7.4). The radius of curvature R of the phase fronts changes from R_1 to R_2 in the same way as for a spherical wave, so that

$$\frac{1}{R_2} = \frac{1}{R_1} - \frac{1}{f} . \tag{7.16}$$

The beam parameter q therefore satisfies the imaging equation

$$\frac{1}{q_2} = \frac{1}{q_1} - \frac{1}{f} . \tag{7.17}$$

If q_1 and q_2 are measured at the distances d_1 and d_2 from the lens, we obtain from (7.17) and (7.7a)–(7.7c) the relation

$$q_2 = \frac{(1 - d_2/f)q_1 + (d_1 + d_2 - d_1 d_2/f)}{(1 - d_1/f) - q_1/f} , \tag{7.18}$$

which allows the spot size w and radius of curvature R at any distance d_2 behind the lens to be calculated.

If, for instance, the laser beam is focused into the interaction region with absorbing molecules, the beam waist of the laser resonator has to be transformed into a beam waist located in this region. The beam parameters in the waists are purely imaginary, because in the focal plane is $R = \infty$; that is, from (7.4) we obtain

$$q_1 = i\pi w_1^2/\lambda , \quad q_2 = i\pi w_2^2/\lambda . \tag{7.19}$$

The beam diameters in the waists are $2w_1$ and $2w_2$, and the radius of curvature is infinite. Inserting (7.19) into (7.18) and equating the imaginary and the real parts

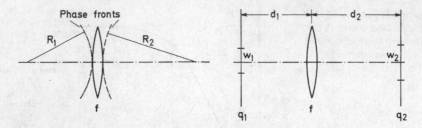

Figure 7.4 Imaging of a Gaussian beam by a thin lens

yields the two equations

$$\frac{d_1 - f}{d_2 - f} = \frac{w_1^2}{w_2^2} , \tag{7.20}$$

$$(d_1 - f)(d_2 - f) = f^2 - f_0^2 , \quad \text{with} \quad f_0 = \pi w_1 w_2 / \lambda . \tag{7.21}$$

Since $d_1 > f$ and $d_2 > f$, this shows that any lens with $f > f_0$ can be used. For a given f, the position of the lens is determined by solving the two equations for d_1 and d_2,

$$d_1 = f \pm \frac{w_1}{w_2} \sqrt{f^2 - f_0^2} , \tag{7.22}$$

$$d_2 = f \pm \frac{w_2}{w_1} \sqrt{f^2 - f_0^2} . \tag{7.23}$$

From (7.20) we obtain the beam waist radius w_2 in the collimated region

$$w_2 = w_1 \left(\frac{d_2 - f}{d_1 - f} \right)^{1/2} . \tag{7.24}$$

When the Gaussian beam is mode-matched to another resonator, the beam parameter q_2 at the mirrors of this resonator must match the curvature R of the mirror and the spot size w in (5.39a), (5.39b). From (7.18), the correct values of f, d_1, and d_2 can be calculated.

We define the collimated or *waist region* as the range $|z| \leq z_R$ around the beam waist at $z = 0$, where at $z = \pm z_R$ the spot size $w(z)$ has increased by a factor of $\sqrt{2}$ compared with the value w_0 at the waist. Using (7.8) we obtain

$$w(z) = w_0 \left[1 + \left(\frac{\lambda z_R}{\pi w_0^2} \right)^2 \right]^{1/2} = \sqrt{2} w_0 , \tag{7.25}$$

which yields for the *waist length* or *Rayleigh length*

$$z_R = \pi w_0^2 / \lambda . \tag{7.26}$$

Figure 7.5 Beam waist region and Rayleigh length z_R of a Gaussian beam

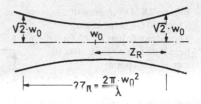

The waist region extends about one Rayleigh distance on either side of the waist (Fig. 7.5). The length of the Rayleigh distance depends on the spot size and therefore on the focal length of the focusing lens. Figure 7.6 depicts the dependence of the full Rayleigh length $2z_R$ on w_0 for two different wavelengths.

Gaussian beams do not diverge linearly as conventional light beams. Within the Raleigh length around the focus the divergence is very small (near field). Farther away from the focus the beam becomes more and more a spherical wave (far field) and the divergence angel approaches the constant asymptotic limit θ_G (Fig. 7.8)

If the beam is not a pure Gaussian beam but contains admixtures of higher order modes, the beam quality can be defined by the parameter

$$M^2 = w(R) \cdot \theta / (w_0 \theta_G)$$

where w is the actual spot size of the beam and w_0 that of a pure Gaussian beam. For a pure Gaussian beam is $M = 1$, for admixtures of higher order modes is $M > 1$. At large distances $z \gg z_R$ from the waist, the Gaussian beam wavefront is essentially a spherical wave emitted from a point source at the waist. This region is called the *far field*. The divergence angle θ (far-field half angle) of the beam can be

Figure 7.6 Full Rayleigh lengths $2z_R$ as a function of the beam waist w_0 for two different wavelengths $\lambda_1 = 632.8\,\text{nm}$ (HeNe laser) and $\lambda_2 = 10.6\,\mu\text{m}$ (CO_2 laser)

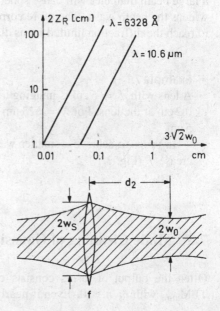

Figure 7.7 Focusing of a Gaussian beam by a lens

Figure 7.8 Divergence of
a focused Gaussian beam.
The extrapolations of the far-
field light rays towards $x = 0$
merge into a point

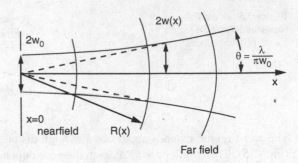

obtained from (7.8) and Fig. 7.1 with $z \gg z_R$ as

$$\theta = \frac{w(z)}{z} = \frac{\lambda}{\pi w_0} . \tag{7.27}$$

Note, however, that in the near-field region the center of curvature *does not* coincide
with the center of the beam waist (Fig. 7.1). When a Gaussian beam is focused by
a lens or a mirror with focal length f, the spot size in the beam waist is for $f \gg w_s$

$$w_0 = \frac{f\lambda}{\pi w_s} , \tag{7.28}$$

where w_s is the spot size at the lens (Fig. 7.7).

To avoid diffraction losses the diameter of the lens should be $d \geq 3w_s$.

In order to reach a minimum beam waist, diffraction should be small. This means
the beam should have a large diameter before the lens. With uncorrected lenses such
a large beam diameter will cause spherical aberration which distorts the focus and
widens the beam waist. Therefore corrected microscope objectives should be used
to reach the diffraction limited focus diameter.

Example 7.1
A lens with $f = 5\,\text{cm}$ is imaging a Gaussian beam with a spot size of $w_s = 0.2\,\text{cm}$ at the lens. For $\lambda = 623\,\text{nm}$ the focal spot has the waist radius $w_0 = 5\,\mu\text{m}$.

In order to achieve a smaller waist radius, one has to increase w_s or decrease f (Fig. 7.7).

7.3 Mode Cleaning of Gaussian Beams

Often the output of lasers consists of an overlap of several transverse modes
TEM_{mnq} with $m, n > 0$. Even if nearly all of the output power is contained in the

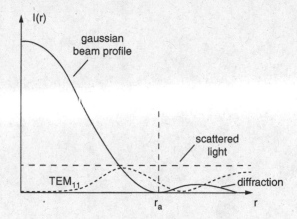

Figure 7.9 Suppression of higher-order modes and scattered light by an aperture with radius r_a. The intensity of the TEM$_{11}$ mode and of scattered light are not to scale but enlarged

fundamental mode with $m = n = 0$, spurious admixtures of higher order modes will detoriate the quality of the output beam. Other perturbations of Gaussian beams are due to dust particles on lenses or mirrors which give rise to scattered light which shows interference patterns and overlaps the pure Gaussian beam. Also the diffraction by lenses or apertures causes an intensity distribution which contains besides the central diffraction maximum higher diffraction orders which spoil the pure Gaussian intensity distribution. In order to obtain the minimum spot size in the focus the beam has to be cleaned from all unwanted admixtures. This can be achieved with spatial filtering by a narrow circular aperture in the focal area of the imaging lens. This aperture transmits most of the Gaussian beam but suppresses higher order modes and scattered light (Fig. 7.9). If the radius r_a of the aperture is chosen correctly to be equal to the minimum in the diffraction pattern of the Gaussian beam, the diffraction of the Gaussian beam is minimized while that of the overlapping light is larger. The best filtering of the Gaussian beam with the radial intensity distribution $I(r) = I_0 \cdot \exp(-2r^2/w^2)$ and $w = f \cdot \lambda/(\pi w_s)$ can be achieved for $r_a = 1/e^2 I_0$, i.e. the radius r must be equal to the spot size w, which depends on the focal length f of the imaging lens and the beam radius w_s at the lens.

Solutions

Chapter 2

1. **a)** With $\beta = 1/(kT) \Rightarrow$

$$\sum_{q=0}^{\infty} q \cdot h v e^{-q h v \cdot \beta} = -\frac{\partial}{\partial \beta} \left(\sum_{q=0}^{\infty} e^{-q h v \beta} \right)$$

$$= -\frac{\partial}{\partial \beta} \left(\frac{1}{1 - e^{-h v \beta}} \right) = \frac{h v e^{-h v \beta}}{\left(1 - e^{-h v \beta}\right)^2} .$$

b)

$$\sum_{q=0}^{\infty} e^{-q h v \beta} = \frac{1}{1 - e^{-h v \beta}} \qquad \frac{a}{b} = \frac{h v}{e^{h v/(kT)} - 1} .$$

2. **a)** The spot size on the output mirror is

$$dA = \pi\, w_s^2 = \pi (0.1)^2 \, \text{cm}^2 = 3 \times 10^{-2} \, \text{cm}^2 .$$

The irradiance at the mirror is then

$$I_1 = \frac{1}{\pi\, 10^{-2}} \, \text{W/cm}^2 \approx 30 \, \text{W/cm}^2 = 3 \times 10^5 \, \text{W/m}^2 .$$

The solid angle $d\Omega$ into which the laser beam is emitted is:

$$d\Omega = \left(4 \times 10^{-3}\right)^2 / 4\pi = 1.3 \times 10^{-6} \, \text{sr} .$$

The radiance L of the laser is:

$$L = \frac{1}{dA\, d\Omega} = 2 \times 10^{11} \, \text{W m}^{-2} \, \text{sr}^{-1} .$$

W. Demtröder, *Laser Spectroscopy 1*, DOI 10.1007/978-3-642-53859-9,
© Springer-Verlag Berlin Heidelberg 2014

At a surface at a distance $z = 1\,\mathrm{m}$ from the mirror, the spot size is:

$$A_2 = \mathrm{d}A + z^2\mathrm{d}\Omega = 4.4 \times 10^{+2}\,\mathrm{cm}^2 \ .$$

The intensity at the surface is:

$$I_2 = \frac{1}{4.4 \times 10^{-2}} \frac{\mathrm{W}}{\mathrm{cm}^2} = 23\,\mathrm{W/cm}^2 = 2.3 \times 10^5\,\mathrm{W/m}^2 \ .$$

b) For a spectral width $\delta\nu = 1\,\mathrm{MHz}$, the spectral power density at the mirror is:

$$\rho_1 = (I_1/c)/\delta\nu = 10^{-9}\,\mathrm{Ws}^2/\mathrm{m}^3 \ .$$

This should be compared with the visible part of the solar radiation on Earth, $I \approx 10^3\,\mathrm{W/m}^2$, $\delta\nu = 3 \times 10^{16}\,\mathrm{s}^{-1} \Rightarrow \rho_{\mathrm{SR}} = 10^{-22}\,\mathrm{Ws}^2/\mathrm{m}^3$, which is smaller by 13 orders of magnitude.

3. $I = I_0\,\mathrm{e}^{-\alpha d}$

$$I_\parallel = I_0\,\mathrm{e}^{-100 \cdot 0.1} = I_0\,\mathrm{e}^{-10} = 4.5 \times 10^{-5} I_0$$
$$I_\perp = I_0\,\mathrm{e}^{-5 \cdot 0.1} = I_0\,\mathrm{e}^{-0.5} = 0.6 I_0 \ .$$

4. $I = \dfrac{P_0}{4\pi r^2} = \dfrac{100}{4\pi (0.02)^2} \dfrac{\mathrm{W}}{\mathrm{m}^2} = 2 \times 10^4\,\mathrm{W/m}^2$

$$I_\nu = \frac{I}{\Delta\nu} \ .$$

$$\text{For } \left.\begin{array}{l} \Delta\lambda = 100\,\mathrm{nm} \\ \lambda = 400\,\mathrm{nm} \end{array}\right\} \Rightarrow |\Delta\nu| = \frac{c}{\lambda^2}\Delta\lambda$$

$$= 1.8 \times 10^{14}\,\mathrm{s}^{-1}$$

$$I_\nu = \frac{2 \times 10^4}{1.8 \times 10^{14}} \frac{\mathrm{Ws}}{\mathrm{m}^2} = 1.1 \times 10^{-10}\,\mathrm{Ws\,m}^{-2}$$
$$\rho_\nu = I_\nu/c = 3.6 \times 10^{-19}\,\mathrm{Ws}^2\,\mathrm{m}^{-3} \ .$$

The spectral mode density is

$$n(\nu) = \frac{8\pi\nu^2}{c^3} \ .$$

Within the volume of the sphere with $r = 2\,\mathrm{cm}$

$$V = \frac{4}{3}\pi r^3 = 3.3 \times 10^{-5}\,\mathrm{m}^3$$

are

$$N = n(\nu) \, V \, \Delta \nu = \frac{8\pi \nu^2}{c^3} \Delta \nu \, V = \frac{8\pi}{c \, \lambda^2} \Delta \nu \, V = 3 \times 10^{15} \, \text{modes} \, .$$

The energy per mode is

$$W_{\text{m}} = \frac{\rho_\nu \, \Delta \nu \, V}{N} = 7 \times 10^{-25} \, \text{Ws/mode} \, .$$

The energy of a photon at $\lambda = 400 \, \text{nm}$ is

$$E = h\nu = h\frac{c}{\lambda} = 4.95 \times 10^{-19} \, \text{Ws} = 3.1 \, \text{eV}$$

$\Rightarrow$ The average number of photons per mode is

$$n_{\text{ph}} = \frac{W_{\text{m}}}{h\nu} = 1.5 \times 10^{-6} \, .$$

The average number of photons per mode is therefore very small.

5. $I = I_0 \, e^{-\alpha x} = 0.9 I_0$

$$\Rightarrow \alpha x = -\ln 0.9 \Rightarrow \alpha x = 0.1 \, .$$

With $x = 5 \, \text{cm} \Rightarrow \alpha = 0.02 \, \text{cm}^{-1}$

$$\alpha = N\sigma \Rightarrow N = \frac{\alpha}{\sigma} = \frac{0.02}{10^{-14}} \, \text{cm}^{-3} = 2 \times 10^{12} \, \text{cm}^{-3} \, .$$

6. **a)** $\tau_i = \dfrac{1}{\sum A_{in}} = \dfrac{1}{13 \times 10^7} \, \text{s} = 7.7 \, \text{ns}$

$$\frac{\text{d}N_n}{\text{d}t} = N_i \, A_{in} - N_n \, A_n \, .$$

For stationary conditions $\text{d}N_n/\text{d}t = 0$

$$\Rightarrow \frac{N_n}{N_i} = \frac{A_{in}}{A_n} = A_{in} \tau_n :$$

$$\frac{N_1}{N_i} = 3 \times 10^7 \cdot 5 \times 10^{-7} = 15$$

$$\frac{N_2}{N_i} = 1 \times 10^7 \cdot 6 \times 10^{-9} = 0.06$$

$$\frac{N_3}{N_i} = 5 \times 10^7 \cdot 10^{-8} = 0.5 \, .$$

b) With $g_0 = 1$, $g_i = 3$ we obtain:

$$B_{0i}^{(0)} = \frac{g_i}{g_0} B_{i0} = 3B_{i0} = \frac{3c^3}{8\pi h \nu^3} A_{i0}$$
$$= 4.6 \times 10^{20} \, \text{m}^3 \, \text{W}^{-1} \, \text{s}^{-3} \, .$$

If the absorption rate and total emission rate of level $|i\rangle$ should be equal, we obtain:

$$B_{0i}^{(\nu)} \rho_\nu = A_i = 1.3 \times 10^8 \, \text{s}^{-1}$$
$$\Rightarrow \rho_\nu = \frac{1.3 \times 10^8}{4.6 \times 10^{20}} \, \text{Ws}^2/\text{m}^3 = 2.8 \times 10^{-13} \, \text{Ws}^2/\text{m}^3 \, .$$

With a laser bandwidth of $\Delta\nu_2 = 10 \, \text{MHz}$, the energy density is

$$\rho = \int \rho_\nu \, d\nu \approx \rho_\nu \, \Delta\nu_2 = 2.8 \times 10^{-6} \, \text{Ws/m}^3$$
$$\Rightarrow I = c\rho = 6.3 \times 10^2 \, \text{W/m}^2 = 63 \, \text{mW/cm}^2 \, .$$

c) $B_{0i}^{(\nu)} = \frac{c}{h\nu} \int \sigma_{0i} \, d\nu \approx \frac{c}{h\nu} \overline{\sigma_{0i}} \, \Delta\nu_a$.

With $\Delta\nu_a = 1/\tau_i \Rightarrow \Delta\nu_a = 1/(2\pi\tau_i)$ for the absorption linewidth, the absorption cross-section becomes:

$$\sigma_{0i} = 4.3 \times 10^{-14} \, \text{m}^2 = 4.3 \times 10^{-10} \, \text{cm}^2 \, .$$

7. The Rabi flopping frequency for the resonance case $\omega = \omega_{i2}$ is

$$\Omega = \sqrt{(D_{i2}E_0/\hbar)^2 + (\gamma/2)^2}$$

where D_{i2} is the dipole matrix element and $\gamma = (\gamma_i + \gamma_2)/2$.
The relation between D_{i2} and the spontaneous transition probability A_{i2} is

$$A_{i2} = \frac{16\pi^2\nu^3}{3\epsilon_0 hc^3} |D_{i2}|^2 = \frac{16\pi^2}{3\epsilon_0 h\lambda^3} |D_{i2}|^2 \, .$$

This gives for Ω:

$$\Omega^2 = |D_{i2}|^2 E_0^2/\hbar^2 + (\gamma/2)^2 = \frac{3\epsilon_0\lambda^3 A_{i2}}{4h} + (\gamma/2)^2 \, .$$

With $\lambda = 600 \, \text{nm}$, $A_{i2} = 10^{-7} \, \text{s}^{-1}$ $\gamma/2 = \frac{1}{4}(\frac{1}{\tau_i} + \frac{1}{\tau_2}) = 7.7 \times 10^7 \, \text{s}^{-1}$ we obtain:

$$\Omega^2 = \left(2.17 \times 10^9 E_0^2 - 5.5 \times 10^{15}\right) \text{s}^{-2} \geq \frac{1}{\tau_2^2} = 2.8 \times 10^{16} \, \text{s}^{-2}$$
$$\Rightarrow E_0^2 \geq 1.5 \times 10^7 \, \text{V}^2/\text{m}^2 \Rightarrow E_0 \geq 3.9 \times 10^3 \, \text{V/m} \, .$$

The intensity of the inducing field is then

$$I = c\epsilon_0 E_0^2 = 4 \times 10^{14} \, \text{W/m}^2$$

and the energy density

$$\rho = I/c = \epsilon_0 E_0^2 = 1.33 \times 10^6 \, \text{Ws/m}^3 \,.$$

This can be compared with the intensity of the Sun's radiation on Earth, which is $I_{\text{sun}} \approx 10^3 \, \text{W/m}^2$.

8. Dust particles on the lens L_1 cause scattering of light in all directions. This light is not focussed by L_1, and therefore only a tiny fraction can pass through the aperture. The same is true for imperfections of lenses or mirror surfaces. Without the aperture the superposition of scattered light or light with deformed wavefronts with the incident light causes interference patterns. The aperture therefore "cleans" the Gaussian laser beam.

9. For coherent illumination of the slits, the following condition holds:

$$b^2 d^2/r^2 \leq \lambda^2 \Rightarrow d^2 \leq r^2 \lambda^2/b^2$$

where b = source diameter, d = slit separation, and r = distance between source and slits.

a) $b = 1 \, \text{mm}, r = 1 \, \text{m}, \lambda = 400 \, \text{nm}$

$$\Rightarrow d^2 \leq \frac{1 \cdot 16 \times 10^{-14}}{10^{-6}} = 16 \times 10^{-8} \, \text{m}^2$$

$$\Rightarrow d \leq 0.4 \, \text{mm} \,.$$

b) $b = 10^9 \, \text{m}, \lambda = 500 \, \text{nm}, r = 4 \, \text{Ly} = 3.78 \times 10^{16} \, \text{m}$

$$\Rightarrow d^2 \leq 357 \, \text{m}^2 \Rightarrow d \leq 19 \, \text{m} \,.$$

c) Here the maximum slit separation d is limited by the coherence length L_c of the laser beam, which depends on the spectral width $\Delta \nu_L$ of the laser radiation. With $\Delta \nu_L = 1 \, \text{MHz}$ we obtain

$$\Delta s_c = \frac{c}{2\pi \Delta \nu_L} = 47.7 \, \text{m} \,.$$

10. Induced and spontaneous transition probabilities are equal when the radiation field contains one photon per mode. This means:

$$\bar{n} = \frac{1}{e^{h\nu/kT} - 1} = 1 \Rightarrow e^{h\nu/kT} = 2$$

$$\Rightarrow T = \frac{h\nu}{k \ln 2} = \frac{hc}{\lambda k \ln 2} \,.$$

a) For $\lambda = 589\,\text{nm}$ we obtain for a thermal radiation field:

$$T = 3.53 \times 10^4 \,\text{K} \,.$$

If a laser beam is sent through a cavity with $V = 1\,\text{cm}^3$, the condition $B_{ik}\rho = A_{ik}$ can be fulfilled at modest laser intensities. This can be estimated as follows:

The number of modes in the cavity within the frequency interval $\Delta\nu_L = 10\,\text{MHz}$ (natural linewidth of the $3P$–$3S$ transition of Na) is

$$n\,\text{d}\nu = \frac{8\pi}{c\lambda^2}\,\text{d}\nu = 2.4 \times 10^6\,/\text{cm}^3 \,.$$

The energy of a photon at $\lambda = 589\,\text{nm}$ is $h\nu = 3.36 \times 10^{-19}\,\text{Ws}$. With 1 photon per mode, the radiation density in the cavity with $V = 1\,\text{cm}^3$ is:

$$\rho = 8.06 \times 10^{-13}\,\text{Ws/cm}^3 \,.$$

The intensity of a laser beam with a spectral width of $10\,\text{MHz}$ is then inside the cavity

$$I = \rho c = 24 \times 10^{-3}\,\text{W/cm}^2 = 24\,\text{mW/cm}^2 \,.$$

b) For $\nu = 1.77 \times 10^9\,\text{s}^{-1}$ we obtain

$$T = 0.12\,\text{K} \,.$$

The energy density ρ of the thermal field within the natural linewidth $\text{d}\nu = 0.15\,\text{s}^{-1}$ at $T = 0.12\,\text{K}$ is

$$\rho = \rho_\nu\,\text{d}\nu = n(\nu)\,h\nu\,\text{d}\nu \,.$$

With $n(\nu) = \frac{8\pi\nu^2}{c^3} = 2.9 \times 10^{-12}\,/\text{cm}^3$

$$\Rightarrow \rho = 5 \times 10^{-37}\,\text{Ws/cm}^3 \,.$$

This is 24 orders of magnitude smaller than the visible radiation in a).

11. $\dfrac{1}{\tau_{\text{eff}}} = \dfrac{1}{\tau_{\text{sp}}} + n\sigma\overline{v} \,.$

At $p = 10\,\text{mb}$ the atomic density is $n = 3 \times 10^{17}\,\text{cm}^{-3}$,

At $T = 400\,\text{K}$ the mean relative velocity is

$$\overline{v} = \sqrt{\frac{8kT}{\pi\mu}} \quad \text{with } \mu = \frac{m_{N_2} \times m_{N_a}}{m_{N_2} + m_{N_a}} = 12.6\,\text{AMU}$$

$$1\,\text{AMU} = 1.66 \times 10^{-27}\,\text{kg}$$

$$\Rightarrow \bar{v} = 820\,\text{m/s} = 8.2 \times 10^4\,\text{cm/s}$$

$$\Rightarrow \frac{1}{\tau_{\text{eff}}} = \frac{10^9}{16} + 3 \times 10^{17} \times 4 \times 10^{-15} \times 8.2 \times 10^4\,\text{s}^{-1}$$

$$= 1.62 \times 10^8\,\text{s}^{-1}$$

$$\Rightarrow \tau_{\text{eff}} = 6.2\,\text{ns} = 0.388\tau_{\text{sp}} \quad \text{with } \tau_{\text{sp}} = 16\,\text{ns} .$$

Chapter 3

1. The natural linewidth is

$$\Delta \nu_n = \frac{1}{2\pi} \left(\frac{1}{\tau(3s_2)} + \frac{1}{\tau(2p_4)} \right)$$

$$= \frac{1}{2\pi} (1.7 \times 10^7 + 5.6 \times 10^7)\,\text{s}^{-1}$$

$$= 11.6\,\text{MHz} .$$

The Doppler width is

$$\Delta \nu_D = 7.16 \times 10^{-7} \nu_0 \sqrt{T/M}$$

With $\nu_0 = c/\lambda = 4.74 \times 10^{14}\,\text{s}^{-1}$, $T = 400\,\text{K}$, $M = 20\,\text{AMU}$
$\Rightarrow \Delta \nu_D = 1.52 \times 10^9\,\text{s}^{-1} = 1.52\,\text{GHz}$.
The pressure broadening has two contributions:

a) by collisions with He atoms.

$$\Delta \nu_p = \frac{1}{2\pi} (n_{\text{He}} \sigma_B(\text{Ne} - \text{He}) \bar{v} .$$

At $p = 2\,\text{mb}$ and $T = 400\,\text{K}$
$\Rightarrow n_{\text{He}} = p/(kT) = 3.6 \times 10^{16}\,\text{cm}^{-3}$
$\sigma_B(\text{Ne} - \text{He}) = 6 \times 10^{-14}\,\text{cm}^2$, $\bar{v} = 1.6 \times 10^5\,\text{cm/s}$
$\Rightarrow \Delta \nu_p = 5.5 \times 10^7\,\text{s}^{-1} = 55\,\text{MHz}$.

b) by collisions Ne − Ne (resonance broadening)

$$\bar{v}(\text{Ne} - \text{Ne}) = 8.8 \times 10^4\,\text{cm/s}$$

$$\sigma_B(\text{Ne} - \text{Ne}) = 1 \times 10^{-13}\,\text{cm}^2$$

$$n_{\text{Ne}} = 3.6 \times 10^{15}\,\text{cm}^{-3}$$

$$\Rightarrow \Delta \nu_p(\text{Ne} - \text{Ne}) = 5\,\text{MHz} .$$

The line shift is

$$\Delta\nu_s(\text{Ne} - \text{Ne}) = 0.5\,\text{MHz} .$$

The total pressure broadening is

$$\Delta\nu_p = 55 + 5 = 60\,\text{MHz} .$$

The total shift is

$$\Delta\nu_s = 9 + 0.5 = 9.5\,\text{MHz} .$$

2. $n = p/kT$
   with $p = 1\,\text{mb} \triangleq 10^2\,\text{Pa} \Rightarrow n = 2.4 \times 10^{22}\,\text{m}^{-3}$

$$\overline{v} = \sqrt{\frac{8kT}{\pi\mu}} \qquad \mu = \frac{44 \times 14z}{191}\,\text{AMU}$$

$$\Rightarrow \overline{v} = 433\,\text{m/s} .$$

a) The pressure-broadened linewidth is

$$\Delta\nu_p = \frac{1}{2\pi}n\sigma_b\overline{v}$$

with $\sigma_b = 5 \times 10^{-14}\,\text{cm}^2 \Rightarrow \Delta\nu_p = 8.3 \times 10^6\,\text{s}^{-1} = 8.3\,\text{MHz}$.
The saturation broadening of the homogeneous linewidth $\Delta\nu_p$ is

$$\Delta\nu_s = \Delta\nu_p\sqrt{1 + S} .$$

The saturation parameter S is defined as the ratio of induced emission rate $B_{ik}\rho_\nu\,d\nu$ within the spectral interval $d\nu$ to the total relaxation rate $\gamma = 1/\tau_{\text{eff}}$. Because $B_{ik}\rho_\nu\,d\nu = I\sigma_a/h\nu$ we can write:

$$S = \frac{I\sigma_a}{h\nu\gamma} = \frac{I\sigma_a}{h\nu\,2\pi\,\Delta\nu_p}$$

where

$$I = \frac{50\,\text{W}}{\pi\,\frac{1}{4}\,10^{-2}\,\text{cm}^2} = 6.4 \times 10^3\,\text{W/cm}^2$$

is the laser intensity in the focal plane.
With $\sigma_a = 10^{-14}\,\text{cm}^2$, $\gamma = 2\pi\Delta\nu_p = 2\pi \cdot 8.3 \times 10^6\,\text{s}^{-1} = 5.2 \times 10^7\,\text{s}^{-1}$
$h\nu = 1.9 \times 10^{-20}\,\text{Ws} \Rightarrow$

$$S = 64 .$$

The saturation broadening is then:

$$\Delta\nu_s = \Delta\nu_p\sqrt{65} = 8.06\Delta\nu_p = 66.9\,\text{MHz} .$$

The Doppler width is

$$\Delta\nu_D = 7.16 \times 10^{-7}\,(c/\lambda)\sqrt{T/M} \quad (T/K \text{ and } M/\text{AMU}).$$

With $M = 32 + 6 \times 19 = 146\,\text{AMU}$ for SF_6 and $T = 300\,\text{K}$ we obtain

$$\Delta\nu_D = 30\,\text{MHz}.$$

Saturation broadening is dominant.
b) At the temperature $T = 10\,\text{K}$, the Doppler width is, for $\lambda = 21\,\text{cm}$ and $M = 1\,\text{AMU}$,

$$\Delta\nu_D = 7.16 \times 10^{-7}(c/\lambda)\sqrt{T/M}$$
$$= 3.23 \times 10^3\,\text{s}^{-1} = 3.23\,\text{kHz}.$$

The natural linewidth is

$$\Delta\nu_n = A_{ik}/2\pi + (4/2\pi)10^{-15}\,\text{s}^{-1} = 6.4 \times 10^{-16}\,\text{s}^{-1}.$$

For the Lyman-α transition at $\lambda = 121.6\,\text{nm}$ it is

$$\Delta\nu_D = 5.6 \times 10^9\,\text{s}^{-1} = 5.6\,\text{GHz}; \quad \Delta\nu_n = 1.5 \times 10^8\,\text{s}^{-1}.$$

The absorption coefficient is $\alpha = n\sigma_{ik}$. The absorption cross-section is related to the spontaneous transition probability by

$$\sigma_{ik} = \frac{\pi}{8}\lambda^2 A_{ik}/\Delta\nu_n = \frac{\pi^2}{4}\lambda^2 = 1.09 \times 10^3\,\text{cm}^2 \approx 1 \times 10^3\,\text{cm}^2.$$

We can assume the star radiation to consist of many spectral intervals with width $\Delta\nu_n$. Each of these spectral parts is absorbed only by H atoms within the velocity group $v_z = (\nu - \nu_0) \cdot \lambda \pm \Delta v_z$ with $\Delta v_z = \lambda \cdot \Delta\nu_n$ inside the Doppler-absorption profile with width $\Delta\nu_D$. This is the fraction $\Delta\nu_n/\Delta\nu_D$ of all H atoms.
The absorption coefficient is therefore

$$\alpha = n\sigma_{ik}\Delta\nu_{n/\Delta\nu_D}$$
$$= 10 \times 10^3 \times 6.4 \times 10^{-16}/3.23 \times 10^3$$
$$= 2 \times 10^{-16}\,\text{cm}^{-1}.$$

The radiation has decreased to $10\,\%\ I_0$ for

$$e^{-\alpha L} = 0.1 \Rightarrow \alpha L = 2.3 \Rightarrow L = \frac{2.3}{2 \times 10^{-16}}\,\text{cm} = 1.15 \times 10^{16}\,\text{cm}$$
$$L = 1.15 \times 10^{11}\,\text{km} = 0.012\,\text{Ly}.$$

For the Lyman-α radiation the absorption cross-section is

$$\sigma_{ik} = \frac{\pi^2}{4}\lambda^2 = 3.7 \times 10^{-10}\,\mathrm{cm}^2$$
$$\Rightarrow \alpha = n\sigma_{ik}\Delta\nu_n/\Delta\nu_D = 1 \times 10^{-10}\,\mathrm{cm}^{-1}$$
$$L = \frac{2.3}{\alpha} = 2.3 \times 10^{10}\,\mathrm{cm} = 2.3 \times 10^5\,\mathrm{km}\ .$$

c) With $\tau = 20\,\mu s$, the natural linewidth is:

$$\Delta\nu_n = \frac{1}{2\pi\tau} = 8\,\mathrm{kHz}\ .$$

With $\lambda = 3.39 \times 10^{-6}\,\mathrm{m}$, $M = 16\,\mathrm{AMU}$ the Doppler width is

$$\Delta\nu_D = 7.16 \times 10^{-7}(c/\lambda)\sqrt{T/M} = 270\,\mathrm{MHz}\ .$$

The pressure broadened linewidth is

$$\Delta\nu_p = n\sigma_b\overline{v} = (p/kT)\sigma_b\overline{v} = 17\,\mathrm{MHz}\ .$$

The transit time broadening is

$$\Delta\nu_{tr} = 0.4\overline{v}/w$$

with $w = 0.5\,\mathrm{cm}$, $\overline{v} = 700\,\mathrm{m/s} \Rightarrow \Delta\nu_{tr} = 56\,\mathrm{kHz}$.
d) In order to fulfill $\Delta\nu_{tr} < \Delta\nu_n \Rightarrow$

$$0.4\overline{v}/w < \frac{1}{2\pi\tau} \Rightarrow w > 0.8\pi\tau v = 3.51\,\mathrm{cm}$$
$$\Rightarrow \text{diameter } 2w > 7\,\mathrm{cm}\ .$$

The saturation broadening is

$$\Delta\nu_S = \Delta\nu_p\sqrt{1+S}\ .$$

For $\sigma_a = 10^{-10}\,\mathrm{cm}^2$ and with $I = \frac{10^{-2}}{0.5^2\pi}\,\frac{\mathrm{W}}{\mathrm{cm}^2} = 1.27 \times 10^{-2}\,\frac{\mathrm{W}}{\mathrm{cm}^2}$ we obtain (see Problem 3.2a)

$$S = \frac{I\sigma_a}{h\nu\,2\pi\,\Delta\nu_p} = 2.2 \times 10^{-1} = 0.22$$
$$\Rightarrow \Delta\nu_S = 17\,\mathrm{MHz} \times 1.09 = 18.62\,\mathrm{MHz}\ .$$

Saturation broadening plays here a minor role.

3. **a)** The Lorentzian and the Gaussian profiles intersect for $I_L(\omega) = I_G(\omega)$. The normalization $I_L(\omega_0) = I_G(\omega_0) = I_0$ requires:

$$\frac{I_0(\gamma/2)^2}{(\omega - \omega_0)^2 + (\gamma/2)^2} = I_0 e^{\frac{-(\omega-\omega_0)^2}{0.368\omega_0^2}}$$

$$\Rightarrow \ln\left[(\omega - \omega_0)^2 + (\gamma/2)^2\right] - 2\ln(\gamma/2) = \frac{(\omega - \omega_0)^2}{0.368\omega_0^2} .$$

With: $\delta\omega_0 = 2\pi\delta\nu_D = 1 \times 10^{10}\,\text{s}^{-1}$

$$\gamma = 2\pi \times 10^7\,\text{s}^{-1} = 6.3 \times 10^7\,\text{s}^{-1}$$

we obtain

$$\ln\left[(\omega - \omega_0)^2 + 9.9 \times 10^{14}\right] - 34.5 = \frac{(\omega - \omega_0)^2}{0.36 \times 10^{20}}$$

$$\Rightarrow (\omega - \omega_0) = 2.18 \times 10^{10}\,\text{s}^{-1} ; \quad \nu - \nu_0 = 3.47\,\text{GHz} .$$

This is 347 times the natural linewidth.

b) At the intersection point ω_c the intensity has decreased to

$$I = I_0 e^{\frac{-2.18^2 \times 10^{20}}{0.36 \times 10^{20}}} = I_0 \times 1.85 \times 10^{-6} .$$

c) At $(\omega - \omega_0) = 0.1(\omega - \omega_c)$ the Lorentzian profile has decreased to

$$I_L = I_0 \frac{(\gamma/2)^2}{[0.1(\omega - \omega_c)]^2 + (\gamma/2)^2} = I_0 \frac{3.15^2 \times 10^{14}}{2.18^2 \times 10^{18} + 3.15^2 \times 10^{14}}$$

$$= 2 \times 10^{-4} I_0 .$$

The Doppler profile has only decreased to

$$I_D = 0.876 I_0 .$$

d) $\Delta\omega_S = \Delta\omega_n\sqrt{1 + S} = 0.58\omega_D$
with $\Delta\omega_n = 2\pi \times 10^7\,\text{s}^{-1}$ and $\delta\omega_D = 1 \times 10^{10}\,\text{s}^{-1}$
$\Rightarrow \sqrt{1 + S} = 80 \Rightarrow S = 7.9$.
The saturation parameter is related to the absorption cross-section by

$$S = \frac{\sigma_a I / h\nu}{\gamma} \quad \text{with } \sigma_a = \frac{\pi^2}{4}\lambda^2$$

$$\Rightarrow I = \gamma S h\nu / \sigma_a$$

with $\lambda = 589\,\text{nm} \Rightarrow \sigma_a = 8.56 \times 10^{-13}\,\text{m}^2 = 8.56 \times 10^{-9}\,\text{cm}^2$
$\Rightarrow I = 195\,\text{W/m}^2 = 19.5\,\text{mW/cm}^2 .$

4. a) At 1 bar the atomic density is

$$n = p/kT = 1.4 \times 10^{19}\,\text{cm}^{-3}\,.$$

According to Fig. 3.12 the line broadening is

$$\Delta\nu_\text{p} = \frac{1}{2\pi} \times 2\,\text{cm}^{-1} \triangleq 10\,\text{GHz}\,.$$

Table 3.1 gives 9.1 GHz.
b) For resonant broadening (Li+Li collisions) the linewidth is

$$\gamma_\text{res} = 2\pi\Delta\nu_\text{p} = \frac{ne^2 f_{ik}}{4\pi\,\epsilon_0 m_0 \omega_{ik}}\,.$$

For $n(\text{Li}) = 1.4 \times 10^{16}\,\text{cm}^{-3}$, $f_{ik} = 0.65$, $\omega_{ik} = 2\pi c/\lambda = 2.8 \times 10^{15}\,\text{s}^{-1}$
$\Rightarrow \Delta\nu_\text{p} = 1.3 \times 10^8\,\text{s}^{-1} = 130\,\text{MHz}$. At $n = 1.4 \times 10^{19} \Rightarrow \Delta\nu_\text{p} = 130\,\text{GHz}$.
This is about 13 times larger then for $\text{Li} + \text{Ar}$ collisions.
5. The mean flight time between two collisions is $\bar{t}_\text{c} = \Lambda/\bar{v}$ where $\Lambda = \frac{1}{n\sigma}$ is
the mean free pathlength, and $\bar{v} = \sqrt{8kT/\pi\mu}$ is the mean relative velocity
between the collision partners.
The effective lifetime is

$$\frac{1}{\tau_\text{eff}} = \frac{1}{\tau_\text{sp}} n\sigma\bar{v} \Rightarrow \gamma_\text{eff} = \gamma_\text{sp} + n\sigma\bar{v}\,.$$

The natural linewidth is doubled for

$$n\sigma\bar{v} = \gamma_\text{sp} = 1/\tau_\text{sp}$$

$$\Rightarrow \bar{t}_\text{c} = \frac{1}{n\sigma\bar{v}} = \tau_\text{sp}\,.$$

For $\bar{v} = 820\,\text{m/s}$ (see Problem 2.11); $\sigma = 4 \times 10^{-15}\,\text{cm}^2$

$$\tau_\text{sp} = 16\,\text{ns} \Rightarrow n = 1.9 \times 10^{17}\,\text{cm}^{-3}$$

$$\Rightarrow p = nkT = 1 \times 10^3\,\text{Pa} = 10\,\text{mbar}\,.$$

At a pressure $p = 10\,\text{mbar}$ of N_2 the linewidth of the $\text{Na}(3S\text{–}3P)$ transition
is doubled; i.e., the homogeneous linewidth is then 20 MHz, compared to the
much larger inhomogeneous Doppler width of about 1 GHz.
6. The Doppler width is

$$\Delta\nu_\text{D} = 7.16 \times 10^{-7}(c/\lambda)\sqrt{T/M} = 1.6 \times 10^9\,\text{s}^{-1}\,.$$

The pressure broadening is, according to Table 3.1

$$\Delta\nu_\text{p}/p = 8\,\text{MHz/torr}\,.$$

At 10 mbar $\hat{=}$ 7.6 torr $\Rightarrow$

$\Delta \nu_p(10 \text{ mbar}) = 60.8 \text{ MHz}$.

On the other hand is $\Delta \nu_p = \frac{1}{2\pi} n \sigma_b \overline{v}$

$\Rightarrow \sigma_b = 2\pi \Delta \nu_p / (n \overline{v})$.

At $p = 10 \text{ mbar} \Rightarrow n = 2.1 \times 10^{19} \text{ m}^{-3}$

The broadening cross-section is (due to elastic and inelastic collisions)

$$\sigma_b = 2.6 \times 10^{-15} \text{ cm}^2 .$$

If the broadening of the upper level $|k\rangle$ is twice as large as that of the lower level $|i\rangle$, we obtain with

$$\Delta \nu_p = \frac{1}{2\pi}(\gamma_i + \gamma_k)$$

the relaxation parameters

$$\gamma_i = \frac{2\pi}{3} \Delta \nu_p = 1.27 \times 10^8 \text{ s}^{-1} , \quad \gamma_k = \frac{4\pi}{3} \Delta \nu_p = 2.5 \times 10^8 \text{ s}^{-1} .$$

The saturation broadening is at low pressures

$$\Delta \nu_S = \Delta \nu_n \sqrt{1 + S} .$$

In order to exceed the pressure broadening at a Ne pressure of 10 mb

$$\Delta \nu_S > \Delta \nu_p \Rightarrow \sqrt{1 + S} > \frac{\Delta \nu_p}{\Delta \nu_n} = \frac{60.8 \times 10^6}{6.4 \times 10^6} = 9.5$$

since the natural linewidth is $\Delta \nu_n = \frac{1}{2\pi \tau_{sp}} = 6.4 \text{ MHz} \Rightarrow S \geq 8.5$

$$S = \frac{\sigma_a I / h \nu}{\gamma} = \frac{\pi^2 \lambda^2 I / h \nu}{4 \gamma} \Rightarrow I = \frac{4 \gamma h \nu}{\pi^2 \lambda^2} S$$

$$I = \frac{4 \times 3.7 \times 10^8 \times 2 \times 1.6 \times 10^{-19}}{\pi^2 \times 7.69^2 \times 10^{-14}} \times 8.5 \approx 690 \text{ W/m}^2 = 69 \text{ mW/cm}^2 .$$

The saturation broadening exceeds the Doppler width for $p = 10 \text{ mbar}$, when

$$\Delta \nu_S = \Delta \nu_p \sqrt{1 + S} > \Delta \nu_D$$

$$\Rightarrow \sqrt{1 + S} > \Delta \nu_D / \Delta \nu_p = 1.6 \times 10^9 / 6.08 \times 10^7 = 26 \Rightarrow S \geq 691$$

$$I = 588 \text{ mW/cm}^2 .$$

The laser beam has to be focussed to a cross-section

$$\pi w_S^2 = \frac{100}{588} \text{ cm}^2 = 0.17 \text{ cm}^2 .$$

Chapter 4

1. From the equation

$$\frac{\lambda}{\Delta\lambda} = mN$$

we obtain with $N = 1800 \times 100 = 1.8 \times 10^5, m = 1$
$\lambda/\Delta\lambda = 1.8 \times 10^5$.

However, this does not take into account the finite width b of the entrance slit s_1. The two spectral lines at λ_1 and λ_2 can be resolved if the images $s_2(\lambda_1)$ and $s_2(\lambda_2)$ can be resolved. The width of these slit images is

$$\Delta s_2 = f_2\lambda/a + bf_2/f_1 .$$

For $a = 10\,\text{cm}$, $f_2 = f_1 = 2\,\text{m} \Rightarrow \Delta s_2 = 20\lambda + 10\,\mu\text{m}$.
For $\lambda = 500\,\text{nm} \Rightarrow \Delta s_2 = 20\,\mu\text{m}$.
The separation of $s_2(\lambda_1)$ and $s_2(\lambda_2)$ is:

$$\delta s_2 = f_2(\text{d}\beta/\text{d}\lambda)\Delta\lambda \quad \text{with } \beta = \text{diffraction angle}.$$

From the grating equation for $m = 1$:

$$d(\sin\alpha + \sin\beta) = \lambda$$

$$\Rightarrow \frac{\text{d}\beta}{\text{d}\lambda} = \left(\frac{\text{d}\lambda}{\text{d}\beta}\right)^{-1} = \frac{1}{d\,\cos\beta}$$

$$\cos\beta = \sqrt{1 - \sin^2\beta} = \sqrt{1 - \left(\frac{\lambda}{d} - \sin\alpha\right)^2}$$

$$\Rightarrow \delta s_2 = \frac{f_2\Delta\lambda}{d\,\cos\beta} \geq \Delta s_2 \Rightarrow \Delta\lambda \geq \frac{\Delta s_2\, d\,\cos\beta}{f_2} .$$

For $\alpha = 45°$, $\lambda = 500\,\text{nm}$, $d = (1/18{,}000)\,\text{cm} = 5.6 \times 10^{-5}\,\text{cm} = 0.56\,\mu\text{m}$
$\cos\beta = 0.9825 \Rightarrow \beta = 11°$

$$\Delta\lambda \geq 1.1 \times 10^{-11}\,\text{m} \Rightarrow \frac{\lambda}{\Delta\lambda} = \frac{500 \times 10^{-9}}{1.1 \times 10^{-11}} = 4.5 \times 10^4 .$$

This is three times smaller than mN.
The useful minimum entrance slit width is given by

$$b_{\text{min}} = \frac{2f_1}{d}\lambda = \frac{2}{0.1} \times 5 \times 10^{-7}\,\text{m}$$

$$= 10^{-5}\,\text{m} = 10\,\mu\text{m} .$$

2. The optimum blaze angle is

$$\theta = (\alpha - \beta)/2$$

with $\alpha = 20°$, $\lambda = 500\,\text{nm}$, β can be obtained from the grating equation with $m = 1$;

$$d(\sin\alpha + \sin\beta) = \lambda$$

$$\Rightarrow \sin\beta = +\sin\alpha - \lambda/d \quad \text{where } d = \frac{1}{18,000}\,\text{cm} = 560\,\text{nm}$$

$$= +0.34 - 0.89 = -0.55 \Rightarrow \beta = -33.5°$$

$$\Rightarrow \theta = (20 + 33.5)/2 = 26.7°$$

3. The condition for a Littrow grating to first order is:

$$2d\sin\alpha = \lambda$$

$$\Rightarrow d = \frac{\lambda}{2\sin\alpha} = \frac{488\,\text{nm}}{2 \times 0.42} = 580.9\,\text{nm}$$

$$\Rightarrow \text{number of grooves: } 1721\,/\text{mm} .$$

4. $d_1/\cos\alpha = d_2/\cos\epsilon$

$$\Rightarrow \frac{d_2}{d_1} = \frac{\cos\epsilon}{\cos\alpha} .$$

For $\epsilon = 60° \Rightarrow \cos\alpha = 0.1\cos\epsilon = 0.05$

$$\Rightarrow \alpha = 87° .$$

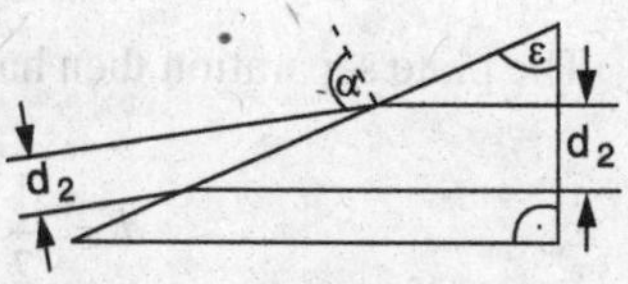

Fig. A1. Beam expanding prism

The incident beam has an angle of $90° - \alpha = 3°$ against the prism surface.

5. The spectral resolution is

$$\frac{\lambda}{\Delta\lambda} = \frac{600}{10^{-4}} = 50\frac{\Delta s}{\lambda} \Rightarrow \Delta s = \frac{6 \times 10^6}{50}\lambda = 7.2 \times 10^{-2}\,\text{m} = 7.2\,\text{cm} .$$

6. The maximum transmission is

$$I_T/I_0 = \frac{T^2}{(T + A)^2} = \frac{(1 - R - A)^2}{(1 - R)^2} .$$

With $R = 0.98$, $A = 0.003 \Rightarrow$

$$I_T/I_0 = \frac{0.017^2}{0.02^2} = 0.72 .$$

The reflectivity finesse is $F_R^* = \frac{\pi\sqrt{R}}{1-R} = 155.5$.

The flatness finesse is: $F_f^* = 50$. According to (4.57)

$$\frac{1}{F_{total}^{*2}} = \frac{1}{F_R^{*2}} + \frac{1}{F_f^{*2}} = 4.4 \times 10^{-4} \Rightarrow F_{tot}^* = 47.6.$$

The spectral resolution is

$$\frac{\lambda}{\Delta\lambda} = F^* \frac{\Delta s}{\lambda}.$$

For $d = 5\,\mathrm{nm} \Rightarrow \Delta s = 1\,\mathrm{cm}$

$$\Rightarrow \frac{\lambda}{\Delta\lambda} = 47.6 \times \frac{10^{-2}}{5 \times 10^{-7}} = 9.5 \times 10^5.$$

7. For $\Delta\lambda = 10^{-2}\,\mathrm{nm}$ and $\lambda = 500\,\mathrm{nm}$ the spectral resolution has to be at least:

$$\frac{\lambda}{\Delta\lambda} \geq \frac{500}{10^{-2}} = 5 \times 10^4.$$

The effective finesse of the FPI in Problem 4.6 is

$$F_{total}^* = 47.6.$$

The plate separation then has to be

$$d = \frac{1}{2}\Delta s = \frac{1}{2}\frac{\lambda^2}{\Delta\lambda F^*} = 0.26\,\mathrm{mm}.$$

The free spectral range is

$$\delta\nu = \frac{c}{2d} \Rightarrow |\delta\lambda| = +\frac{c}{\nu^2}|\delta\nu| = \frac{\lambda^2}{2d} = 3.8 \times 10^{-10}\,\mathrm{m} = 0.38\,\mathrm{nm}.$$

The spectral interval $\Delta\lambda$ transmitted by the spectrograph should be smaller than $\delta\nu$ in order to avoid the overlap of different orders. This means that the spectral resolution of the spectrograph

$$\Delta\lambda = \frac{d\lambda}{dx}\Delta s \leq 0.38\,\mathrm{nm}$$

with a linear dispersion of $d\lambda/dx = 5 \times 10^{-2}\,\mathrm{nm/mm}$

$$\Rightarrow \Delta s \leq \frac{0.38}{5 \times 10^{-1}}\,\mathrm{mm} = 0.76\,\mathrm{mm}.$$

8. The free spectral range must be

$$\delta\lambda > 200\,\text{nm}$$

$$\Rightarrow \delta\lambda = \frac{\lambda^2}{2d} \geq 200\,\text{nm} \Rightarrow d \leq 625\,\text{nm} .$$

If the bandwidth is 5 nm, the finesse must be

$$F^* = \delta\nu/\Delta\nu = |\delta\lambda/\Delta\lambda| = \frac{625}{5} = 125 .$$

If the finesse is solely determined by the reflectivity R

$$\Rightarrow F^* = \frac{\pi\sqrt{R}}{1-R} \Rightarrow R = 0.9753 .$$

9. For $\rho \ll r$ the free spectral range is

$$\delta\nu = \frac{c}{4d} \Rightarrow d = \frac{c}{48\nu} = \frac{3\times10^8}{4\times3\times10^9}\,\text{m} = 2.5\times10^{-2}\,\text{m} = 2.5\,\text{cm}$$

$$F^* = \frac{\delta\nu}{\Delta\nu} = \frac{3\times10^9}{10^7} = 300$$

$$\frac{1}{F^{*2}} = \frac{1}{F_R^2} + \frac{1}{F_f^2} \Rightarrow F_R = \frac{F^* \cdot F_f}{\sqrt{F_f^2 - F^{*2}}} = 375$$

$$\Rightarrow R = 0.9916 = 99.16\,\% .$$

10. $T(\lambda) = T_0 \cos^2\left(\frac{\pi\Delta n L_1}{\lambda}\right) \cos^2\left(\frac{\pi\Delta n L_2}{\lambda}\right)$

with 2 % absorption losses $T_0 = 0.98$.

a) $T(\lambda) = 0.98 \cos^2\left(\frac{0.05\pi\times10^{-3}}{\lambda\,[\text{m}]}\right) \cos^2\left(\frac{0.05\pi\times4\times10^{-3}}{\lambda\,[\text{m}]}\right)$

Transmission peaks appear for the condition

$$\frac{5\times10^{-5}\pi}{\lambda} = m_1\pi \quad\text{and}\quad \frac{2\times10^{-4}\pi}{\lambda} = m_2\pi \quad (m_1, m_2 \in N)$$

$$\Rightarrow \lambda_1 = \frac{5\times10^{-5}}{m_1} \quad\text{and}\quad \lambda_2 = \frac{2\times10^{-4}}{m_2} .$$

For $\lambda = 500\,\text{nm}$ we obtain:

$$m_1 = 100 \quad\text{and}\quad m_2 = 400 .$$

For $m_1 = 101 \Rightarrow \lambda = 495\,\text{nm}$.
The thin plate has a free spectral range $\Delta\lambda = 5\,\text{nm}$.

$$\text{For } m_2 = 401 \Rightarrow \lambda = 498.75\,\text{nm}.$$
$$\text{The thick plate has } \Delta\lambda = 1.25\,\text{nm}.$$

b) $T(\alpha, \lambda) = T_0 \left[1 - \sin^2 \left(\frac{2\pi}{\lambda} \Delta n\, L \right) \sin^2 2\alpha \right]$

Where $\lambda = \frac{2\Delta n\, L}{m}$ is the first factor 0 and $T(\alpha, \lambda_{\max})$ has a maximum transmission T_0, independent of α. For $\lambda = \frac{2\Delta n\, L}{m + \frac{1}{2}}$ this factor becomes 1 and the transmission is

$$T(\alpha) = T_0 (1 - \sin^2 2\alpha)\,.$$

The contrast is then:

$$\frac{T_{\max}}{T_{\min}} = \frac{1}{1 - \sin^2 2\alpha}\,.$$

11. The output voltage V_S is

$$V_S = \frac{R}{R + R_1} V_0 = \frac{1}{1 + R_1/r} V_0\,.$$

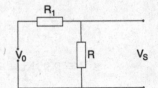

R is the parallel circuit of R_2 and C:

$$\frac{1}{R} = \frac{1}{1/i\omega C} + \frac{1}{R_2} = \frac{1}{R_2} - i\omega C \Rightarrow R = \frac{R_2}{1 - i\omega R_2 C}$$

$$\Rightarrow V_S = \frac{V_0}{1 + \frac{R_1}{R_2}(1 - i\omega C R_2)} = \frac{1}{\left(1 + \frac{R_1}{R_2}\right) - i\omega C R_1} V_0$$

$$\Rightarrow |V_S| = \frac{R_2/(R_1 + R_2)}{\sqrt{1 + \omega^2 C^2 \frac{R_1^2/R_2^2}{(R_1 + R_2)^2}}} V_0\,.$$

For $\omega = 0 \Rightarrow |V_S(0)| = \frac{R_2}{R_1 + R_2} V_0$

$$\Rightarrow |V_S(\omega)| = \frac{V_S(0)}{\sqrt{1 + \left(\omega C \frac{R_1 R_2}{R_1 + R_2}\right)^2}} = \frac{V_S(0)}{\sqrt{1 + (\omega\tau)^2}}\,.$$

The phase shift between V_S and V_0 is

$$\tan \varphi = \frac{\Im(V_S)}{\Re(V_S)} = \frac{\omega C R_1}{1 + R_1/R_2} = \frac{\omega C R_1 R_2}{R_1 + R_2} = \omega \tau .$$

12. $\Delta I = \dfrac{\beta P_0}{G}$

with $\beta = 0.8;\ P_0 = 10^{-9}\,\text{W};\ G = 10^{-9}\,\text{W/K}$

$$\Rightarrow \Delta T = 0.8\,\text{K} .$$

$$T = T(0) + \frac{\beta P_0}{G}\left(1 - \mathrm{e}^{-(G/H)t}\right) .$$

For $\Delta T = 0.9 \Delta T_\infty = 0.9\dfrac{\beta P_0}{G} \Rightarrow 1 - \mathrm{e}^{-(G/H)t} = 0.9 \Rightarrow \mathrm{e}^{-(G/H)t} = 0.1$

$$\Rightarrow \frac{G}{H}t = -\ln 0.1 \Rightarrow t = \frac{H}{G} \times 2.3 = \frac{10^{-8}}{10^{-9}} \times 2.3\,\text{s} = 23\,\text{s} .$$

The time constant is

$$\tau = H/G = 10\,\text{s} .$$

The frequency dependence of ΔT is

$$\Delta T = \frac{a\beta P_0 G}{\sqrt{G^2 + \Omega^2 H^2}} .$$

For $G^2 + \Omega^2 H^2 = 4G^2$ is $\Delta T(\Omega) = 0.5\Delta T(\Omega = 0)$

$$\Rightarrow \Omega^2 = \frac{3G^2}{H^2} = \frac{3 \times 10^{-18}}{10^{-16}}\,\text{s}^{-2} = 3 \times 10^{-2}\,\text{s}^{-2}$$

$$\Rightarrow \Omega = 1.73 \times 10^{-1} = 0.173\,\text{s}^{-1} .$$

13. The heating current $i = 1\,\text{mA}$ produces at $R = 10^{-3}\,\Omega$ a power of $P = R \times i \times i = Ri^2 = 10^{-3} \times 10^{-6}\,\text{W} = 10^{-9}\,\text{W}$. If the incident radiation brings an additional power of $10^{-10}\,\text{W}$ to the bolometer, the heating power must be reduced by this amount.

$$\Rightarrow \Delta i = (di/dP)\Delta P = \frac{\Delta P}{2Ri} = \frac{10^{-10}}{2 \times 10^{-3} \times 10^{-3}}\,\text{A} = 5 \times 10^{-5}\,\text{A}$$

$$\Rightarrow \Delta i = 50\,\mu\text{A}$$

14. The anode voltage pulse is

$$U_a(t) = \frac{Q(t)}{C} = \left(\frac{1}{C} \int_0^{\Delta t} i_{\mathrm{ph}}(t)\, dt \right) \times e^{-t/RC} \ .$$

a) The time constant $\tau = RC = 10^3 \times 10^{-11}\,\mathrm{s} = 10^{-8}\,\mathrm{s}$, which governs the decay of the voltage at C, is long compared with the rise time $\Delta t = 1.5\,\mathrm{ns}$. Therefore we can neglect the decay during the rise time and obtain for the pulse maximum

$$U_a = \frac{1}{C} \times 10^6 e = \frac{1}{C} \times 1.6 \times 10^{-13}\ \text{Coulombs}$$

with $C = 10^{-11}$ Farads we obtain

$$U_a(t) = 1.6 \times 10^{-2} \times e^{-t \times 10^8}\,\mathrm{V}\ .$$

The peak amplitude is $16\,\mathrm{mV} = U_{\mathrm{max}}$.

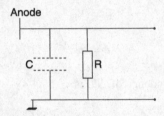

The halfwidth of the pulse is obtained from

$$e^{-10^8 t} = \tfrac{1}{2} \Rightarrow \Delta t_1 = 10^{-8} \ln 2 = 6.9 \times 10^{-9}\,\mathrm{s}\ .$$

b) For $10^{-12}\,\mathrm{W}$ cw radiation at $\lambda = 500\,\mathrm{nm}$, the number of photoelectrons per second is

$$n_{\mathrm{PE}} = \eta \frac{10^{-12}\,\mathrm{W}}{h\nu}\,\mathrm{s}^{-1} = 0.2 \times 2.2 \times 10^6\,\mathrm{s}^{-1} = 4.5 \times 10^5\,\mathrm{s}^{-1}\ .$$

With an amplification factor M, the anode current is:

$$i_a = n_{\mathrm{PE}} \times e \times M\ .$$

The voltage across the anode resistor R is

$$U_a = i_a R = R n_{\mathrm{PE}} e M = 10^3 \times 4.5 \times 10^5 \times 1.6 \times 10^{-19} \times 10^6$$
$$= 7.2 \times 10^{-5}\,\mathrm{V} = 72\,\mu\mathrm{V}\ .$$

Note: For cw measurements, a larger resistance of $R \approx 1\,\mathrm{M\Omega}$ is used because the time resolution is not important here.

For $R = 10^6\,\Omega \Rightarrow U_a = 72\,\mathrm{mV}$.

In order to produce 1 V output pulses for single photoelectrons, an amplification of M_L or $6 \cdot 9$ of the preamplifier is required.

15. For $10^{-17}\,\mathrm{W}$ at $\lambda = 500\,\mathrm{nm}$, 25 photons/s fall onto the first cathode. The human eye can see 20 photons/s $\hat{=}\ 8 \times 10^{-18}\,\mathrm{W}$; with a collection efficiency of 0.1 the last phosphor screen has to emit at least $8 \times 10^{-17}\,\mathrm{W}$. With a conversion efficiency of 0.2, the intensity amplification V_I has to be

$$V_I = \frac{8 \times 10^{-17}}{1 \times 10^{-17} \times 0.2^3} = 1000 = 10^3 \, .$$

16. $U_{ph}(i = 0) = \dfrac{kT}{e}\left[\ln\left(\dfrac{i_{ph}}{i_d}\right) + 1\right]$

with $i_{ph} = 50\,\mu\mathrm{A}$ and $i_d = 50\,\mathrm{nA}$

$$\Rightarrow U_{ph}(i = 0) = 0.2\,\mathrm{V} \, .$$

Chapter 5

1. The threshold inversion is

$$\Delta N_{thr} = \frac{\gamma}{2\sigma L}$$

$\gamma = 5\,\%$, round trip length $= 2 \times 20\,\mathrm{cm} = 40\,\mathrm{cm}$.
The absorption cross-section is related to the Einstein coefficient B_{ik} by

$$B_{ik} = \frac{c}{h\nu}\int \sigma \, d\nu \approx \frac{c}{h\nu}\overline{\sigma}\Delta\nu$$

with $\Delta\nu = 20\,\mathrm{MHz}$.

With $B_{ik} = \dfrac{c^3}{8\pi h\nu^3}A_{ik} \Rightarrow$

$$\overline{\sigma} = \frac{h\nu}{c\Delta\nu}B_{ik} = \frac{\lambda^2}{8\pi\Delta\nu}A_{ik}$$

$$\Rightarrow \Delta N_{thr} = \frac{8\pi\Delta\nu\gamma}{2\lambda^2 L A_{ik}} = \frac{8\pi \times 2 \times 10^7 \times 5 \times 10^{-2}}{2 \times 2.25 \times 10^{-14} \times 0.4 \times 5 \times 10^7} = 2.5 \times 10^{12}\,\mathrm{m}^{-3}$$

$$= 2.5 \times 10^6\,\mathrm{cm}^{-3} \, .$$

2. The spacing of the longitudinal modes is

$$\Delta v = \frac{c}{2d} = \frac{3 \times 10^8}{0.8} = 375\,\text{MHz} .$$

The population density

$$N(v_z) = N(v_z = 0)e^{-(v-v_0)^2/8v^2}$$

with $\delta v = 2 \times 10^9\,\text{s}^{-1}$, $(v - v_0) = 375\,\text{MHz}$ the population density for the adjacent modes has decreased to

$$N_1 = N_0 e^{-0.1875} = 0.83 N_0 .$$

If $N_{\text{thr}}(v_z = 0) = N_0 \Rightarrow N_1 = 0.83 N_0$.
According to Problem 5.1 the threshold inversion is

$$\Delta N_{\text{thr}} = \frac{8\pi \Delta v \gamma}{2\lambda^2 L\, A_{ik}} = \frac{8\pi \times 5 \times 10^7 \times 0.1}{2 \times 6.33^2 \times 10^{-14} \times 0.8 \times 10^8} = 1.96 \times 10^{12}\,\text{m}^{-3}$$

$\Rightarrow$ Oscillation begins at the adjacent modes if threshold is reached for this mode. Then the inversion at the central mode is (without saturation) $\Delta N_0 = \Delta N_1/0.83$.

3. $v_a = v_r + \dfrac{\Delta v_r}{\Delta v_m}(v_0 - v_r)$

with $\Delta v_r = 2\,\text{MHz}$, $\Delta v_m = \Delta v_D = 1\,\text{GHz}$; $(v_0 - v_r) = 0.5\Delta v_D$

$$\Rightarrow v_a = v_r + 10^6\,\text{s}^{-1} .$$

The mode is pulled by 1 MHz.

4. $\delta v_{\text{spa}} = \dfrac{2d}{ap}\delta v$; $p = 2, 3, 4, \dots$

$$\delta v = \frac{c}{2d} = 150\,\text{MHz} ; \ d = 2\,\text{m} ; \ a = 0.2\,\text{m}$$

$$\Rightarrow \delta v_{\text{spa}} = 1.5 \times 10^9\,\text{s}^{-1} = 1.5\,\text{GHz} \quad \text{for } p = 2 .$$

For $p = 3 \Rightarrow \delta v_{\text{spa}} = 1.0\,\text{GHz}$

For a Doppler width of $\Delta v_D = 1\,\text{GHz}$ the gain at the first adjacent spatial hole burning mode is $g = g_0 e^{-1/0.36} = 0.06 g_0$. This mode does not reach the threshold.

The adjacent resonator mode is 150 MHz away from the line center. Its unsaturated gain is

$$g = g_0 e^{-\frac{0.15^2}{0.36 \times 1}} = 0.94 g_0 .$$

Here the net gain is $0.94 \times 1.1 = 1.03$ without mode competition. The two adjacent resonator modes reach the threshold. Therefore three longitudinal modes can oscillate.

5. The total output intensity of a laser with unsaturated gain g_0, internal cavity losses γ_0, length L of the active medium, and mirror losses $T + A = 1 - R = \gamma_M$ is.

$$I_{\text{out}} = \gamma_M \left[\frac{2g_0 L}{\gamma_0 + \gamma_M} - 1 \right] \frac{I_{\text{sat}}}{2} \ .$$

Differentiating gives

$$\frac{dI_{\text{out}}}{d\gamma_M} = \left[\left(\frac{2g_0 L}{\gamma_0 + \gamma_M} - 1 \right) - \gamma_M \frac{2g_0 L}{(\gamma_0 + \gamma_M)^2} \right] \frac{I_{\text{sat}}}{2} = 0$$

$$\Rightarrow \gamma_M^{\text{opt}} = \sqrt{2g_0 L \gamma_0} - \gamma_0 \ .$$

With $\gamma_0 = 0.1$ and $2g_0 L = 2 \Rightarrow$

$$\gamma_M^{\text{opt}} = 0.347 = 34.7\,\% = 1 - R \ .$$

The output mirror should have a reflectivity of $R = 65.3\,\%$.

6. The spot size at the center of the resonator is

$$w_0 = \sqrt{\frac{\lambda L}{2\pi}} = \sqrt{\frac{6.33 \times 10^{-7} \times 0.3}{2\pi}} \ \text{m} = 1.7 \times 10^{-4}\,\text{m} = 0.17\,\text{mm} \ .$$

The spot size at the mirror is

$$w(L/2) = \sqrt{2} \times w_0 = 0.24\,\text{mm} \ .$$

The diameter of the beam (distance between $1/e$ points) is $2w_0$ and $2w$ respectively.

The divergence angle of the laser beam is

$$\theta = \frac{w(L/2)}{L/2} = 1.6 \times 10^{-3}\,\text{rad} \ .$$

The spot size on the lens is

$$w_s = 30\,\text{cm} \times 1.6 \times 10^{-3} + w(L/2)$$

$$= 4.8 \times 10^{-2}\,\text{cm} + 2.4 \times 10^{-2}\,\text{cm} = 7.2 \times 10^{-2}\,\text{cm} = 0.72\,\text{mm}$$

and the beam diameter $2w_s = 1.44\,\text{mm}$.

The location of the focus can be calculated from the lens equation

$$\frac{1}{a} + \frac{1}{b} = \frac{1}{f} \quad \text{with } a = 50 + 15\,\text{cm}; \ b = ?; \ f = 30\,\text{cm}$$

$$\Rightarrow b = 55.7\,\text{cm} \ .$$

The focus is 55.7 cm away from the lens, 105.7 cm away from the output mirror.

The beam waist at the focus is

$$w_0 = f\lambda/(\pi w_s) = \frac{30 \times 6.3 \times 10^{-5}}{\pi \times 0.072} \, \text{cm} = 0.84 \times 10^{-2} \, \text{cm}$$

$$= 0.084 \, \text{mm} = 84 \, \mu\text{m} \, .$$

The Rayleigh length is

$$z_R = \frac{\pi w_0^2}{\lambda} = 3.5 \, \text{cm} \, .$$

7. The beam waist at the focus is

$$w_0 = \frac{f_1 \lambda}{\pi w} = \frac{1 \times 5 \times 10^{-5}}{\pi \times 0.1} \, \text{cm} = 1.59 \, \mu\text{m} \, .$$

The power transmitted through the aperture with radius a is

$$P_t = P_i \left(1 - e^{-2a^2/w_0^2}\right) \, .$$

For $P_t/P_i = 0.95$ we obtain

$$0.05 = e^{-2a^2/(1.59^2 \times 10^{-6})} \quad \text{with } a \text{ in mm}$$
$$\Rightarrow a^2 = -(\ln 0.05) \times \tfrac{1}{2} \times 1.59^2 \times 10^{-6} = 3.79 \times 10^{-6} \, \text{mm}^2$$
$$\Rightarrow a = 1.95 \times 10^{-3} \, \text{mm} \Rightarrow 2a = 39 \, \mu\text{m} = 2.45 w_0 \, .$$

8. The axial modes are separated by

$$\Delta \nu = \frac{c}{2d} = 300 \, \text{MHz} \quad \text{for } d = 50 \, \text{cm} \, .$$

The gain factor G follows the Doppler profile

$$G(\nu) = G(\nu_0) e^{-(\nu - \nu_0)^2/(0.36 \Delta \nu_D^2)} \, .$$

with $\Delta \nu_D = 1.5 \, \text{GHz}$ and $\nu_1 - \nu_0 = 300 \, \text{MHz} \Rightarrow$

$$G(\nu_1) = G(\nu_0) e^{-0.11} = 0.896 \, .$$

with $G(\nu_0) = 1.3 \Rightarrow G(\nu_1) = 1.16$.
With 4 % losses the net gain at ν_1 is 1.12
$\Rightarrow$ the losses of the etalon at ν_1 must be at least 12 % in order to prevent laser oscillation at ν_1.

The transmission of the etalon with thickness t
and refractive index $n = 1.4$

$$T = \frac{1}{1 + F \sin^2 \phi/2} \quad \text{with } \phi = \frac{2\pi v \Delta s}{c} = \frac{2\pi v}{c} 2nt \ .$$

For $v = v_0 \to T = 1 \Rightarrow \phi/2 - m\pi = \frac{2\pi v_0}{c} nt$.

For $v = v_1 \Rightarrow T \leq 0.88 \Rightarrow F \sin^2 \phi/2 \geq 0.12$

$$\Rightarrow \sin \phi_1/2 \geq \sqrt{0.12/F} \ .$$

Since $v_1 = v_0 + 300\,\text{MHz} \Rightarrow \phi_1 = \phi_0 + \Delta\phi$

$$\Rightarrow \Delta\phi = \frac{4\pi(v_1 - v_0)}{c} nt$$

$$\Rightarrow \Delta\phi = 2\pi \frac{3 \times 10^8}{3 \times 10^{10}} nt \quad \text{with } t \text{ in cm}$$

$$= 2\pi \times 10^{-2} nt \ .$$

The thickness t of the etalon should be small in order to minimize walk-off
losses by the tilted etalon. If we assume as a reasonable number $t = 0.5\,\text{cm}$,
$n = 1.4$

$$\Rightarrow \Delta\phi = 2\pi \times 7 \times 10^{-3} = 2.5°$$

$$\Rightarrow \sin \Delta\phi/2 = \sin \phi_1/2 = 0.044$$

$$\Rightarrow F \geq \frac{0.12}{0.044^2} = 63 \ .$$

With $F^* = \frac{\pi}{2}\sqrt{F}$, we obtain for the necessary finesse F^* the relation

$$F^* \geq 12.5 \ .$$

Since $F^* = \frac{\pi\sqrt{R}}{1-R} \Rightarrow R_E \geq 0.78$, the etalon reflectivity should be larger than
78%.

9. The resonator with $R_1 = \infty$ and $R_2 = 400\,\text{cm}$ and $d = 100\,\text{cm}$ is equivalent
to a spherically symmetric resonator with $d = 200\,\text{cm}$ and

$$R = R_1 = R_2 = 400\,\text{cm} \ .$$

The spot sizes w_s on the mirrors are

$$w_s = \left(\frac{\lambda d}{\pi}\right)^{1/2} \left[\frac{2d}{R} - \left(\frac{d}{R}\right)^2\right]^{-1/4}$$

$$= 5.96 \times 10^{-4}\,\text{m} = 0.596\,\text{mm} \ .$$

The transmission of the spherical aperture with radius a in the center of the resonator is, for the fundamental modes,

$$T = 1 - e^{-2a^2/w_s^2} > 0.99$$

$$\Rightarrow e^{-2a^2/w_s^2} \le 0.01 \Rightarrow a^2 \ge \frac{w_s^2}{2}\ln 100$$

$$\Rightarrow a \ge 0.904\,\text{mm}\,.$$

According to Fig. 5.12 the Fresnel number N_F should be smaller than 0.8, in order to increase the losses of the TEM_{10} mode above 10 %. The Fresnel number is defined as $N_F = \frac{1}{\pi}\frac{\pi a^2}{\pi w_s^2}$, where w_s is the beam waist of the fundamental mode $\Rightarrow$

$$a^2 < 0.8 \times \pi w_s^2 = 0.8\pi \times 0.596^2\,\text{mm}^2 = 0.89\,\text{mm}^2$$

$$\Rightarrow a \le 0.944\,\text{mm}\,.$$

The radius a of the aperture therefore must lie between $0.904 \le a \le 0.944\,\text{mm}$.

10. With $L = 15\,\text{cm}$ the free spectral range is

$$\delta\nu = \frac{c}{2d} = 10^9\,\text{s}^{-1}$$

$\Rightarrow$ only one mode can oscillate if this mode is close to the center of the gain profile.

The unsaturated gain at ν_0 is 10 %. With losses of 3 % the net gain is 7 % $\Rightarrow G(\nu_0) = 1.07$.

When tuning away from the gain center, the net gain factor should always be > 1.

$$\Rightarrow G = 1.1 \times e^{-(\nu-\nu_0)^2/(0.36\Delta\nu_D^2)} - 0.03 \ge 1$$

$$\Rightarrow e^{-(\nu-\nu_0)^2/0.3\Delta\nu_D^2} \ge \frac{1.03}{1.1} = 0.936$$

$$\Rightarrow (\nu-\nu_0)^2 \ge 0.3\Delta\nu_D^2\ln\frac{1}{0.936}\,.$$

With $\Delta\nu_D = 1.5 \times 10^9\,\text{s}^{-1} \Rightarrow$

$$\nu - \nu_0 \le 2.13 \times 10^8\,\text{c}^{-1} = 213\,\text{MHz}\,.$$

The maximum tuning range is from $\nu_0 - 213\,\text{MHz}$ up to $\nu_0 + 213\,\text{MHz}$.
In order to tune over one free spectral range, the mirror separation must change

by $\lambda/2 \Rightarrow \delta\nu = 10^9\,\mathrm{s}^{-1}$, requiring $\Delta d = \lambda/2$

$$\Rightarrow \nu - \nu_0 = 426\,\mathrm{MHz}\ \text{requiring}\ \Delta d = (\lambda/2)\frac{(\nu - \nu_0)}{\delta\nu} = \frac{\lambda}{2} \times 0.426$$

$$= 0.213\lambda = 0.135\,\mu\mathrm{m} \quad \text{at} \quad \lambda = 632\,\mathrm{nm}$$

$$\Rightarrow \Delta V = \frac{dV}{dx}\Delta x = \left(\frac{10^{-9}\,\mathrm{m}}{V}\right)^{-1} \times 1.35 \times 10^{-7}\,\mathrm{m} = 135\,\mathrm{V}.$$

11. $\dfrac{\Delta\nu}{\nu} = \dfrac{\Delta d}{d} = \alpha\,\Delta T$.

A temperature drift of $1\,^\circ\mathrm{C/h}$ gives, for invar rods ($\alpha = 1.2 \times 10^{-6}\,\mathrm{K}^{-1}$), a frequency drift per hour of

$$\frac{\Delta\nu}{\nu} = 1.2 \times 10^{-6}\ .$$

For $\nu = c/\lambda = 6 \times 10^{14}\,\mathrm{s}^{-1} \Rightarrow \Delta\nu = 7.2 \times 10^8\,\mathrm{s}^{-1}/\mathrm{h} = 720\,\mathrm{MHz/h}$.
For fused quartz ($\alpha = 0.4\text{--}0.5 \times 10^{-6}\,\mathrm{K}^{-1}$ the drift is three times smaller, while for Cerodur it is more than 12 times smaller.

12. With $L = 100\,\mathrm{cm}$ the mode spacing is $\delta\nu = 150\,\mathrm{MHz}$.
   a) For a solid etalon with $t = 1\,\mathrm{cm}$, $n = 1.4 \Rightarrow$

$$\frac{\Delta\nu}{\nu} = \frac{\Delta t}{t} + \frac{\Delta n}{n}\ .$$

The second term is small and can be neglected

$$\Rightarrow \frac{\Delta\nu}{\nu} = \alpha\Delta T = 2 \times 4 \times 10^{-7} = 8 \times 10^{-7}$$

$$\Rightarrow \Delta\nu = 4.9 \times 10^7\,\mathrm{s}^{-1}\ .$$

b) For an air-spaced etalon we can neglect the first term if the spacers are made of cerodur or the distance is temperature-compensated.
The optical path due to air at a pressure p is for a length d equal to $s = nd$ with n (air at $p = 1\,\mathrm{bar}$) $= 1.00028$.
The change Δs is

$$\Delta s = (n - 1)\,d\,\frac{\Delta p}{p}$$

$$\Rightarrow \left|\frac{\Delta\nu}{\nu}\right| = \frac{\Delta s}{s} = \frac{n-1}{n} \cdot \frac{\Delta p}{p} = 0.00028 \times \frac{4}{1000} = 1.12 \times 10^{-6}\ .$$

For $\nu = 6 \times 10^{14}\,\mathrm{s}^{-1} \Rightarrow \Delta\nu = 6.72 \times 10^8\,\mathrm{s}^{-1} = 672\,\mathrm{MHz}$.
This illustrates that an air-spaced etalon is less stable than a solid etalon.

c) For a temperature drift of $1\,°C/h$ the solid etalon has a frequency drift of $336\,MHz/h$.

13. The transmission of the Pockels cell is

$$T = T_0 \cos^2 aV$$

where V is the applied voltage and a is a constant which depends on the electro-optic coefficient and the dimensions of the modulator.

For $V = 0 \Rightarrow T = T_0$, for $V = 600 \Rightarrow T = 0 \Rightarrow aV = \pi/2$.

The system should operate at the maximum slope of dT/dV.

$$\Delta T = \frac{dT}{dV} \Delta V = -2aT_0 \cos aV \sin aV \; \Delta V \;.$$

For a fluctuation in intensity of $5\,\%$ the transmission must change by $\Delta T = -0.05T_0$ in order to compensate for the fluctuations.

$$\Rightarrow \Delta V = \frac{0.05}{2a \cos aV \sin aV} \quad \text{with } a = \frac{\pi}{2 \times 600} \; V^{-1} \;.$$

The maximum slope is realized for $aV = 45°$

$$\Rightarrow \cos aV = \sin aV = \tfrac{1}{2}\sqrt{2}$$

$$\Rightarrow \Delta V = \frac{2 \times 0.05 \times 600}{\pi} V = 19\,V \;.$$

14. The free spectral range of the etalon is

$$\delta\nu_E = \frac{c}{2d} = 8 \times 10^9\,s^{-1} \Rightarrow d = 1.8\,cm \;.$$

a) The change of d with temperature is for invar ($\alpha = 1.2 \times 10^{-6}\,K^{-1}$)

$$\Delta d = d\alpha\Delta T$$

$$\Rightarrow \frac{\Delta d}{d} = 1.2 \times 10^{-6} \times 10^{-2} = 1.2 \times 10^{-8}$$

$$\Rightarrow \left|\frac{\Delta\nu}{\nu}\right| = \frac{\Delta d}{d} = 1.2 \times 10^{-8} \;.$$

For $\nu = 5 \times 10^{14}\,s^{-1}$ ($\lambda = 600\,nm$) $\Rightarrow \Delta\nu = 6 \times 10^6\,s^{-1} = 6\,MHz$.

b) If d changes by $1\,mm$ due to acoustic vibrations

$$\Rightarrow \frac{\Delta d}{d} = \frac{10^{-7}}{1.8} = 5.6 \times 10^{-8} = \left|\frac{\Delta\nu}{\nu}\right|$$

$$\Rightarrow \Delta\nu = 5.6 \times 10^{-8} \times 5 \times 10^{14} = 28\,MHz \;.$$

c) With a free spectral range $\delta\nu_{\text{FPI}} = 10\,\text{GHz}$ of the FPI and a finesse $F^* = 50$, the full halfwidth of the transmission peak is

$$\Delta\nu_{\text{FPI}} = \delta\nu/F^* = 200\,\text{MHz}\ .$$

The transmitted intensity is

$$I_t = I_0 T = I_0 \frac{1}{1 + F\sin^2(\phi/2)} \quad \text{with } F = \left(\frac{2}{\pi}F^*\right)^2 = 1 \times 10^3\ .$$

The stabilization system interprets an intensity change of 1 % as a transmission change ΔT, i.e., a change $\Delta\phi$ of ϕ, and because

$$\phi = \frac{2\pi}{\lambda}\Delta s = \frac{2\pi\nu}{c}\Delta s \Rightarrow \Delta\phi = \frac{2\pi}{c}\Delta s\,\Delta\nu$$

also as a change of ν.

A rough estimation of $\Delta\nu$ proceeds as follows.

A frequency change of $100\,\text{MHz}$ changes (at a fixed plate separation $d = 0.5\Delta s$) the transmission by 100 % from 0 to 1. A transmission change of 1 % therefore corresponds to a frequency change of $0.01 \times 100\,\text{MHz} = 1\,\text{MHz}$.

A more elaborate calculation uses the relation

$$\Delta T = \frac{dT}{d\phi}\frac{d\phi}{d\nu}\Delta\nu \Rightarrow \Delta\nu = \frac{0.01}{\frac{dT}{d\phi}\frac{d\phi}{d\nu}} \quad \text{because } \Delta T = 0.01$$

$$\frac{dT}{d\phi} = \frac{F\sin(\phi/2)\cos(\phi/2)}{(1 + F\sin^2\phi/2)^2}$$

$$\frac{d\phi}{d\nu} = \frac{2\pi\,\Delta s}{c}\ .$$

References

Chapter 1

1. *Laser Spectroscopy I–XXI*, Proc. Int. Confs. 1973–2013,
   I, Vale 1973, ed. by R.G. Brewer, A. Mooradian (Plenum, New York 1974);
   II, Megeve 1975, ed. by S. Haroche, J.C. Pebay-Peyroula, T.W. Hänsch, S.E. Harris, Lecture Notes Phys., Vol. 43 (Springer, Berlin, Heidelberg 1975);
   III, Jackson Lake Lodge 1977, ed. by J.L. Hall, J.L. Carlsten, Springer Ser. Opt. Sci., Vol. 7 (Springer, Berlin, Heidelberg 1977);
   IV, Rottach-Egern 1979, ed. by H. Walther, K.W. Rothe, Springer Ser. Opt. Sci., Vol. 21 (Springer, Berlin, Heidelberg 1979);
   V, Jaspers 1981, ed. by A.R.W. McKellar, T. Oka, B.P. Stoichef, Springer Ser. Opt. Sci., Vol. 30 (Springer, Berlin, Heidelberg 1981);
   VI, Interlaken 1983, ed. by H.P. Weber, W. Lüthy, Springer Ser. Opt. Sci., Vol. 40 (Springer, Berlin, Heidelberg 1983);
   VII, Maui 1985, ed. by T.W. Hänsch, Y.R. Shen, Springer Ser. Opt. Sci., Vol. 49 (Springer, Berlin, Heidelberg 1985);
   VIII, Are 1987, ed. by W. Persson, S. Svanberg, Springer Ser. Opt. Sci., Vol. 55 (Springer, Berlin, Heidelberg 1987);
   IX, Bretton Woods 1989, ed. by M.S. Feld, J.E. Thomas, A. Mooradian (Academic, New York 1989);
   X, Font Romeau 1991, ed. by M. Ducloy, E. Giacobino, G. Camy (World Scientific, Singapore 1992);
   XI, Hot Springs, VA 1993, ed. by L. Bloomfield, T. Gallagher, D. Larson, AIP Conf. Proc. **290** (AIP, New York 1993);
   XII, Capri, Italy 1995, ed. by M. Inguscio, M. Allegrini, A. Sasso (World Scientific, Singapore 1995);
   XIII, Hangzhou, P.R. China 1997, ed. by Y.Z. Wang, Y.Z. Wang, Z.M. Zhang (World Scientific, Singapore 1997);
   XIV, Innsbruck, Austria 1999, ed. by R. Blatt, J. Eschner, D. Leihfried, F. Schmidt-Kaler (World Scientific, Singapore 1999);
   XV, Snowbird, USA 2001, ed. by St. Chu (World Scientific, Singapore 2002);
   XVI, Palmcove, Australia 2003, ed. by P. Hannaford, A. Sidoven, H. Bachor, K. Baldwin (World Scientific, Singapore 2004);

XVII, Scotland 2005, ed. by E.A. Hinds, A. Ferguson, E. Riis (World Scientific, Singapore 2005);
XVIII, Telluride, Colorado 2007, Proceedings ed. by L. Holberg, J. Bergquist, M. Casevich (World Scientific, Singapore 2008);
XIX, Hokaido, Japan 2009, Proceedings ed. by H. Katori (World Scientific, Singapore 2010);
XX, Hannover 2011, Proceedings ed by W. Ertmer, R. Scholz (Logos Verlag, Berlin 2011);
XXI, Berkely, California 2013, Proceedings ed. by R. Falcone, L. Holberg. IOP J. Phys.: Conference Series (Open Access)

2. Int. Conferences on Photonics, Optics and Laser Technology

3. *Advances in Laser Sciences I–IV*, Int. Conf. 1985–1989,
   I, Dallas 1985, ed. by W.C. Stwally, M. Lapp (Am. Inst. Phys., New York 1986);
   II, Seattle 1986, ed. by W.C. Stwalley, M. Lapp, G.A. KennedyWallace (AIP, New York 1987);
   III, Atlantic City 1987, ed. by A.C. Tam, J.L. Gale, W.C. Stwalley (AIP, New York 1988);
   IV, Atlanta 1988, ed. by J.L. Gole et al. (AIP, New York 1989)

4. M. Feld, A. Javan, N. Kurnit (Eds.): *Fundamental and Applied Laser Physics*, Proc. Esfahan Symposium 1971 (Wiley, London 1973)

5. A. Mooradian, T. Jaeger, P. Stokseth (Eds.): *Tunable Lasers and Applications*, Springer Ser. Opt. Sci., Vol. 3 (Springer, Berlin, Heidelberg 1976)

6. R.A. Smith (Ed.): *Very High Resolution Spectroscopy* (Wiley Interscience, New York 1970)

7. *Int. Colloq. on Doppler-Free Spectroscopic Methods for Simple Molecular Systems, Aussois, May 1973* (CNRS, Paris 1974)

8. S. Martellucci, A.N. Chester (Eds.): *Analytical Laser Spectroscopy*, Proc. NATO ASI (Plenum, New York 1985)

9. Y. Prior, A. Ben-Reuven, M. Rosenbluth (Eds.): *Methods of Laser Spectroscopy* (Plenum, New York 1986)

10. A.C.P. Alves, J.M. Brown, J.M. Hollas (Eds.): *Frontiers of Laser Spectroscopy of Gases*, NATO ASI Series, Vol. 234 (Kluwer, Dordrecht 1988)
    T.W. Hänsch, M. Inguscio (Eds.): *Frontiers in Laserspectroscopy* (North Holland, Amsterdam 1994)

11. W. Demtröder, M. Inguscio (Eds.): *Applied Laser Spectroscopy*, NATO ASI Series, Vol. 241 (Plenum, New York 1991)

12. H. Walther (Ed.): *Laser Spectroscopy of Atoms and Molecules*, Topics Appl. Phys., Vol. 2 (Springer, Berlin, Heidelberg 1976)

13. K. Shimoda (Ed.): *High-Resolution Laser Spectroscopy*, Topics Appl. Phys., Vol. 13 (Springer, Berlin, Heidelberg 1976)

14. A. Corney: *Atomic and Laser Spectroscopy*, new edition (Oxford Univ. Press, Oxford 2006)

15. V.S. Letokhov: *Laserspektroskopie* (Vieweg, Braunschweig 1977);
    V.S. Letokhov (Ed.): *Laser Spectroscopy of Highly Vibrationally Excited Molecules* (Hilger, Bristol 1989)

16. J.M. Weinberg, T. Hirschfeld (Eds.): *Unconventional Spectroscopy*, SPIE Proc. **82** (1976)

17. S. Jacobs, M. Sargent III, J. Scott, M.O. Scully (Eds.): *Laser Applications to Optics and Spectroscopy* (Addison-Wesley, Reading, MA 1975)

18. D.C. Hanna, M.A. Yuratich, D. Cotter: *Nonlinear Optics of Free Atoms and Molecules*, Springer Ser. Opt. Sci., Vol. 17 (Springer, Berlin, Heidelberg 1979)

19. M.S. Feld, V.S. Letokhov (Eds.): *Coherent Nonlinear Optics*, Topics Curr. Phys., Vol. 21 (Springer, Berlin, Heidelberg 1980)

20. S. Stenholm: *Foundations of Laser Spectroscopy* (Dover Publ., New York 2005)

21. J.I. Steinfeld (Ed.): *Laser and Coherence Spectroscopy* (Plenum, New York 1978)

22. W.M. Yen, M.D. Levenson (Eds.): *Lasers, Spectroscopy and New Ideas*, Springer Ser. Opt. Sci., Vol. 54 (Springer, Berlin, Heidelberg 1987)

23. I.S. Kliger (Ed.): *Ultrasensitive Laser Spectroscopy* (Academic, New York 1983)

24. B.A. Garetz, J.R. Lombardi (Eds.): *Advances in Laser Spectroscopy, Vols. I–III* (Heyden, London 1982, 1983, 1984);
K.F. Renk: *Basics of Laser Physics* (Springer, Berlin, Heidelberg 2012)

25. S. Svanberg: *Atomic and Molecular Spectroscopy*, 4th edn., Springer Ser. Atoms Plasmas, Vol. 6 (Springer, Berlin, Heidelberg 2004, reprint 2012)

26. D.L. Andrews, A.A. Demidov: *An Introduction to Laser Spectroscopy*, 2nd edn. (Plenum, New York 2002);
D.L. Andrews (Ed.): *Applied Laser Spectroscopy* (Wiley-VCH, Weinheim 1992)

27. L.J. Radziemski, R.W. Solarz, J. Paissner: *Laser Spectroscopy and its Applications* (Dekker, New York 1987)

28. V.S. Letokhov (Ed.): *Lasers in Atomic, Molecular and Nuclear Physics* (World Scientific, Singapore 1989)

29. E.R. Menzel: *Laser Spectroscopy* (Dekker, New York 1994);
H. Abramczyk: *Introduction to Laser Spectroscopy* (Elsevier, Amsterdam 2005);
D.L. Andrews, A.A. Demidov: *An Introduction to Laser Spectroscopy*, 2nd edn. (Springer, Berlin, Heidelberg 2002)

30. Z.-G. Wang, H.-R. Xia: *Molecular and Laser Spectroscopy*, Springer Ser. Chem. Phys., Vol. 50 (Springer, Berlin, Heidelberg 1991);
R. Blatt, W. Neuhauser (Eds.): *High Resolution Laser Spectroscopy*. Appl. Phys. B **59** (1994)

31. J. Sneddon (Ed.): *Lasers in Analytical Atomic Spectroscopy* (Wiley, New York 1997)

32. R. Menzel: *Photonics: Linear and Nonlinear Interaction of Laser Light and Matter*, 2nd edn. (Springer, Heidelberg 2007)

33. J. Hecht: *Laser Pioneers* (Academic, Boston 1992)

34. Ch.H. Townes: *How the Laser Happened. Adventures of a Scientist* (Oxford Univ. Press, Oxford 1999)

35. J.C. Lindon, G.E. Trauter, J.L. Holmes: *Encyclopedia of Spectroscopy and Spectrometry, Vols. I–III* (Academic, London 2000)

36. F. Träger (Ed.): *Springer Handbook of Lasers and Optics*, 2nd edn. (Springer, Berlin, Heidelberg 2012)

Chapter 2

37. A. Corney: *Atomic and Laser Spectroscopy* (Oxford Univ. Press, Oxford 2006)

38. A.P. Thorne, U. Litzén, S. Johansson: *Spectrophysics* (Springer, Heidelberg 1999)

39. I.I. Sobelman: *Atomic Spectra and Radiative Transitions*, 2nd edn., Springer Ser. Atoms Plasmas, Vol. 12 (Springer, Berlin, Heidelberg 1992)

40. H.G. Kuhn: *Atomic Spectra*, 2nd edn. (Prentice Hall, Upper Saddle River, N.J. 1970)

41. M. Born, E. Wolf: *Principles of Optics*, 5th edn. (Pergamon, Oxford 1999)

42. R. Loudon: *The Quantum Theory of Light*, 3rd edn. (Clarendon, Oxford 2000)

43. W. Schleich: *Quantum Optics in Phase Space* (Wiley-VCH, Weinheim 2001)

44. S. Suter: *The Physics of Laser–Atom Interaction* (Cambridge Studies in Modern Optics, Cambridge 1997)

45. J.W. Robinson (Ed.): *Handbook of Spectroscopy, Vols. I–III* (CRC, Cleveland, Ohio 1974–81)
    Atomic Spectroscopy (Dekker, New York 1996)

46. L. May (Ed.): *Spectroscopic Tricks, Vols. I–III* (Plenum, New York 1965–73)

47. J.O. Hirschfelder, R. Wyatt, R.D. Coulson (Eds.): *Lasers, Molecules and Methods*, Adv. Chem. Phys., Vol. 78 (Wiley, New York 1986)

48. M. Cardona, G. Güntherodt (Eds.): *Light Scattering in Solids I–VI*, Topics Appl. Phys., Vols. 8, 50, 51, 54, 66, 68 (Springer, Berlin, Heidelberg 1983–91)

49. A. Stimson: *Photometry and Radiometry for Engineers* (Wiley-Interscience, New York 1974)

50. W.L. Wolfe: 'Radiometry'. In: *Appl. Optics and Optical Engineering, Vol. 8*, ed. by J.C. Wyant, R.R. Shannon (Academic, New York 1980);
    W.L. Wolfe: *Introduction to Radiometry* (SPIE, Bellingham, WA 1998)

51. D.S. Klinger, J.W. Lewis, C.E. Randull: *Polarized Light in Optics and Spectroscopy* (Academic, Boston 1997)

52. S. Huard, G. Vacca: *Polarization of Light* (Wiley, Chichester 1997)

53. E. Collet: *Polarized Light: Fundamentals and Applications* (Dekker, New York 1993)

54. D. Eisel, D. Zevgolis, W. Demtröder: Sub-Doppler laser spectroscopy of the NaK-molecule. J. Chem. Phys. **71**, 2005 (1979)

55. W.L. Wiese: 'Transition probabilities'. In: *Methods of Experimental Physics, Vol. 7a*, ed. by B. Bederson, W.L. Fite (Academic, New York 1968) p. 117;
    W.L. Wiese, M.W. Smith, B.M. Glennon: Atomic Transition Probabilities. Nat'l Standard Reference Data Series NBS4 and NSRDS-NBS22 (1966–1969), see Data Center on Atomic Transition Probabilities and Lineshapes, NIST Homepage (www.nist.org);
    P.L. Smith, W.L. Wiese: *Atomic and Molecular Data for Space Astronomy* (Springer, Berlin 1992)

56. C.J.H. Schutte: *The Wave Mechanics of Atoms, Molecules and Ions* (Arnold, London 1968);
    R.E. Christoffersen: *Basic Principles and Techniques of Molecular Quantum Mechanics* (Springer, Heidelberg 1989)

57. M.O. Scully, W.E. Lamb Jr., M. Sargent III: *Laser Physics* (Addison Wesley, Reading, MA 1974)

58. P. Meystre, M. Sargent III: *Elements of Quantum Optics*, 2nd edn. (Springer, Berlin, Heidelberg 1991)

59. G. Källen: *Quantum Electrodynamics* (Springer, Berlin, Heidelberg 1972)

60. C. Cohen-Tannoudji, B. Diu, F. Laloe: *Quantum Mechanics, Vols. I, II* (Wiley-International, New York 1977);
    C. Cohen-Tannoudji, J. Dupont-Roche, G. Grynberg: *Atom–Photon Interaction* (Wiley, New York 1992)

61. L. Mandel, E. Wolf: Coherence properties of optical fields. Rev. Mod. Phys. **37**, 231 (1965);
    L. Mandel, E. Wolf: *Optical Coherence and Quantum Optics* (Cambridge University Press, Cambridge 1995)

62. G.W. Stroke: *An Introduction to Coherent Optics and Holography* (Academic, New York 1969)

63. J.R. Klauder, E.C.G. Sudarshan: *Fundamentals of Quantum Optics* (Benjamin, New York 1968)

64. A.F. Harvey: *Coherent Light* (Wiley Interscience, London 1970)

65. H. Kleinpoppen: 'Coherence and correlation in atomic collisions'. In: *Adv. Atomic and Molecular Physics, Vol. 18, ed. by D.R. Bates, B. Bederson (Academic, ...,* H.J. Beyer, K. Blum, R. Hippler: *Coherence in Atomic Collision Physics* (Plenum, New York 1988)

66. R.G. Brewer: 'Coherent optical spectroscopy'. In: *Frontiers in Laser Spectroscopy, Vol. 1,* ed. by R. Balian, S. Haroche, S. Liberman (North-Holland, Amsterdam 1977) p. 342

67. B.W. Shore: *The Theory of Coherent Excitation* (Wiley, New York 1990)

Chapter 3

68. I.I. Sobelman, L.A. Vainstein, E.A. Yukov: *Excitation of Atoms and Broadening of Spectral Lines*, 2nd edn., Springer Ser. Atoms Plasmas, Vol. 15 (Springer, Berlin, Heidelberg 1995)

69. R.G. Breene: *Theories of Spectral Line Shapes* (Wiley, NewYork 1981)

70. K. Burnett: *Lineshapes Laser Spectroscopy* (Cambridge University Press, Cambridge 2000)

71. See, for instance, *Proc. Int. Conf. on Spectral Line Shapes*,
    Vol. 1, ed. by B. Wende (De Gruyter, Berlin 1981);
    Vol. 2, 5th Int. Conf., Boulder 1980, ed. by K. Burnett (De Gruyter, Berlin 1983);
    Vol. 3, 7th Int. Conf., Aussois 1984, ed. by F. Rostas (De Gruyter, Berlin 1985);
    Vol. 4, 8th Int. Conf., Williamsburg 1986, ed. by R.J. Exton (Deepak Publ., Hampton, VA 1987);
    Vol. 5, 9th Int. Conf.; Torun, Poland 1988, ed. by J. Szudy (Ossolineum, Wroclaw 1989);
    Vol. 6, Austin 1990, ed. by L. Frommhold, J.W. Keto (AIP Conf. Proc. No. 216, 1990);
    Vol. 7, Carry Le Rovet 1992, ed. by R. Stamm, B. Talin (Nova Science, Paris 1994);
    Vol. 8, Toronto 1994, ed. by A.D. May, J.R. Drummond (AIP, New York 1995);
    Vol. 9, Florence 1996, ed. by M. Zoppi, L. Olivi (AIP, New York 1997);
    Vol. 10, State College, PA, USA, ed. by R.M. Herrmann (AIP, New York 1999);
    Vol. 11, Berlin 2000, ed. by J. Seidel (AIP, New York 2001);
    Vol. 12, Berkeley, CA, 2002, ed. by C.A. Back (AIP, New York 2002);
    Vol. 13, Paris, 2004, ed. by E. Dalimier (Frontier Group 2004);
    Vol. 14, Auburn, AL, 2006, ed. by E. Oks, M. Pindzola (At. Mol. Chem. Phys. Vol. 874, Springer 2006);
    Vol. 15–19, AIP Conf. Proc. Spectral Line Shapes (AIP, New York 2009)

72. C. Cohen-Tannoudji: *Quantum Mechanics* (Wiley, New York 1977)

73. S.N. Dobryakov, Y.S. Lebedev: Analysis of spectral lines whose profile is described by a composition of Gaussian and Lorentz profiles. Sov. Phys. Dokl. **13**, 9 (1969)

74. A. Unsöld: *Physik der Sternatmosphären* (Springer, Berlin, Heidelberg 1955);
    A. Unsöld, B. Baschek: *The New Cosmos*, 5th edn. (Springer, Berlin, Heidelberg 2005)

75. E. Lindholm: Pressure broadening of spectral lines. Ark. Mat. Astron. Fys. **32**A, 17 (1945)

76. A. Ben Reuven: The meaning of collisional broadening of spectral lines. The classical oscillation model. Adv. Atom. Mol. Phys. **5**, 201 (1969)

77. F. Schuler, W. Behmenburg: Perturbation of spectral lines by atomic interactions. Phys. Rep. C **12**, 274 (1974)

78. D. Ter Haar: *Elements of Statistical Mechanics* (Pergamon, New York 1977)

79. A. Gallagher: 'The spectra of colliding atoms'. In: *Atomic Physics, Vol. 4*, ed. by G. zu Putlitz, E.W. Weber, A. Winnaker (Plenum, New York 1975)

80. K. Niemax. G. Piehler: Determination of van der Waals constants from the red wings of self-broadened Cs principal series lines. J. Phys. B **8**, 2718 (1975)

81. N. Allard, J. Kielkopf: The effect of neutral nonresonant collisions on atomic spectral lines. Rev. Mod. Phys. **54**, 1103 (1982)

82. U. Fano, A.R.P. Rau: *Atomic Collisions and Spectra* (Academic, New York 1986)

83. K. Sando, Shi-I.: Pressure broadening and laser-induced spectral line shapes. Adv. At. Mol. Phys. **25**, 133 (1988)

84. A. Gallagher: Noble-gas broadening of the Li resonance line. Phys. Rev. **A12**, 133 (1975)

85. J.N. Murrel: *Introduction to the Theory of Atomic and Molecular Collisions* (Wiley, Chichester 1989)

86. R.J. Exton, W.L. Snow: Line shapes for satellites and inversion of the data to obtain interaction potentials. J. Quant. Spectrosc. Radiat. Transfer. **20**, 1 (1978)

87. H. Griem: *Principles of Plasma Spectroscopy* (Cambridge University Press, Cambridge 1997)

88. A. Sasso, G.M. Tino, M. Inguscio, N. Beverini, M. Francesconi: Investigations of collisional line shapes of neon transitions in noble gas mixtures. Nuov. Cimento D **10**, 941 (1988)

89. C.C. Davis, I.A. King: 'Gaseous ion lasers'. In: *Adv. Quantum Electronics, Vol. 3*, ed. by D.W. Godwin (Academic, New York 1975)

90. W.R. Bennett: *The Physics of Gas Lasers* (Gordon and Breach, New York 1977)

91. R. Moore: 'Atoms in dense plasmas'. In: *Atoms in Unusual Situations*, ed. by J.P. Briand, Nato ASI, Ser. B, Vol. 143 (Plenum, New York 1986)

92. H. Motz: *The Physics of Laser Fusion* (Academic, London 1979)

93. T.P. Hughes: *Plasmas and Laser Light* (Hilger, Bristol 1975)

94. A.S. Katzantsev, J.C. Hénoux: *Polarization Spectroscopy of Ionized Gases* (Kluwer Academ., Dordrecht 1995)

95. I.R. Senitzky: 'Semiclassical radiation theory within a quantum mechanical framework'. In: *Progress in Optics* **16** (North-Holland, Amsterdam 1978) p. 413

96. W.R. Hindmarsh, J.M. Farr: 'Collision broadening of spectral lines by neutral atoms'. In: *Progr. Quantum Electronics, Vol. 2, Part 4*, ed. by J.H. Sanders, S. Stenholm (Pergamon, Oxford 1973)

97. N. Anderson, K. Bartschat: *Polarization, Alignment and Orientation in Atomic Collisions* (Springer, Heidelberg 2001)

98. R.G. Breen: 'Line width'. In: *Handbuch der Physik, Vol. 27*, ed. by S. Flügge (Springer, Berlin 1964) p. 1

99. J. Hirschfelder, Ch.F. Curtiss, R.B. Bird: *Molecular Theory of Gases and Liquids* (Wiley, New York 1954)

100. S. Yi Chen, M. Takeo: Broadening and shift of spectral lines due to the presence of foreign gases. Rev. Mod. Phys. **29**, 20 (1957)

101. K.M. Sando, Shih-I. Chu: Pressure broadening and laser-induced spectral line shapes. Adv. At. Mol. Phys. **25**, 133 (1988)

102. R.H. Dicke: The effect of collisions upon the Doppler width of spectral lines. Phys. Rev. **89**, 472 (1953)

103. R.S. Eng, A.R. Calawa, T.C. Harman, P.L. Kelley: Collisional narrowing of infrared water vapor transitions. Appl. Phys. Lett. **21**, 303 (1972)

104. A.T. Ramsey, L.W. Anderson: Pressure Shifts in the ^{23}Na Hyperfine Frequency. J. Chem. Phys. **43**, 191 (1965)

105. K. Shimoda: 'Line broadening and narrowing effects' In: *High-Resolution Spectroscopy* Topics Appl. Phys., Vol. 13, ed. by K. Shimoda (Springer, Berlin, Heidelberg 1976) p. 11

106. J. Hall: 'The line shape problem in laser saturated molecular absorptions'. In: *Lecture Notes in Theor. Phys., Vol. 12A*, ed. by K. Mahanthappa, W. Brittin (Gordon and Breach, New York 1971)

107. V.S. Letokhov, V.P. Chebotayev: *Nonlinear Laser Spectroscopy*, Springer Ser. Opt. Sci., Vol. 4 (Springer, Berlin, Heidelberg 1977)

108. K.H. Drexhage: 'Structure and properties of laser dyes'. In: *Dye Lasers*, 3rd edn., Topics Appl. Phys., Vol. 1, ed. by F.P. Schäfer (Springer, Berlin, Heidelberg 1990)

109. D.S. McClure: 'Electronic spectra of molecules and ions in crystals'. In: *Solid State Phys., Vols. 8 and 9* (Academic, New York 1959)

110. W.M. Yen, P.M. Selzer (Eds.): *Laser Spectroscopy of Solids*, Springer Ser. Opt. Sci., Vol. 14 (Springer, Berlin, Heidelberg 1981)

111. A.A. Kaminskii: *Laser Crystals*, 2nd edn., Springer Ser. Opt. Sci., Vol. 14 (Springer, Berlin, Heidelberg 1991)

112. C.H. Wei, K. Holliday, A.J. Meixner, M. Croci, U.P. Wild: Spectral hole-burning study of BaFClBrSm$^{(2+)}$. J. Lumin. **50**, 89 (1991)

113. W.E. Moerner: *Persistent Spectral Hole-Burning: Science and Applications*, Topics Curr. Phys., Vol. 44 (Springer, Berlin, Heidelberg 1988)

Chapter 4

114. R. Kingslake, B.J. Thompson (Eds.): *Applied Optics and Optical Engineering, Vols. 1–10* (Academic, New York 1969–1985);
M. Bass, E. van Skryland, D. Williams, W. Wolfe (Eds.): *Handbook of Optics, Vols. I and II* (McGraw-Hill, New York 1995)

115. E. Wolf (Ed.): *Progress in Optics, Vols. 1–42* (North-Holland, Amsterdam 1961–2001)

116. M. Born, E. Wolf: *Principles of Optics*, 7th edn. (Pergamon, Oxford 1999)

117. A.P. Thorne, U. Litzen, S. Johansson: *Spectrophysics*, 2nd edn. (Springer, Berlin 1999);
G.L. Clark (Ed.): *The Encyclopedia of Spectroscopy* (Reinhold, New York 1960)

118. (a) L. Levi: *Applied Optics* (Wiley, London 1980);
(b) D.F. Gray (Ed.): *Am. Inst. Phys. Handbook* (McGraw-Hill, New York 1980)

119. R.D. Guenther: *Modern Optics* (Wiley, New York 1990)

120. F. Graham-Smith, T.A. King: *Optics and Photonics* (Wiley, London 2000)

121. H. Lipson: *Optical Physics*, 4th edn. (Cambridge University Press, Cambridge 2010)

122. K.I. Tarasov: *The Spectroscope* (Hilger, London 1974)

123. S.P. Davis: *Diffraction Grating Spectrographs* (Holt, Rinehard & Winston, New York 1970)

124. A.B. Schafer, L.R. Megil, L. Dropleman: Optimization of the Czerny-Turner spectrometer. J. Opt. Soc. Am. **54**, 879 (1964)

125. *Handbook of Diffraction Gratings, Ruled and Holographic* (Jobin Yvon Optical Systems, Metuchen, NJ 1970);
    Bausch and Lomb Diffraction Grating Handbook (Bausch & Lomb, Rochester, NY 1970)

126. G.W. Stroke: 'Diffraction gratings'. In: *Handbuch der Physik, Vol. 29*, ed. by S. Flügge (Springer, Berlin, Heidelberg 1967)

127. E.G. Loewen, E. Popov: *Diffraction Gratings and Applications* (CRC Press 1997)

128. M.C. Hutley: *Diffraction Gratings* (Academic, London 1982);
    E. Popov, E.G. Loewen: *Diffraction Gratings and Applications* (Dekker, New York 1997)

129. See, for example, E. Hecht: *Optics*, 4th edn. (Addison-Wesley, London 2002)

130. G. Schmahl, D. Rudolph: 'Holographic diffraction gratings'. In: *Progress in Optics* **14**, 195 (North-Holland, Amsterdam 1977)

131. E. Loewen: 'Diffraction gratings: ruled and holographic'. In: *Applied Optics and Optical Engineering, Vol. 9* (Academic, New York 1980)

132. M.D. Perry, et al.: High-efficiency multilayer dielectric diffraction gratings. Opt. Lett. **20**, 940 (1995)

133. Basic treatments of interferometers may be found in general textbooks on optics. A more detailed discussion has, for instance, been given in S. Tolansky: *An Introduction to Interferometry* (Longman, London 1973);
    W.H. Steel: *Interferometry* (Cambridge Univ. Press, Cambridge 2009);
    J. Dyson: *Interferometry* (Machinery Publ., Brighton 1970);
    M. Francon: *Optical Interferometry* (Academic, New York 1966)

134. H. Polster, J. Pastor, R.M. Scott, R. Crane, P.H. Langenbeck, R. Pilston, G. Steingerg: New developments in interferometry. Appl. Opt. **8**, 521 (1969)

135. K.M. Baird, G.R. Hanes: 'Interferometers'. In: [114], Vol. 4, pp. 309–362

136. P. Hariharan: *Optical Interferometry*, 2nd edn. (Academic, New York 2010);
    W.S. Gornall: The world of Fabry–Perots. Laser Appl. **2**, 47 (1983)

137. M. Francon, J. Mallick: *Polarisation Interferometers* (Wiley, London 1971)

138. H. Welling, B. Wellingehausen: High resolution Michelson interferometer for spectral investigations of lasers. Appl. Opt. **11**, 1986 (1972)

139. P.R. Saulson: *Fundamentals of Interferometric Gravitational Wave Detectors* (World Scientific, Singapore 1994)

140. R.W.P. Drever, J.L. Hall, F.V. Kowalski, J. Hough, G.M. Ford, A.J. Munley, H. Ward: Laser phase and frequency stabilization using an optical resonator. Appl. Phys. B **31**, 97 (1983);
    A. Wicht, K. Danzmann, M. Fleischhauer, M. Scully, G. Müller, R.-H. Rinkleff: White-light cavities, atomic phase coherence and gravitational wave detectors. Opt. Commun. **134**, 431 (1997)

141. P. Griffiths, J.A. de Haseth: *Fourier-Transform Infrared Spectroscopy* (Wiley, New York 2007)

142. S.P. Davies, M.C. Abrams, J.W. Brault: *Fourier Transform Spectroscopy* (Academic Press, New York 2001)

143. V. Grigull, H. Rottenkolber: Two beam interferometer using a laser. J. Opt. Soc. Am. **57**, 149 (1967);
    W. Schumann, M. Dubas: *Holographic Interferometry*, Springer Ser. Opt. Sci., Vol. 16 (Springer, Berlin, Heidelberg 1979);
    W. Schumann, J.-P. Zürcher, D. Cuche: *Holography and Deformation Analysis*, Springer Ser. Opt. Sci., Vol. 46 (Springer, Berlin, Heidelberg 1986);
    G.R. Toker: *Holographic Interferometry* (CRC Press 2012)

144. W. Marlow: Hakenmethode. Appl. Opt. **6**, 1715 (1967)

145. I. Meroz (Ed.): *Optical Transition Probabilities. A Representative Collection of Russian Articles* (Israel Program for Scientific Translations, Jerusalem 1962)

146. D.S. Rozhdestvenski: Anomale Dispersion im Natriumdampf. Ann. Phys. **39**, 307 (1912)

147. S. Ezekiel, H.J. Arditty (Eds.): *Fiber-optic rotation sensors*, Springer Ser. Opt. Sci., Vol. 32 (Springer, Berlin, Heidelberg 1982);
G.E. Stedman: Ring laser tests of fundamental physics and geophysics. Rep. Prog. Phys. **60**, 615 (1997);
E. Udd, W.B. Spillman, Jr.: *Fiber Optic Sensors* (Wiley 2011)

148. J.P. Marioge, B. Bonino: Fabry–Perot interferometer surfacing. Opt. Laser Technol. **4**, 228 (1972);
J.V. Ramsay: Aberrations of Fabry–Perot Interferometers. Appl. Opt. **8**, 569, (1969)

149. M. Hercher: Tilted etalons in laser resonators. Appl. Opt. **8**, 1103 (1969)

150. W.R. Leeb: Losses introduced by tilting intracavity etalons. Appl. Phys. **6**, 267 (1975)

151. W. Demtröder, M. Stock: Molecular constants and potential curves of Na_2 from laser-induced fluorescence. J. Mol. Spectrosc. **55**, 476 (1975)

152. P. Connes: L'etalon de Fabry–Perot spherique. Phys. Radium **19**, 262 (1958);
P. Connes: *Quantum Electronics and Coherent Light*, ed. by P.H. Miles (Academic, New York 1964) p. 198

153. D.A. Jackson: The spherical Fabry–Perot interferometer as an instrument of high resolving power for use with external or with internal atomic beams. Proc. Roy. Soc. (London) A **263**, 289 (1961)

154. J.R. Johnson: A high resolution scanning confocal interferometer. Appl. Opt. **7**, 1061 (1968)

155. M. Hercher: The spherical mirror Fabry–Perot interferometer. Appl. Opt. **7**, 951 (1968)

156. R.L. Fork, D.R. Herriot, H. Kogelnik: A scanning spherical mirror interferometer for spectral analysis of laser radiation. Appl. Opt. **3**, 1471 (1964)

157. F. Schmidt-Kaler, D. Leibfried, M. Weitz, T.W. Hänsch: Precision measurements of the isotope shift of the $1s$–$2s$ transition of atomic hydrogen and deuterium. Phys. Rev. Lett. **70**, 2261 (1993)

158. J.R. Johnson: A method for producing precisely confocal resonators for scanning interferometers. Appl. Opt. **6**, 1930 (1967)

159. P. Hariharan: *Optical Interferometry* (Academic, New York 2010);
G.W. Hopkins (Ed.): *Interferometry*. SPIE Proc. **192** (1979);
R.J. Pryputniewicz (Ed.): *Industrial Interferometry*. SPIE Proc. **746** (1987);
R.J. Pryputniewicz (Ed.): *Laser Interferometry*. SPIE Proc. **1553** (1991);
J.D. Briers: Interferometric testing of optical systems and components. Opt. Laser Techn. (February 1972) p. 28

160. J.M. Vaughan: *The Fabry–Perot Interferometer* (Hilger, Bristol 1989);
Z. Jaroscewicz, M. Pluta (Eds.): *Interferometry 89: 100 Years after Michelson: State of the Art and Applications*. SPIE Proc. **1121** (1989)

161. J. McDonald: *Metal Dielectric Multilayer* (Hilger, London 1971)

162. A. Thelen: *Design of Optical Interference Coatings* (McGraw-Hill, New York 1988);
Z. Knittl: *Optics of Thin Films* (Wiley, New York 1976)

163. V.R. Costich: 'Multilayer dielectric coatings'. In: *Handbook of Lasers*, ed. by R.J. Pressley (CRC, Cleveland, Ohio 1972);
D. Ristau, H. Ehlers: 'Thin Film Optical Coatings'. In: F. Träger (Ed.): *Springer Handbook of Lasers and Optics* (Springer, Heidelberg 2007)

164. H.A. MacLeod (Ed.): Optical interference coatings. Appl. Opt. **28**, 2697–2974 (1989);
     R.E. Hummel, K.H. Guenther (Eds.): *Optical Properties, Vol. 1: Thin Films for Optical Coatings* (CRC, Cleveland, Ohio 1995);
     A. Musset, A. Thelen: 'Multilayer antireflection coatings'. In: *Progress in Optics* **3**, 203 (North-Holland, Amsterdam 1970)

165. See, for instance: Newport Research Corp.: *Ultralow loss supermirrors* (www.newport.com)

166. J.T. Cox, G. Hass: In: *Physics of Thin Films, Vol. 2*, ed. by G. Hass (Academic, New York 1964)

167. E. Delano, R.J. Pegis: 'Methods of synthesis for dielectric multilayer filters'. In: *Progress in Optics, Vol. 7*, 69 (North-Holland, Amsterdam 1969)

168. H.A. Macleod: *Thin Film Optical Filter*, 3rd edn. (Inst. of Physics Publ., London 2001)

169. J. Evans: The birefringent filter. J. Opt. Soc. Am. **39**, 229 (1949)

170. H. Walther, J.L. Hall: Tunable dye laser with narrow spectral output. Appl. Phys. Lett. **17**, 239 (1970)

171. M. Okada, S. Iliri: Electronic tuning of dye lasers by an electro-optic birefringent Fabry–Perot etalon. Opt. Commun. **14**, 4 (1975)

172. B.H. Billings: The electro-optic effect in uniaxial crystals of the type XH_2PO_4. J. Opt. Soc. Am. **39**, 797 (1949)

173. R.L. Fork, D.R. Herriot, H. Kogelnik: A scanning spherical mirror interferometer for spectral analysis of laser radiation. Appl. Opt. **3**, 1471 (1964)

174. V.G. Cooper, B.K. Gupta, A.D. May: Digitally pressure scanned Fabry–Perot interferometer for studying weak spectral lines. Appl. Opt. **11**, 2265 (1972)

175. J.M. Telle, C.L. Tang: Direct absorption spectroscopy, using a rapidly tunable cw-dye laser. Opt. Commun. **11**, 251 (1974)

176. P. Cerez, S.J. Bennet: New developments in iodine-stabilized HeNe lasers. IEEE Trans. IM-**27**, 396 (1978)

177. K.M. Evenson, J.S. Wells, F.R. Petersen, B.L. Danielson, G.W. Day, R.L. Barger, J.L. Hall: Speed of light from direct frequency and wavelength measurements of the methane-stabilized laser. Phys. Rev. Lett. **29**, 1346 (1972)

178. K.M. Evenson, D.A. Jennings, F.R. Petersen, J.S. Wells: 'Laser frequency measurements: a review, limitations and extension to 197 THz'. In: *Laser Spectroscopy III*, ed. by J.L. Hall, J.L. Carlsten, Springer Ser. Opt. Sci., Vol. 7 (Springer, Berlin, Heidelberg 1977)

179. K.M. Evenson, J.S. Wells, F.R. Petersen, B.L. Davidson, G.W. Day, R.L. Barger, J.L. Hall: The speed of light. Phys. Rev. Lett. **29**, 1346 (1972)

180. A. DeMarchi (Ed.): *Frequency Standards and Metrology* (Springer, Berlin, Heidelberg 1989)

181. P.R. Bevington: *Data Reduction and Error Analysis for the Physical Sciences* (McGraw-Hill, New York 1969)

182. J.R. Taylor: *An Introduction to Error Analysis* (Univ. Science Books, Mill Valley 1982)

183. P. Potuluri, M.E. Gehm, M.E. Sullivan, D.J. Brady: Measurement efficient optical wavemeters. Opt. Expr. **12**, 6219 (2004), http://www.disp.dukwe.edu

184. J.L. Hall, S.A. Lee: Interferometric real time display of CW dye laser wavelength with sub-Doppler accuracy. Appl. Phys. Lett. **29**, 367 (1976)

185. J.J. Snyder: 'Fizeau wavelength meter'. In: *Laser Spectroscopy III*, ed. by J.L. Hall, J.L. Carlsten, Springer Ser. Opt. Sci., Vol. 7 (Springer, Berlin, Heidelberg 1977) p. 419 and Laser Focus May 1982 p. 55

186. R.L. Byer, J. Paul, M.D. Duncan: 'A wavelength meter'. In: *Laser Spectroscopy III*, ed. by J.L. Hall, J.L. Carlsten, Springer Ser. Opt. Sci., Vol. 7 (Springer, Berlin, Heidelberg 1977) p. 414

187. A. Fischer, H. Kullmer, W. Demtröder: Computer-controlled Fabry–Perot-wavemeter. Opt. Commun. 20, 277 (1981)

188. N. Konishi, T. Suzuki, Y. Taira, H. Kato, T. Kasuya: High precision wavelength meter with Fabry–Perot optics. Appl. Phys. 25, 311 (1981)

189. K. Quedraogo et al.: Compact and accurate concept of laser wavemeters based on Ellipsometry. Rev. Scient. Instr. 82, 055102 (2011)

190. F.V. Kowalski, R.E. Teets, W. Demtröder, A.L. Schawlow: An improved wavemeter for CW lasers. J. Opt. Soc. Am. 68, 1611 (1978)

191. R. Best: *Theorie und Anwendung des Phase-Locked Loops* (AT Fachverlag, Stuttgart 1976)

192. F.M. Gardner: *Phase Lock Techniques* (Wiley, New York 1966);
Phase-Locked Loop Data Book (Motorola Semiconductor Prod., Inc. 1973)

193. B. Edlen: Dispersion of standard air. J. Opt. Soc. Am. 43, 339 (1953)

194. J.C. Owens: Optical refractive index of air: Dependence on pressure, temperature and composition. Appl. Opt. 6, 51 (1967)

195. R. Castell, W. Demtröder, A. Fischer, R. Kullmer, K. Wickert: The accuracy of laser wavelength meters. Appl. Phys. B 38, 1 (1985)

196. P.J. Fox et al.: A reliable compact and low cost Michelson wavemeter for laser wavelength measurements. Am. J. Phys. 67, 624 (1999)

197. J. Cachenaut, C. Man, P. Cerez, A. Brillet, F. Stoeckel, A. Jourdan, F. Hartmann: Description and accuracy tests of an improved lambdameter. Rev. Phys. Appl. 14, 685 (1979)

198. J. Viqué, B. Girard: A systematic error of Michelson's type lambdameter. Rev. Phys. Appl. 21, 463 (1986)

199. J.J. Snyder: 'An ultrahigh resolution frequency meter'. *Proc. 35th Ann. Freq. Control US-AERADCOM* May 1981. Appl. Opt. 19, 1223 (1980)

200. P. Juncar, J. Pinard: Instrument to measure wavenumbers of CW and pulsed laser lines: The sigma meter. Rev. Sci. Instrum. 53, 939 (1982);
P. Jacquinot, P. Juncar, J. Pinard: 'Motionless Michelson for high precision laser frequency measurements'. In: *Laser Spectroscopy III*, ed. by J.L. Hall, J.L. Carlsten, Springer Ser. Opt. Sci., Vol. 7 (Springer, Berlin, Heidelberg 1977) p. 417

201. J.J. Snyder: Fizeau wavemeter. SPIE Proc. 288, 258 (1981)

202. M.B. Morris, T.J. McIllrath, J. Snyder: Fizeau wavemeter for pulsed laser wavelength measurement. Appl. Opt. 23, 3862 (1984)

203. J.L. Gardner: Compact Fizeau wavemeter. Appl. Opt. 24, 3570 (1985)

204. J.L. Gardner: Wavefront curvature in a Fizeau wavemeter. Opt. Lett. 8, 91 (1983)

205. http://www.rp-photonics.com/wavemeters.html

206. J.J. Keyes (Ed.): *Optical and Infrared Detectors*, 2nd edn., Topics Appl. Phys., Vol. 19 (Springer, Berlin, Heidelberg 1980)

207. A. Rogalski: *Infrared Detectors*, 2nd edn. (CRC Press, Boca Raton, Florida 2010)

208. P.N. Dennis: *Photodetectors* (Plenum, New York 1986)

209. G.R. Osche: *Optical Detection Theory for Laser Applications* (Wiley Interscience, Hoboken, NJ 2002)

210. J. Haus: *Optical Sensors* (Wiley-VCH, Weinheim 2010)

211. E.L. Dereniak, G.D. Boreman: *Infrared Detectors and Systems* (Wiley, New York 1996)

212. G.H. Rieke: *Detection of Light: From the Ultraviolet to the Submillimeter* (Cambridge University Press, Cambridge 1994)

213. J. Wilson, J.F.B. Hawkes: *Optoelectronics* (Prentice Hall, London 1998)

214. R. Paul: *Optoelektronische Halbleiterbauelemente* (Teubner, Stuttgart 1985)

215. T.S. Moss, G.J. Burell, B. Ellis: *Semiconductor Opto-Electronics* (Butterworth, London 1973)

216. R.W. Boyd: *Radiometery and the Detection of Optical Radiation* (Wiley, New York 1983)

217. E.L. Dereniak, D.G. Crowe: *Optical Radiation Detectors* (Wiley, New York 1984)

218. F. Stöckmann: Photodetectors, their performance and limitations. Appl. Phys. **7**, 1 (1975)

219. F. Grum, R.L. Becher: *Optical Radiation Measurements, Vols. 1 and 2* (Academic, New York 1979 and 1980)

220. R.H. Kingston: *Detection of Optical and Infrared Radiation*, Springer Ser. Opt. Sci., Vol. 10 (Springer, Berlin, Heidelberg 1978)

221. R. Fischer: *Optical System Design*, 2nd edn. (McGraw Hill, Columbus, Ohio 2008)

222. E.H. Putley: 'Thermal detectors'. In: [206], p. 71

223. C. Kuenzer, St. Dech: *Thermal Infrared Remote Sensing* (Springer. Berlin, Heidelberg 2013)

224. T.E. Gough, R.E. Miller, G. Scoles: Infrared laser spectroscopy of molecular beams. Appl. Phys. Lett. **30**, 338 (1977);
M. Zen: Cryogenic bolometers, in *Atomic and Molecular Beam Methods*, ed. by G. Scoles (Oxford Univ. Press, New York 1988) Vol. 1

225. D. Bassi, A. Boschetti, M. Scotoni, M. Zen: Molecular beam diagnostics by means of fast superconducting bolometer. Appl. Phys. B **26**, 99 (1981);
A.T. Lee et al.: Superconducting bolometer with strong electrothermal feedback. Appl. Phys. Lett. **69**, 1801 (1996)

226. J. Clarke, P.L. Richards, N.H. Yeh: Composite superconducting transition edge bolometer. Appl. Phys. Lett. B **30**, 664 (1977)

227. M.J.E. Golay: A Pneumatic Infra-Red Detector. Rev. Scient. Instrum. **18**, 357 (1947)

228. B. Tiffany: Introduction and review of pyroelectric detectors. SPIE Proc. **62**, 153 (1975)

229. C.B. Boundy, R.L. Byer: Subnanosecond pyroelectric detector. Appl. Phys. Lett. **21**, 10 (1972)

230. G. Gautschi: *Piezoelectric Sensorics*, ISBN 3-540-42259-5 (Springer, Berlin, Heidelberg 2002)

231. L.E. Ravich: Pyroelectric detectors and imaging. Laser Focus **22**, 104 (1986)

232. H. Melchior: 'Demodulation and photodetection techniqes'. In: *Laser Handbook, Vol. 1*, ed. by F.T. Arrecchi, E.O. Schulz-Dubois (North-Holland, Amsterdam 1972) p. 725

233. H. Melchior: Sensitive high speed photodetectors for the demodulation of visible and near infrared light. J. Lumin. **7**, 390 (1973)

234. D. Long: 'Photovoltaic and photoconductive infrared detectors'. In: [206], p. 101

235. E. Sakuma, K.M. Evenson: Characteristics of tungsten nickel point contact diodes used as a laser harmonic generation mixers. IEEE J. QE-**10**, 599 (1974)

236. K.M. Evenson, M. Ingussio, D.A. Jennings: Point contact diode at laser frequencies. J. Appl. Phys. **57**, 956 (1985);
H.D. Riccius, K.D. Siemsen: Point-contact diodes. Appl. Phys. A **35**, 67 (1984);
H. Rösser: Heterodyne spectroscopy for submillimeter and far-infrared wavelengths. Infrared Phys. **32**, 385 (1991)

237. H.-U. Daniel, B. Maurer, M. Steiner: A broad band Schottky point contact mixer for visible light and microwave harmonics. Appl. Phys. B **30**, 189 (1983);
T.W. Crowe: GaAs Schottky barrier mixer diodes for the frequency range from 1–10 THz. Int. J. IR and Millimeter Waves **10**, 765 (1989);
H.P. Röser, R.V. Titz, G.W. Schwab, M.F. Kimmitt: Current-frequency characteristics of submicron GaAs Schottky barrier diodes with femtofarad capacitances. J. Appl. Phys. **72**, 3194 (1992)

238. F. Capasso: Band-gap engineering via graded-gap structure: Applications to novel photodetectors. J. Vac. Sci. Techn. B**12**, 457 (1983);
K.-S. Hyun, C.-Y. Park: Breakdown characteristics in InP/InGaAs avalanche photodiodes with p-i-n multiplication layer. J. Appl. Phys. **81**, 974 (1997)

239. F. Capasso (Ed.): *Physics of Quantum Electron Devices*, Springer Ser. Electron. Photon., Vol. 28 (Springer, Berlin, Heidelberg 1990)

240. J. Kataoka et al.: Recent progress of avalanche photodiodes in high resolution X-ray and γ-ray detectors. Xiv:astro-ph/0602391

241. F. Capasso: Multilayer avalanche photodiodes and solid state photomultipliers. Laser Focus **20**, 84 (July 1984)

242. http://www.hamamatsu.com/jp/en/product/category/3100/4003/index.html

243. G.A. Walter, E.L. Dereniak: Photodetectors for focal plane arrays. Laser Focus **22**, 108 (March 1986)

244. A. Tebo: IR detector technology. Arrays. Laser Focus **20**, 68 (July 1984);
E.L. Dereniak, R.T. Sampson (Eds.): *Infrared Detectors, Focal Plane Arrays and Imaging Sensors*, SPIE Proc. **1107** (1989);
E.L. Dereniak (Ed.): Infrared Detectors and Arrays. SPIE Proc. **930** (1988)

245. D.F. Barbe (Ed.): *Charge-Coupled Devices*, Topics Appl. Phys., Vol. 38 (Springer, Berlin, Heidelberg 1980)

246. http://www.specinst.com/What_Is_A_CCD.html

247. http://en.wikipedia.org/wiki/Charge-coupled_device

248. see special issue on CCDs of Berkeley Lab **23**, 3 (Fall 2000) and G.C. Holst: *CCD Arrays, Cameras and Display* (Sofitware, ISBN 09640000024, 2000);
K.P. Proll, J.M. Nivet, C. Voland: Enhancement of the dynamic range of the detected intensity in an optical measurement system by a three channel technique. Appl. Opt. **41**, 130 (2002)

249. H. Zimmermann: *Integrated Silicon Optoelectronics* (Springer, Berlin, Heidelberg 2000)

250. R.B. Bilborn, J.V. Sweedler, P.M. Epperson, M.B. Denton: Charge transfer device detectors for optical spectroscopy. Appl. Spectrosc. **41**, 1114 (1987)

251. I. Nin, Y. Talmi: CCD detectors record multiple spectra simultaneously. Laser Focus **27**, 111 (August 1991)

252. H.R. Zwicker: Photoemissive detectors. In *Optical and Infrared Detectors*, 2nd edn., ed. by J. Keyes, Topics Appl. Phys., Vol. 19 (Springer, Berlin, Heidelberg 1980)

253. C. Gosh: Photoemissive materials. SPIE Proc. **346**, 62 (1982)

254. R.L. Bell: *Negative Electron Affinity Devices* (Clarendon, Oxford 1973)

255. L.E. Wood, T.K. Gray, M.C. Thompson: Technique for the measurement of photomultiplier transit time variation. Appl. Opt. **8**, 2143 (1969)

256. J.D. Rees, M.P. Givens: Variation of time of flight of electrons through a photomultiplier. J. Opt. Soc. Am. **56**, 93 (1966)

257. (a) B. Sipp. J.A. Miehé, R. Lopes Delgado: Wavelength dependence of the time resolution of high speed photomultipliers used in single-photon timing experiments. Opt. Commun. **16**, 202 (1976)
     (b) G. Beck: Operation of a 1P28 photomultipier with subnanosecond response time. Rev. Sci. Instrum. **47**, 539 (1976)
     (c) B.C. Mongan (Ed.): *Adv. Electronics and Electron Physics, Vol. 74* (Academic, London 1988)

258. S.D. Flyckt, C. Marmonier: *Photomultiplier tubes* (Photonics Brive, France 2002)

259. A. van der Ziel: *Noise in Measurements* (Wiley, New York 1976)

260. A.T. Young: Undesirable effects of cooling photomultipliers. Rev. Sci. Instrum. **38**, 1336 (1967);
     Hamamatsu: *Photomultiplier handbook*
     http://hamamatsu.com/photomultiplier

261. J. Sharpe, C. Eng: Dark Current in Photomultiplier Tubes (EMI Ltd. information document, ref. R-P021470)

262. Phototubes and Photocells. In: *An Introduction to the Photomultiplier* (RCA Manual, EMI Ltd. information sheet, 1966);
     Hamamatsu Photonics: *Photomultiplier Tubes: Basis and Applications*, 3rd edn. (Hamamatsu City, Japan 2006)

263. Photomultiplier Handbook. psec.uchicago.edu/links/Photomultiplier_Handbook.pdf

264. R.W. Engstrom: *Photomultiplier Handbook* (RCA Corp. 1980)

265. G. Pietri: Towards picosecond resolution. Contribution of microchannel electron multipiers to vacuum tube design. IEEE Trans. NS-**22**, 2084 (1975);
     J.L. Wiza: Microchannel plate detectors (Galileo information sheet, Sturbridge, MA, 1978)

266. I.P. Csonba (Ed.): *Image Intensification*, SPIE Proc. **1072** (1989)

267. *Proc. Topical Meeting on Quantum-Limited Imaging and Image Processing* (Opt. Soc. Am., Washington, DC 1986)

268. T.P. McLean, P. Schagen (Eds.): *Electronic Imaging* (Academic, London 1979)

269. H.K. Pollehn: 'Image intensifiers'. In: [114], Vol. 6 (1980) p. 393

270. S. Jeffers, W. Weller: 'Image intensifier optical multichannel analyser for astronomical spectroscopy'. In: *Adv. Electronics and Electron Phys. B* **40** (Academic, New York 1976) p. 887

271. L. Perko, J. Haas, D. Osten: Cooled and intensified array detectors for optical spectroscopy. SPIE Proc. **116**, 64 (1977)

272. J.L. Hall: 'Arrays and charge coupled devices'. In: [114], Vol. 8 (1980) p. 349

273. J.L. Weber: Gated optical multichannel analyzer for time resolved spectroscopy. SPIE Proc. **82**, 60 (1976)

274. R.G. Tull: A comparison of photon-counting and current measuring techniques in spectrometry of faint sources. Appl. Opt. **7**, 2023 (1968)

275. J.F. James: On the use of a photomultiplier as a photon counter. Monthly Notes R. Astron. Soc. **137**, 15 (1967);
     W. Becker, A. Bergmann: Detectors for High-Speed Photon Counting
     (www.becker-bickl.de)

276. D.V. O'Connor, D. Phillips: *Time-Correlated Photon-Counting* (Academic, London 1984);
     G.F. Knoll: *Radiation Detectors and Measurement* (Wiley, New York 1979);
     S. Kinishita, T. Kushida: High performance time-correlated single photon counting apparatus, using a side-on type photon multiplier. Rev. Sci. Instrum. **53**, 469 (1983)

277. P.W. Kruse: 'The photon detection process'. In: *Optical and Infrared Detectors*, 2nd edn., ed. by J.J. Keyes, Topics Appl. Phys., Vol. 19 (Springer, Berlin, Heidelberg 1980)

278. Signal Averagers. (Information sheet, issued by Princeton Appl. Res., Princeton, NJ, 1978)

279. Information sheet on transient recorders, Biomation, Palo Alto, CA

280. C. Morgan. Digital signal processing: Laser Focus **10**, 59 (Nov. 1977)

281. Handshake: Information sheet on waveform digitizing instruments (Tektronic, Beaverton, OR 1979)

282. Hamamatsu photonics information sheet (February 1989)

283. H. Mark: *Principles and Practice of Spectroscopic Calibration* (Wiley, New York 1991)

284. A.C.S. van Heel (Ed.): *Advanced Optical Techniques* (North-Holland, Amsterdam 1967)

285. W. Göpel, J. Hesse, J.N. Zemel (Eds.): *Sensors, A Comprehensive Survey* (Wiley-VCH, Weinheim 1992);
W. Göpel: *Sensors Update* (Wiley-VCH, Weinheim 1998)

286. D. Dragoman, M. Dragoman: *Advanced Optical Devices* (Springer, Heidelberg 1999)

287. F. Grum, R.L. Becherer (Eds.): *Optical Radiation Measurements, Vols. I, II* (Academic, New York 1979, 1980)

288. C.H. Lee: *Picosecond Optoelectronics Devices* (Academic, New York 1984)

289. *The Photonics and Application Handbook* (Laurin, Pittsfield, MA 1990)

290. F. Träger (Ed.): *Springer Handbook of Lasers and Optics*, 2nd edn. (Springer, Berlin, Heidelberg 2012)

Chapter 5

291. A.E. Siegman: *An Introduction to Lasers and Masers* (McGraw-Hill, New York 1971);
A.E. Siegman: *Lasers* (Oxford Univ. Press, Oxford 1986)

292. I. Hecht: *The Laser Guidebook*, 2nd edn. (McGraw-Hill, New York 1999)

293. *Lasers, Vols. 1–4*, ed. by A. Levine (Dekker, New York 1966–76)

294. A. Yariv: *Quantum Electrons*, 3rd edn. (Wiley, New York 1989);
A. Yariv: *Optical Electronics*, 4th edn. (Sounders College Publishing, Harcourt Brace 1991)

295. O. Svelto: *Principles of Lasers*, 5th edn. (Springer, Heidelberg 2010)

296. *Laser Handbook, Vols. I–V* (North-Holland, Amsterdam 1972–1985);
M.J. Weber: *Handbook of Lasers* (CRC, New York 2001);
M.J. Weber: *Handbook of Laser Wavelengths* (CRC, New York 1999)

297. A. Maitland, M.H. Dunn: *Laser Physics* (North-Holland, Amsterdam 1969);
P.W. Milonni, J.H. Eberly: *Lasers* (Wiley, New York 2010)

298. F.K. Kneubühl, M.W. Sigrist: *Laser*, 7th edn. (Teubner, Stuttgart 2008)

299. I.T. Verdeyen: *Laser Electronics*, 3rd edn. (Prentice Hall, Englewood Cliffs, NJ 1994)

300. C.C. Davis: *Lasers and Electro-Optic*, 2nd edn. (Cambridge University Press, Cambridge 2014)

301. M.O. Scully, W.E. Lamb Jr., M. Sargent III: *Laser Physics* (Addison Wesley, Reading, MA 1974);
P. Meystre, M. Sargent III: *Elements of Quantum Optics*, 2nd edn. (Springer, Berlin, Heidelberg 1991), *Quantum Optics* (Cambridge Univ. Press, Cambridge 2013)

302. H. Haken: *Laser Theory* (Springer, Berlin, Heidelberg 1984)

303. D. Eastham: *Atomic Physics of Lasers* (Taylor & Francis, London 1986)

304. R. Loudon: *The Quantum Theory of Light*, 3rd edn. (Clarendon, Oxford 1997)

305. A.F. Harvey: *Coherent Light* (Wiley, London 1970)

306. E. Hecht: *Optics*, 3rd edn. (Addison Wesley, Reading, MA 1997), 4th edn. (Pearson Education 2013)

307. M. Born, E. Wolf: *Principles of Optics*, 7th edn. (Cambridge Univ. Press, Cambridge 2013)

308. G. Koppelmann: Multiple beam interference and natural modes in open resonators. *Progress in Optics* **7** (North-Holland, Amsterdam 1969) pp. 1–66

309. A.G. Fox, T. Li: Resonant modes in a maser interferometer. Bell System Techn. J. **40**, 453 (1961)

310. G.D. Boyd, J.P. Gordon: Confocal multimode resonator for millimeter through optical wavelength masers. Bell Syst. Techn. J. **40**, 489 (1961)

311. G.D. Boyd, H. Kogelnik: Generalized confocal resonator theory. Bell Syst. Techn. J. **41**, 1347 (1962)

312. A.G. Fox, T. Li: Modes in maser interferometers with curved and tilted mirrors. Proc. IEEE **51**, 80 (1963)

313. H.K.V. Lotsch: The Fabry–Perot resonator. Optik **28**, 65, 328, 555 (1968);
H.K.V. Lotsch: Optik **29**, 130, 622 (1969);
H.K.V. Lotsch: The confocal resonator system. Optik **30**, 1, 181, 217, 563 (1969/70)

314. H. Kogelnik, T. Li: Laser beams and resonators. Proc. IEEE **54**, 1312 (1966)

315. N. Hodgson, H. Weber: *Optical Resonators* (Springer, Berlin, Heidelberg, New York 1997), *Laser Resonators and Beam Propagation* (Springer 2005)

316. A.E. Siegman: Unstable optical resonators. Appl. Opt. **13**, 353 (1974)

317. W.H. Steier: 'Unstable resonators'. In: *Laser Handbook III*, ed. by M.L. Stitch (North-Holland, Amsterdam 1979)

318. R.L. Byer, R.L. Herbst: The unstable-resonator YAG laser. Laser Focus **14**, 48 (July 1978)

319. Y.A. Anan'ev: *Laser Resonators and the Beam Divergence Problem* (Hilger, Bristol 1992)

320. N. Hodgson, H. Weber: Unstable resonators with excited converging wave. IEEE J. QE-**26**, 731 (1990);
N. Hodgson, H. Weber: High-power solid state lasers with unstable resonators. Opt. Quant. Electron. **22**, 39 (1990)

321. W. Magnus, F. Oberhettinger, R.P. Soni: *Formulas and Theories for the Special Functions of Mathematical Physics* (Springer, Berlin, Heidelberg 1966);
M. Abramowitz, I. Stegun: *Handbook of Mathematical Functions* (Dover Publication 2012)

322. T.F. Johnston: Design and performance of broadband optical diode to enforce one direction travelling wave operation of a ring laser. IEEE J. QE-**16**, 483 (1980);
T.F. Johnson: Focus on Science 3, No. 1, (1980) (Coherent Radiation, Palo Alto, Calif.);
G. Marowsky: A tunable flash lamp pumped dye ring laser of extremely narrow bandwidth. IEEE J. QE-**9**, 245 (1973)

323. I.V. Hertel, A. Stamatovic: Spatial hole burning and oligo-mode distance control in CW dye lasers. IEEE J. QE-**11**, 210 (1975)

324. D. Kühlke, W. Diehl: Mode selection in cw-laser with homogeneously broadened gain. Opt. Quant. Electron. **9**, 305 (1977)

325. W.R. Bennet Jr.: *The Physics of Gas Lasers* (Gordon and Breach, New York 1977)

326. R. Beck, W. Englisch, K. Gürs: *Table of Laser Lines in Gases and Vapors*, 2nd edn., Springer Ser. Opt. Sci., Vol. 2 (Springer, Berlin, Heidelberg 1978);
M.J. Weber: *Handbook of Laser Wavelengths* (CRC, New York 1999)

327. K. Bergmann, W. Demtröder: A new cascade laser transition in a He-Ne mixture. Phys. Lett. 29 A, 94 (1969)

328. B.J. Orr: A constant deviation laser tuning device. J. Phys. E **6**, 426 (1973)

329. L. Allen, D.G.C. Jones: The helium-neon laser. Adv. Phys. **14**, 479 (1965)

330. C.E. Moore: Atomic Energy Levels, Nat. Stand. Ref. Ser. **35**, NBS Circular 467 (U.S. Dept. Commerce, Washington, DC 1971)

331. P.W. Smith: On the optimum geometry of a 6328 Å laser oscillator. IEEE J. QE-**2**, 77 (1966)

332. W.B. Bridges, A.N. Chester, A.S. Halsted, J.V. Parker: Ion laser plasmas. IEEE Proc. **59**, 724 (1971)

333. A. Ferrario, A. Sirone, A. Sona: Interaction mechanisms of laser transitions in argon and krypton ion-lasers. Appl. Phys. Lett. **14**, 174 (1969)

334. C.C. Davis, T.A. King: 'Gaseous ion lasers'. In: *Adv. Quantum Electronics, Vol. 3*, ed. by D.W. Goodwin (Academic, London 1975)

335. G. Herzberg: *Molecular Spectra and Molecular Structure, Vol. II* (Van Nostrand Reinhold, New York 1945)

336. D.C. Tyle: 'Carbon dioxyde lasers". In: *Adv. Quantum Electronics, Vol. 1*, ed. by D.W. Goodwin (Academic, London 1970)

337. W.J. Witteman: *The CO_2 Laser*, Springer Ser. Opt. Sci., Vol. 53 (Springer, Berlin, Heidelberg 1987)

338. H.W. Mocker: Rotational level competition in CO_2-lasers. IEEE J. QE-**4**, 769 (1968)

339. J. Haisma: Construction and properties of short stable gas lasers. Phillips Res. Rpt., Suppl. No. 1 (1967) and Phys. Lett. **2**, 340 (1962)

340. M. Hercher: Tunable single mode operation of gas lasers using intra-cavity tilted etalons. Appl. Opt. **8**, 1103 (1969)

341. P.W. Smith: Stabilized single frequency output from a long laser cavity. IEEE J. QE-**1**, 343 (1965)

342. P. Zory: Single frequency operation of argon ion lasers. IEEE J. QE-**3**, 390 (1967)

343. V.P. Belayev, V.A. Burmakin, A.N. Evtyunin, F.A. Korolyov, V.V. Lebedeva, A.I. Odintzov: High power single-frequency argon ion laser. IEEE J. QE-**5**, 589 (1969)

344. P.W. Smith: Mode selection in lasers. Proc. IEEE **60**, 422 (1972)

345. W.W. Rigrod, A.M. Johnson: Resonant prism mode selector for gas lasers. IEEE J. QE-**3**, 644 (1967)

346. R.E. Grove, E.Y. Wu, L.A. Hackel, D.G. Youmans, S. Ezekiel: Jet stream CW-dye laser for high resolution spectroscopy. Appl. Phys. Lett. **23**,. 442 (1973)

347. H.W. Schröder, H. Dux, H. Welling: Single mode operation of CW dye lasers. Appl. Phys. **1**, 347 (1973)

348. T.W. Hänsch: Repetitively pulsed tunable dye laser for high resolution spectroscopy. Appl. Opt. **11**, 895 (1972)

349. J.P. Goldsborough: 'Design of gas lasers'. In: *Laser Handbook I*, ed. by F.T. Arrecchi, E.O. Schulz-Dubois (North-Holland, Amsterdam 1972) p. 597

350. B. Peuse: 'New developments in CW dye lasers'. In: *Physics of New Laser Sources*, ed. by N.B. Abraham, F.T. Annecchi, A. Mooradian, A. Suna (Plenum, New York 1985)

351. M. Pinard, M. Leduc, G. Trenec, C.G. Aminoff, F. Laloc: Efficient single mode operation of a standing wave dye laser. Appl. Phys. **19**, 399 (1978)

352. E. Samal, W. Becker: *Grundriß der praktischen Regelungstechnik* (Oldenbourg, München 1996)

353. P. Horrowitz, W. Hill: *The Art of Electronics*, 2nd edn. (Cambridge Univ. Press, Cambridge 1989). http://en.wikipedia.org/wiki/PID_controller

354. Schott Information Sheet (Jenaer Glaswerk Schott & Gen., Mainz 1972)

355. R.W. Cahn, P. Haasen, E.J. Kramer (Eds.): *Materials Science and Technology, Vol. 11* (Wiley-VCH, Weinheim 1994)

356. J.J. Gagnepain: *Piezoelectricity* (Gordon & Breach, New York 1982);
Ch.R. Bowen: *Piezo-active Composites*, Springer Ser. Mater. Sci., Vol. 185 (Springer, Berlin, Heidelberg 2013)

357. W. Jitschin, G. Meisel: Fast frequency control of a CW dye jet laser. Appl. Phys. **19**, 181 (1979)

358. D.P. Blair: Frequency offset of a stabilized laser due to modulation distortion. Appl. Phys. Lett. **25**, 71 (1974)

359. J. Hough, D. Hills, M.D. Rayman, L.-S. Ma, L. Holbing, J.L. Hall: Dye-laser frequency stabilization using optical resonators. Appl. Phys. B **33**, 179 (1984)

360. F. Paech, R. Schmiedl, W. Demtröder: Collision-free lifetimes of excited NO_2 under very high resolution. J. Chem. Phys. **63**, 4369 (1975)

361. S. Seel, R. Storz, G. Ruosa, J. Mlynek, S. Schiller: Cryogenic optical resonators: a new tool for laser frequency stabilization at the 1 kHz lLevel. Phys. Rev. Lett. **78** 4741 (1997)

362. G. Camy, B. Decomps, J.-L. Gardissat, C.J. Bordé: Frequency stabilization of argon lasers at 582.49 THz using saturated absorption in $^{127}I_2$. Metrologia **13**, 145 (1977)

363. F. Spieweck: Frequency stabilization of Ar^+ lasers at 582 THz using expanded beams in external $^{127}I_2$ cells. IEEE Trans. IM-**29**, 361 (1980)

364. S.N. Bagayev, V.P. Chebotajev: Frequency stability and reproducibility at the 3.39 nm He-Ne laser stabilized on the methane line. Appl. Phys. **7**, 71 (1975)

365. J.L. Hall: 'Stabilized lasers and the speed of light'. In: *Atomic Masses and Fundamental Constants, Vol. 5*, ed. by J.H. Sanders, A.H. Wapstra (Plenum, New York 1976) p. 322

366. M. Niering et al.: Measurement of the hydrogen 1S–2S transition frequency by phase coherent comparison with a microwave cesium fountain clock. Phys. Rev. Lett. **84**, 5496 (2000)

367. Y.T. Zhao, J.M. Zhao, T. Huang, L.T. Xiao, S.T. Jia: Frequency stabilization of an external cavity diode laser with a thin Cs vapour cell. J. Phys. D: Appl. Phys. **37**, 1316 (2004)

368. T.W. Hänsch: 'High resolution spectroscopy of hydrogen'. In: *The Hydrogen Atom*, ed. by G.F. Bassani, M. Inguscio, T.W. Hänsch (Springer, Berlin, Heidelberg 1989)

369. P.H. Lee, M.L. Skolnik: Saturated neon absorption inside a 6328 Å laser. Appl. Phys. Lett. **10**, 303 (1967)

370. H. Greenstein: Theory of a gas laser with internal absorption cell. J. Appl. Phys. **43**, 1732 (1972)

371. T.N. Niebauer, J.E. Faller, H.M. Godwin, J.L. Hall, R.L. Barger: Frequency stability measurements on polarization-stabilized HeNe lasers. Appl. Opt **27**, 1285 (1988)

372. C. Salomon, D. Hils, J.L. Hall: Laser stabilization at the millihertz level. J. Opt. Soc. Am. B **5**, 1576 (1988)

373. D.G. Youmans, L.A. Hackel, S. Ezekiel: High-resolution spectroscopy of I_2 using laser-molecular-beam techniques. J. Appl. Phys. **44**, 2319 (1973)

374. D.W. Allan: 'In search of the best clock: An update'. In: *Frequency Standards and Metrology* (Springer, Berlin, Heidelberg 1989)

375. F. Lei Hong et al.: Proc. SPIE, Vol. 4269, p. 143 (2001)

376. E.A. Gerber, A. Ballato (Eds.): *Precision Frequency Control* (Academic, New York 1985)

377. D. Hils, J.L. Hall: 'Ultrastable cavity stabilized laser with subhertz linewidth'. In [378], p. 162

378. *Frequency Standards and Metrology*, ed. by A. De Marchi (Springer, Berlin, Heidelberg 1989);
M. Zhu, J.L. Hall: Short and long term stability of optical oscillators. J. Opt. Soc. Am. B **10**, 802 (1993)

379. K.M. Baird, G.R. Hanes: Stabilisation of wavelengths from gas lasers. Rep. Prog. Phys. **37**, 927 (1974)

380. T. Ikegami, S. Sudo, Y. Sakai: *Frequency Stabilization of Semiconductor Laser Diodes.* (Artech House, Boston 1995)

381. J. Hough, D. Hils, M.D. Rayman, L.S. Ma, L. Hollberg, J.L. Hall: Dye-laser frequency stabilization using optical resonators. Appl. Phys. B **33**, 179 (1984)

382. E. Peik, T. Schneider, C. Tamm: Laser frequency stabilization to a single ion. J. Phys. B. At. Mol. Phys. **39**, 149 (2006)

383. J.C. Bergquist (Ed.): Proc. 5th Symposium on Frequency Standards and Metrology (World Scientific, Singapore 1996)

384. M. Ohtsu (Ed.): *Frequency Control of Semiconductor Lasers* (Wiley, New York 1991);
E.D. Black: An Introduction to Pound-Drever-Hall laser-frequency stabilization. Am. J. Phys. **69**, 79 (2001)

385. Laser Frequency Stabilization, Standards, Measurements and Applications, SPIE Vol. 4269 (Soc. Photo-Opt. Eng., Bellingham, USA 2001)

386. L.F. Mollenauer, J.C. White, C.R. Pollock (Eds.): *Tunable Lasers*, 2nd. edn., Topics Appl. Phys., Vol. 59 (Springer, Berlin, Heidelberg 1992)

387. H.G. Danielmeyer: Stabilized efficient single-frequency Nd:YAG laser. IEEE J. QE-**6** 101 (1970)

388. J.P. Goldsborough: Scanning single frequency CW dye laser techniques for high-resolution spectroscopy. Opt. Eng. **13**, 523 (1974)

389. S. Gerstenkorn, P. Luc: *Atlas du spectre d'absorption de la molecule d'iode* (Edition du CNRS, Paris 1978) with corrections in Rev. Phys. Appl. **14**, 791 (1979)

390. A Giachetti, R.W. Stanley, R. Zalibas: Proposed secondary standard wavelengths in the spectrum of thorium. J. Opt. Soc. Am. **60**, 474 (1970)

391. H. Kato et al.: *Doppler-free high-resolution Spectral Atlas of Iodine Molecule* (Society for the Promotion of Science, Japan 2000)

392. M. Kabir, S. Kasahara, W. Demtröder, Y. Tamamitani, A. Doi, H. Kato: Doppler-free laser polarization spectroscopy and optical–optical double resonances of a large molecule: Naphthalene. J. Chem. Phys. **119**, 3691 (2003)

393. B.A. Palmer, R.A. Keller, F.V. Kovalski, J.L. Hall: Accurate wave-number measurements of uranium spectral lines. J. Opt. Soc. Am. **71**, 948 (1981)

394. (a) W. Jitschin, G. Meisel: 'Precise frequency tuning of a single mode dye laser'. In: *Laser'77, Opto-Electronics*, ed. by W. Waidelich (IPC Science and Technology, Guildford, Surrey 1977)
(b) J.L. Hall: 'Saturated absorption spectroscopy'. In: *Atomic Physics, Vol. 3*, ed. by S. Smith, G.K. Walters (Plenum, New York 1973)

395. H.M. Nussenzweig: *Introduction to Quantum Optics* (Gordon & Breach, New York 1973)

396. W. Brunner, W. Radloff, K. Junge: *Quantenelektronik* (VEB Deutscher Verlag der Wissenschaften, Berlin 1975) p. 212

397. A.L. Schawlow, C.H. Townes: Infrared and optical masers. Phys. Rev. **112**, 1940 (1958)

398. C.J. Bordé, J.L. Hall: 'Ultrahigh resolution saturated absorption spectroscopy'. In: *Laser Spectroscopy*, ed. by R.G. Brewer, H. Mooradian (Plenum, New York 1974) pp. 125–142

399. J.L. Hall, M. Zhu, P. Buch: Prospects focussing laser-prepared atomic fountains for optical frequency standards applications. J. Opt. Soc. Am. B **6**, 2194 (1989)

400. R.W.P. Drever, J.L. Hall, F.V. Kowalski, J. Hough, G.M. Ford, A.J. Munley, H.W. Ward: Laser phase and frequency stabilization using an optical resonator. Appl. Phys. B **31**, 97 (1983)

401. S.N. Bagayev, A.E. Baklarov, V.P. Chebotayev, A.S. Dychkov, P.V. Pokasov: 'Super high resolution laser spectroscopy with cold particles'. In: *Laser Spectroscopy VIII*, ed. by W. Persson, J. Svanberg, Springer Ser. Opt. Sci., Vol. 55 (Springer, Berlin, Heidelberg 1987); V.P. Chebotayev: 'High resolution laser spectroscopy'. In: *Frontier of Laser Spectroscopy*, ed. by T.W. Hänsch, M. Inguscio (North-Holland, Amsterdam 1994)

402. K. Ueda, N. Uehara: Laser diode pumped solid state lasers for gravitational wave antenna. Proc. SPIE **1837**, 337 (1992)

403. G. Ruoso, R. Storz, S. Seel, S. Schiller, J. Mlynek: Nd:YAG laser frequency stabilization to a supercavity at the 0.1 Hz instability level. Opt. Comm. **133**, 259 (1997)

404. F.J. Duarte: *Tunable Laser Handbook* (Academic Press, New York 1995)

405. P.F. Moulton: 'Tunable paramagnetic-ion lasers'. In: *Laser Handbook, Vol. 5*, ed. by M. Bass, M.L. Stitch (North-Holland, Amsterdam 1985) p. 203

406. R.S. McDowell: 'High resolution infrared spectroscopy with tunable lasers'. In: *Advances in Infrared and Raman Spectroscopy, Vol. 5*, ed. by R.J.H. Clark, R.E. Hester (Heyden, London 1978)

407. L.E. Mollenauer, J.L. White, C.R. Pollack: *Tunable Lasers*, 2nd edn. (Springer, Heidelberg 1993)

408. E.D. Hinkley, K.W. Nill, F.A. Blum: 'Infrared spectroscopy with tunable lasers'. In: *Laser Spectroscopy of Atoms and Molecules*, ed. by H. Walther, Topics Appl. Phys. Vol. 2 (Springer, Berlin, Heidelberg 1976)

409. G.P. Agraval (Ed.): *Semiconductor Lasers* (AIP, Woodbury 1995)

410. F. Duarte, F.J. Duarte: *Tunable Laser Optics* (Academic Press, New York 2007)

411. A. Mooradian: 'High resolution tunable infrared lasers'. In: *Very High Resolution Spectroscopy*, ed. by R.A. Smith (Academic, London 1976)

412. I. Melgailis, A. Mooradian: 'Tunable semiconductor diode lasers and applications'. In: *Laser Applications in Optics and Spectroscopy* ed. by S. Jacobs, M. Sargent, M. Scully, J. Scott (Addison Wesley, Reading, MA 1975) p. 1

413. H.C. Casey, M.B. Panish: *Heterostructure Lasers* (Academic, New York 1978)

414. C. Vourmard: External-cavity controlled 32 MHz narrow band CW GaAs-diode laser. Opt. Lett. **1**, 61 (1977)

415. W. Fleming, A. Mooradian: Spectral characteristics of external cavity controlled semiconductor lasers. IEEE J. QE-**17**, 44 (1971)

416. W. Fuhrmann, W. Demtröder: A widely tunable single-mode GaAs-diode laser with external cavity. Appl. Phys. B **49**, 29 (1988)

417. H. Tabuchi, H. Ishikawa: External grating tunable MQW laser with wide tuning range at 240 nm. Electron. Lett. **26**, 742 (1990);
M. de Labachelerie, G. Passedat: Mode-hop suppression of Littrow grating-tuned lasers. Appl. Opt. **32**, 269 (1993)

418. H. Wenz, R. Großkloß, W. Demtröder: Kontinuierlich durchstimmbare Halbleiterlaser. Laser & Optoelektronik **28**, 58 (Febr. 1996);
C.J. Hawthorne, K.P. Weber, R.E. Schulten: Littrow configuration tunable external cavity diode laser with fixed direction output beam. Rev. Sci. Instrum. **72**, 4477 (2001)

419. Sh. Nakamura, Sh.F. Chichibu: *Introduction to Nitride Semiconductor Blue Lasers* (Taylor and Francis, London 2000)

420. N.W. Carlson: *Monolythic Diode Laser Arrays* (Springer, Berlin, Heidelberg, New York 1994);
R. Diehl (Ed.): *High Power Diode Lasers* (Springer, Berlin, Heidelberg, New York 2000)

421. R.C. Powell: *Physics of Solid-State Laser Materials* (Springer, Berlin, Heidelberg, New York 1998)

422. A.A. Kaminskii: *Laser Crystals*, 2nd edn., Springer Ser. Opt. Sci., Vol. 14 (Springer, Berlin, Heidelberg 1990); A.A. Kaminsky: *Crystalline Lasers* (CRC, New York 1996)

423. M. Inguscio, R. Wallenstein (Eds.): *Solid State Lasers; New Developments and Applications* (Plenum, New York 1993)

424. U. Dürr: Vibronische Festkörperlaser: Der Übergangsmetallionen-Laser. Laser Optoelectr. **15**, 31 (1983)

425. A. Miller, D.M. Finlayson (Eds.): *Laser Sources and Applications* (Institute of Physics, Bristol 1996)

426. F. Gan: *Laser Materials* (World Scientific, Singapore 1995)

427. S.T. Lai: Highly efficient emerald laser. J. Opt. Soc. Am. B **4**, 1286 (1987)

428. G.T. Forrest: Diode-pumped solid state lasers have become a mainstream technology. Laser Focus **23**, 62 (1987)

429. W.P. Risk, W. Lenth: Room temperature continuous wave 946 nm Nd:YAG laser pumped by laser diode arrays and intracavity frequency doubling to 473 nm. Opt. Lett. **12**, 993 (1987)

430. P. Hammerling, A.B. Budgor, A. Pinto (Eds.): *Tunable Solid State Lasers I*, Springer Ser. Opt. Sci., Vol. 47 (Springer, Berlin, Heidelberg 1984)

431. A.B. Budgor, L. Esterowitz, L.G. DeShazer (Eds.): *Tunable Solid State Lasers II*, Springer Ser. Opt. Sci., Vol. 52 (Springer, Berlin, Heidelberg 1986)

432. W. Koechner: *Solid-State Laser Enginering*, 6th edn., Springer Ser. Opt. Sci., Vol. 1 (Springer, Berlin, Heidelberg 2006)

433. N.P. Barnes: 'Transition Metal Solid State Lasers'. In: Tunable Laser Handbook, ed. by F.J. Duarte (Academic, San Diego 1995)

434. W.B. Fowler (Ed.): *Physics of Color Centers* (Academic, New York 1968)

435. F. Lüty: 'F_A-Centers in Alkali Halide Crystals'. In: *Physics of Color Centers*, ed. by W.B. Fowler (Academic, New York 1968)

436. L.F. Mollenauer, D.H. Olsen: Broadly tunable lasers using color centers. J. Appl. Phys. **46**, 3109 (1975)

437. L.F. Mollenauer: 'Color center lasers'. In: *Laser Handbook IV*, ed. by M. Bass, M.L. Stitch (North-Holland, Amsterdam 1985) p. 143

438. V. Ter-Mikirtychev: Diode pumped tunable room-temparature LiF:F_2^- color center laser. Appl. Opt. **37**, 6442 (1998)

439. H.W. Kogelnik, E.P. Ippen, A. Dienes, C.V. Shank: Astigmatically compensated cavities for CW dye lasers. IEEE J. QE-**8**, 373 (1972)

440. R. Beigang, G. Litfin, H. Welling: Frequency behavior and linewidth of CW single mode color center lasers. Opt. Commun. **22**, 269 (1977)

441. G. Phillips, P. Hinske, W. Demtröder, K. Möllmann, R. Beigang: NaCl-color center laser with birefringent tuning. Appl. Phys. B **47**, 127 (1988)

442. G. Litfin: Color center lasers. J. Phys. E **11**, 984 (1978)

443. L.F. Mollenauer, D.M. Bloom, A.M. Del Gaudio: Broadly tunable CW lasers using F_2^+-centers for the 1.26–1.48 μm and 0.82–1.07 μm bands. Opt. Lett. **3**, 48 (1978)

444. L.F. Mollenauer: Room-temperature stable F_2^+-like center yields CW laser tunable over the 0.99–1.22 μm range. Opt. Lett. **5**, 188 (1980)

445. W. Gellermann, K.P. Koch, F. Lüty: Recent progess in color center lasers. Laser Focus **18**, 71 (1982)

446. M. Stuke (Ed.): *25 Years Dye Laser*, Topics Appl. Phys., Vol. 70 (Springer, Berlin, Heidelberg 1992)

447. F.P. Schäfer (Ed.): *Dye Lasers*, 3rd edn., Topics Appl. Phys., Vol. 1 (Springer, Berlin, Heidelberg 1990)

448. F.J. Duarte, L.W. Hillman: *Dye Laser Principles* (Academic, Boston 1996) and (Elsevier 2012);
F.J. Duarte: *Tunable Lasers Handbook* (Academic, New York 1996);
F.J. Duarte: Tunable Organic Dye Lasers. Prog. Quantum Electron. **36**, 29 (2012)

449. G. Marowsky, R. Cordray, F.K. Tittel, W.L. Wilson, J.W. Keto: Energy transfer processes in electron beam excited mixtures of laser dye vapors with rare gases. J. Chem. Phys. **67**, 4845 (1977)

450. B. Steyer, F.P. Schäfer: Stimulated and spontaneous emission from laser dyes in the vapor phase. Appl. Phys. **7**, 113 (1975)

451. F.J. Duarte (Ed.): *High-Power Dye Lasers*, Springer Ser. Opt. Sci., Vol. 65 (Springer, Berlin, Heidelberg 1991)

452. W. Schmidt: Farbstofflaser. Laser **2**, 47 (1970)

453. G.H. Atkinson, M.W. Schuyler: A simple pulsed laser system, tunable in the ultraviolet. Appl. Phys. Lett. **27**, 285 (1975)

454. A. Hirth, H. Fagot: High average power from long pulse dye laser. Opt. Commun. **21**, 318 (1977)

455. J. Jethwa, F.P. Schäfer, J. Jasny: A reliable high average power dye laser. IEEE J. QE-**14**, 119 (1978)

456. J. Kuhl, G. Marowsky, P. Kunstmann, W. Schmidt: A simple and reliable dye laser system for spectroscopic investigations. Z. Naturforsch. **27a**, 601 (1972)

457. H. Walther, J.L. Hall: Tunable dye laser with narrow spectral output. Appl. Phys. Lett. **6**, 239 (1970)

458. P.J. Bradley, W.G.I. Caugbey, J.I. Vukusic: High efficiency interferometric tuning of flashlamp-pumped dye lasers. Opt. Commun. **4**, 150 (1971)

459. M. Okada, K. Takizawa, S. Ieieri: Tilted birefringent Fabry–Perot etalon for tuning of dye lasers. Appl. Opt. **15** 472 (1976)

460. G.M. Gale: A single-mode flashlamp-pumped dye laser. Opt. Commun. **7**, 86 (1973);
F.J. Duarte, R.W. Conrad: Single-mode flashlamp-pumped dye laser oscillators. Appl. Opt. **25**, 663 (1986);
F.J. Duarte, T.S. Taylor, A. Costella, I. Garcia-Moreno, R. Sastre: Long-pulse narrow-linewidth dispersion solid-state dye laser oscillator. Appl. Opt. **37**, 3987 (1998)

461. J.J. Turner, E.I. Moses, C.L. Tang: Spectral narrowing and electro-optical tuning of a pulsed dye-laser by injection-locking to a CW dye laser. Appl. Phys. Lett. **27**. 441 (1975)

462. M. Okada, S. Ieiri: Electronic tuning of dye-lasers by an electro-optical birefringent Fabry–Perot etalon. Opt. Commun. **14**, 4 (1975)

463. J. Kopainsky: Laser scattering with a rapidly tuned dye laser. Appl. Phys. **8**, 229 (1975)

464. F.P. Schäfer, W. Schmidt, J. Volze: Appl. Phys. Lett. **9** 306 (1966)

465. P.P. Sorokin, J.R. Lankard: IBM J. Res. Develop. **10**, 162 (1966);
P.P. Sorokin: Organic lasers. Sci. Am. **220**, 30 (February 1969)

466. F.B. Dunnings, R.F. Stebbings: The efficient generation of tunable near UV radiation using a N_2-pumped dye laser. Opt. Commun. **11**, 112 (1974)

467. T.W. Hänsch: Repetitively pulsed tunable dye laser for high resolution spectroscopy. Appl. Opt. **11**, 895 (1972)

468. R. Wallenstein: Pulsed narrow band dye lasers. Opt. Acta **23**, 887 (1976)

469. I. Soshan, N.N. Danon, V.P. Oppenheim: Narrowband operation of a pulsed dye laser without intracavity beam expansion. J. Appl. Phys. **48**, 4495 (1977)

470. S. Saikan: Nitrogen-laser-pumped single-mode dye laser. Appl. Phys. **17**, 41 (1978)

471. K.L. Hohla: Excimer-pumped dye lasers – the new generation. Laser Focus **18**, 67 (1982)

472. O. Uchino, T. Mizumami, M. Maeda, Y. Miyazoe: Efficient dye lasers pumped by a XeCl excimer laser. Appl. Phys. **19**, 35 (1979)

473. R. Wallenstein: 'Pulsed dye lasers'. In: *Laser Handbook Vol. 3*, ed. by M.L. Stitch (North-Holland, Amsterdam 1979) pp. 289–360

474. W. Hartig: A high power dye laser pumped by the second harmonics of a Nd:YAG laser. Opt. Commun. **27**, 447 (1978)

475. F.J. Duarte, J.A. Piper: Narrow linewidths, high pulse repetition frequency copper-laser-pumped dye-laser oscillators. Appl. Opt. **23**, 1991 (1984)

476. M.G. Littman: Single-mode operation of grazing-incidence pulsed dye laser. Opt. Lett. **3**, 138 (1978)

477. K. Liu, M.G. Littmann: Novel geometry for single-mode scanning of tunable lasers. Opt. Lett. **6**, 117 (1981)

478. M. Littmann, J. Montgomery: Grazing incidence designs improve pulsed dye lasers: Laser Focus **24**, 70 (February 1988)

479. F.J. Duarte, R.W. Conrad: Diffraction-limited single-longitudinal-mode multiple-prism flashlamp-pumped dye laser oscillator. Apl. Opt. **26**, 2567 (1987);
F.J. Duarte, R.W. Conrad: Multiple-prism Littrow and grazing incidence pulsed CO_2 lasers. Appl.Opt. **24**, 1244 (1985)

480. F.J. Duarte, J.A. Piper: Prism preexpanded gracing incidence grating cavity for pulsed dye lasers. Appl. Opt. **20**. 2113 (1981)

481. F.J. Duarte: Multipass dispersion theory of prismatic pulsed dye lasers. Opt. Acta **31**, 331 (1984)

482. Lambda Physik information sheet, Göttingen 2000
and http://www.lambdaphysik.com

483. S. Leutwyler, E. Schumacher, L. Wöste: Extending the solvent palette for CW jet-stream dye lasers. Opt. Commun. **19**, 197 (1976)

484. P. Anliker, H.R. Lüthi, W. Seelig, J. Steinger, H.P. Weber: 33 watt CW dye laser. IEEE J. QE-**13**, 548 (1977)

485. P. Hinske: Untersuchung der Prädissoziation von Cs_2-Molekülen. Diploma Thesis, F.B. Physik, University of Kaiserslautern (1988)

486. A. Bloom: Modes of a laser resonator, containing tilted birefringent plates. J. Opt. Soc. Am. **64**, 447 (1974)

487. H.W. Schröder, H. Dux, H. Welling: Single-mode operation of CW dye lasers. Appl. Phys. **7**, 21 (1975)

488. H. Gerhardt, A. Timmermann: High resolution dye-laser spectrometer for measurements of isotope and isomer shifts and hyperfine structure. Opt. Commun. **21**, 343 (1977)

489. H.W. Schröder, L. Stein, D. Fröhlich, F. Fugger, H. Welling: A high power single-mode CW dye ring laser. Appl. Phys. **14**, 377 (1978)

490. D. Kühlke, W. Diehl: Mode selection in CW laser with homogeneously broadened gain. Opt. Quantum Electron. **9**, 305 (1977)

491. J.D. Birks: Excimers. Rep. Prog. Phys. **38**, 903 (1977)

492. C.K. Rhodes (Ed.): *Excimer Lasers*, 2nd edn., Topics Appl. Phys., Vol. 30 (Springer, Berlin, Heidelberg 1984)

493. H. Scheingraber, C.R. Vidal: Discrete and continuous Franck–Condon factors of the $Mg_2 A\ ^1\Sigma_\mu^+ - X\ ^1\Sigma_g^+$ system. J. Chem. Phys. **66**, 3694 (1977)

494. D.J. Bradley: 'Coherent radiation generation at short wavelengths'. In: *High-Power Lasers and Applications*, ed. by K.L. Kompa, H. Walther, Springer Ser. Opt. Sci., Vol. 9 (Springer, Berlin, Heidelberg 1978) pp. 9–18

495. R.C. Elton: *X-Ray Lasers* (Academic, New York 1990)

496. H.H. Fleischmann: High current electron beams. Phys. Today **28**, 34 (May 1975)

497. C.P. Wang: Performance of XeF/KrF lasers pumped by fast discharges. Appl. Phys. Lett. **29**, 103 (1976)

498. H. Pummer, U. Sowada, P. Oesterlin, U. Rebban, D. Basting: Kommerzielle Excimerlaser. Laser u. Optoelektronik **17**, 141 (1985)

499. S.S. Merz: Switch developments could enhance pulsed laser performance. Laser Focus **24**, 70 (May 1988)

500. P. Klopotek, V. Brinkmann, D. Basting, W. Mückenheim: A new excimer laser producing long pulses at 308 nm. Lambda-Physik Mitteilung (Göttingen 1987)

501. D. Basting: Excimer lasers – new perspectives in the UV. Laser and Elektro-Optic **11**, 141 (1979)

502. K. Miyazak: T. Fukatsu, I. Yamashita, T. Hasama, K. Yamade, T. Sato: Output and picosecond amplifcation characteristics of an efficient and high-power discharge excimer laser. Appl. Phys. B **52**, 1 (1991)

503. J.M.J. Madey: Stimulated emission of Bremsstrahlung in a periodic magnetic field. J. Appl. Phys. **42**, 1906 (1971)

504. E.L. Saldin, E. Schneidmiller, M. Yurkow: *The Physics of Free Electron Lasers* (Springer, Berlin, Heidelberg, New York 2000)

505. T.C. Marshall: *Free-Electron Lasers* (MacMillan, New York 1985)

506. H.P. Freund, T.M. Antonsen: *Principles of Free-electron Lasers* (Chapmand Hall, London 1992) and (Springer 1996)

507. *Report of the Committee of the National Academy of Science on Free Electron Lasers and Other Advanced Sources of Light* (National Academy Press, Washington 1994);

508. J. Dunn et al.: Demonstration of X-ray amplification in transient gain nickel-like palladium scheme. Phys. Rev. Lett. **80**, 2825 (1998);
Ch. Brau: *Free-Electron Lasers* (Academic Press, Boston 1990)

509. R. Lee: Science on High Energy Lasers From Today to NIF. In: Energy and Technology Review UCRL-ID 119 170 (1996);
P. Luchini, H. Motz: *Undulators and Free-electron Lasers* (Oxford University Press, Oxford 1990);
Ch. Brau: *Free Electron Lasers* (Academic, Boston 1990)

510. B. Rus et al.: Multi-millijoule highly coherent X-ray laser at 21 mm operating in deep saturation through double pass amplification. Phys. Rev. **A66**, 063806 (2002);
G. Dattoli, A. Renieri, A. Torre: *Lectures in Free-Electron Laser theory and related topics* (World Scientific, Singapore 1995)

511. J. Bokon, P.H. Bucksbaum, R.R. Freeman: Generation of 35.5 nm coherent radiation. Opt. Lett. **8**, 217 (1983);
P. Luchini, H. Motz: *Undulators and Free-Electron Lasers* (Clarendon Press, Oxford 1990)

512. C. Yamanaka (Ed.): *Short-Wavelength Lasers*, Springer Proc. Phys., Vol. 30 (Springer, Berlin, Heidelberg 1988)

513. D.L. Matthews, M.D. Rosen: Soft X-ray lasers. Sci. Am. **259**, 60 (1988)

514. T.J. McIlrath (Ed.): *Laser Techniques for Extreme UV Spectroscopy*, AIP Conf. Proc. **90** (1986)

515. B. Wellegehausen, M. Hube, F. Jim: Investigation on laser plasma soft X-ray sources generated with low-energy laser systems. Appl. Phys. B **49**, 173 (1989)

516. E. Fill (Guest Ed.): X-ray lasers. Appl. Phys. B **50**, 145–226 (1990);
E. Fill: Appl. Phys. B **58**, 1–56 (1994)

517. A.L. Robinson: Soft X-ray laser at Lawrence Livermore Lab. Science **226**, 821 (1984)

518. E.E. Fill (Ed.): *X-Ray Lasers* (Institute of Physics, Bristol 1992)

519. P.V. Nickler, K.A. Janulewicz (Eds.) X-Ray Lasers 2006. Proc. of 10th Int. Conf., Berlin (Springer Proceedings in Physics Vol. 115, Springer, Berlin, Heidelberg 2007)

520. J. Dunn: The X-ray Laser: From Underground to Tabletop. https://www.llnl.gov/str/Dunn. html

521. Y. Wang et al.: Demonstration of high repetition rate tabletop soft X-ray laser with saturated output at wavelengths down to 13.9 mm. Phys. Rev. A **72**, 053807 (2003)

522. J. Hecht: The history of X-ray lasers. Opt. Photonics News (May 2008)

Chapter 6

523. A. Yariv, P. Yeh: *Optical waves in crystals* (Wiley, New York 1984);
C.R. Vidal: Coherent VUV sources for high resolution spectroscopy. Appl. Opt. **19**, 3897 (1980);

524. J.R. Reintjes: 'Coherent ultraviolet and VUV-sources'. In: *Laser Handbook V*, ed. by M. Bass, M.L. Stitch (North-Holland, Amsterdam 1985)

525. N. Bloembergen: *Nonlinear Optics*, 4th edn. (World Scientific, Singapore 1996)

526. A. Newell, J.V. Moloney: *Nonlinear Optics* (Addison-Wesley, Reding, MA 1992)

527. D.L. Mills: *Nonlinear Optics*, 2nd edn. (Springer, Berlin, Heidelberg 1998);

528. G.S. He, S.H. Liv: *Physics of Nonlinear Optics* (World Scientific, Singapore 1999);
     G.I. Stegeman, R.A. Stegeman: *Nonlinear Optics, Phenomena, Materials and Devices* (Wiley 2012)

529. G.C. Baldwin: *An Introduction to Nonlinear Optics* (Springer 2013)

530. F. Zernike, J.E. Midwinter: *Applied Nonlinear Optics* (Academic, New York 1973; Dover Publ. 2006)

531. A.C. Newell, J.V. Moloney: *Nonlinear Optics* (Addison Wesley, Redwood City 1992);

532. Y.P. Svirko, N.I. Zheluder: *Polarization of Light in Nonlinear Optics* (Wiley, Chichester 1998)

533. P.G. Harper, B.S. Wherrett (Eds.) *Nonlinear Optics* (Academic, London 1977);
     Laser Analytical Systems GmbH, Berlin, now available at Spectra Physics, Palo Alto, USA

534. P. Günther (Ed.): *Nonlinear Optical Effects and Materials* (Springer, Berlin, Heidelberg, New York 2000)

535. R.W. Boyd: *Nonlinear Optics*, 3rd edn. (Elsevier, Amsterdam 2008);
     P.E. Powers: *Fundamentals of Nonlinear Optics* (CRC Press 2011)

536. M.G. Kuzyk: *Nonlinear Optics Lecture Notes*. http://nlosource.com/LectureNotesBook.pdf

537. Y.G. Di, G.G. Gurzadian, D.N. Nikogosian: *Handbook of Nonlinear Optical Crystals*, 3rd edn., Springer Ser. Opt. Sci., Vol. 64 (Springer, Berlin, Heidelberg 2013)

538. J.T. Lin, C. Chen: Choosing a nonlinear crystal. Laser Optoelectron. **6**, 59 (Nov. 1987)

539. D.A. Kleinman, A. Ashkin, G.D. Boyd: Second harmonic generation of light by focussed laser beams. Phys. Rev. **145**, 338 (1966)

540. P. Lokai, B. Burghardt, S.D. Basting: W. Mückenheim: Typ-I Frequenzverdopplung und Frequenzmischung in β-BaB$_2$O$_4$. Laser Optoelektr. **19**, 296 (1987)

541. R.S. Adhav, S.R. Adhav, J.M. Pelaprat: BBO's nonlinear optical phase-matching properties. Laser Focus **23**, 88 (1987)

542. K. Nakamura et al.: Periodic poling of magnesium oxyde-doped lithium niobate. J. Appl. Phys. **21**, 4528 (2002)

543. G.G. Gurzadian et al.: *Handbook of Nonlinear Optical Crystals* (Springer, Berlin, Heidelberg, New York 2013);
     D. Nikogosyan: *Nonlinear Optical Crystals: A Complete Survey* (Springer, Berlin. Heidelberg 2005)

544. H. Schmidt, R. Wallenstein: Beta-Bariumbort: Ein neues optisch-nichtlineares Material. Laser Optoelektr. **19**, 302 (1987)

545. Ch. Chuangtian, W. Bochang, J. Aidong, Y. Giuming: A new-type ultraviolet SHG crystal: β-BaB$_2$O$_4$. Scientia Sinica B **28**, 235 (1985)

546. I. Shogi, H. Nakamura, K. Obdaira, T. Kondo, R. Ito: Absolute measurement of second order nonlinear-optical cefficient of beta-BaB$_2$O$_4$ for visible and ultraviolet second harmonic wavelengths. J. Opt. Soc. Am. B **16**, 620 (1999)

547. H. Kouta, Y. Kuwano: Attaining 186 nm light generation in cooled beta BaB$_2$O$_4$ crystal. Opt Lett. **24**, 1230 (1999)

548. J.C. Baumert, J. Hoffnagle, P. Günter: High efficiency intracavity frequency doubling of a styril-9 dye laser with KNbO$_3$ crystals. Appl. Opt. **24**, 1299 (1985)

549. T.J. Johnston Jr.: 'Tunable dye lasers'. In: *The Encyclopedia of Physical Science and Technology*, Vol. 14 (Academic, San Diego 1987)

550. W.A. Majewski: A tunable single frequency UV source for high resolution spectroscopy. Opt. Commun. **45**, 201 (1983)

551. K.J. Unstern, M.H. Dunns: The generation of tunable UV radiation from 718-789 nm by intracavity frequency doubling of a coumarin 102 dye laser. Opt. Commun. **35**, 259 (1980)

552. D. Fröhlich, L. Stein, H.W. Schröder, H. Welling: Efficient frequency doubling of CW dye laser radiation. Appl. Phys. **11**, 97 (1976)

553. A. Renn, A. Hese, H. Busener: Externer Ringresonator zu Erzeugung kontinuierliche UV-Strahlung. Laser Optoelektr. **3**, 11 (1982)

554. F.V. Moers, T. Hebert, A. Hese: Theory and experiment of CW dye laser injection locking and its application to second harmonic generation. Appl. Phys. B **40**, 67 (1986)

555. N. Wang, V. Gaubatz: Optical frequency doubling of a single-mode dye laser in an external resonator. Appl. Phys. B **40**, 43 (1986)

556. J.T. Lin, C. Chen: Chosing a nonlinear crystal. Laser and Optronics **6**, 59 (November 1987)

557. Castech-Phoenix Inc. Fujian, China, information sheets

558. L. Wöste: 'UV-generation in CW dye lasers'. In: *Advances in Laser Spectroscopy* ed. by F.T. Arrechi, F. Strumia, H. Walther (Plenum, New York 1983)

559. H.J. Müschenborn, W. Theiss, W. Demtröder: A tunable UV-light source for laser spectroscopy using second harmonic generation in β-BaB_2O_4. Appl. Phys. B **50**, 365 (1990)

560. H. Dewey: Second harmonic generation in $KB_4OH \cdot 4H20$ from 217 to 315 nm. IEEE J. QE-**12**, 303 (1976)

561. E.U. Rafailov et al.: Second harmonic generation from a first-order quasi-phase-matched GaAs–AlGaAs waveguide crystal. Opt. Lett. **26**, 1984 (2001)

562. M. Feger, G.M. Magel, D.H. Hundt, R.L. Byer: Quasi-phase-matched second harmonic generation: tuning and tolerances. IEEE J. Quant. Electr. **28**, 2631 (1992)

563. M. Pierrou, F. Laurell, H. Karlsson, T. Kellner, C. Czeranowsky, G. Huber: Generation of 740 mW of blue light by intracavity frequency doubling with a first order quasi-phase-matched $KTiOPO_4$ crystal. Opt. Lett. **24**, 205 (1999)

564. F.B. Dunnings: Tunable ultraviolet generation by sum-frequency mixing. Laser Focus **14**, 72 (1978)

565. S. Blit, E.G. Weaver, F.B. Dunnings, F.K. Tittel: Generation of tunable continuous wave ultraviolet radiation from 257 to 329 nm. Opt. Lett. **1**, 58 (1977)

566. R.F. Belt, G. Gashunov, Y.S. Liu: KTP as an harmonic generator for Nd:YAG lasers. Laser Focus **21**, 110 (1985)

567. J. Halbout, S. Blit, W. Donaldson, Ch.L. Tang: Efficient phase-matched second harmonic generation and sum-frequency mixing in urea. IEEE J. QE-**15**, 1176 (1979)

568. G. Nath, S. Haussühl: Large nonlinear optical coefficient and phase-matched second harmonic generation in $LiIO_3$. Appl. Phys. Lett. **14**, 154 (1969)

569. F.B. Dunnings, F.K. Tittle, R.F. Stebbings: The generation of tunable coherent radiation in the wavelength range 230 to 300 nm using lithium formate monohydrate. Opt. Commun. **7**, 181 (1973)

570. U. Simon, S. Waltman, I. Loa, L. Holberg, T.K. Tittel: External cavity difference frequency source near 3.2 μm based on mixing a tunable diode laser with a diode-pumped Nd:YAG laser in $AgGaS_2$. J. Opt. Soc. Am. **B12**, 323 (1995)

571. F.B. Dunnings: Tunable ultraviolet generation by sum-frequency mixing. Laser Focus **14**, 72 (May 1978)

572. G.A. Massey, J.C. Johnson: Wavelength tunable optical mixing experiments between 208 and 259 nm. IEEE J. QE-**12**, 721 (1976)

573. T.A. Paul, F. Merkt: High-resolution spectroscopy of xenon using a tunable Fourier-transform-limited all-solid-state vacuum-ultraviolet laser system. J. Phys. B: At. Mol. Opt. Phys. **38**, 4145 (2005)

574. A. Borsutzky, R. Brünger, Ch. Huang, R. Wallenstein: Harmonic and sum-frequency generation of pulsed laser radiation in BBO, LBO and KD$^+$P. Appl. Phys. B **52**, 55 (1991)

575. G.A. Rines, H.H. Zenzie, R.A. Schwarz, Y. Isanova, P.F. Moulton: Nonlinear conversion of Ti:Sapphire wavelengths. IEEE J. of Selected Topics in Quant. Electron. **1**, 50 (1995)

576. K. Matsubara, U. Tanaka, H. Imago, M. Watanabe: All-solid state light source for generation of continious-wave coherent radiation near 202 nm. J. Opt. Soc. Am. B **16**, 1668 (1999)

577. C.R. Vidal: Third harmonic generation of mode-locked Nd:glass laser pulses in phase-matched Rb-Xe-mixtures. Phys. Rev. A **14**, 2240 (1976)

578. R. Hilbig, R. Wallenstein: Narrow band tunable VUV-radiation generated by nonresonant sum- and difference-frequency mixing in xenon and krypton. Appl. Opt. **21**, 913 (1982)

579. C.R. Vidal: 'Four-wave frequency mixing in gases'. In: *Tunable Lasers*, 2nd edn., ed. by I.F. Mollenauer, J.C. White, C.R. Pollock, Topics Appl. Phys., Vol. 59 (Springer, Berlin, Heidelberg 1992)

580. G. Hilber, A. Lago, R. Wallenstein: Broadly tunable VUV/XUV-adiation generated by resonant third-order frequency conversion in Kr. J. Opt. Soc. Am. B **4**, 1753 (1987);

581. A. Lago, G. Hilber, R. Wallenstein: Optical frequency conversion in gaseous media. Phys. Rev. A **36**, 3827 (1987)

582. S.E. Harris, J.F. Young, A.H. Kung, D.M. Bloom, G.C. Bjorklund: 'Generation of ultraviolet and VUV-radiation'. In: *Laser Spectroscopy I*, ed. by R.G. Brewer, A. Mooradian (Plenum, New York 1974)

583. A.H. Kung, J.F. Young, G.C. Bjorklund, S.E. Harris: Generation of Vacuum Ultraviolet Radiation in Phase-matched Cd Vapor. Phys. Rev. Lett. **29**, 985 (1972)

584. W. Jamroz, B.P. Stoicheff: 'Generation of tunable coherent vacuum-ultraviolet radiation'. In: *Progress in Optics* **20**, 324 (North-Holland, Amsterdam 1983)

585. A. Timmermann, R. Wallenstein: Generation of tunable single-frequency CW coherent vacuum-ultraviolet radiation. Opt. Lett. **8**, 517 (1983)

586. T.P. Softley, W.E. Ernst, L.M. Tashiro, R.Z. Zare: A general purpose XUV laser spectrometer: Some applications to N_2, O_2 and CO_2. Chem. Phys. **116**, 299 (1987);
C. Fischer M.W. Sigrist: *Mid-IR Difference Frequency Generation*. Topics Appl. Phys., Vol. 89 S. 97–143 (Springer, Berlin, Heidelberg 2003)

587. P.F. Levelt, W. Ubachs: XUV-laser spectroscopy in the Cu $^1\Sigma_u^+$, $v = 0$ and c_3 $^1\Pi_u$, $v = 0$ Rydberg states of N_2. Chem. Phys. **163**, 263 (1992)

588. H. Palm, F. Merkt: Generation of tunable coherent XUV radiation beyond 19 eV by resonant four wave mixing in argon. Appl. Phys. Lett. **73**, 157 (1998)

589. U. Hollenstein, H. Palm, F. Merkt: A broadly tunable XUV laser source with a 0.008 cm^{-1} bandwidth. Rev. Sci. Instrum. **71**, 4023 (2000)

590. D.L. Matthews, R.R. Freeman (Eds.): The generation of coherent XUV and soft X-Ray radiation. J. Opt. Soc. Am. B **4**, 533 (1987)

591. A.S. Pine: 'IR-spectroscopy via difference-frequency generation'. In: *Laser Spectroscopy III*, ed. by J.L. Hall, J.L. Carlsten, Springer Ser. Opt. Sci., Vol. 21 (Springer, Berlin, Heidelberg 1977) p. 376

592. A.S. Pine: High-resolution methane ν_3-band spectra using a stabilized tunable difference-frequency laser system. J. Opt. Soc. Am. **66**, 97 (1976).
A.S. Pine: J. Opt. Soc. Am. **64**, 1683 (1974)

593. A.H. Hilscher, C.E. Miller, D.C. Bayard, U. Simon, K.P. Smolka, R.F. Curl, F.K. Tittel: Optimization of a midinfrared high-resolution difference frequency laser spectrometer. J. Opt. Soc. Am. B **9**, 1962 (1992)

594. U. Simon, S. Waltman, I. Loa, L. Holberg, T.K. Tittel: External cavity difference frequency source near 3.2 μm based on mixing a tunable diode laser with a diode-pumped Nd:YAG-laser in AgGaS$_2$. J. Opt. Soc. Am. B **12**, 322 (1995)

595. M. Seiter, D. Keller, M.W. Sigrist: Broadly tunable difference-frequency spectrometry for trace gas detection with noncollinear critical phase-matching in LiNbO$_3$. Appl. Phys. B **67**, 351 (1998)

596. W. Chen, J. Buric, D. Boucher: A widely tunable cw laser difference frequency source for high resolution infrared spectroscopy. Laser Physics **10**, 521 (2000)

597. S. Stry, P. Hering, M. Mürtz: Portable difference-frequency laser-based cavity leak-out spectrometer for trace gas analysis. Appl. Phys. **B75**, 297 (2002)

598. M.H. Chou, J. Hauden, M.A. Arhore, M.M. Feger: 1.5 μm band wavelength conversion based on difference frequency generation in LiNbO$_3$ waveguide with integrated coupling structure. Opt. Lett. **23**, 1004 (1998)

599. R.Y. Shen (Ed.): *Nonlinear Infrared Generation*, Topics Appl. Phys., Vol. 16 (Springer, Berlin, Heidelberg 1977)

600. K.M. Evenson, D.A. Jennings, K.R. Leopold, L.R. Zink: 'Tunable far infrared spectroscopy'. In: *Laser Spectroscopy VII*, ed. by T.W. Hänsch, Y.R. Shen, Springer Ser. Opt. Sci., Vol. 49 (Springer, Berlin, Heidelberg 1985) p. 366

601. M. Inguscio: Coherent atomic and molecular spectroscopy in the far-infrared. Physica Scripta **37**, 699 (1988)

602. L.R. Zink, M. Prevedelti, K.M. Evenson, M. Inguscio: 'High resolution far-infrared spectroscopy'. In: *Applied Laser Spectroscopy X*, ed. by W. Demtröder, M. Inguscio (Plenum, New York 1990)

603. M. Inguscio, P.R. Zink, K.M. Evenson, D.A. Jennings: Sub-Doppler tunable far-infrared spectroscopy. Opt. Lett. **12**, 867 (1987)

604. S.E. Harris: Tunable optical parametric oscillators. Proc. IEEE **57**, 2096 (1969);
M.A. Belkin et al.: Room temperature terahertz quantum cascade laser source based on intracavity difference frequency generation. Appl. Phys. Lett. **92**, 201101 (2008)

605. R.L. Byer: 'Parametric oscillators and nonlinear materials'. In: *Nonlinear Optics*, ed. by P.G. Harper, B.S. Wherret (Academic, London 1977)

606. R.L. Byer, R.L. Herbet, R.N. Fleming: 'Broadly tunable IR-source'. In: *Laser Spectroscopy II*, ed. by S. Haroche, J.C. Pebay-Peyroula, T.W. Hänsch, S.E. Harris, Lect. Notes Phys., Vol. 43 (Springer, Berlin, Heidelberg 1974) p. 207

607. C.L. Tang, L.K. Cheng: *Fundamentals of Optical Parametric Processes and Oscillators* (Harwood, Amsterdam 1995)

608. A. Yariv: 'Parametric processes'. In: *Progress in Quantum Electronics Vol. 1, Part 1*, ed. by J.H. Sanders, S. Stenholm (Pergamon, Oxford 1969)

609. R. Fischer: Vergleich der Eigenschaften doppelt-resonanter optischer parametrischer Oszillatoren. Exp. Technik der Physik **21**, 21 (1973)

610. C.L. Tang, W.R. Rosenberg, T. Ukachi, R.J. Lane, L.K. Cheng: Advanced materials aid parametric oscillator development. Laser Focus **26**, 107 (1990)

611. A. Fix, T. Schröder, R. Wallenstein: New sources of powerful tunable laser radiation in the ultraviolet, visible and near infrared. Laser und Optoelektronik **23**, 106 (1991)

612. R.L. Byer, M.K. Oshamn, J.F. Young, S.E. Harris: Visible CW parametric oscillator. Appl. Phys. Lett. **13**, 109 (1968)

613. J. Pinnard, J.F. Young: Interferometric stabilization of an optical parametric oscillator. Opt. Commun. **4**, 425 (1972)

614. J.G. Haub, M.J. Johnson, B.J. Orr, R. Wallenstein: Continuously tunable, injection-seeded β-barium borate optical parametric oscillator. Appl. Phys. Lett. **58**, 1718 (1991)

615. S. Schiller, J. Mlynek (Eds.): Special issue on cw optical parametric oscillators. Appl. Phys. B **25** (June 1998) and J. Opt. Soc. Am. B **12**, (11) (1995)

616. M.E. Klein, M. Scheidt, K.J. Boller, R. Wallenstein: Dye laser pumped, continous-wave KTP optical parametric oscillators. Appl. Phys. B **25**, 727 (1998)

617. R. Al-Tahtamouni, K. Bencheik, R. Sturz, K. Schneider, M. Lang, J. Mlynek, S. Schiller: Long-term stable operation and absolute frequency stability of a doubly resonant parametric oscillator. Appl. Phys. B **25**, 733 (1998)

618. Information leaflet Coherent Laser Group, Palo Alto, Calif.; and Linos Photonics GmbH, Göttingen, Germany

619. V. Wilke, W. Schmidt: Tunable coherent radiation source covering a spectral range from 185 to 880 nm. Appl. Phys. **18**, 177 (1979)

620. W. Hartig, W. Schmidt: A broadly tunable IR waveguide Raman laser pumped by a dye laser. Appl. Phys. **18**, 235 (1979)

621. J.D. Kafka, T. Baer: Fiber Raman laser pumped by a Nd:YAG laser. Hyperfine Interactions **37**, 1–4 (Dec. 1987)

622. V. Karpov, E.M. Dianov, V.M. Paramonov, O.I. Medvedkov: Laser-diode pumped phospho-silicate-fiber Raman laser with an output of 1 W at 1.48 μm. Opt. Lett. **24**, 1 (1999)

623. I.K. Ilev, H. Kumagai, K. Toyoda: A widely tunable (0.54–1.01 μm) double-pass fiber Raman laser. Appl. Phys. Lett. **69**, 1846 (1996)

624. S.A. Babin et al.: All-fiber widely tunable Raman fiber laser with controlled output spectrum. Optics Express **15**, 8438 (2007)

625. K. Zhao, S. Jackson: Highly efficient free-running cascaded Raman fiber laser that uses broadband pumping. Optics Express **13**, 4731 (2005)

626. C. Headley, G. Agraval (Eds.): *Raman Amplification in Fiber Optical Communication Systems* (Academic Press, New York 2004)

627. C. Lin, R.H. Stolen, W.G. French, T.G. Malone: A CW tunable near-infrared (1.085–1.175 μm) Raman oscillator. Opt. Lett. **1**, 96 (1977)

628. D.J. Brink, D. Proch: Efficient tunable ultraviolet source based on stimulated Raman scattering of an excimer-pumped dye laser. Opt. Lett. **7**, 494 (1982)

629. A.Z. Grasiuk, I.G. Zubarev: High power tunable IR Raman lasers. Appl. Phys. **17**, 211 (1978)

630. K. Suto: *Semiconductor Raman Lasers* (Boston Artech House, Boston 1994)

631. B. Wellegehausen, K. Ludewigt, H. Welling: Anti-Stokes Raman lasers. SPIE Proc. **492**, 10 (1985)

632. A. Weber (Ed.): *Raman Spectroscopy of Gases and Liquids*, Topics Curr. Phys., Vol. 11 (Springer, Berlin, Heidelberg 1979)

633. K. Suto, J. Nishizawa: *Semiconductor Raman Lasers* (Artech House, Boston 1994)

Index

Printed in the United States
By Bookmasters